AF581791

Special Issue of the Manufacturing Engineering Society (MES)

Special Issue of the Manufacturing Engineering Society (MES)

Volume 2

Editors

Eva M. Rubio
Ana M. Camacho

MDPI • Basel • Beijing • Wuhan • Barcelona • Belgrade • Manchester • Tokyo • Cluj • Tianjin

Editors
Eva M. Rubio
Universidad Nacional de
Educación a Distancia (UNED)
Spain

Ana M. Camacho
Universidad Nacional de Educación a
Distancia (UNED)
Spain

Editorial Office
MDPI
St. Alban-Anlage 66
4052 Basel, Switzerland

This is a reprint of articles from the Special Issue published online in the open access journal *Materials* (ISSN 1996-1944) (available at: https://www.mdpi.com/journal/materials/special_issues/Manufacturing_Society).

For citation purposes, cite each article independently as indicated on the article page online and as indicated below:

LastName, A.A.; LastName, B.B.; LastName, C.C. Article Title. *Journal Name* **Year**, *Article Number*, Page Range.

Volume 2
ISBN 978-3-03936-766-5 (Hbk)
ISBN 978-3-03936-767-2 (PDF)

Volume 1-2
ISBN 978-3-03936-768-9 (Hbk)
ISBN 978-3-03936-769-6 (PDF)

© 2020 by the authors. Articles in this book are Open Access and distributed under the Creative Commons Attribution (CC BY) license, which allows users to download, copy and build upon published articles, as long as the author and publisher are properly credited, which ensures maximum dissemination and a wider impact of our publications.
The book as a whole is distributed by MDPI under the terms and conditions of the Creative Commons license CC BY-NC-ND.

Contents

About the Editors

Eva M. Rubio is a Professor at the Department of Manufacturing Engineering of the National Distance Education University (UNED, Spain). She received her MSc in Aeronautical Engineering (1997) from the Polytechnic University of Madrid and her PhD in Industrial Engineering (2002) from the National University of Distance Education (UNED). Her research activity is dedicated to the field of Manufacturing Engineering, with the following main lines of work: analysis of machining processes; analysis of metal forming processes; industrial metrology; and teaching and innovation in engineering.

Ana M. Camacho is a Professor at the Department of Manufacturing Engineering of the National Distance Education University (UNED, Spain). She received her MSc in Industrial Engineering from the University of Castilla-La Mancha (UCLM) in 2001 and her PhD in Industrial Engineering from the UNED in 2005. Her main research interest is innovation in manufacturing engineering and materials technology, especially focused on analysis of metal forming and additive manufacturing techniques through computer-aided engineering tools and experimental testing, and development of methodologies for materials selection in demanding applications.

Article

Study on Near-Net Forming Technology for Stepped Shaft by Cross-Wedge Rolling Based on Variable Cone Angle Billets

Sutao Han [1], Xuedao Shu [1,2,*] and Chang Shu [3]

[1] Faculty of Mechanical Engineering & Mechanics, Ningbo University, Ningbo 315211, China; hansutao@foxmail.com

[2] Department of Mechanical Engineering, Northwestern University, Evanston, IL 60208, USA

[3] Department of Mechanical and Aerospace Engineering, University of Florida, Gainesville, FL 32608, USA; changshu@ufl.edu

* Correspondence: xuedao.shu@northwestern.edu or shuxuedao@nbu.edu.cn

Received: 30 June 2018; Accepted: 18 July 2018; Published: 25 July 2018

Abstract: Considering problems about concaves at the stepped shaft ends, this paper established the plastic flow kinetic theories about metal deforming during the cross-wedge rolling (CWR) process. By means of the DEFORM-3D finite element software and the point tracing method, the forming process of stepped shafts and the forming mechanism of concaves at shaft ends were studied. Based on the forming features of stepped shafts, rolling pieces were designed using variable cone angle billets. Single-factor tests were conducted to analyze the influence law of the shape parameters of billet with variable cone angle on end concaves, and rolling experiments were performed for verification. According to the results, during the rolling process of stepped shafts, concaves will come into being in stages, and the increasing tendency of its depth is due to the wave mode, the parameters of cone angle $\boldsymbol{\alpha}$, the first cone section length **n**. Furthermore, the total cone section length **m** has an increasingly weaker influence on the end concaves. Specifically, cone angle $\boldsymbol{\alpha}$ has the most significant influence on the quality of shaft ends, which is about twice the influence of the total cone section length **m**. The concave depth will decrease at the beginning, and then increase with the increasing of the cone angle $\boldsymbol{\alpha}$ and the first cone section length **n**, and it will decrease with the increasing of the total cone section length **m**. Finite element numerical analysis results are perfectly consistent with experimental results, with the error ratio being lower than 5%. The results provide a reliable theoretical basis for effectively disposing of end concave problems during CWR, rationally confirming the shape parameters of billets with a variable cone angle, improving the quality of stepped shaft ends, and realizing the near-net forming process of cross-wedge rolling without a stub bar.

Keywords: stepped shaft; cross wedge rolling; plastic flow kinetic theories; variable cone angle billets; near-net forming

1. Introduction

The process of cross-wedge rolling (CWR) plays a vital role in the thermoplastic forming of steels thanks to its many advantages including high efficiency, material saving, energy saving, and environmental protection [1,2]. In the process of CWR, plastic deformation including axial extension and radial compression occurs on rolling pieces. As the flow velocity of surface metal is faster than that of core metal, in order to ensure the quality of rolling ends, stub bars of a certain length (about 8–10% the mass of the rolling piece), together with sufficient machining allowance, need to be reserved. As a result, the utilization of material is usually lower than 85% during the CWR process, and this turns out to be the bottleneck of CWR development, and cannot meet the requirements of the conservation-minded society [3–5].

Considering the stub bars generated in the CWR process, research studies has been done by experts all over the world from different perspectives. Among them, Danno and Awano [6] studied the influence laws of different process parameters on the volume of stub bars formed during CWR. Pater [7,8] obtained the forming laws of concaves by analyzing the stress–strain and metal flow during CWR, and proposed to erect clamps on rolling ends to control their quality. In China, Wei Xinhong [9] illustrated the formation mechanism of end concaves from different perspectives using the displacement method, grid method, stress and strain analysis, etc. Zhenhai Ma [10] studied the influence laws of process parameters on the volume of stub bars, and put forward the theory that the concave size would decrease with the increasing of the forming angle and the constant size of the rolling piece's end, and would increase with the increasing of the spreading angle and deformation quantity. Wangming Zu and Xinhong Wei [11,12] erected side wedges on the die in order to restrain the flow of the surface metal at the rolling end, thus decreasing the size of the concave. Wenwei Gong [13] put forward the method to design press blocks on the die in order to restrain the flow of the surface metal of the rolling piece, thus constraining the formation of concaves. Chuan Liu and Jie Wei [14,15] designed a closed–open cross-wedge rolling die in order to change the metal flow direction of rolling piece, and thus enhancing the quality at shaft ends. Refering to Pater's cross-wedge rolling technology based on chamfered end billets [16], Xiaoying Zhang and Bin Hu [17,18], by analyzing the reasons of end concave formation, put forward the method of changing the shape of billet ends to decrease concave size. However, stepped shafts that have been widely applied are not covered in the above research studies, which were only limited to the rolling of simple shafts such as oil pump shafts, and research on the shaft end quality forming with conical billets has not been deepened. The research work of the above experts provides a reference for the research ideas and methods of the influence of the billet shape on the end quality.

This paper, by means of the DEFORM-3D 11.0 finite element software, established the finite element model of stepped shaft rolling. Based on the study on the forming mechanism of stepped shaft rolling, the method to design billet ends to the shape with a variable cone angle was raised and aimed at the staged deformation of stepped shafts during CWR, for the control of shaft end quality. Single-factor tests were conducted to study the influence laws of cone parameters on the quality of stepped shaft ends. In combination with rolling experiments for verification, it illustrated that billets with a variable cone angle could effectively reduce the size of stub bars at the rolling end. The results provide a reliable theoretical basis for effectively disposing of end concave problems during CWR, rationally confirming the shape parameters of billets with variable cone angles and improving the quality of stepped shaft ends.

2. Materials and Methods

2.1. Plastic Flow Kinetic Theories of Metal Deformation in CWR

The CWR process is as illustrated in Figure 1. The rolling piece rotates clockwise under the driving of the roller that rotates anticlockwise. At the contact surface with the rolling piece, the roller has the minimum circular velocity at point **A** and the maximum circular velocity at point **B**, which is located on the wedge top of the die. At point **K**, the roller and the rolling piece have identical velocity. In the figure, ω_1 refers to the rotational angular velocity of the roller, ω_2 is the rotational angular velocity of the rolling piece, V_K is the linear velocity of the roller die at point **K**, R_K is the rolling radius, which is defined by the distance from point **K** to the rotation center of the roller, $\mathbf{H}_1$ is the length of the rolled section, and $\mathbf{H}_2$ is the length of the unrolled section. The rolling piece is fed along the die and the rotational angular velocity of the rolling piece is smaller than its velocity at the rigid contact with the wedge of die. The relationship between the rotational angular velocity of the rolled piece and the linear velocity of the wedge [1] is determined by the following parameters:

$$\omega_2 = \frac{\mathbf{V}_K}{r_K} \tag{1}$$

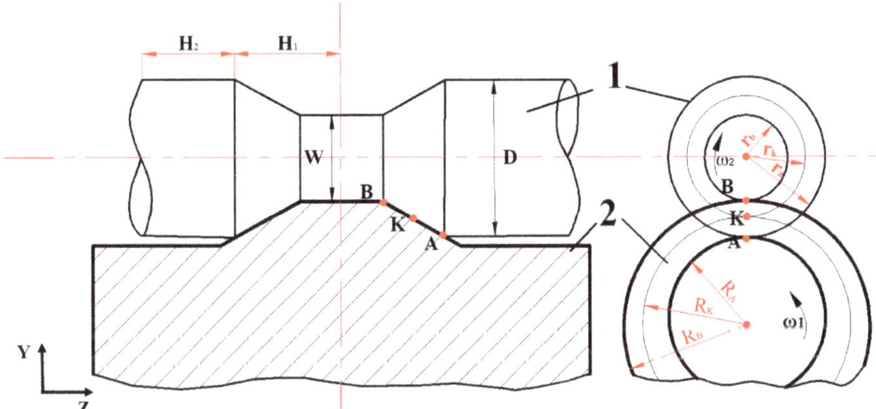

Figure 1. The front view and side view of the typical cross-wedge rolling (CWR) process. 1 is the rolling piece, 2 is the roller.

With increasing the length of the rolled section, the length of unrolled section will decrease. Based on the incompressibility of metallic bulk and the stable process of rolling, the below formula is obtained:

$$\delta = \frac{\mathrm{D}}{\mathrm{W}} \tag{2}$$

$$\mathbf{\Delta H_1} = \mathbf{\Delta H_2}\delta^2 \tag{3}$$

wherein $\mathbf{\Delta H_1}$ is the increment of the rolled section, $\mathbf{\Delta H_2}$ is the increment of the unrolled section, δ is the compression ratio, D is the original diameter of the rolling piece, and W is the diameter after rolling.

Due to the forming wedge pressure, metal in the rolled section would deform into an oval shape. The rotational angular velocity of the rolling piece increased gradually from the inlet to the outlet of the deformation area. As a result, fiber crimp occurred at the axial direction on the rolling piece. In this paper, metallic particles were regarded as the analysis objects, to study the displacement velocity of metal in the rolling area. As shown in Figure 2, the transient variable of the velocity of metallic particles at the direction of axis OX is:

$$\Delta \mathbf{V}_{\mathrm{x}} = (\omega_{\mathrm{BBIX}} - \omega_{\mathrm{BX}})\mathrm{r_b} \tag{4}$$

The displacement velocity at axis OZ is:

$$\Delta \mathbf{V}_{\mathrm{Z}} = (\mathbf{V}_{\mathrm{HZ}} - \mathbf{V}_{\mathrm{Z}}) \tan\beta = \mathbf{V}_{\mu\mathrm{H}} \tan\beta \tag{5}$$

where in ω_{BBIX} is the angular velocity at the outlet of the deformation area of the rolling piece, ω_{BX} is the angular velocity at the inlet of the deformation area of the rolling piece, $\mathrm{r_b}$ is the radius of rolling piece at the shaping completion area, $\mathbf{V}_{\mathrm{HZ}}$ is the axial velocity of spreading of the wedge, $\mathbf{V}_{\mathrm{Z}}$ is the axial velocity of extension of the rolling piece, $\mathbf{V}_{\mu\mathrm{H}}$ is the sliding velocity, and β is the spreading angle.

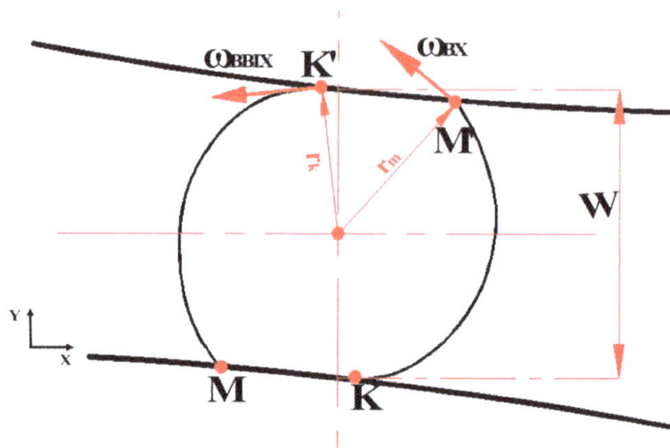

Figure 2. Schematic diagram deformation of simple shafts under load during CWR.

By means of Formulas (4) and (5), it is able to easily obtain the inclination angle between the crimped metal fiber of the rolling piece and the vertical axis:

$$\varphi = \arctan\frac{\Delta V_X}{\Delta V_Z} = \arctan\frac{(\omega_{BBIX} - \omega_{BX})d}{2V_{\mu H}\tan\beta} \tag{6}$$

Due to ignoring the difference between the rolling velocity in the rolling area and the velocity of metal deformation, as well as the longitude curve of metal deformation during calculation, the crimping angle in the rolling experiment is relatively larger. The results of finite element simulation are as shown in Figure 3.

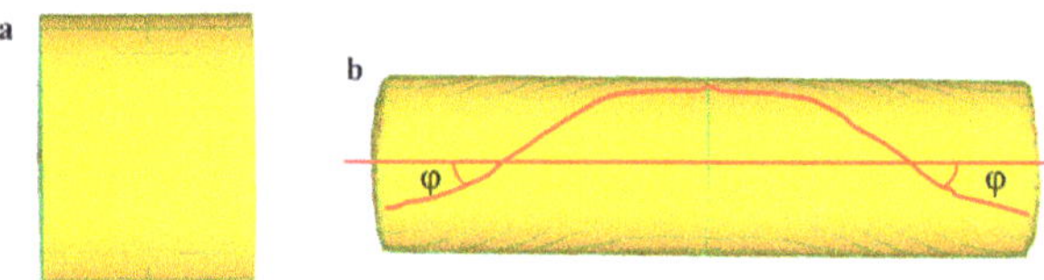

Figure 3. Results of finite element simulation of metal crimping of simple shaft: (**a**) billet; (**b**) rolled piece.

As for metal in the unrolled section, which is the metal at the end face, the displacement velocity at the stable rolling stage is composed of two elements, including the circular motion in the cross-section and the linear motion at the axial direction. See the formula below:

$$\begin{aligned} x &= a\cos\omega_2 t \\ y &= a\sin\omega_2 t \\ z &= bt \end{aligned} \tag{7}$$

Wherein a is the radial distance from the metallic particle to the axis of the rolling piece.

At the spreading stage of the rolling process, the rotary radius of the rolling piece and the geometrical shape of the deformation area did not change any more, and spreading length became the exclusive element of change. At this moment, rolling and shaping became stable. Considering the incompressibility of deforming metal, the displacement curve of the end metal at the stable rolling stage is expressed as follows:

$$
\begin{aligned}
x &= a\cos\omega_2 t \\
y &= a\sin\omega_2 t \\
z &= \frac{(r_0+r_1)^3(r_0-r_1)\sin\beta\tan\beta}{8r_0^2\tan\alpha}t
\end{aligned}
\tag{8}
$$

2.2. Confirmation of the Finite Element Simulation Scheme

The rolling process of multi-step shafts always starts from a certain plane in the middle to both ends for metal shaping and the influence of plastic metal on the concave size of the rolling end will decrease with the increasing of the distance from the shaft end [19,20]. Consequently, in order to reduce the quantity of calculation and enhance calculation efficiency, when using DEFORM-3D software for finite element simulation, a three-diameter shaft was selected as the research object, and contact surface constraint was applied to the middle of a certain plane, as shown in Figure 4. The dimension parameters of the product were as follows: D = 50 mm, L1 = 10 mm, L2 = 37.4 mm, L3 = 25.9 mm, d1 = 40 mm, d2 = 35 mm, the spreading angle of the die = 8°, forming angle = 30°, the finite element model was of 20CrMnTi material. The roller radius is 500 mm, the parameters of friction between the workpiece and rollers are 0.2, the parameters of friction between the workpiece and guide plates are 0.01, and the rolling temperature was 1050 °C. The billet was of plastic material and four-dimension mesh was applied, with the number of grid points as 150,000. The finite element model established is as shown in Figure 5.

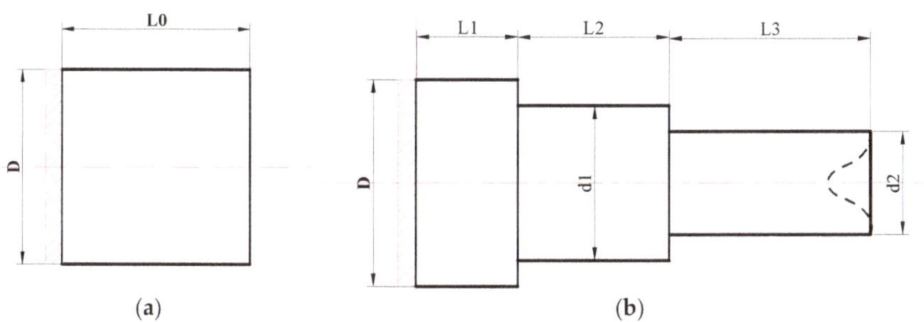

Figure 4. Diagrams of traditional billets and product: (**a**) traditional billet; (**b**) rolled piece.

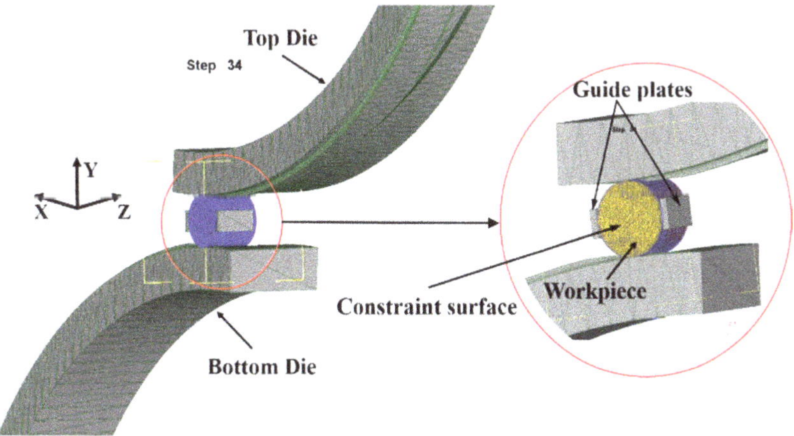

Figure 5. Finite element model.

2.3. Forming Features of Stepped Shaft Rolling and Forming Mechanisms of Concaves

The features of stepped shafts rolling are large axial and radial plastic deformation, different radial compression of metal at different stages, an uneven distribution of axial and radial deformation, etc. As a result, the end concave of a rolling piece made from stepped shaft is larger than that made from common shafts. The point tracing method was used to study the forming features of stepped shafts and the forming mechanisms of end concave. As shown in Figure 6, **P1** is at the circular edge, **P3** is at the circular center and **P2** is at the 1/2 radius of the rolling piece. The axial displacement results are as illustrated below.

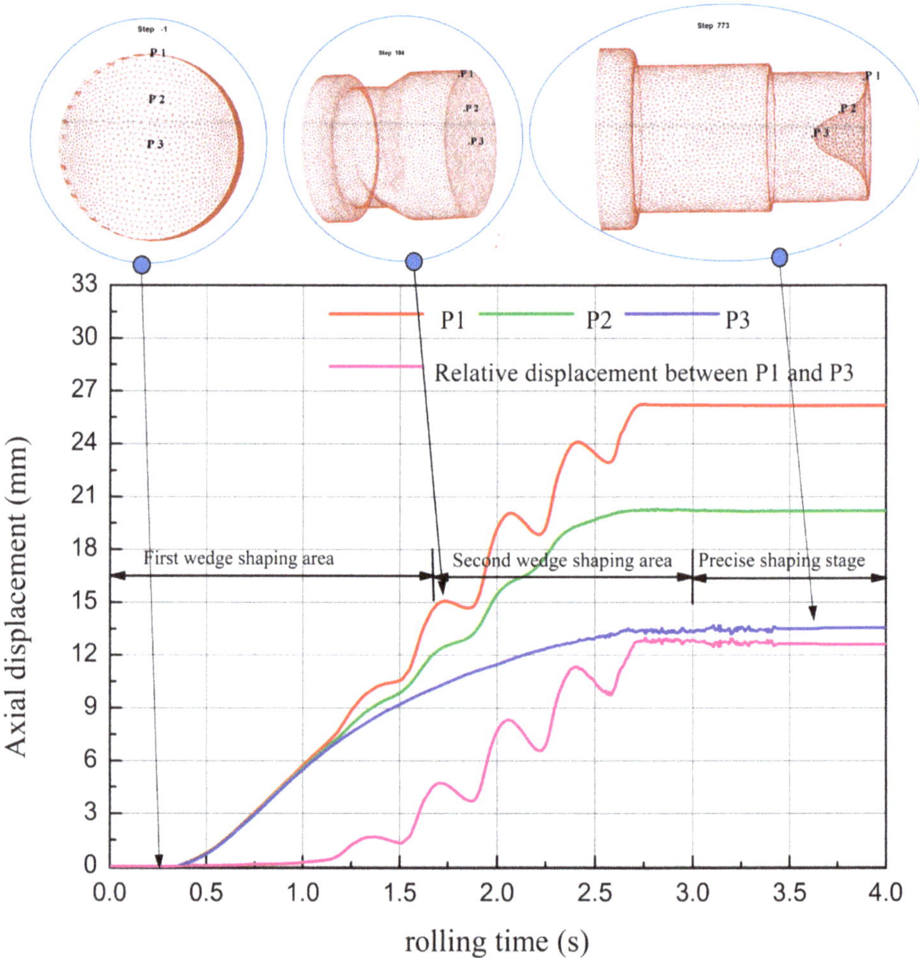

Figure 6. Process of forming of end concaves of stepped shaft based on the point tracing method.

As illustrated in the figure, the forming process of stepped shaft is divided into three stages, including: the first wedge-shaping stage (0–1.71 s), the second wedge-shaping stage (1.71–3 s) and the precise shaping stage (3–4 s), and the entire process is complete within 4 s. During 0–1 s, the end metal moves axially and rises perpendicularly. During 1–2.75 s, the axial displacement velocity of **P3** gradually slows down due to the influence of the formed concave; meanwhile, **P2** and **P3** have horizontal displacement under the influence of fiber crimp and their movement path matches with the spiral curve equation. Due to the variances of deformation, **P2** at 1/2 radius is selected for calculating the movement path of the shaft end, during 0–1.71 s, the simulation result of the displacement curve's slope is 7.2, and the theoretically calculated result is 6.9, with the error at 4%, and by 1.71–2.47 s, the simulation result of displacement curve's slope is 8.76, and the theoretically calculated result is 9.2, with the error at 5%. Based on these results, the movement laws of end metal during stepped shaft rolling basically match with plastic flow kinetic theories.

Based on the forming of end concave in the process of rolling, within 0–1 s, the first wedge starts to wedge into and spread the billet. At the moment, the percentage reduction of area is relatively small, and the quantity of metal deformation is not huge. As a large quantity of metal exits at the ends, metal flow is restrained to a certain extent. As a result, the overall deformation is uniform and concave does not occur. At the stable rolling stage, metal moves axially and rises perpendicularly. At 1.72 s, the second wedge shaping stage starts. At this moment, the quantity of transient metal deformation increases with increasing the ratio of reduction of area. Due to the crimping of metal fibers, the surface metal moves as per the spiral path, the axial motion of core metal is constrained, and the concave can be clearly observed at the rolling ends. This happens because the quantity of metal at the ends is less, the function of inhibition is weakened, and the flow velocity of the surface metal is obviously faster than that of the core metal. Besides, the accumulation effect of uneven deformation starts to occur. The axial displacement slope of core metal gradually reduces and that of surface metal increases on the contrary. Since 1.83 s, the second wedge spreading stage starts. At the moment, the three steps have been initially formed, and relatively deep concave can be seen at shaft ends. This is because in this stage, the radial compression is large, making the end "bowl-like": the surface metal is thin, the core is hollow, and the end metal loses its function of inhibition. As for the end metal, the radial compression and lateral extension are extremely uneven, and the phenomena of folding occurs; as for the core, the axial flow gradually slows down, and the surface metal moves perpendicularly at the axial direction. The helix lead and amplitude of the movement path gradually become larger from the outlet to the inlet point of the deforming area, and the increasing rate of concave depth is accelerated.

Take the rotation period within 1.83–2.23 s as an example. As shown in Figure 7, during 1.83–2.06 s, the roller starts to compress the **P1–P3** section. At the moment, the radial compression gradually becomes larger, and the surface metal starts to fold to the middle. The axial extension of the core metal is restrained, and the flow velocity slows down gradually; on the contrary, the axial extension of the surface metal is free of restraining, and the flow velocity accelerates, making the concave enlarges gradually. By 2.02 s, the roller is vertical with the **P1–P3** section, and radial compression reaches the peak value. However, due to the crimping of metal fibers during the CWR process, the forming of the φ angle is hysteretic, that is to say, by 2.06 s or so, the concave is the largest. Within 2.06–2.23 s, metal at the **P1–P3** section extends laterally, the **P1–P3** vertical section gradually starts its radial compression stage, and the folding effect occurs. Metal at the outer edge of the **P1–P3** section moves axially to the inner side, and the concave becomes smaller. The effect of folding is gradually weaker with the approaching of the core, and the core metal will not be affected. By 2.23 s, the folding effect has the most significant effect, and consequently, concave reaches its minimum value, as shown in Figure 8.

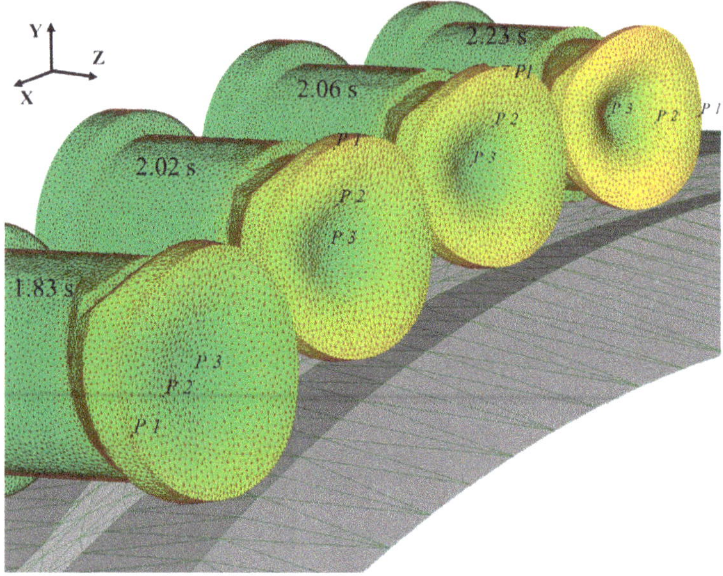

Figure 7. Variation of the rolling piece in the rotation period of 1.83–2.23 s.

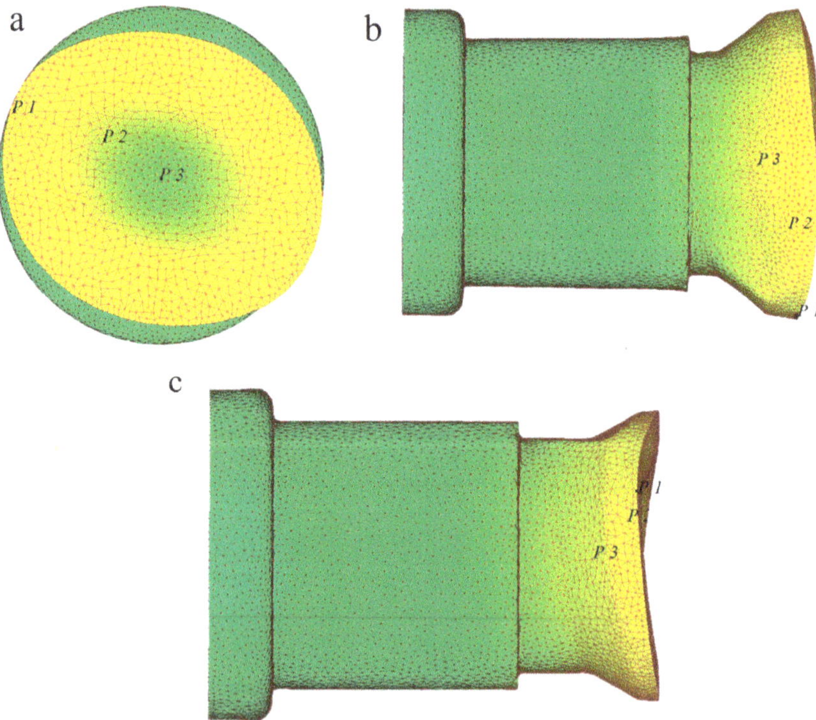

Figure 8. Three views of the rolled piece when rolling time is 2.23 s: (**a**) *Z* axis direction view of rolled piece; (**b**) *Y* axis direction view of rolled piece; (**c**) *X* axis direction view of rolled piece.

2.4. Design of Billet Shape and Features of Shaft Ends Made of Billets with Variable Cone Angle

As the rolling of stepped shafts features large metal deformation, which is in stages and uneven, the billet end is designed to the shape of the variable cone angle, to gradually weaken or even eliminate the effect of uneven axial and radial deformation, and maintain the levelness of the rolling end. The forming process of the stepped shaft by cross-wedge rolling with the variable cone angle billet is shown in Figure 9. Figure 10 shows the diagrams of the billet with the variable cone angle, and the shape of a typical rolling piece made from such a billet.

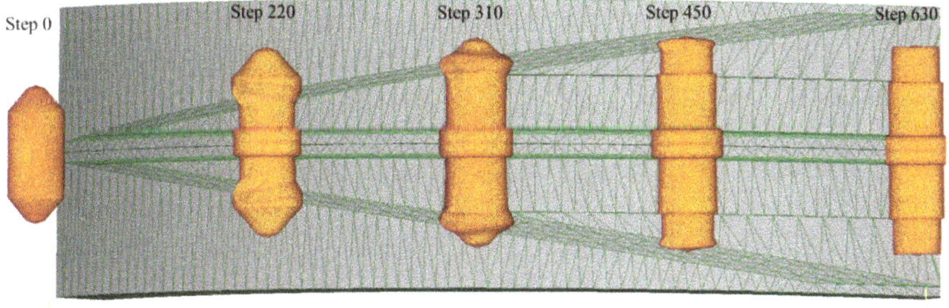

Figure 9. Finite element simulation of a variable cone angle billet forming to be stepped shaft by cross-wedge rolling.

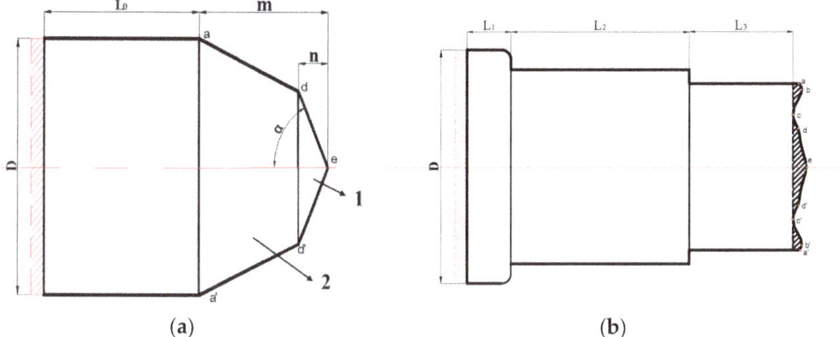

Figure 10. Shapes of billet with a variable cone angle and its product: (**a**) shape of billet with a variable cone angle; 1 is the first cone section, 2 is the second cone region; (**b**) shape of a typical rolling piece from such a billet.

As shown in Figure 10b, the rolling end made by the billet with a variable cone angle is "W"-shaped. The depth of the concave is the distance from point **c** to **a**. This is because during the forming of a stepped shaft, the core metal of the billet will be ahead of the metal on the edge during moving. As a result, in the process of forming, hysteretic metal is between point **a** on the edge and point **e** in the core. Due to the unevenness of deformation and the crimping of metal fibers, the metal on the edge would fold along those points with the minimum displacement, coming into being the deepest area of the concave, that is, point **c**.

In order to improve the quality of rolling ends, this paper, based on the results of single-factor tests, studied the influence laws of dimensions of the billet with a variable cone angle on the quality of end concave of the stepped shaft. The parameters and results of single-factor tests are shown in Table 1.

Table 1. Test design and results of single-factor test.

Number	m/mm	α/°	n/mm	Concave Depth/mm
1	15	50	3	4.946
2	15	60	3	4.337
3	15	70	3	2.452
4	15	76.5	3	5.972
5	15	80	3	6.828
6	15	70	1.5	4.672
7	15	70	3	2.452
8	15	70	4.5	3.075
9	15	70	6	3.856
10	15	70	7.5	5.249
11	12.5	70	3	4.111
12	15	70	3	2.452
13	17.5	70	3	2.263
14	20	70	3	2.294
15	22.5	70	3	1.745

3. Results and Discussion

3.1. Influence of the Total Cone Section Length m on Concave Depth

When **n** = 3 mm and **α** = 70°, the influence of **m** on concave depth is as shown in Figure 11. As illustrated, the depth will decrease obviously with the increasing of **m**. When **m** = 12.5–15 mm, the depth of concave changes tremendously with the increasing of **m**, in which the rate of change is −0.66. When **m** = 15–20 mm, the change of the depth with the increasing of **m** slows down, approximately at the rate of −0.03, and when **m** = 20–22.5 mm, the change is faster, with the rate increasing to −0.22. Within the range from 12.5 mm to 15 mm, **m** makes the most significant difference on concave depth, which is about three times compared to the range [20, 22.5].

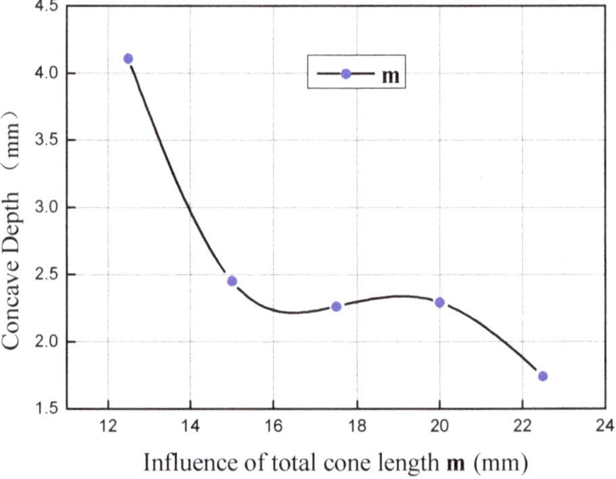

Figure 11. Influence of total cone length **m** on concave depth.

Within the range of 12.5–17.5, when **m** = 12.5 mm, as the displacement hysteresis point **c** is close to point **d**, the depth of concave is mainly subject to the difference between the axial displacement of the metal at point **d** and those on the edge. At the beginning of rolling, as the displacement of the core metal compared to those on the edge is subtle, the axial displacement of metal within the 1/4 radius range nearby point **e** is smaller than that of the metal on the edge during the second wedge-spreading stage. As a result, point **d** is relatively hysteretic to point **a** and **e** to a certain extent, that is to say, the distribution from the core to the edge is U-shaped, and the folding effect occurs, making point **d** the core of concave. With the increasing of **m**, the second cone section becomes larger, the conicity becomes smaller, and the core metal at shaft end moves much more ahead of the metal on the edge. As a result, the axial displacement difference between the metal within the 1/4 radius range near point **e** and the metal on the edge is smaller. Meanwhile, point **c** slowly moves from point **d** to the range between point **d** and **a**, the radial distance of folding becomes shorter, the folding effect becomes weaker, and the concave is shallower.

Within the range of **m** = 17.5–20 mm, due to the distance advantage of metal at point **e**, it is flush with point **a** at the axial direction upon the completion of rolling. With the increasing of L, as the conicity difference between the second and the first cone sections becomes larger, at the second wedge-spreading stage, the resistance, due to the core metal of the cone, against the axial flow of the metal between point **a** and **d** is gradually larger, point **c** continues to move to point **b**, the cone point angle ∠ded′ gradually becomes smaller, and the first cone section as a whole starts to move ahead of the metal on the edge. Within the range from 17.5 mm to 20 mm, due to the restriction of cone angle against metal flow, the folding effect should be more significant, and consequently, the concave depth slightly increases.

Within the range of **m** = 20–22.5 mm, with the increasing of **m**, the impediment effect of cone angle becomes larger, the hysteretic distance of point **c** increasingly becomes larger, the phenomena of folding occurs earlier, and the folding effect is more obvious. Meanwhile, with the proceeding of the rolling process, the core metal of the cone moves laterally due to the influence of hysteretic metal and torsional effect. The metal at point **c** is compensated, and the complementary effect offsets the folding effect. As a result, concave depth decreases.

3.2. *Influence of the Length of the First Cone Section n on Concave Depth*

The curve of change of concave depth with **n** is as shown in Figure 12. As illustrated, when **m** =15 mm and $\alpha = 70^\circ$, concave depth will firstly decrease and then increase with the increasing of **n**. When **n** = 1.5–3 mm, the larger **n** is, the smaller the concave depth will be, and the change tendency is fast, with the rate at −1.48. When **n** = 3–7.5 mm, the larger **n** is, the larger the concave depth will be, and the change tendency slows down, at the rate of 0.62.

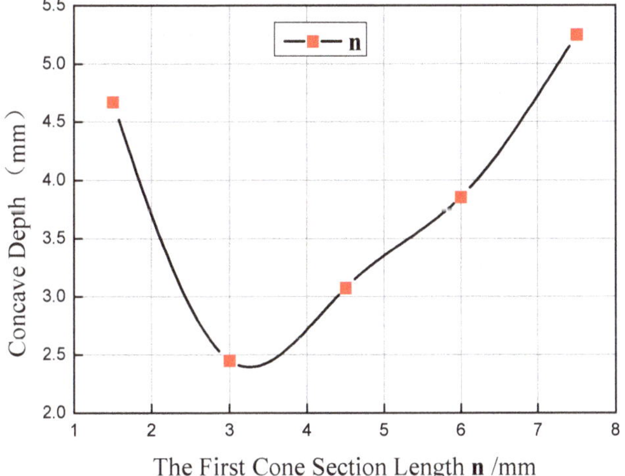

Figure 12. Influence of the first cone section length **n** on concave depth.

Based on the forming mechanism of concave in stepped shaft, when **n** changes from 1.5 mm to 3 mm, the depth of concave is subject to the deformation unevenness degree of metal on the edge and between point **a** and **b**. With the increasing of **n**, the volume of the first cone section will enlarge, the difference of initial displacement between the metal in the core and that on the edge becomes larger, and the radial distance between point **a** and **d** becomes shorter. Consequently, the deformation unevenness of the metal between point **a** and **c** decreases, the axial displacement difference between point **a** and **c** becomes smaller, and the concave depth decreases accordingly. Within the range of **n** = 3–7.5 mm, the depth of the concave is subject to the deformation unevenness of the metal on the edge and that within the 1/4 radius range. At the moment, the radial distance between point **a** and **d** becomes shorter, that between point **d** and **e** becomes longer, and point **c** moves from the middle of point **a** and **d** to the middle of point **d** and **e**. With the increasing of **n**, the difference of initial displacement at the axial direction of the metal at point **d** and **e** becomes larger, the deformation unevenness of metal becomes larger, and the concave depth increases accordingly.

3.3. Influence of Cone Angle α on Concave Depth

The curve of change of concave depth with α is as shown in Figure 13. As illustrated, when **m** = 15 mm and **n** = 3 mm, concave depth will firstly decrease, and then increase with the increasing of α. When α = 50–68°, the larger α is, the smaller the concave depth will be, with the change rate at −0.125 mm/°. When α = 68–80°, the larger α is, the larger the concave depth will be, with the change rate at 0.44 mm/°, which is about 3.5 times the change rate within the range of 50° to 68°.

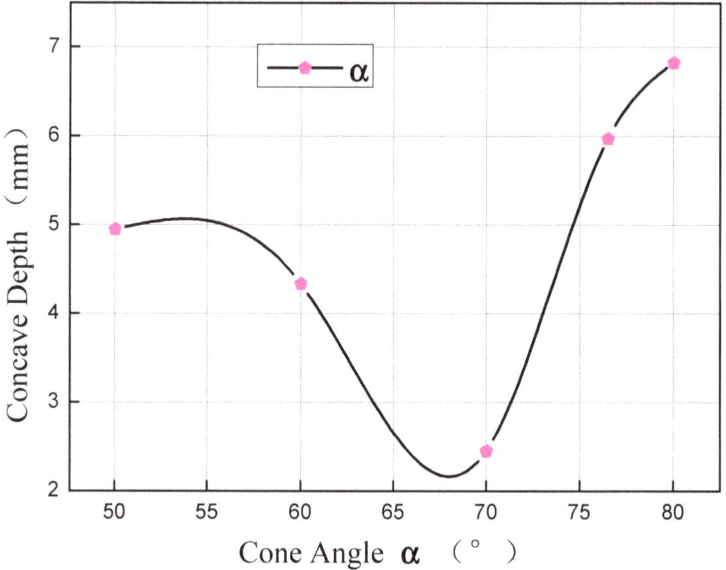

Figure 13. Influence of cone angle α on concave depth.

Within the range from 50° to 70°, the concave depth is subject to the deformation unevenness of metal on the edge and metal at point **d**. When α = 50°, point **c** and point **d** overlap. As the cone angle is small and the displacement advantage of point **e** relatively to point **d** is obvious, the concave is a U-shaped distribution from the core to the edge during rolling, and the metal will fold along point **d**. With the increasing of the angle, the volume of the first cone section increases, the difference of the initial displacement at the axial direction between point **e** and point **d** becomes gradually smaller, the flow line of metal is improved, the folding effect at point **d** gradually becomes weaker, and the concave depth becomes smaller. Within the range from 70° to 80°, the depth of concave is mainly subject to the axial displacement difference between point **d** and point **e**. With the gradual increasing of the cone angle, the initial displacement difference at the axial direction between point **d** and **e** gradually becomes smaller, the radial distance enlarges, the deformation unevenness increases, and the depth of the concave increases accordingly.

3.4. Comprehensive Analysis of the Influence Degree of the Shape Parameters of Billets

Different shape parameters of the billet have different impacts on the quality of concave at rolling end, and each factor has a different dimension. It cannot directly compare the importance of all factors to the end quality. In order to further confirm the priority sequence thereof, influence factor μ and influence coefficient λ, both with the dimension as 1, are introduced herein. μ refers to the relative variation of factors, and λ refers to the relative variation of the concave depth. Through the analysis of these data, the influence laws of different influence factors on the quality of the shaft end are analyzed and obtained, as shown in Figure 14.

$$\mu = \frac{A_i}{A_0};\ \lambda = \frac{B_i}{B_0} \tag{9}$$

where in A_i is the level value of the influence factor, A_0 is the level value of the third group of experiment factors, B_i is the depth of the end concave, and B_0 is the depth of the end concave of the third group of experiment.

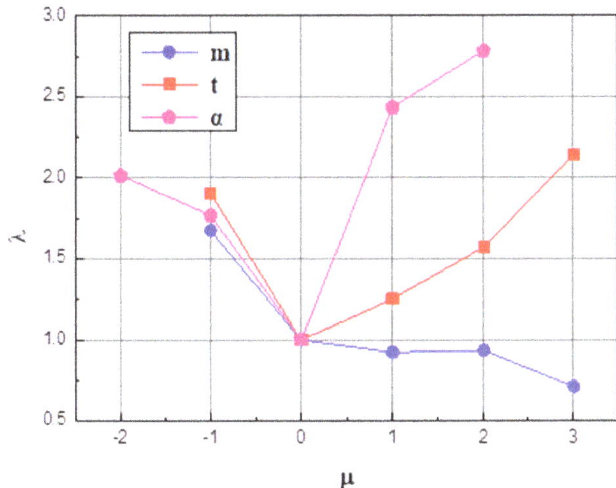

Figure 14. Comparison of influence of different factors on the concave quality of the rolling end.

As illustrated in Figure 11, the parameters of cone angle **α**, the height of the first cone section **n**, and the total cone height **m** have an increasingly weaker influence on the end concaves. Specifically, core angle **α** has the most significant impact on the end concave's quality, which is about two times the influence compared to the impact on total cone height **m**. When μ < 0, the difference of influence from these three parameters on concave is not obvious, so the sequence of influences from huge to small is the height of the first cone section, the cone angle, and the total cone height. When μ > 0, the cone angle **α** and the height of the first cone section **n** start to have a direct proportional relationship to the influence on the concave; the influence ratio of the cone angle **α** increases, and is 56% larger than that of the height of the first cone section **n**. Meanwhile, the total cone height continues to have an inverse proportional relationship to its influence ratio on the concave, but the influence level is weakened, and is about 50% of the influence within the range from −1 to 0.

3.5. Experimental Verification

Experiments were fulfilled on H630 rolling mills and Figure 15a–c shows respectively the rolling mill, rolling die, and a picture of the infrared thermometer. Stepped shaft products from a traditional billet and from a billet with a variable cone angle are shown in Figure 16. Figure 17 is the comparison between simulation and experimental results. Table 2 compares the statistical data of the simulation and experimental concaves.

Figure 15. Diagram of the rolling site: (**a**) the H630 type rolling mill; (**b**) the cross-wedge rolling die; (**c**) heating device.

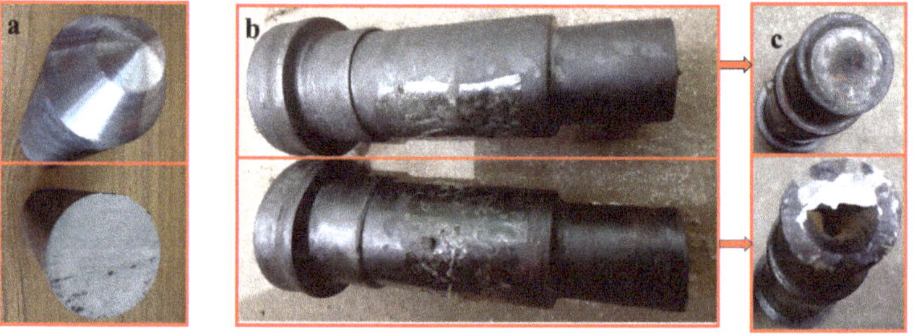

Figure 16. Billets and rolled products: (**a**) comparison of billets; (**b**) lateral comparison of rolled products; (**c**) axial comparison of rolled products.

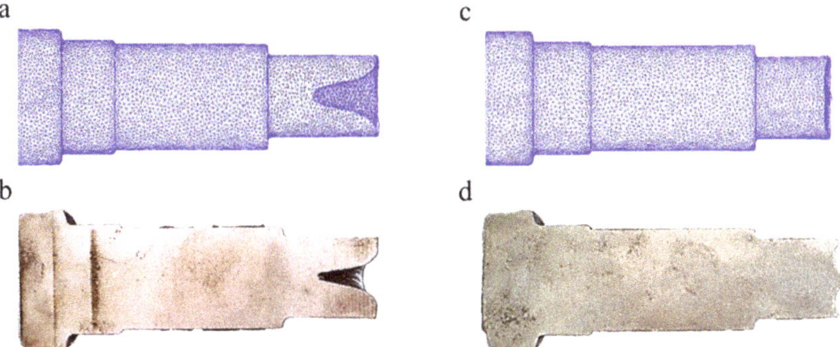

Figure 17. Comparison of Simulation (**a**,**c**) and Experimental (**b**,**d**) rolling effects of stepped shafts: (**a**) simulation results of rolling by a traditional billet; (**b**) experiment results of rolling by traditional billet; (**c**) simulation results of rolling by a billet with a variable cone angle; (**d**) experiment results of rolling by a billet with a variable cone angle.

Table 2. Concave depth statistics of finite element simulation and experimental results of stepped shaft by cross-wedge rolling.

Concave Depth / Billet Type	Finite Element Simulation Results/mm	Experimental Results/mm	Material Utilization
traditional billet	25.44 mm	24.95 mm	86.8%
billet with variable cone angle	1.93 mm	1.85 mm	99%

Through comparison, the simulation and experimental results are basically consistent. Specifically, the simulated concave depth using traditional billet is 25.44 mm, and in practical rolling, the figure is 24.95 mm, with the error ratio at 1.9%. By means of billets with a variable cone angle, the simulated depth is 1.93 mm, and the practical result is 1.85 mm, with the error at 4%. The depth of concave of the rolling pieces using a cone-shaped billet is about 92.5% smaller than those using traditional billets, and the utilization rate is 14% higher comparatively. Based on the results of finite element simulation and rolling experiment, the rolling of stepped shafts using billets with a variable cone angle has the outstanding impact on resolving end concave problems and enhancing the utilization rate of the material.

The billet end can be preformed to a variable cone angle shape before cross-wedge rolling by the hot roller shearing equipment [21,22] developed by our research group, and the near-net forming technology for stepped shaft by cross-wedge rolling based on variable cone angle billets has also reached the industrial application level.

4. Conclusions

1. In the process of cross-wedge rolling (CWR), metal in the shaping area will crimp like fibers, and the axial movement path of the end metal is in line with the spiral curve equation. During the rolling of the stepped shaft, subject to the impact of the changes of the percentage reduction of area, fiber crimp of metal, etc., the shaping process of the concave develops by stages, and the depth is increased by the wave mode. The adoption of a billet with a variable cone angle can effectively enhance the quality of stepped shaft ends and increase the utilization rate to more than 99%.
2. The parameters of the cone angle α, the first cone section length $\mathbf{n}$, and the total cone section length $\mathbf{m}$ have increasingly weaker influence on end concaves. Specifically, the cone angle α makes the most significant impact on the end concave's quality, which is about two times the

influence of the total cone section length **m**. The depth of concave will firstly decrease and then increase with the increasing of the cone angle **α** and of the first cone section length **n**, and will decrease with the increasing of the total cone section length **m**.

3. Through comparison, the error ratio between the experimental and simulation results of rolling stepped shafts using billets with a variable cone angle is lower than 5%. It means that the results of finite element simulation are reliable, and can reflect the changes of shaping of the rolling piece during practical production.

Author Contributions: S.H. designed the experiment scheme, carried out the cross wedge rolling experiment and wrote the manuscript. X.S. participated in the design of the experiment scheme, the cross wedge rolling experiment, the writing and the revision of the manuscript. C.S. participated in the revision of the manuscript and cross wedge rolling experiment.

Funding: This research was funded by K.C. Wong Education Foundation, Hong Kong, the National Natural Science Foundation of China [grant number 51475247], and The Natural Science Foundation of Zhejiang [grant number LZ17E050001].

Conflicts of Interest: The authors declare no conflicts of interest.

References

1. Hu, Z.H.; Zhang, K.S.; Wang, B.Y.; Shu, X.D. *Formed Technology and Simulation of Parts about the Cross-Wedge Rolling*, 1st ed.; Metallurgy Press: Beijing, China, 2005; pp. 1–64, ISBN 7-5024-3518-2.
2. Pater, Z. A study of cross wedge rolling process. *J. Mater. Process. Technol.* **1998**, *80–81*, 370–375. [CrossRef]
3. Shu, X.D.; Zhao, P. Dynamic characteristics of automobile semi-axes cross wedge rolling mills. *J. Univ. Sci. Technol. Beijing* **2007**, *29*, 1254. [CrossRef]
4. Shu, X.D.; Valery, Y.S.; Kozhevnikova, G.; Sun, B.S.; Peng, W.F. *Cross Wedge Rolling Theory and Forming Technology*, 1st ed.; Science Press: Beijing, China, 2014; pp. 1–10, ISBN 978-7-03-040397-1.
5. Pater, Z. Theoretical and experimental analysis of cross wedge rolling. *Int. J. Mach. Tools Manuf.* **2000**, *40*, 49–63. [CrossRef]
6. Danno, A.; Awano, T. Effect of rolling conditions on formation of central cavity in 2-roll cross rolling. *J. JSTP* **1976**, *17*, 117–124.
7. Pater, Z. Finite element analysis of cross wedge rolling. *J. Mater. Process. Technol.* **2006**, *173*, 201–208. [CrossRef]
8. Pater, Z.; Tomczak, J.; Bulzak, T. New forming possibilities in cross wedge rolling processes. *Arch. Civ. Mech. Eng.* **2018**, *18*, 149–161. [CrossRef]
9. Wei, X.H. Study on the End Rolling Quality of Cross Wedge Rolling. Master's Thesis, Ningbo University, Ningbo, China, 2011.
10. Ma, Z.H.; Yang, C.P.; Hu, Z.H. Study on the influence factors of the end concave of cross wedge rolling. *Forg. Stamp. Technol.* **2002**, *27*, 29–30.
11. Zu, W.M. Study on the Key Technology of Multi-Wedge Cross Wedge Rolling Forming Automobile Semi-Axis. Ph.D. Thesis, University of Science and Technology Beijing, Beijing, China, 2008.
12. Shu, X.D.; Wei, X.H.; Chen, L.P. Influence analysis of block wedge on rolled-piece end quality in cross wedge rolling. *Appl. Mech. Mater.* **2011**, *101*, 1055–1058. [CrossRef]
13. Gong, W.W.; Shu, X.D.; Peng, W.F.; Sun, B.S. The research on the microstructure evolution law of cross wedge rolling asymmetric shaft-parts based on parity wedge. *Appl. Mech. Mater.* **2012**, *201*, 1121–1125. [CrossRef]
14. Liu, C.; Shu, X.D.; Chao, C.; Sun, B.S. Influence rules of technological parameters on end quality in closed cross wedge rolling (CWR). *Adv. Mater. Res.* **2014**, *1095*, 756–761. [CrossRef]
15. Wei, J.; Shu, X.D.; Tian, D.Y.; Xiang, W.; Wei, Y.; Han, S.; Peng, W.; Liu, C.; Wang, Y.; Niu, B. Study in shaft end forming quality of closed-open cross wedge rolling shaft using a wedge block. *Int. J. Adv. Manuf. Technol.* **2017**, *93*, 1095–1105. [CrossRef]
16. Pater, Z. Cross-wedge rolling. In *Comprehensive Materials Processing*; Button, S.T., Ed.; Elsevier Ltd.: Amsterdam, The Netherlands, 2014; Volume 3, pp. 211–279, ISBN 0080965326.
17. Zhang, X.Y. Simulation of deformation principle of near net shape cross wedge rolling process to form shafts without concaves. *Forg. Stamp. Technol.* **2013**, *38*, 156–159.

18. Hu, B.; Shu, X.D.; Hong, Z. Study on influence of end concavity of rolled piece by billet circular arc angle with CWR. *Appl. Mech. Mater.* **2014**, *599*, 478–481. [CrossRef]
19. Duan, W.Z. Research on Rules of Material Loss on the Ends of Cross Wedge Rolling Workpiece. Ph.D. Thesis, University of Science and Technology Beijing, Beijing, China, 2010.
20. Zeng, J.; Xu, C.; Ren, W.; Li, P. Study on the deformation mechanism for forming shafts without concavity during the near-net forming cross wedge rolling process. *Int. J. Adv. Manuf. Technol.* **2017**, *91*, 127–136. [CrossRef]
21. Shu, X.D.; Hu, B.; Li, C.B.; Lv, Z.T.; Peng, W.F.; Sun, B.S. One Kind of Billet Shearing Devices. ZL201310583234.9, 22 July 2015.
22. Shu, X.D.; Wang, Y.; Li, Z.X.; Wei, J.; Zhou, Z.R. Crank Slider Type Hot Billet Shearing Mechanism Driven by Denatured Eccentric Circular Gear. ZL 201510414990.8, 31 May 2017.

© 2018 by the authors. Licensee MDPI, Basel, Switzerland. This article is an open access article distributed under the terms and conditions of the Creative Commons Attribution (CC BY) license (http://creativecommons.org/licenses/by/4.0/).

Article

Process Parameter Effects on Biocompatible Thermoplastic Sheets Produced by Incremental Forming

Marc Sabater [1], M. Luisa Garcia-Romeu [1,*], Marina Vives-Mestres [2], Ines Ferrer [1] and Isabel Bagudanch [1]

1 Department of Mechanical Engineering & Industrial Construction, University of Girona, 17071 Girona, Spain; marc.sabater@udg.edu (M.S.); ines.iferrer@udg.edu (I.F.); isabel.bagudanch@udg.edu (I.B.)

2 Department of Computer Science, Applied Mathematics & Statistics, University of Girona, 17003 Girona, Spain; marina.vives@udg.edu

* Correspondence: mluisa.gromeu@udg.edu; Tel.: +34-972-418-265

Received: 6 July 2018; Accepted: 6 August 2018; Published: 8 August 2018

Abstract: There has been increasing interest in the processes that enable part customization and small-batch production in recent years. The prosthetic sector, in which biocompatible materials are used, is one of the areas that requires these types of processes; Incremental Sheet Forming (ISF) technology can meet these requirements. However, the biocompatible thermoplastic polymers formed by this technology have not yet been tested. Hence, the aim of this paper is to cover this gap in our knowledge by analyzing the effects of process parameters on the ISF process with the aim of optimizing these parameters before the actual production of, in this case, customized prostheses. Tests with polycaprolactone (PCL) and ultra-high molecular weight polyethylene (UHMWPE) were performed. Maximum force, surface roughness and maximum depth were statistically analyzed by means of response surface methodology and survival analysis. Spindle speed and tool diameter were shown to be the most influential process parameters in terms of maximum forming force and surface roughness for both materials. In contrast, survival analysis applied to maximum depth showed a greater influence of tool diameter in PCL sheets and a greater influence of spindle speed in the case of UHMWPE.

Keywords: single; point; incremental; forming; thermoplastics; biocompatible; temperature; process; parameters

---

1. Introduction

The paradigms of manufacturing have evolved from craft production to mass production and then from mass customization to what S. J. Hu [1] calls personalization or personalized production. There are several important concepts and technologies that have facilitated the development of personalized production, including open-architecture products, personalization design, on-demand manufacturing systems and cyber-physical systems.

In the context of on-demand manufacturing systems, Incremental Sheet Forming (ISF) emerged to meet the demand for rapid prototyping and small-batch production. The process consists of a sheet being formed by means of a round-tipped tool (or punch), which makes a series of small incremental deformations in the sheet on a predefined path that is governed by a numerical control. It is a simple process that can be applied in a number of different fields, ranging from the automotive and aeronautic sectors to architecture. However, it is struggling to find a place in industrial production beyond the various research efforts carried out in recent years. More recently, the biotechnology sector has been attracting most of these research efforts, which have involved a variety of raw materials (metals, such as titanium or polymers, such as polycaprolactone) as well as different variants of the ISF process (such as Single Point Incremental Forming, or SPIF, i.e., without the presence of a die or Two Point Incremental Forming, TPIF, which fully or partially employs a die).

Within the biotechnology sector, the manufacture of prostheses using ISF process involves two of the key elements of personalized production: design personalization, since customized geometry required is for each patient and on-demand manufacturing systems, since ISF technology makes small-batch production possible. The ISF process also offers certain flexibility since it does not require dedicated machinery that would entail heavy investment. In fact, an ISF production system can be adapted from one of the most common machines found in the workshops: a computer numerically controlled (CNC) milling machine fitted with a clamping system for the sheet. In addition, if a TPIF system is required, the die can be made of wood or resin. Hence, transforming a milling machine into an ISF system is a very affordable solution.

Initially, the raw materials being formed in ISF systems were metallic, such as aluminum alloys (AA1050 or AA3003, for example) and some steels (DC04 and AISI 304, among others). Such materials were widely used because of their good formability. Other metallic materials, such as magnesium alloys and titanium alloys [2] need to be heated to be formed incrementally, resulting in increased complexity and cost. Recently, however, some researchers have focused on polymeric materials and thermoplastics [3], because these have the advantage of being able to be formed at room temperature using the ISF process.

With regard to the process itself, a common concern for researchers from the traditional literature review for metallic and polymer materials is how the process parameters (step down, sheet thickness, tool diameter and wall angle, among others) affect the finished product; that is to say, how they affect the various specific response variables. The aim of such work is to establish the optimal combination of process parameters for achieving the desired effect on these different response variables. Three of the response variables that have garnered most attention in ISF studies are: (i) the maximum axial force (FZ max), to ensure that the maximum capacity is not exceeded, especially when a machining center has been adapted; (ii) the attained final roughness (Ra), which serves as an indicator of the quality of the finished product; and (iii) the maximum depth (of penetration) (Z) of the tool, which serves as an indicator of the material's maximum formability before any tearing, wrinkling or breakage occurs.

Aerens et al. [4], working with aluminum and steel alloys, estimated the steady state axial forming force (FZs) by means of an analytical model that accounts for tensile strength, initial sheet thickness, tool diameter, scallop height, and initial wall angle. The works of Li et al. [5,6] investigated further by seeking an efficient model for tangential force prediction whereas Bahloul et al. [7] focused on minimizing the sheet thinning rate and the tool (or punch) loads. Recently, Centeno et al. [8] shed light on the importance of spindle speed in considering the variation force for metallic materials—a factor which is even more important for polymeric materials [9–11].

Final roughness is another aspect that has captured attention. Recently, Liu et al. [12] aimed to provide a predictive model and the optimal process parameters for minimizing surface roughness in a study in which they incrementally formed a sheet of AA 7075 O-temper aluminum and investigated four process parameters: step down, feed rate, sheet thickness, and tool diameter. Radu and Cristea [13] discarded sheet thickness and introduced spindle speed for three materials: DC01 steel, 304 stainless steel and AA1050 aluminum alloy. Echrif and Hrairi [14] carried out a similar study on an AA1050-O aluminum alloy sheet.

Formability analysis, regardless of the material, is mainly focused on obtaining forming limit diagrams, e.g., Silva et al. [3], where dedicated equipment is required. However, since the initial stages of the development of ISF technology, alternative and simple formability indicators have been used, such as maximum reachable wall angle and its corresponding maximum reachable depth [2,15].

It is well known that using trial and error methodology for determining the best combination of process parameters in manufacturing processes is expensive and time consuming. Having reviewed the current status of ISF, we have noted that before addressing optimization, a Box-Behnken design (BBD) of response surface methodology can provide a systematic approach to examining the main effects of the process parameters—and the interactions between them—on the response variables. We have also observed that studies involving polymeric materials are scarce and, in the case of biocompatible materials, non-existent. Therefore, the aim of this paper is to investigate the effects of four process parameters: tool diameter, spindle speed, feed rate and step down on three response variables: forming axial force, surface roughness and final depth in the ISF process using two biocompatible thermoplastic materials: polycaprolactone (PCL) and ultra-high molecular weight polyethylene (UHMWPE).

2. Materials and Methods

2.1. Geometry and Materials

Tests were performed on 2 mm-thick sheets of the biocompatible polymers, UHMWPE and PCL. The test geometry in this experiment is a pyramidal frustum (Figure 1a), the features of which are:

- varying wall angle along the part's depth.
- 105 mm in length.
- 45° for the initial wall angle
- 80 mm generatrix radius.

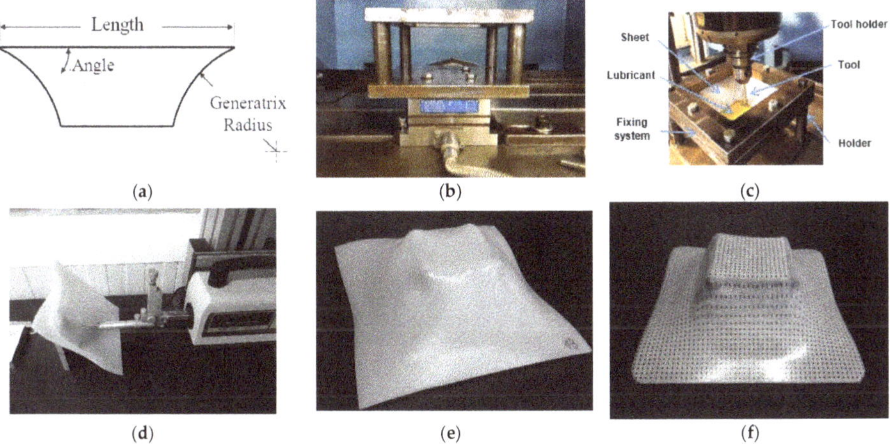

Figure 1. (**a**) Test geometry (**b**) Dynamometer with tooling (**c**) Experimental setup (**d**) Roughness measurement on ultra-high molecular weight polyethylene (UHWMPE) (**e**) UHMWPE part (**f**) polycaprolactone (PCL) part.

From a general point of view, both polymers present low density and high ductility but differ in their thermal properties, e.g., Vicat and melting temperature are lower for PCL (Table 1). The commercially available UHMWPE sintered sheets were initially around 10 mm thick and were sliced by a CNC saw machine and converted into 150 × 150 × 2 mm^3 sheets. The PCL sheets were produced in our laboratory by compression molding. Around 55 g of PCL pellets (Sigma Aldrich, Saint Louis, MO, USA, ≈3 mm, average Mn = 80,000) were positioned into the cavity (150 × 150 × 2 mm^3) of a stainless-steel cast which was previously warmed to a set temperature between 60 and 80 °C inside a heating hydraulic press. A low load was applied for a fixed time to guarantee the melt of the material. Subsequently, the load was increased, thus keeping the sheet in place for a few more minutes to complete the final compaction of the fused polymer. The pressure was then retired, and the cast cooled to room temperature by placing it in a cooling press.

Table 1. Mechanical properties.

Material	Vicat Softening Temperature (°C)	Tensile Strength (MPa)	Elastic Modulus (MPa)
PCL	44.3	15.2	375
UHMWPE	80	19	700

The selection of these two materials is made according to two aspects, their mechanical behavior and their final application. They represent, in both cases, two confronted or extreme cases. From the point of view of mechanical behavior, PCL shows a decrease in strength after the yield point, although it is maintained in a stable value. Whereas for UHMWPE a strain hardening behavior is appreciated (in [16] where self-made PCL and UHMWPE sheets are compared) demonstrating that is a more rigid material. While under the point of view of the final biomedical application, they also respond to two different possible sectors where ISF technology can develop products. The characteristic of biodegradability is very important for PCL; this is why at present it is highly valued by doctors. While the basic properties of UHMWPE have been significant in the orthopedic sector for years.

2.2. *Experimental Setup*

The ISF tests were carried out using a Kondia®HS1000 3-axis milling machine (Kondia, Elgoibar, Spain). The details of the clamping system (Figure 1b) and the setup are described in detail in previous works [10]. The forming forces were measured by a table-type dynamometer Kistler® 9257B (Kistler ibérica SL, Granollers, Spain); surface roughness was determined by means of a Mitutoyo Surftest SV-2000 profilometer (Sariki, Cerdanyola del Vallès, Spain) (Figure 1c) and the maximum depth was recorded by a direct reading off the Kondia milling machine (Kondia, Elgoibar, Spain).

2.3. *Design of Experiments*

Box-Behnken designs (BBD) [17] are three-level designs that allow second order response surfaces to be fitted efficiently. The design for the four factors consists of 27 experimental runs (Table 2) with 24 unique experimental settings (one replicate) plus three replicates at the central point. The explanatory variables came from the classical approach for metal ISF studies: thickness was kept constant for comparison and spindle speed was added because of its importance in polymer forming. The levels of the chosen variable were as follows:

- Dt: Tool diameter (6, 10, 14 mm)
- S: Spindle speed (Free*, 1000, 2000 rpm)
- F: Feed rate (1500, 2250, 3000 mm/min)
- Δz: Step down (0.2, 0.35, 0.5 mm).

* Rotation is considered to be free when any rotation of the tool is due solely to the friction between the sheet and the tool itself.

Table 2. Design of experiments and results.

ID	Tool Diameter, Dt (mm)	Spindle Speed, S (rpm)	Feed Rate, F (mm/min)	Step Down, Δz (mm)	PCL			UHMWPE		
					Fz Max (N)	Ra (μm)	Zmax * (mm)	Fz max (N)	ΔRa (μm)	Zmax * (mm)
1	6	Free	2250	0.35	208.72	0.498	27.7 (0)	485.33	0.437	42.7 (1)
2	14	Free	2250	0.35	439.14	0.627	29.1 (0)	1027.50	0.750	42.7 (1)
3	6	2000	2250	0.35	190.95	0.41	42.7 (1)	414.58	0.916	42.0 (0)
4	14	2000	2250	0.35	329.64	2.23	43.0 (1)	697.15	−0.194	35.7 (0)
5	10	1000	1500	0.20	314.38	0.608	41.4 (0)	635.68	0.511	43.0 (1)
6	10	1000	3000	0.20	309.16	0.622	43.0 (1)	596.08	0.242	43.0 (1)
7	10	1000	1500	0.50	293.05	1.393	42.0 (0)	636.66	0.391	43.0 (1)
8	10	1000	3000	0.50	291.63	0.509	43.0 (1)	591.00	0.324	43.0 (1)
9	10	1000	2250	0.35	325.57	0.585	38.2 (0)	643.44	0.477	42.7 (1)
10	6	1000	2250	0.20	214.14	0.453	43.0 (1)	491.68	0.508	43.0 (1)
11	14	1000	2250	0.20	425.94	1.114	39.0 (0)	765.37	0.739	43.0 (1)
12	6	1000	2250	0.50	197.63	0.484	37.0 (0)	399.84	0.373	43.0 (1)
13	14	1000	2250	0.50	390.87	1.549	24.0 (0)	858.88	0.635	43.0 (1)
14	10	Free	1500	0.35	320.78	1.023	40.6 (0)	818.05	0.332	42.7 (1)
15	10	2000	1500	0.35	275.32	1.880	41.3 (0)	581.63	0.230	38.5 (0)
16	10	Free	3000	0.35	343.10	0.716	40.3 (0)	747.58	0.198	42.7 (1)
17	10	2000	3000	0.35	282.30	1.735	40.6 (0)	558.92	0.420	39.2 (0)
18	10	1000	2250	0.35	296.22	0.464	42.7 (1)	595.00	0.524	42.7 (1)
19	6	1000	1500	0.35	227.42	0.527	42.7 (1)	486.95	0.354	42.7 (1)
20	14	1000	1500	0.35	418.57	1.385	30.1 (0)	802.43	0.863	42.7 (1)
21	6	1000	3000	0.35	240.99	0.330	42.7 (1)	449.30	0.231	42.7 (1)
22	14	1000	3000	0.35	381.09	0.902	31.5 (0)	830.85	0.956	42.7 (1)
23	10	Free	2250	0.20	330.25	0.859	36.0 (0)	727.72	0.115	42.8 (0)
24	10	2000	2250	0.20	281.23	1.775	37.0 (0)	554.77	0.549	38.8 (0)
25	10	Free	2250	0.50	355.89	0.579	43.0 (1)	774.19	0.801	43.0 (1)
26	10	2000	2250	0.50	266.34	2.102	43.0 (1)	587.48	0.407	36.5 (0)
27	10	1000	2250	0.35	280.30	0.598	42.7 (1)	611.80	0.440	42.7 (1)

Note: * In parentheses: 1 = maximum depth accomplished, 0 = sheet fracture.

2.4. Analysis Procedure

The methodology used to estimate the response surface models is the one described by Myers [18]. The main steps are summarized in Figure 2 and are as follows:

- Estimation of the full model with first order, two-way interactions and pure quadratic terms.
- Sequentially removal of the non-significant terms based on the tests on individual regression and groups of coefficients. Each model was evaluated in terms of the fit statistics: R^2, R^2-adjusted, R^2-predicted and RMSE. In addition, the test for significance of regression (p-value associated to Model in the ANOVA table) was observed: a p-value $< \alpha$ indicated that the regression was significant. The lack of fit test was also examined as an indicator of the tentative model satisfactorily describing the data when its p-value was high.
- Model adequacy checking: last squares regression assumptions.

The approach we followed was to retain, in the model, the smallest subset of explanatory variables providing a significant regression test and a non-significant lack-of-fit test as well as good fit statistics (high R^2, adj-R^2, pred-R^2 and low RMSE) together with an appropriate model adequacy. The reason for using a subset of explanatory variables, rather than all of them, is that the estimates of the coefficients will have smaller variance and the predictions will be more precise. The Shapiro Wilk normality test (SWNT) was used to check the normality of the residuals.

Note that coded variables (−1, 0 and 1, indicating low, medium and high level, respectively) were used to compare the size of the coefficients and that the significance level in all cases is $\alpha = 0.05$. The statistical analysis was conducted using R software [19].

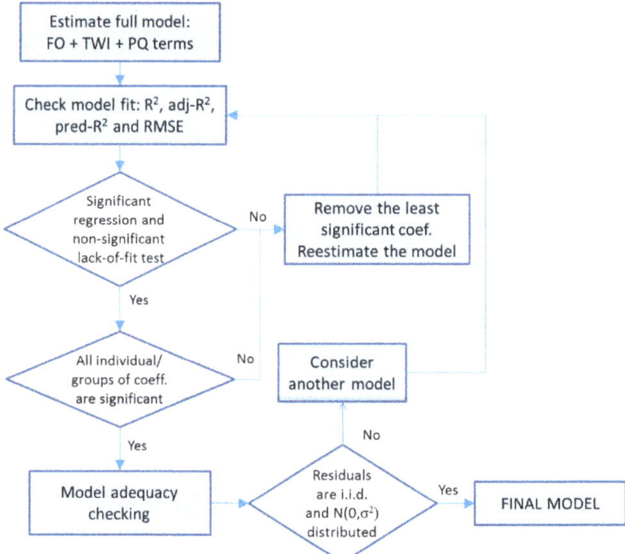

Figure 2. Model selection scheme.

3. Results and Discussion

The effects of varying various forming parameters on the three response variables under study (forming force, roughness and maximum depth) during an ISF process have been explored previously in a number of studies ranging from normal ISF to hot ISF, in experiments or simulations, and with regard to metals as well as polymers. We shall now go on to compare the results of these studies with the results we obtained for the two biocompatible materials in this work.

The factor levels (i.e., the process parameters) as well as response values for all 27 experimental runs are shown in Table 2. Note that the response value ΔRa has been introduced; we shall discuss the reasons for this in the following subsections, in which the statistical models proposed for each material are shown and briefly discussed.

3.1. Maximum Axial Force (Fz Max)

3.1.1. PCL

The model that best fits Fz max for PCL is as follows:

$$\hat{y} = 304.83 + 92.12 \cdot \text{Dt} - 31.01 \cdot \text{S} - 22.93 \cdot \text{Dt} \cdot \text{S}, \quad (1)$$

It is a very good model which describes 94% of the variability of Fz max and has a predicted R^2 of 92%. The residuals of the model are homoscedastic, independent and identically normally distributed (SWNT p-value = 0.06).

Two of the first order factors, tool diameter (Dt) and spindle speed (S), are significant, as is the interaction between the two (Table 3) with Dt being the most influential because its coefficient is higher (three times higher than the coefficient of S). In general, the higher Dt (14 mm) produces higher Fz max values. It is clear that the portion of sheet to be formed using a greater tool diameter requires higher forces, as has been pointed out previously for metals (Al7075-O in Li et al. [6], Al3003-O in Bahoul et al. [7] and AISI304 in Centeno et al. [8]) as well as for polymers (PVC in Bagudanch et al. [9]). In contrast, an increase in spindle speed reduces the forces required. The mechanism able to explain this fact involves the change in the friction conditions, as the heat originating from the increase in friction decreases the various forming force values and, as a result, the mechanical behavior of the polymer material may change. This is also consistent with the work presented by Centeno et al. [8,20] and Baharudin et al. [21] for metal materials and, in the case of polymers, by Davarpanah et al. [11] for PLA and PVC, by Lozano-Sánchez et al. [16] for PCL and UHMWPE and by Bagudanch et al. [22] for PVC, PC, PP, UHMWPE and PCL.

The interpretation of the interaction effect (Dt·S) is as follows: a reduction in S (from 2000 rpm to 0 rpm) slightly increases Fz max when Dt = 6 mm, while the increment is more acute when Dt = 14 mm. As we have said, the increase in spindle speed increases the friction and temperature, causing lower force values, which agrees with results independently observed. Hence, it would be expected that, at higher tool diameters, the heat would concentrate even more in the forming region resulting in a temperature increase [22] which, in turn, would decrease the value of the required forming force; however, this does not occur according to the results we obtained. In the Dt·S interaction, it seems that the effect of the increased contact zone at higher tool diameters—which causes an increase in the forming force [4]—is a more dominant influence than the increase in temperature caused by the higher spindle speed.

Table 3. Maximum force (Fz max) model results.

PCL					UHWMPE				
Parameter estimates					**Parameter estimates**				
	Coefficient		*p*-value			Coefficient		Pr (>|t|)	
(Intercept)	304.84		<0.001		(Intercept)	626.33		<0.001	
Dt	92.12		<0.001		Dt	187.88		<0.001	
S	−31.01		<0.001		S	−98.82		<0.001	
Dt·S	−22.93		0.009		F	−15.64		0.033	
					Δz	6.40		0.359	
					Dt·S	−64.90		<0.001	
					Dt·Δz	46.34		<0.001	
					S^2	38.24		<0.001	
Analysis of variance					**Analysis of variance**				
	Df	Sum Sq	Mean Sq	F value	*p*-value		Df	Sum Sq	Mean Sq	F value	*p*-value
Model	3	115,468	38,489	149.76	<0.001	Model	7	579,363	82,766	148.86	<0.001
Residuals	23	5918	257			Residuals	19	10,570	556		
Lack of fit	5	1185	237	0.90	0.502	Lack of fit	17	9360	551	0.91	0.644
Pure Error	18	4733	263			Pure Error	2	1210	605		
Summary of fit					**Summary of fit**				
R^2	0.95	RMSE	16.04		R^2	0.98	RMSE	23.59	
Adj, R^2	0.94	Pred. R^2	0.93		Adj, R^2	0.98	Pred. R^2	0.96	

Table 3. *Cont.*

PCL	UHWMPE

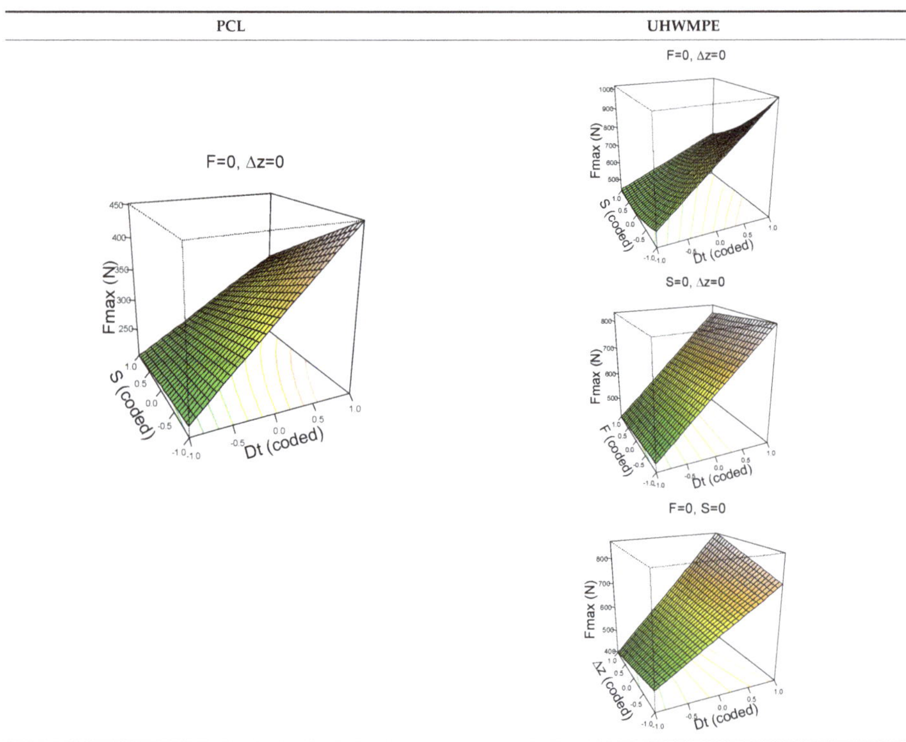

3.1.2. UHMWPE

The model that best explains Fz max on UHMWPE material is as follows:

$$\hat{y} = 626.33 + 187.88 \cdot Dt - 98.82 \cdot S - 15.64 \cdot F + 6.40 \cdot \Delta z - 64.90 \cdot Dt \cdot S + 46.34 \cdot Dt \cdot \Delta z + 38.24 \cdot S^2 \quad (2)$$

This is a very good model able to explain 98% of the variability of Fz max and with a high capability of predicting new response values (pred R^2 = 96%). The residuals of the retained model are i.i.d. distributed (SWNT *p*-value = 0.3303).

Like the PCL model, the UHMWPE model depends highly on Dt which is the most influential first order factor and also appears in the two significant two-way interactions. In general, high values of Dt result in higher values of Fz max. It is already known that an increase in tool diameter entails an increase in force. The second most influential factor is spindle speed, S, and it has a significant quadratic effect and interacts with Dt. The feed rate, F, is the least influential factor because of its lower coefficient; however, it is still significant. The effect of F can be appreciated in the (F, Dt) surface plot in Table 3; when F moves from 1500 to 3000 mm/min, Fz max is reduced on average by 15.64 N.

The interaction effects can be better understood from the surface plots. With regard to Dt·S, it can be appreciated how changing from S = 2000 rpm to S = 0 rpm increases Fmax in all cases, however this increment is more acute when Dt = 14 mm than it is when Dt = 6 mm.

The other significant interaction (Dt·Δz) has the opposite effect: on the one hand, when Dt = 6 mm, a reduction in Δz implies an increase in Fz max; on the other hand, when Dt = 14 mm, a reduction in Δz implies a reduction in Fz max. The reduction in step size decreases the portion of sheet being formed, which should also lead to a reduction in the force: the model shows this trend for high tool diameters. However, it is not the case for low tool diameters for which the forming force slightly increases.

3.2. Surface Roughness (Ra)

Roughness assessment needed to be conducted in a different way for each material, since the blank sheet was produced in different ways in each case. Furthermore, since it is known that sheet metal mark orientation can affect the results of roughness measures [23], it should be pointed out that the roughness values were obtained, in both cases, perpendicular to the step-down direction.

Tool diameter and step down have been identified as parameters that influence surface roughness, Ra, [13,14] mainly in metals, whereas for PVC and PC, step down and spindle speed with some interactions were significant [23]. The manner in which the tool diameter and step down parameters improve or worsen the surface roughness can be seen in the diagrams in Figure 3, which shows how increased Dt reduces Ra, while increased Δz increases it.

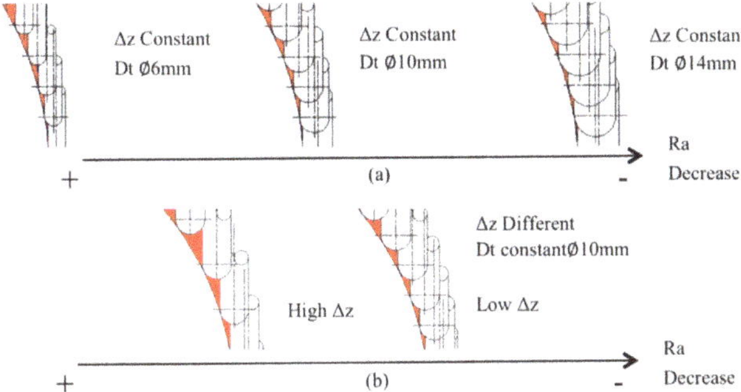

Figure 3. (**a**) Effect of different tool diameters, Dt, on surface roughness, Ra (**b**) Effect of different step down, Δz, on surface roughness, Ra.

3.2.1. PCL

The model that best explains Ra for the PCL material includes all four factors of the experiment:

$$\hat{y} = 0.57 + 0.43{\cdot}Dt + 0.49{\cdot}S - 0.17{\cdot}F + 0.1{\cdot}\Delta z + 0.42{\cdot}Dt{\cdot}S - 0.22{\cdot}F{\cdot}\Delta z + 0.51{\cdot}S^2 + 0.16{\cdot}F^2 + 0.21{\cdot}\Delta z^2 \quad (3)$$

The model explains 88% of the variability of the response Ra and has a good predicted R^2 of 74%. Residuals are independent, homoscedastic and normally distributed (SWNT *p*-value = 0.3082). The model is complex because of the high numbers of terms included in it, but the surface plots shown in Table 4 can help in interpreting it.

Table 4. Surface Roughness (Ra) model results.

PCL			UHWMPE		
Parameter estimates			**Parameter estimates**		
	Coefficient	*p*-value		Coefficient	*p*-value
(Intercept)	0.57	<0.001	(Intercept)	0.46	<0.001
Dt	0.43	<0.001	Dt	0.08	0.259
S	0.49	<0.001	S	−0.03	0.707
F	−0.17	0.010	F	−0.03	0.703
Δz	0.10	0.105	Dt·S	−0.36	0.006
Dt·S	0.42	0.001			
F· Δz	−0.22	0.038			
S^2	0.51	<0.001			
F^2	0.16	0.070			
Δz^2	0.21	0.020			

Analysis of variance (PCL)						**Analysis of variance** (UHWMPE)					
	Df	Sum Sq	Mean Sq	F value	*p*-value		Df	Sum Sq	Mean Sq	F value	*p*-value
Model	9	7.97	0.89	22.24	<0.001	Model	4	0.59	0.15	2.77	<0.001
Residuals	17	0.68	0.04			Residuals	22	1.18	0.05		
Lack of fit	15	0.67	0.04	8.14	0.115	Lack of fit	14	0.91	0.06	1.89	0.184
Pure Error	2	0.01	0.01			Pure Error	8	0.27	0.03		

Summary of fit (PCL)				**Summary of fit** (UHWMPE)			
R^2	0.92	RMSE	0.20	R^2	0.33	RMSE	0.23
Adj, R^2	0.88	Pred. R^2	0.74	Adj, R^2	0.21	Pred. R^2	−0.33

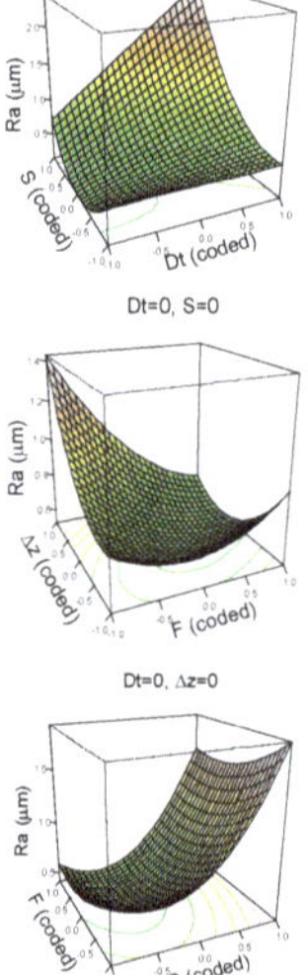

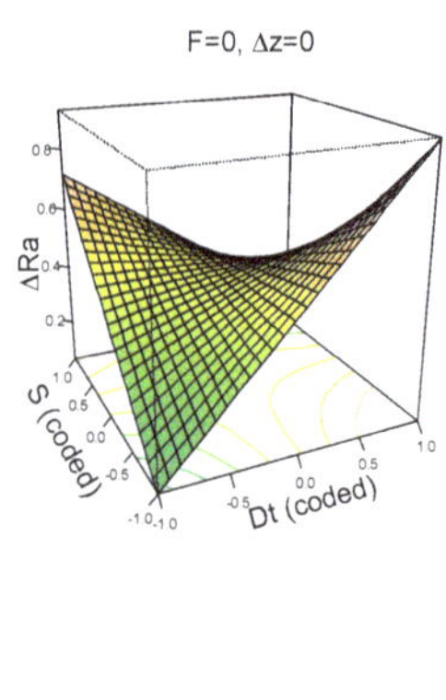

In general, and in contrast to previous studies, a higher Dt is associated with a higher Ra (worse surface quality). However, this effect is more acute when S is at its high level (S = 2000 rpm), due to the friction and the heat that the surface is receiving [9]. Similarly, a higher Δz level is associated with a higher Ra, which is as expected, but again, this effect is more acute when the feed rate is low, F = 1500 mm/min, since at lower feed rates, contact time with the surface is higher, therefore there is more heat, which worsens surface quality. The effect of the quadratic terms S and F show a simple minimum pattern on the response surface. The stationary point of the response surface is Dt = 1.3,

S = −1, F = 0.6, Δz = 0.1 in coded units or Dt = 15 mm, S = 0 rpm, F = 2688 mm/min, Δz = 0.36 mm. This stationary point represents a minimum for the Ra value.

3.2.2. UHMWPE

The parameter of roughness for UHMWPE analyzed here will be the difference between the roughness before and after processing (ΔRa = Rabefore processing–Raafter processing). In contrast to PCL in which the sheets proceed direct from a mold, in the case of UHMWPE, the sheets were sliced in layers from a thicker molded sheet, which meant the surface already began with a texture whose starting roughness was high, at around 1.05 μm on average. With one exception, the ΔRa value is always positive, which means that the processing operation improves the roughness of the material. The roughness is reduced on average by 0.464 μm with 95% CI = [0.36, 0.57].

The model selected to explain the differences in roughness in UHMWPE is as follows:

$$\hat{y} = 0.46 + 0.08 \cdot Dt - 0.03 \cdot S - 0.03 \cdot F - 0.36 \cdot Dt \cdot S \quad (4)$$

The model, however, explains only 21% of the variability of the response variable and has a bad predicted R^2. It was decided to keep F in the model to achieve a non-significant lack-of-fit test. The general regression test is significant. Residuals are normally distributed (SWNT *p*-value = 0.87).

The most important factor explaining the differences in roughness in UHMWPE is the interaction between Dt and S which has a negative coefficient. The surface plots shown in Table 4 can help interpret the model. They show a saddle point (the stationary point) near the center of the plot. From the center point, increasing or decreasing Dt and S at the same time produces a reduction in ΔRa, while increasing one factor and decreasing the other leads to an increase in ΔRa. In other words, Ra after processing is slightly reduced when Dt and S are both at their most positive or most negative values (low ΔRa values), whereas Ra is highly reduced when Dt and S are at opposite levels (very low ΔRa values).

3.3. *Maximum Achieved Depth (Zmax)*

The response variable, maximum achieved depth, Zmax, both on UHMWPE and PCL, cannot be analyzed using a response surface model because the data shows a highly right-skewed distribution which is truncated at different depths (42.7 mm for a step down of 0.35 mm or 43.0 mm in the other cases) depending on the experimental settings. For example, for UHMWPE, 20 out of 27 experiments reached the maximum specified depth (Zmax = 100%), five were in the interval [90,100) and two in the interval [80,90). All tested models have a significant lack of fit test and show a dependence on the residuals vs. fitted values, mainly when Zmax = 100%.

The reason for analyzing the Zmax response is to find the point, z, at which the material breaks and to find which factors explain the breaking depth (Zmax). Generally, this can be modeled using survival (or reliability) analysis which aims to analyze the relationship of time to an event. In this context, the event is defined as the breaking (or failure) of the polymer and time is represented by the depth, z.

The experimental data shows many censored observations, that is, experiments in which the material did not break. Each experiment is an observation of the type (zi,ci) where zi is Zmax and ci = 1 if there has been a failure, and ci = 0 (censored) if otherwise. These ci values are shown in parentheses in Table 2 alongside the Zmax values.

The aim of this section is to explain the probability of the material surviving at depth z, that is, the survival, or reliability, function: $S(z) = 1 - F(z) = P(Z \geq z)$–being careful not to confuse S(z) with the factor S = Spindle-speed. Another important function in survival analysis is the hazard function, $\lambda(z)$. It assesses the instantaneous risk of demise at next mm, conditional on survival to that depth. In other words, it is the expected number of events that will occur in the next mm given that there has not been an event to that depth (z).

In this analysis, the Kaplan Meyer method (a non-parametric approach) is used to estimate S(z). Moreover, the Mantel-Haenszel test (log-rank test) is used to check for differences between survival curves (factor levels).

It is finished by fitting a Cox proportional hazard regression model (semi parametric approach) to determine whether the model including the significant factors is significant itself and to look for significant interactions. It is used with the likelihood ratio, Wald and logrank tests.

Note that the number of experiments at each level is not balanced. For example, there are six experiments at S = −1, 15 at S = 0 and 6 at S = 1. This is because data was not collected to carry out a survival analysis, but rather a BBD of experiments.

3.3.1. PCL

Kaplan Meyer curves (Figure 4) can be interpreted as follows: at z = 20 mm, all observations/experiments are without event and the survival (S(z)) is 1, or, equivalently, 100%. The solid lines, colored according to factor levels, show the events with vertical drops (at each z). Colored curves showing different patterns indicate that there are differences in survival between factor levels, such as is the case of factor Dt (Mantel-Haenszel test p-value = 0.033): it is the only significant factor in PCL survival.

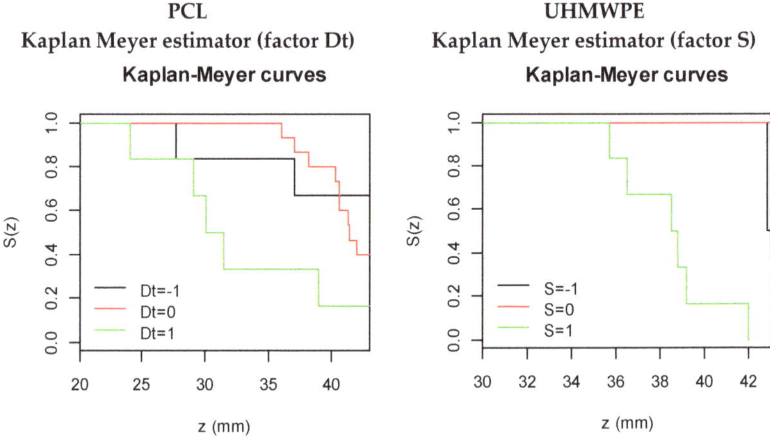

Figure 4. Significant survival curves according to factor levels on PCL and UHMWPE.

Note that the survival curves are not balanced: there are more experiments on Dt = 10 mm than on Dt = 6 or 14 mm (15 vs. 6). However, it is noticeable that survival is lower for Dt = 14 mm.

The Cox proportional hazard regression model has a unique significant factor: Dt. The coefficient of S (b = 0.95) is interpretable in its exponential form ($e^{0.95}$ = 2.59) as the multiplicative effects of the hazard. That is, increasing Dt one level (e.g., from −1 to 0) increases the danger of breaking by, on average, a factor of 2.59. The overall tests of significance p-value (likelihood ratio, Wald and logrank) are significant, indicating that the model is appropriate.

Nevertheless, when S is included in the Cox model together with Dt, the individual regression coefficient of S has a p-value of 0.07 and the overall model is significant. In this case, the coefficient associated with S is b = −0.70 meaning that increasing S by one level reduces the danger of breaking by $e^{-0.70}$ = 0.50. Note, however, that the interaction between Dt and S is not significant. More experiments should be carried out in order to confirm the significance of S in the survival function.

3.3.2. UHMWPE

Kaplan Meyer curves for Zmax on UHMWPE show significant differences only across S levels (Mantel-Haenszel test p-value < 0.001). From Figure 4, it can be seen that S(z) is lower when S = 2000 rpm. No breaks were observed for S = 1000.

The Cox proportional hazard regression model has S as a unique significant factor. Its coefficient is $\hat{\beta} = 3.38$ meaning that increasing S one level increases, on average, the danger of breaking by $e^{3.38} = 29.49$.

The Kaplan-Meyer curves are shown globally without being separated by factor levels in Figure 5. In the graphics comparing PCL versus UHMWPE, it can be observed that PCL sheets break earlier (z around 25) and in greater quantity than UHMWPE sheets in which the first breakage appears at around z = 35.

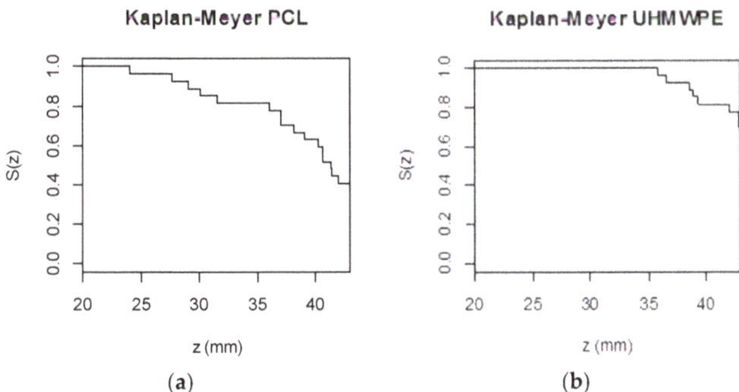

Figure 5. Kaplan-Meyer curves without separating by factor levels for (**a**) PCL and (**b**) UHMWPE.

4. Conclusions

In this work, SPIF experimental tests using two different polymers biocompatible have been carried out following a Box-Behnken design for four factors and a survival analysis. The maximum forming force, surface roughness and Zmax response achieved in the experiments has been statistically analyzed and empirical models for each material have been obtained. Using the proposed models, it is possible to control the Fz max, Ra and Zmax.

Among the process parameters analyzed and according to the data summarized in Table 5, spindle speed and tool diameter have been found to be the most influential parameters in terms of maximum forming force variation for both materials. From the roughness analysis, one can observe again the importance of spindle speed and tool diameter on these biocompatible polymers, as both involve an increase in temperature due to either the friction at higher spindle speeds or the increase in surface contact and contact time between tool and sheet, both of which worsen the surface because the materials flow, degrade and lose their properties.

Finally, the response Zmax, on UHMWPE and PCL, cannot be analyzed using a response surface model because data is highly right-skewed and truncated at different depths. However, taking into account the objective of the paper, the data can be analyzed using a novel method in this field: survival analysis. The results have shown that the most important factor is spindle speed for UHMWPE and tool diameter tool for PCL.

Table 5. Summary of the coefficients of the selected models (bold indicates significance (α = 0.05)).

	Intercept	Dt	S	F	Δz	Dt·S	Dt·Δz	F·Δz	S^2	F2	Δz^2
Fz max											
PCL	**304.84**	**92.12**	**−31.01**			**−22.93**					
UHMWPE	**626.33**	**187.88**	**−98.82**	**−15.64**	6.40	−64.90	**+46.34**		**38.24**		
Ra											
PCL	**0.57**	**0.43**	**0.49**	**−0.17**	0.1	**0.42**		**−0.22**	**0.51**	0.16	**0.21**
UHMWPE (ΔRa)	**0.46**	0.08	−0.03	−0.03		**−0.36**					
Z max (Survival analysis)											
PCL		×									
UHMWPE			×								

Author Contributions: M.S., M.L.G.-R. and I.B. conceived and designed the experiments. M.S. and M.V.-M. carried out data and the statistical analysis. M.S. and I.B. performed the SPIF experiments. M.L.G.-R. and I.F. directed the research. M.S. wrote the manuscript. The manuscript was finalized through contributions from all authors, and all authors approve the final manuscript.

Funding: This research has received funding from the University of Girona (MPCUdG2016/036), the Spanish Ministry of Education (DPI2016-77156-R) and the Catalan Agency for Management of University and Research Grants (2017-SGR-0385).

Acknowledgments: The authors would like to thank the Centro de Investigación de Química Aplicada (CIQA) for supplying the PCL sheets. Finally, we are grateful to Marc López for his collaboration during the experimental work.

Conflicts of Interest: The authors declare no conflict of interest.

References

1. Hu, S.J. Evolving Paradigms of Manufacturing: From Mass Production to Mass Customization and Personalization. *Procedia CIRP* **2013**, *7*, 3–8. [CrossRef]
2. Honarpisheh, M.; Abdolhoseini, M.J.; Amini, S. Experimental and numerical investigation of the hot incremental forming of Ti-6Al-4V sheet using electrical current. *Int. J. Adv. Manuf. Technol.* **2015**, *83*, 2027–2037. [CrossRef]
3. Silva, M.B.; Martinho, T.M.; Martins, P.A.F. Incremental forming of hole-flanges in polymer sheets. *Mater. Manuf. Process.* **2013**, *28*, 330–335. [CrossRef]
4. Aerens, R.; Eyckens, P.; Van Bael, A.; Duflou, J.R. Force prediction for single point incremental forming deduced from experimental and FEM observations. *Int. J. Adv. Manuf. Technol.* **2010**, *46*, 969–982. [CrossRef]
5. Li, Y.; Liu, Z.; Lu, H.; Daniel, W.J.T.; Liu, S.; Meehan, P. Efficient force prediction for incremental sheet forming and experimental validation. *Int. J. Adv. Manuf. Technol.* **2014**, *73*, 571–587. [CrossRef]
6. Li, Y.; Daniel, W.J.T.; Liu, Z.; Lu, H.; Meehan, P. Deformation mechanics and efficient force prediction in single point incremental forming. *J. Mater. Process. Technol.* **2015**, *221*, 100–111. [CrossRef]
7. Bahloul, R.; Arfa, H.; BelHadjSalah, H. A study on optimal design of process parameters in single point incremental forming of sheet metal by combining Box–Behnken design of experiments, response surface methods and genetic algorithms. *Int. J. Adv. Manuf. Technol.* **2014**, *74*, 163–185. [CrossRef]
8. Centeno, G.; Bagudanch, I.; Martínez-Donaire, A.J.; García-Romeu, M.L.; Vallellano, C. Critical analysis of necking and fracture limit strains and forming forces in single-point incremental forming. *Mater. Des.* **2014**, *63*, 20–29. [CrossRef]
9. Bagudanch, I.; Garcia-Romeu, M.L.; Centeno, G.; Elías-Zúñiga, A.; Ciurana, J. Forming force and temperature effects on single point incremental forming of polyvinylchloride. *J. Mater. Process. Technol.* **2015**, *219*, 221–229. [CrossRef]

10. Bagudanch, I.; Sabater, M.; Garcia-Romeu, M.L. Incremental forming of polymers: Process parameters selection from the perspective of electric energy consumption and cost. *J. Clean. Prod.* **2016**, *112*, 1013–1024. [CrossRef]
11. Davarpanah, M.A.; Mirkouei, A.; Yu, X.; Malhotra, R.; Pilla, S. Effects of incremental depth and tool rotation on failure modes and microstructural properties in single point incremental forming of polymers. *J. Mater. Process. Technol.* **2015**, *222*, 287–300. [CrossRef]
12. Liu, Z.; Liu, S.; Yanle, L.; Meehan, P.A. Modeling and Optimization of Surface Roughness in Incremental Sheet Forming using a Multi-objective Function. *Mater. Manuf. Process.* **2014**, *29*, 808–818. [CrossRef]
13. Radu, M.C.; Cristea, I. Processing Metal Sheets by SPIF and Analysis of Parts Quality. *Mater. Manuf. Process.* **2013**, *28*, 287–293. [CrossRef]
14. Echrif, S.B.M.; Hrairi, M. Significant Parameters for the Surface Roughness in Incremental Forming Process. *Mater. Manuf. Process.* **2014**, *29*, 697–703. [CrossRef]
15. Attanasio, A.; Fiorentino, A.; Mazzoni, L.; Ceretti, E.; Giardini, C. Design of an Equipment for Forces Measurement in TPIF Process. *Int. Conf. Technol. Plast.* **2008**, *9*, 1783–1788.
16. Lozano-Sánchez, L.M.; Bagudanch, I.; Sustaita, A.O.; Iturbe-Ek, J.; Elizalde, L.E.; Garcia-Romeu, M.L.; Elías-Zúñiga, A. Single-Point Incremental Forming of Two Biocompatible Polymers: An insight into their thermal and structural properties. *Polymers* **2018**, *10*, 391. [CrossRef]
17. Box, G.E.P.; Behnken, D.W. Some New Three Level Designs for the Study of Quantitative Variables. *Technometrics* **1960**, *2*, 455–475. [CrossRef]
18. Myers, R.H.; Montgomery, D.C.; Anderson-Cook, C.M. *Response Surface Methodology: Process and Product Optimization Using Designed Experiments*, 3rd ed.; Wiley Series in Probability and Statistics; Wiley: New York, NY, USA, 2009.
19. R Core Team. *R: A Language and Environment for Statistical Computing*; R Foundation for Statistical Computing: Vienna, Austria, 2014. Available online: http://www.R-project.org/ (accessed on 1 January 2016).
20. Centeno, G.; Martínez-Donaire, A.J.; Bagudanch, I.; Morales-Palma, D.; Garcia-Romeu, M.L.; Vallellano, C. Revisiting formability and failure of AISI304 sheets in SPIF: Experimental approach and numerical validation. *Metals* **2017**, *7*, 531. [CrossRef]
21. Baharudin, B.T.H.T.; Azpen, Q.M.; Sulaima, S.; Mustapha, F. Experimental investigation of forming forces in frictional stir incremental forming of aluminum alloy AA6061-T6. *Metals* **2017**, *7*, 484. [CrossRef]
22. Bagudanch, I.; Vives-Mestres, M.; Sabater, M.; Garcia-Romeu, M.L. Polymer Incremental Sheet Forming Process: Temperature Analysis Using Response Surface Methodology. *Mater. Manuf. Process.* **2016**, *32*, 44–53. [CrossRef]
23. Powers, B.M.; Ham, M.; Wilkinson, M.G. Small Data Set Analysis in Surface Metrology: An Investigation Using a Single Point Incremental Forming Case Study. *Scanning* **2010**, *32*, 199–211. [CrossRef] [PubMed]

© 2018 by the authors. Licensee MDPI, Basel, Switzerland. This article is an open access article distributed under the terms and conditions of the Creative Commons Attribution (CC BY) license (http://creativecommons.org/licenses/by/4.0/).

Article

Force Prediction for Incremental Forming of Polymer Sheets

Gustavo Medina-Sanchez, Alberto Garcia-Collado, Diego Carou * and Rubén Dorado-Vicente

Department of Mechanical and Mining Engineering, University of Jaén, EPS de Jaén, Campus LasLagunillas, 23071 Jaén, Spain; gmedina@ujaen.es (G.M.-S.); acollado@ujaen.es (A.G.-C.); rdorado@ujaen.es (R.D.-V.)
* Correspondence: dcarou@ujaen.es; Tel.: +34-953-211-712; Fax: +34-953-212-420

Received: 29 June 2018; Accepted: 30 August 2018; Published: 3 September 2018

Abstract: Incremental sheet forming (ISF) is gaining attention as a low cost prototyping and small batch production solution to obtain 3D components. In ISF, the forming force is key to define an adequate setup, avoiding damage and reducing wear, as well as to determine the energy consumption and the final shape of the part. Although there are several analytical, experimental and numerical approaches to estimate the axial forming force for metal sheets, further efforts must be done to extend the study to polymers. This work presents two procedures for predicting axial force in Single Point Incremental Forming (SPIF) of polymer sheets. Particularly, a numerical model based on the Finite Element Model (FEM), which considers a hyperelastic-plastic constitutive equation, and a simple semi-analytical model that extends the known specific energy concept used in machining. A set of experimental tests was used to validate the numerical model, and to determine the specific energy for two polymer sheets of polycarbonate (PC) and polyvinyl chloride (PVC). The approaches provide results in good agreement with additional real examples. Moreover, the numerical model is useful for accurately predicting temperature and thickness.

Keywords: incremental forming; FEM; force prediction; numerical model; semi-analytical model; specific energy

---

1. Introduction

Sheet forming by means of local deformations or incremental sheet forming (ISF) is a prototyping and small batch production solution to obtain 3D components developed in the late 20th [1]. During ISF, a tool follows a 3D path deforming a flat sheet clamped to a rigid frame. This technique is noted for their low cost and good forming capabilities, beyond the forming limit curve, compared to conventional forming processes [2]. The process has applications in automotive industry [3], aeronautical industry [4] and biomedicine [5].

The main ISF solutions are punctual incremental forming and doubled sided incremental forming assisted by a partial or full die (Figure 1). Single Point Incremental Forming (SPIF) is flexible, easy to implement low cost solution, and has been in the research spotlight in the last years. However, this process induces complex deformation mechanisms (shear, stretch and bending), which cause poor dimensional results [6], and limit its use in industrial applications [7].

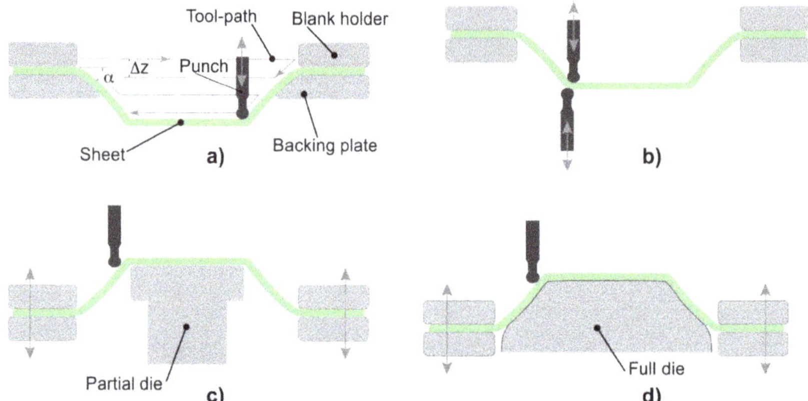

Figure 1. Main incremental sheet forming (ISF) techniques. (**a**) Single point incremental forming; (**b**) Two point incremental forming; (**c**) Double-sided incremental forming with partial die; (**d**) Double-sided incremental forming with full die.

SPIF is usually accomplished by a conventional Computer Numerical Control (CNC) machine equipped with a clamping frame and a punch that follows a G-code program of instructions. The deformation mechanism is related to the forming forces and, therefore, a good knowledge about forming forces allows tackling several key SPIF problems: How to improve dimensional accuracy; how to avoid machine damage and wear; how to extend the technique to different materials, etc.

Despite the industrial interest of low cost small batch production of polymer parts at room temperature [8], the extension of SPIF to these materials is currently a challenge. In addition to the SPIF drawbacks mentioned above, it is difficult to model the deformation behavior of polymer materials. Most research efforts are focused on developing prediction models for metals of known mechanical properties. When coming to glassy polymers, they suffer relaxation after yielding and, therefore, they have a complex constitutive equation with strain rate and temperature material dependence.

The study of the forming force can be an adequate strategy to advance in forming of polymer sheets. Thus, progress can be made from the extension of forming force approaches for metal sheets to polymer sheets or by developing new forming force models. To the best knowledge of the authors, existing analytical and experimental forming force models were specifically developed for metals, and the scarce numerical models for polymers are not comparable to those for metals.

Regarding the force approaches for metals, analytical models define relations between process parameters, material mechanical properties and force. For example, Bansal et al. [9] relates the forming forces with thickness, meridional and circumferential stresses, contact surface and thickness. Besides experimental models, such as the regression equation as defined by Aerens et al. [10], estimate the forming force by means of experimental data. When considering numerical simulations, they are accurate for simple geometries but computationally expensive because of the nonlinearities produced by contact area changes. Behera et al. [11] shows an extensive classification of numerical simulations works focused on metal sheets.

The existing studies about incremental forming of different polymeric materials are limited and force prediction is not their main goal. In this regard, experimental works check new applications and asses how process parameters influence response variables such as force, roughness and incremental depth [12]. On the other hand, analytical and numerical approaches try to determine deformation mechanics and failure modes [13]. Concerning these solutions, it is worth mention the theoretical model based on membrane analysis developed by Silva et al. [14], the constitutive equation, based on overstress proposed by Alkas et al. [15,16], to model viscoplastic materials with only seven material parameters, and the numerical study of Nguyen et al. [17], who considers a viscoelastic material to estimate sheet thickness and spring-back.

The present work proposes a semi-analytical model based on fitting forming force measurements when forming truncated cones for different values of the deformed volume. This deformed volume is determined as a function of the contact surface and thickness. The novelty of this model is to extend the concept of specific energy, used in the orthogonal cutting model to SPIF. In this regard, a new concept called specific forming energy is introduced. This specific forming energy relates the forming forces with the geometric parameters characteristic of the process. The solution is simpler than analytical approaches.

Moreover, a coupled thermo-mechanical numerical model, with a hyperelastic-plastic constitutive equation suitable for polymer sheets, is presented. Unlike other aforementioned models with viscoplastic and viscoelastic constitutive equations, the proposed approach studies a hyperelastic-plastic material, only defined by six material parameters, which considers mechanical and thermal response. The material model parameters are determined minimizing the differences between experimental and simulated material stress-strain curves at different temperatures.

The structure of the paper comprises: Section 2 describes the considered assumptions, and the developed numerical and semi-analytical models; Section 3 explains the setup and tests carried out; Section 4 presents the main results, and discusses the goodness of the proposed approaches with respect to real measurements; and, finally, Section 5 presents the main conclusions.

2. Forming Force Models

2.1. Semi-Analytical Force Model

A simple method that relates the experimental axial forming forces, with a locally deformed volume, is proposed. Note that the main forming force acts along the z direction, so that is the one analyzed.

Following a similar reasoning such as that used in orthogonal cutting, the local plastic deformation induced during SPIF along the z direction requires a power P, so that it should be held that:

$$P = F_z \cdot \Delta z, \tag{1}$$

where F_z is the axial forming force and Δz is the step down.

It stands to reason that the power needed to deform a sheet volume V should be constant for a specific material and temperature. This constant U, similarly to the machining case in Reference [18], can be called specific forming energy:

$$U = \frac{F_z \cdot \Delta z}{V}, V = S \cdot t, \tag{2}$$

where S is the affected punch-sheet area, and t is the mean thickness beneath the punch.

When knowing S and t at a specific position and U for a specific material, it is possible to solve the Equation (2) for obtaining F_z.

The affected area can be calculated from the expression proposed by Bansal et al. [9] as:

$$S = \frac{\pi D}{4}(D/2 + \Delta z) + \frac{\pi D^2}{8(\gamma - \alpha)}(sin\alpha - sin\gamma). \tag{3}$$

This area depends of the incremental depth Δz, the forming tool diameter D and the angle of the sheet α, since γ can be calculated as:

$$\gamma = arcos\left(1 - \frac{2 \cdot \Delta z}{D}\right). \tag{4}$$

The mean thickness of the deformed volume can be calculated from the sheet area deformed in a step down (Figure 2) and the arc of the punch perimeter in contact with the sheet as:

$$t = \frac{A}{(\alpha' + \gamma') \cdot D/2}, \tag{5}$$

$$\alpha' = arcos(cos(\alpha) - 2 \cdot \Delta z/D), \tag{6}$$

$$\gamma' = arcos(1 - 4 \cdot \Delta z/D). \tag{7}$$

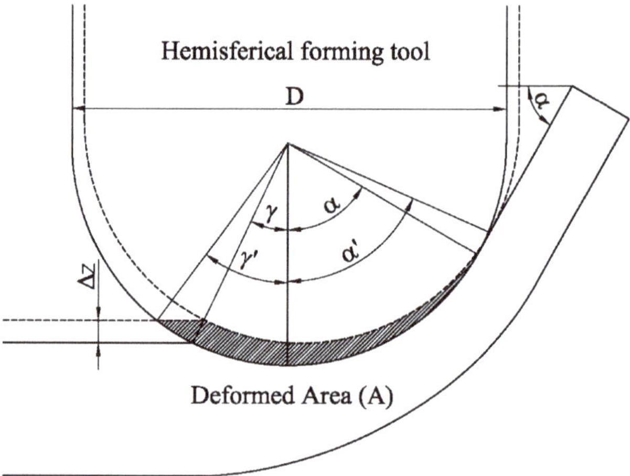

Figure 2. Deformed area defined.

Based on all of the above, the deformed volume calculated in Equation (2) depends exclusively on the geometric parameters: Δz, D, and α. In summary, once the material characteristic U is experimentally determined, the axial forming forces can be estimated from the geometric parameters using Equation (2).

2.2. Numerical Estimation of the Forming Force for Polymer Sheets

The numerical model was carried out in a fully coupled-stress dynamic analysis using the ABAQUS commercial software. The analysis considers the inertia effect, the temperature-dependent of the material response, and the transient thermal response. It also includes the integration of the momentum and the heat flow equations coupling the material energy dissipation during plastic flow, rising the local temperature [19].

The Finite Element (FE) model used for the numerical simulation of the process can be seen in Figure 3. The forming tool and the backing plate were simulated as analytical rigid surfaces in order to perform an efficient numerical contact analysis. The polymer sheet is discretized employing 2052 S4RT elements with reduced integration to avoid the hourglassing, with an active degree of freedom to capture the variation of the temperature at nodes throughout the thickness by bilinear interpolation. The number of integration points used to capture the temperature variation throughout the thickness was 5. The mesh was generated taking into account the tool path that defines the final shape of the part. It was generated uniformly through angle and radius coordinates in order to maintain elements aligned and to generate minimal distortion during the polymer sheet deformation and thickness reduction. The minimum element length in the model was 3 mm, obtaining a minimum stable time increment of 1.2×10^{-6} s that is small enough to avoid instabilities during the entire simulation. Note that, mesh size influences the approach accuracy, nevertheless the authors selected the element size in order to limit the computational time to 10 h (for an Intel Core i8 CPU) maintaining an adequate concordance with the experimental results. The contact between the plastic shell and the forming tool was simulated by surface-to-surface interaction with penalty tangential behavior and hard contact behavior for normal direction.

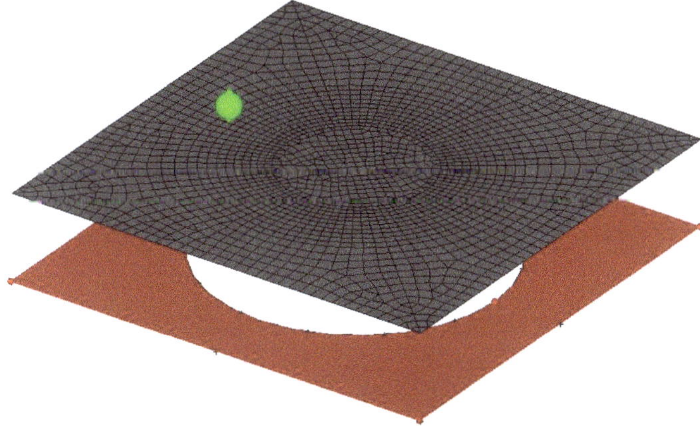

Figure 3. FE model for the SPIF analysis.

The thermal contact conductance Ψ between the metal forming tool and the plastic shell was simulated by gap conductance independent of the contact pressure for polycarbonate (PC) and polyvinyl chloride (PVC) with value of 0.183 W/mK [20]. Other heat transfer mechanisms such as convection or radiation were not taken into consideration due to the low temperature of the plastic shell reached during the entire process. The role of the friction in the formability of the thermoplastics and metals during the SPIF process has been widely discussed by several authors [21–25], concluding that the effect of the friction in the thermoplastic has an important role due to the increase of the temperature of the material. In this work, the values of the dynamic coefficient of friction for the lubricated PC-Shell interface μ_k for PC and PVC are taken from Ludena and Bayer [26]. Thermal properties like conductivity k_t and specific heat C_p are taken from the manufacturer's data sheet. Table 1 summarized all thermal properties employed in the numerical model.

Table 1. Thermal properties for polycarbonate (PC) and polyvinyl chloride (PVC) employed in the numerical model.

Property	PC	PVC
Dynamic friction coefficient, μ_k	0.08	0.12
Thermal contact conductance (W/mK), Ψ	0.183	0.183
Specific heat (kJ/kgK), C_p	1.25	1.18
Thermal conductance (W/mK), k_t	0.2	0.175
Thermal expansion coefficient (m/mK), B	6.5×10^{-5}	7×10^{-5}

The material model employed for the two thermoplastics is a non-linear hyperelastic model combined with J_2-plasticity theory based on isotropic hardening: A simple hardening law that obtains good results with glassy polymers and low material parameters [27]. The hyperelastic component is based on the Arruda-Boyce eight chain model [28] that takes a non-linear Langevin chain statistics into account when deriving the strain energy density function. The predicted stress response of the eight-chain model can be written as follows:

$$\sigma = \frac{\mu}{J\overline{\lambda^*}} \frac{\mathcal{L}^{-1}(\overline{\lambda^*}/\lambda_L)}{\mathcal{L}^{-1}(1/\lambda_L)} \mathrm{dev}[b^*] + \kappa(J-1)\mathbf{I}, \tag{8}$$

where μ is the shear modulus, k the bulk modulus, and λ is the limiting chain stretch. The variable $b^* = J^{-2/3}b$ is the distortional left Cauchy-Green tensor, and $\overline{\lambda^*}$ is the applied chain stretch which can be calculated from:

$$\overline{\lambda^*} = \sqrt{\frac{tr[b^*]}{3}}. \tag{9}$$

The J_2-plasticity component is based on isotropic hardening, that describes the size change of the yield surface σ^0 as a function of the equivalent plastic strain $\bar{\varepsilon}^{pl}$. The model incorporates this effect by an exponential law defined by Equation (10):

$$\sigma^0 = \sigma|_0 + Q_\infty\left(1 - e^{-b\bar{\varepsilon}^{pl}}\right), \tag{10}$$

with, $\sigma|_0$ the yield stress at zero plastic strain, Q_∞ the maximum change of the size in the yield surface and b defines the rate at which the size of the yield surface changes as plastic strain develops. If the material is rate independent, the yield condition is:

$$\sigma^0 = q, \tag{11}$$

with

$$q = \sqrt{\frac{3}{2}S:S}, \tag{12}$$

and S is the deviatoric stress. The yield function is only dependent on the temperature and the equivalent plastic strain ($\bar{\varepsilon}^{pl}$).

The procedure to determinate the values of the needed model parameters (μ, k and λ) is described in Section 3.1.2.

3. Materials and Methods

3.1. Experimental Setup and Polymer Sheets

3.1.1. Setup

The SPIF process is performed using an ALECOP-ODISEA conventional milling machine, with an in-house developed fixing system (Figure 4a). The sheet fixing system is placed on the machining bed, and it consists of a frame made of four aluminum profiles, a die with a hole of 140 mm diameter and an upper die. Eight screws are used to fasten the two dies. Two different tools made of aluminum, with a hemispherical tip of 10 and 12 mm in diameter are used. A lubricant fluid has been used during the experiments.

A 9257BA Kistler dynamometer table is used to measure the value of the force. The dynamometer table is placed between the milling table and the fixing system.

A Flir T335 thermal imaging camera, with a 320 × 240 pixel resolution, is used to measure the temperature reached in the polymer sheet during the process.

Two 200 × 200 mm PVC and PC sheets with a thickness of 3 mm are used. The maximum size of the sheet is the same than that of the working space of the CNC machine to properly fix the sheet. The final shape of the formed sheet is expected to have a 40 mm deep cone, an outer diameter of 128 mm and a cone opening angle α with values between 45° and 60°. Figure 4b shows the shape of the specimen and its theoretical dimensions.

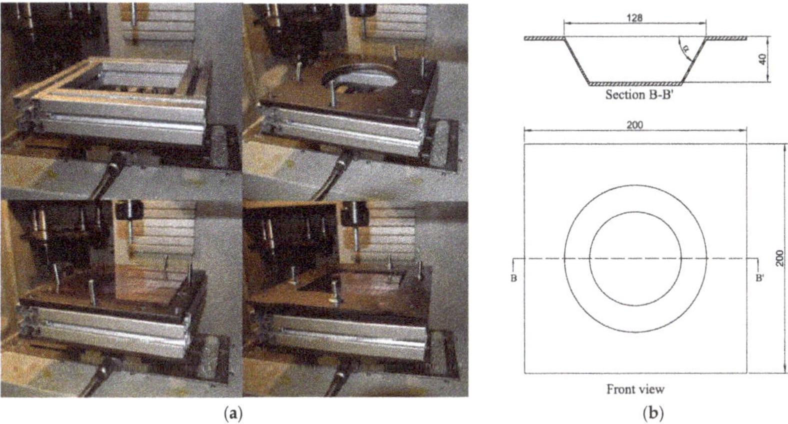

(a) (b)

Figure 4. Setup for SPIF: (**a**) CNC machine with fixing system; (**b**) Shape of processed sheet.

3.1.2. Material Properties

The monotonic stress-strain behavior of both glassy polymers, PVC and PC, at different temperatures were characterized by universal testing machine with 6 kN load cell to calibrate the constitutive model, taking into account the temperature material dependence during the simulation. The tests were carried out in uniaxial tension with three replications of each test by following the ASTM D 638-02a norm. The tests were performed at constant strain rate of 1000 mm/min at 273 and 373 K for PC specimens and 273, 313 and 343 K for PVC specimens. It was noticed that both materials experienced a linear elastic response followed by a yielding, after this point the material undergoes softening behavior (Figure 5). The parameters used for the above described material model (Table 2) were calibrated by Mcalibration® commercial software. Instead of an exponential law, a stress-strain tabular data is used to define the evolution of the yield surface size σ^0. In order to fit PVC and PC material parameters at different temperatures, an optimization method based on Nelder-Mead algorithm was employed. The fitness function was the coefficient of determination R^2, Figure 5 shows the coefficient R^2 for all curves. The calibration was performed with rate independence, for this reason only yield stress values and equivalent plastic strain are provided.

In this work, the experimental tests were conducted with lubricant fluid and, therefore, a low temperature variation was noted. Although it is not the aim of this work, once the numerical simulation is validated, it could be used to predict the temperature evolution since the material model was fitted for different temperatures. Figure 6 shows an example of temperature prediction, the gradient of temperature localized in the plastic sheet along the wake generated by the steel tip was due to the dissipated energy converted into heat due to the frictional sliding. This energy is responsible of the increase of temperature in the thermoplastic sheet that requires a temperature dependent material model to predict the axial force relaxation.

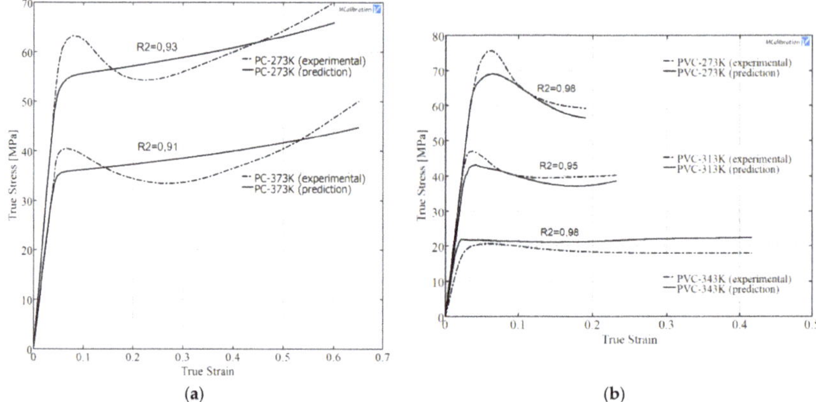

(a) (b)

Figure 5. Experimental behavior and calibrated model results at different temperatures for PC (**a**) and PVC (**b**).

Table 2. Parameters of the material model proposed.

Material	Temperature	Eight Chain Model Components			Isotropic Hardening		
	(K)	μ (MPa)	K	λ (MPa)$^{-1}$	σ_{ys} (MPa)	σ_{ult} (MPa)	ε_{ult}
PC	273	805.3	0.0085	2.06	50.2	65.7	0.61
	373	5398.3	0.02	2.36	34.4	42.1	0.66
PVC	273	630.6	0.006	6.65	64.2	58.1	0.18
	313	1040.5	0.0065	4.18	38.1	39.2	0.24
	343	1480	0.0345	8.2	18.2	22.6	0.42

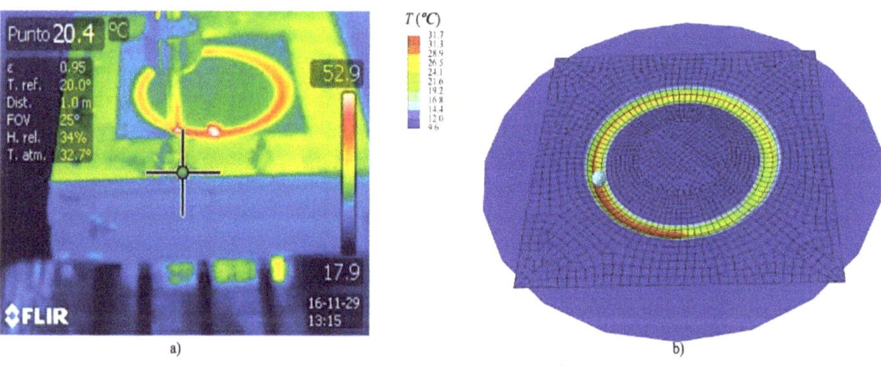

Figure 6. Temperature for a PVC sheet: (**a**) Example of experimental temperature measurements; (**b**) numerical temperature estimation.

3.2. Experimental Procedures

One SPIF test for tuning the numerical model was performed. On the other hand, three tests, changing the deformed volume by means of process parameters, were carried out to determine the semi-analytical approximation. These tests are repeated for each material considered (PVC and PC). In order to validate the models, two additional tests with different conditions to that of the aforementioned tests were carried out.

4. Results and Discussion

4.1. Tuning Tests

The measured axial forming force profile and the numerical estimation, for the tuning tests described in Table 3, are shown in Figure 7. There is an initial transition zone where the forming force grows and a stationary region ($z > 15$ mm) where the force stands approximately constant. The numerical model was fitted to reproduce the stationary region, and the final numerical result is portrayed in Figure 7. For each z, the numerical force is the mean of three values obtained at random punch positions, which can be the reason for the oscillation observed in the estimated curve.

Table 3. Experimental tests conducted to define the semi-analytical and numerical model.

Test	Material	α	Δz	D
Tuning tests	PC	60	0.8	12
	PVC	60	0.6	10
Semi-analytical tests	PC & PVC	60	0.5	10
		45	0.5	10
		60	0.8	12
Validation tests	PC & PVC	60	0.8	10

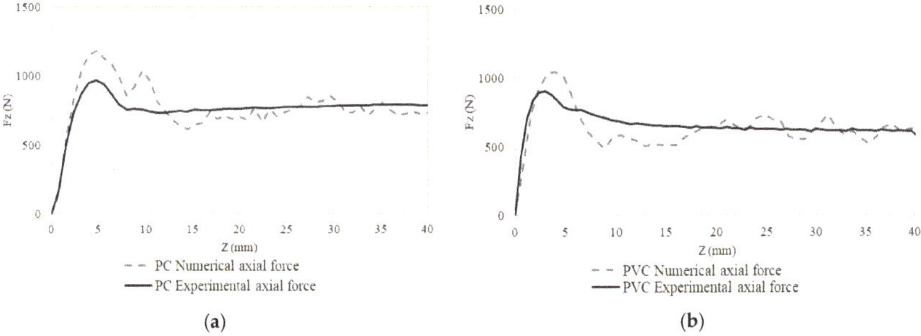

Figure 7. Numerical and experimental axial force profile. (**a**) PC process conditions: Cone shape, $\alpha = 60°$, $D = 12$ mm and $\Delta z = 0.8$ mm; (**b**) PVC process conditions: Cone shape, $\alpha = 60°$, $D = 10$ mm and $\Delta z = 0.6$ mm.

4.2. Specific Energy Equations

Figure 8 shows the forces obtained by the semi-analytical model tests described in Table 3. These forces are the mean value of the axial forming force in the stationary region of each of the experimental tests. Except for the intercept, the experimental measurements agree with Equation (2). The intercept can be interpreted as a minimum forming force to obtain a local deformation.

By fitting the experimental data using a linear function, an equation can be obtained that predicts the value of the axial force as a function of the deformed volume and the step down:

$$Fz = U \cdot \frac{V}{\Delta z} + F_0 \quad (13)$$

where U is the specific forming energy and F_0 is the minimum forming force. For each of the tested polymers, different values of the specific forming force and the minimum force were obtained.

The high value of the determination coefficient shows a good agreement between experimental measurements and the linear approximation.

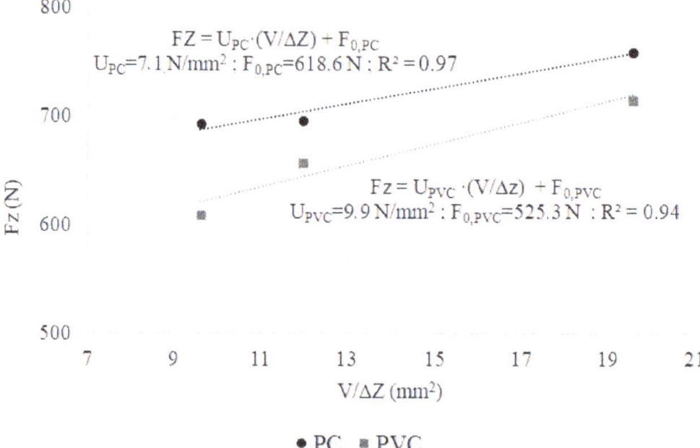

Figure 8. Axial forming forces approximations for PC and PVC.

Only three experimental measurements are enough to obtain the characteristic curve for a specific material. With this curve, any other contact conditions for a material can be computed, predicting the axial forming forces in a simple and accurate way.

4.3. Additional Validation Tests

In order to validate the numerical and semi-analytical models, a test using a different condition to the previous tests was carried out. Figure 9 shows the relative error of the estimation models respect to the stationary axial forming force measurements. The two estimation models agree with the actual forces, and have a relative error below 10%. The best result obtained with the semi-analytical model is due to the deformed volume (in this test) is within the fitted region.

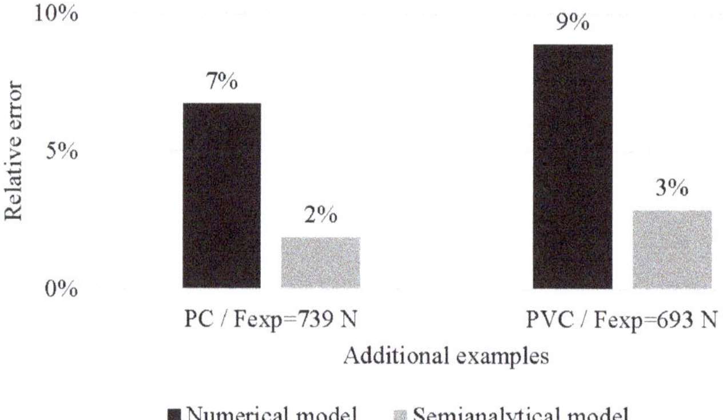

Figure 9. Relative error of the estimation for additional tests. Process conditions: Cone shape, $\alpha = 60^\circ$, $D = 10$ mm and $\Delta z = 0.8$ mm.

The numerical model also provides wall thickness reduction respect to the radial distance (Figure 10). Numerical and experimental thickness measurements are similar for both PVC and PC sheets. The percentage of the mean relative error is always lower than 6% in both cases.

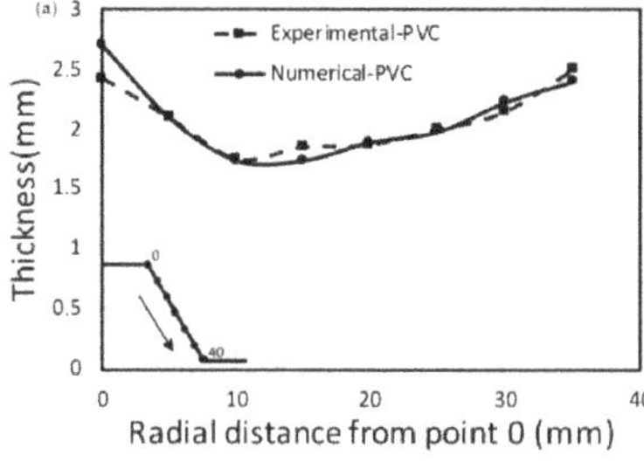

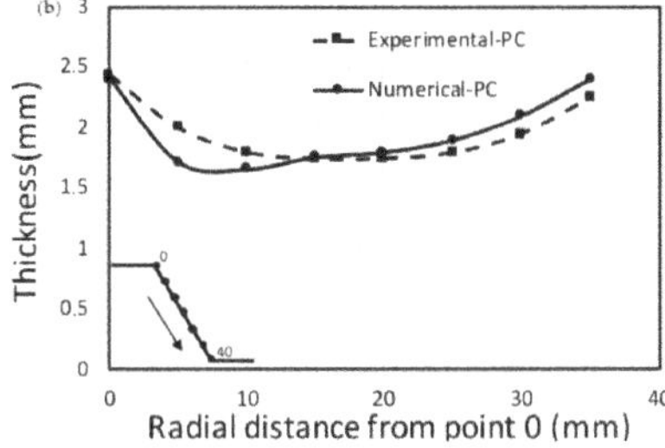

Figure 10. Thickness evolution respect to radial distance, (**a**) PVC and (**b**) PC. Process conditions: Cone shape, $\alpha = 60^\circ$, $D = 10$ mm and $\Delta z = 0.8$ mm.

Minimum thickness values are correctly predicted for both polymers. For the PVC sheet, the numerical model also estimates the radial distance where the minimum thickness occurs (maximum stretching). SPIF process uniformly pushes the same percentage of material to the strain direction. At both wall ends, the thickness is greater than that at the initial component wall: Initial thickness is reduced through the wall by stretching the material to the bottom of the component.

5. Conclusions

The study of forming force in SPIF is key to improve the process. Poor dimensional accuracy is an important SPIF drawback that can be treated knowing the forming force. There are several analytical, experimental and numerical approaches to determine the forming force but, in general, the existing solutions are focused on metals. For polymers, which also have interesting potential applications, it is difficult to accomplish forming force predictions because of their more complex material behavior.

This work shows two axial forming forces approaches: (1) A simple method based on experimental measurements for different deformation volumes and materials, which extends the specific energy concept used in cutting models to incremental forming processes; and (2) a numerical method, that implements a hyperelastic-plastic material.

Material property curves and the axial forming force at steady state for a set of process conditions were used to tuning the numerical model. Regarding to the semi-analytical solution, we conducted experiments with different axial depth, tool diameter and forming angle in order to study the deformation volume influence on the forming force. These tests were repeated for two polymers PVC and PC.

Once the numerical and the semi-analytical models were ready, we compared their estimations with the real measurements of additional forming tests. The results agree with the experimental measurements, a percentage of relative error below 10% was obtained.

Extending the semi-analytical procedure to more materials, and further works on this topic could lead to a fast and robust forming force estimation model. On the other hand, the numerical model not only obtained a good forming force prediction, but also additional variables such as thickness evolution estimation, which together with forming force are two of the main variables in any mechanical forming process.

The two approaches could help to improve the knowledge of the SPIF process on polymers. Each solution addresses a research need: Fast computation (semi-analytical model), or low experimental cost (numerical model).

Author Contributions: Conceptualization, G.M.-S., A.G.-C. and R.D.-V.; Methodology, G.M.-S., A.G.-C. and R.D.-V.; Software, G.M.-S., A.G.-C. and R.D.-V.; Validation, G.M.-S., A.G.-C. and R.D.-V.; Formal Analysis, G.M.-S., A.G.-C. and R.D.-V.; Investigation, G.M.-S., A.G.-C. and R.D.-V.; Resources, G.M.-S., A.G.-C. and R.D.-V.; Data Curation, G.M.-S., A.G.-C. and R.D.-V.; Writing-Original Draft Preparation, G.M.-S., A.G.-C., D.C. and R.D.-V.; Writing-Review & Editing, G.M.-S., A.G.-C., D.C. and R.D.-V.; Visualization, G.M.-S., A.G.-C., and R.D.-V.; Supervision, G.M.-S., A.G.-C., D.C. and R.D.-V.; Project Administration, G.M.-S., A.G.-C., D.C. and R.D.-V.

Acknowledgments: The authors would like to thank the "Mechanical and Energy Engineering" TEP 250 research group.

Conflicts of Interest: No competing financial interests exist.

References

1. Matsubara, M.; Tanaka, S.; Nakamura, T. Development of Incremental Sheet Metal Forming System Using Elastic Tools: Principle of Forming Process and Formation of Some Fundamentally Curved Shapes. *JSME Int. J. Ser. C Dyn. Control Robot. Des. Manuf.* **1996**, *39*, 156–163. [CrossRef]
2. Maqbool, F.; Bambach, M. Revealing the Dominant Forming Mechanism of Single Point Incremental Forming (SPIF) by Splitting Plastic Energy Dissipation. *Procedia Eng.* **2017**, *183*, 188–193. [CrossRef]
3. Jeswiet, J.; Hagan, E. *Rapid Proto-Typing of a Headlight with Sheet Metal*; Canadian Institute of Mining, Metallurgy and Petroleum: Westmount, QC, Canada, 2002.
4. Hussain, G.; Al-Ghamdi, K.A. PEO coating as lubrication means for SPIF of titanium sheet: Characteristics and performance. *Mater. Res. Innov.* **2014**, *18*, S2-727–S2-733. [CrossRef]
5. Lu, B.; Ou, H.; Shi, S.Q.; Long, H.; Chen, J. Titanium based cranial reconstruction using incremental sheet forming. *Int. J. Mater. Form.* **2014**, *9*, 361–370. [CrossRef]
6. Maqbool, F.; Bambach, M. Dominant deformation mechanisms in single point incremental forming (SPIF) and their effect on geometrical accuracy. *Int. J. Mech. Sci.* **2018**, *136*, 279–292. [CrossRef]
7. Al-Ghamdi, K.A.; Hussain, G.; Butt, S.I. Force Variations with Defects and a Force-based Strategy to Control Defects in SPIF. *Mater. Manuf. Process.* **2014**, *29*, 1197–1204. [CrossRef]
8. Franzen, V.; Kwiatkowski, L.; Martins, P.A.F.; Tekkayaa, A.E. Single point incremental forming of PVC. *J. Mater. Process. Technol.* **2009**, *209*, 462–469. [CrossRef]
9. Bansal, A.; Lingam, R.; Yadav, S.K.; Venkata Reddy, N. Prediction of forming forces in single point incremental forming. *J. Manuf. Process.* **2017**, *28*, 486–493. [CrossRef]
10. Aerens, R.; Eyckens, P.; Van Bael, A.; Duflou, J.R. Force prediction for single point incremental forming deduced from experimental and FEM observations. *Int. J. Adv. Manuf. Technol.* **2010**, *46*, 969–982. [CrossRef]
11. Behera, A.K.; de Sousa, R.A.; Ingarao, G.; Oleksik, V. Single point incremental forming: An assessment of the progress and technology trends from 2005 to 2015. *J. Manuf. Process.* **2017**, *27*, 37–62. [CrossRef]
12. Sabater, M.; Garcia-Romeu, M.L.; Vives-Mestres, M.; Ferrer, I.; Bagudanch, I. Process Parameter Effects on Biocompatible Thermoplastic Sheets Produced by Incremental Forming. *Materials* **2018**, *11*, 1377. [CrossRef] [PubMed]
13. Alkas Yonan, S.; Silva, M.B.; Martins, P.A.F.; Tekkaya, A.E. Plastic flow and failure in single point incremental forming of PVC sheets. *Express Polym. Lett.* **2014**, *5*, 301–311. [CrossRef]
14. Silva, M.B.; Alves, L.M.; Martins, P.A.F. Single point incremental forming of PVC: Experimental findings and theoretical interpretation. *Eur. J. Mech. A Solids* **2010**, *29*, 557–566. [CrossRef]
15. Alkas Yonan, S.; Haupt, P.; Kwiatkowski, L.; Franzen, V.; Brosius, A.; Tekkaya, A.E. A Viscoplastic Material Model Based on Overstress for the Simulation of Incremental Sheet Forming of Thermoplastics. *AIP Conf. Proc.* **2011**, *1353*, 803. [CrossRef]
16. Alkas Yonan, S.; Soyarslan, C.; Haupt, P.; Kwiatkowski, L.; Tekkaya, A.E. A simple finite strain non-linear visco-plastic model for thermoplastics and its application to the simulation of incremental cold forming of polyvinylchloride (PVC). *Int. J. Mech. Sci.* **2013**, *66*, 192–201. [CrossRef]
17. Le, V.S.; Nguyen, T.N. A Numerical Simulation of Incremental Forming Process for Polymer Sheets. *Int. J. Model. Simul.* **2012**, *32*, 265–272.
18. Groover, M.P. *Fundamentals of Modern Manufacturing, Binder Ready Version: Materials, Processes, and Systems*; John Wiley & Sons: Hoboken, NJ, USA, 2015; ISBN 9781119128694.
19. Hibbitt, Karlsson & Sorensen, Inc. *ABAQUS/Explicit User's Manual: Version 6.3*; Hibbitt, Karlsson & Sorensen, Inc.: Providence, RI, USA, 2002.
20. Marotta, E.; Fletcher, L.S. Thermal contact conductance of selected polymeric materials. *J. Thermophys. Heat Transf.* **1996**, *10*, 334–342. [CrossRef]
21. Bagudanch, I.; Garcia-Romeu, M.L.; Centeno, G.; Elías-Zúñiga, A.; Ciurana, J. Forming force and temperature effects on single point incremental forming of polyvinylchloride. *J. Mater. Process. Technol.* **2015**, *219*, 221–229. [CrossRef]

22. Durante, M.; Formisano, A.; Langella, A.; Memola Capece Minutolo, F. The influence of tool rotation on an incremental forming process. *J. Mater. Process. Technol.* **2009**, *209*, 4621–4626. [CrossRef]
23. Clavijo-Chaparro, S.L.; Iturbe-Ek, J.; Lozano-Sánchez, L.M.; Sustaita, A.O.; Elías-Zúñiga, A. Plasticized and reinforced poly(methyl methacrylate) obtained by a dissolution-dispersion process for single point incremental forming: Enhanced formability towards the fabrication of cranial implants. *Polym. Test.* **2018**, *68*, 39–45. [CrossRef]
24. Medina-Sánchez, G.; Torres-Jimenez, E.; Lopez-Garcia, R.; Dorado-Vicente, R.; Cazalla-Moral, R. Temperature influence on Single Point Incremental Forming of PVC parts. *Procedia Manuf.* **2017**, *13*, 335–342. [CrossRef]
25. Ward, I.M.; Sweeney, J. *Mechanical Properties of Solid Polymers*, 2nd ed.; John Wiley & Sons: Hoboken, NJ, USA, 2004.
26. Ludema, K.C.; Bayer, R.G. *Tribological Modeling for Mechanical Designers;* ASTM International: West Conshohocken, PA, USA, 1991; ISBN 9780803114128.
27. Bergstrom, J.S. *Mechanics of Solid Polymers: Theory and Computational Modeling;* William Andrew: Norwich, NY, USA, 2015; ISBN 9780323322966.
28. Arruda, E.M.; Boyce, M.C. A three-dimensional constitutive model for the large stretch behavior of rubber elastic materials. *J. Mech. Phys. Solids* **1993**, *41*, 389–412. [CrossRef]

© 2018 by the authors. Licensee MDPI, Basel, Switzerland. This article is an open access article distributed under the terms and conditions of the Creative Commons Attribution (CC BY) license (http://creativecommons.org/licenses/by/4.0/).

Article

Contribution to Reduce the Influence of the Free Sliding Edge on Compression-After-Impact Testing of Thin-Walled Undamaged Composites Plates

Markus Linke [1,*] **and Juan Antonio García-Manrique** [2]

[1] Department of Automotive and Aeronautical Engineering, Hamburg University of Applied Sciences, 20099 Hamburg, Germany

[2] Department of Mechanical Engineering, Universitat Politécnica de Valencia, 46022 Valencia, Spain; jugarcia@mcm.upv.es

* Correspondence: markus.linke@haw-hamburg.de; Tel.: +49-40-428-75-7863

Received: 2 July 2018; Accepted: 6 September 2018; Published: 13 September 2018

Abstract: Standard Compression-After-Impact test devices show a weakening effect on thin-walled specimens due to a free panel edge that is required for compression. As a result, thin-walled undamaged samples do not break in the free measuring area but near the free edge and along the supports. They also show a strength reduction due to the free edge which can become potentially relevant for very weakly damaged panels. In order to reduce the free edge influence on the measured strength, a modified Compression-After-Impact test device has been developed. In an experimental investigation with carbon fiber reinforced plastics, the modified device is compared with a standard device. It is shown that thin-walled undamaged specimens investigated with the modified device now mainly break within the free measuring area and no longer at the free edge and along the bearings as it is the case for standard test devices. The modified device does not cause a free edge weakening effect in comparison to standard devices. The modified device is therefore more suitable for determining the compression strengths of undamaged thin-walled composite plates.

Keywords: Compression-After-Impact testing; Compression-After-Impact strength; carbon fiber reinforced plastics

1. Introduction

Thin-walled fiber-reinforced plastics, in particular, carbon fiber reinforced plastics (CFRP), are very susceptible to reductions in strength due to low velocity impact damages. In aeronautics, this strength loss has to be considered during construction by a sufficient damage tolerance leading to weight penalties. Therefore, it is important to understand the failure mechanisms of impact damaged composite structures in order to improve the damage tolerance of CFRP structures.

The influence of impact damages on the static behavior is typically investigated by Compression-After-Impact (CAI) testing as the residual compressive strength is one major property that significantly decreases due to impact damages. It is a well standardized test (cp. AITM 1-0010 [1], ASTM D 7137 [2], BSS 7260 [3], EN 6038 [4], ISO 18352 [5], SACMA SRM 2R-94 [6]) where the residual strength of damaged composite plates is measured. The residual strength represents a structural strength value which results from the material properties of the plate, the structure of the CFRP, the induced damage mechanisms (such as interlaminar debonding, fiber breakage, matrix cracking, local-sublaminate buckling as well as their interactions) but also from the global geometrically non-linear behavior of the plate under pressure, in particular, as a result of geometric imperfections. Imperfections strongly influence the out-of-plane deformation during the whole compressive load application. Depending on the degree of out-of-plane deformation due to buckling, additional bending moments occur which lead to additional stress levels in the specimens.

The severity of the damage is described by comparing the residual strength in CAI testing with the strength of undamaged specimens. The strength of undamaged specimens is determined either by the in-plane compression strength (e.g., according to ISO 14126 [7]) or by CAI testing (cp. [8,9]).

If the in-plane compression strength of undamaged samples is compared to the residual CAI strength of damaged ones on the one hand, a structural strength value is compared with an in-plane material characteristic. The in-plane material characteristic ideally lacks completely an out-of-plane deformation since the free measuring length in the load direction is small with about 10 to 25 mm according to ISO 14126 [7]. However, the in-plane strength cannot be achieved in CAI testing as an out-of-plane deformation typically occurs during CAI testing resulting in an additional plate bending. The out-of-plane deformation is strongly influenced by geometric imperfections which can become relevant for thin-walled samples due to the free measuring size in the load direction of about 130 mm. As a result, the loss in strength is clearly overestimated.

If the CAI strength is intended to be used for undamaged specimens to quantify the damage severity on the other hand, the measured CAI strength of the undamaged panel is usually reduced due to a free sliding edge which is required for compressing the panel. For a plate damaged with sufficient impact energy, this is generally unproblematic as the damage weakens the plate more than the free edge. The plate then breaks in the area of the damage. However, in the case of an undamaged or only very slightly damaged plate, the weakening of the plate due to the damage may be less than that caused by the free edge. For this reason, the fracture of such samples almost always occurs in the area of the free edge (cp. e.g., [10]). Consequently, the strength in CAI testing is generally underestimated for undamaged samples. This can be relevant also for very weakly damaged samples under specific circumstances. In these cases, standard CAI testing device leads to a systematic error.

Various modifications to CAI test devices are proposed in scientific literature. These differ primarily as a result of the different requirements that CAI testing equipment must meet.

Many articles deal intensively with the prevention of global buckling by the use of adequately modified CAI devices [10–18]. In principle, it is intended to first induce local effects such as local-sublaminate buckling due to delaminated layers in the damaged area. The local buckling results in further matrix cracks and interlaminar debonding finally leading to the collapse of the structure. Since the risk of global buckling is greatest in particular for thin-walled panels, the majority of these articles treats thin composite plates.

Global buckling is basically shifted to higher buckling loads by two main measures. One reduces the free area in the CAI equipment. With the other, additional bearings are applied without changing sample geometry. Typically, the geometry still corresponds to the dimensions according to standardized CAI tests in this case.

The reduction of the free area is achieved either by applying double-sided stabilizing plates (cp. e.g., [10,13–15,17]) to the specimen called anti-buckling plates or by reducing the overall dimensions of CAI test samples through modifications of the equipment (cp. e.g., [11,12]). In both cases, the free measuring area is reduced which is accompanied by an increase in the global buckling load. Moreover, the influence of geometric imperfections is reduced so that out-of-plane deflections under the same load become smaller compared to the unstabilized plate. The stabilizing plates on the sample are either rectangular [10,13,14] or circular [17].

Furthermore, additional lateral supports are introduced on both sides of the samples in comparison to the standardized tests. The bearings are effective as edge knife simple supports. In [16], an additional narrow guide on both sides is applied between the simply supported lateral edges. It acts in the lateral direction over the entire measuring range. The continuous guide is positioned eccentrically so that the guide does not cover the damaged area. Four lateral guides are introduced according to [18] in such a manner that a gap to avoid inference with the damaged area is realized. Due to these guides, the global buckling load is increased.

Research works also deal with the reduction of residual strength in CAI testing due to the potential misalignment with respect to the loaded edges [19]. CAI testing modifications are also reported in the case where the geometry of the specimens under consideration does not fit into the standardized ones (cp. [20] for tapered plates).

Although it is reported in literature that failure in CAI testing occurs under certain conditions at the loaded supports [10], the influence of the weakening effect of the free specimen edge in CAI testing on the residual strength is hardly tackled. Only according to [10] a device is proposed which positions the weakening effect to the plate center. Therefore, the specimens no longer break at the supports but in the plate center. However, the strength reduction due to the free plate edge is not reduced as its length is not changed.

This work deals with the free edge weakening effect of undamaged thin-walled plates in CAI testing. Usually the samples break in the area of the free edge at the bearing and not in the free measuring range. The resulting strength loss which is observed for undamaged samples can also become relevant for very weakly damaged panels as the weakening effect of the free edge can be greater than that of very low energy impact damages. As a result, the influence of strength reduction due to impacts is either not measurable for very weakly damaged panels or is subject to a high degree of uncertainty. In order to illustrate solutions to these difficulties, this work consequently aims at reducing the influence of the free edge in CAI testing on the measured strength of thin-walled samples. For this purpose, an existing CAI testing device is modified in such a way that undamaged samples no longer break in the area of the free edge along the bearings. The comparison with results obtained with a standard CAI device allows to estimate the weakening effect of the free edge. The work focuses on the problem of undamaged samples since the influence of the free edge can be determined more clearly with these than it is possible with damaged samples.

This article is subdivided into six sections. The introduction is followed by Section 2 where typical standards for CAI testing are described. Based on that, the argumentation for the selected modifications is illustrated leading to a modified CAI test device. In Section 3, the specimens and test campaign under investigation are described. The test results are illustrated in Section 4 and discussed in Section 5. Finally, the article concludes with Section 6.

2. Devices for Compression-After-Impact Testing and Test Device Modifications

In aerospace engineering, two different types of fixtures for CAI testing are used. In principle, they differ by the supports at the loaded plate edges. Some CAI test devices exhibit a clamped support. And in contrast to this, other devices realize more or less simply supported edges at the loaded boundaries. In this article, CAI test devices of the former approach, i.e., with a realization of clamped supports at the loaded edges, are investigated. This is due to the fact that a clamped support is practically easier to be realized, in particular, if modifications to the CAI test device are intended to be made.

The standard test device under consideration for the investigation is the Instron "Airbus" Compression Impact Fixture (Instron GmbH, High Wycombe, UK) which agrees with the standards AITM 1-0010 [1] and ISO 18352 [5]. This device is called in the following CAI-standard device or briefly CAI-standard. The geometry and the plate fixture are schematically illustrated in Figure 1a. The nominal dimensions of the specimens to be tested amount to 100 by 150 mm^2 indicated by the surrounding rectangle in Figure 1a. The shaded areas show the clamped part of the coupons. The clamping is achieved by putting specimens between to plane rigid walls of metal. The simply supported bearings are indicated by the dotted lines in the loading direction of the specimen which are parallel to the x-axis. They are applied in the same manner on both sides of the samples. The free measuring area is consequently about 91 by 132 mm^2. The whole edge (100 mm) of the top part of the specimen is loaded whereas the bottom part is completely restrained and does not move.

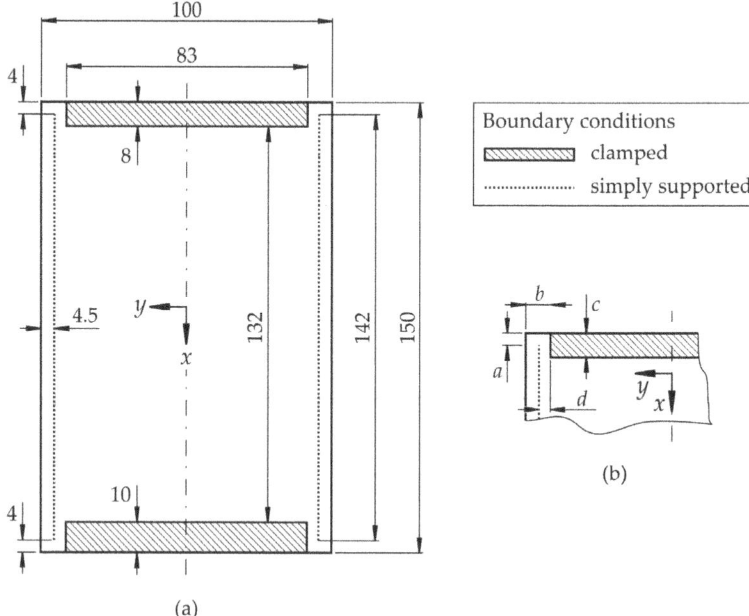

Figure 1. (**a**) Schematic fixture of composite plate (nominal dimension 100 mm × 150 mm) within typical CAI-standard test devices (cp. e.g., standards [1,5]) and (**b**) modified upper corner of CAI test device.

CAI testing using the CAI-standard for undamaged thin-walled composite specimens typically leads to failure at the top clamped boundary starting usually at the free edge of the specimens. Failure does not typically occur in the free range of the specimen like it is the case for damaged specimens. This is due to the fact that the weakening effect of the free edges is greater than the one of plate imperfections. E.g., geometric plate imperfections lead to a theoretical maximum plate bending in the middle of the free measuring area.

Since the free edges at the two upper and two lower corners as well as their support along the width significantly reduce the failure load of undamaged specimens, these areas have been modified in a new CAI testing device which is called CAI-modified below. It is a CAI testing device where the boundaries at all four corners are changed in order to stiffen the free edge area.

The free edges in the lower part of the device are not required for compression. Therefore, they are replaced by clamped supports.

In the upper corners, the two free edges are stiffened by a change in the adjacent bearings. The stiffening effect is primarily caused by constraining the rotational degrees of freedom. This is due to the fact that plate rotations in this area induce additional bending stresses, thereby reducing the failure load. The main dimensions with a stiffening effect are exemplarily shown in Figure 1b. The shorter the free edge a is selected, the stiffer the plate becomes near the free edge. The edge with length b primarily acts as a simple support allowing a rotation around the global y-axis. Clamping consequently leads to the constraint of the rotational degree of freedom and thus to stiffening. The dimension c of the clamped support influences the rotational degrees of freedom around the x- and y-axis in the area where the lateral knife edge support acts. The further the clamping extends into the area between the lateral simple supports (indicated by dotted lines in the Figure 1a,b), the more the free edge area is stiffened. The latter effect increases with a decreasing distance d.

Based on the described relations above between geometry respective constraints with free edge stiffening, the CAI-modified device is derived based on an already existing device manufactured in accordance with [1]. It is illustrated in a schematic sketch in Figure 2a. In the Figure 2b,c, photographs of the device are shown. As a certain length of the free edge is required for specimen compression in the x-direction, the dimension of the free edge has been maintained for about 4 mm (length a according to Figure 1b). But the specimen is now fully clamped along the whole loaded edge (length b according to Figure 1b). As the influence of the free plate edge can be reduced further by a clamped support that extends beyond the simply supported lines in the load direction (indicated by dotted lines in Figure 2a), the upper clamped support has been enlarged in the x-direction in the region between the simply supported lines compared to CAI-standard devices (length c according to Figure 1b). In order not to shorten too much the free measuring area, the support is enlarged by 2 mm in comparison to the standard geometry. The distance d according to Figure 1b is not reduced as the knife edge simple supports of the original device have been used.

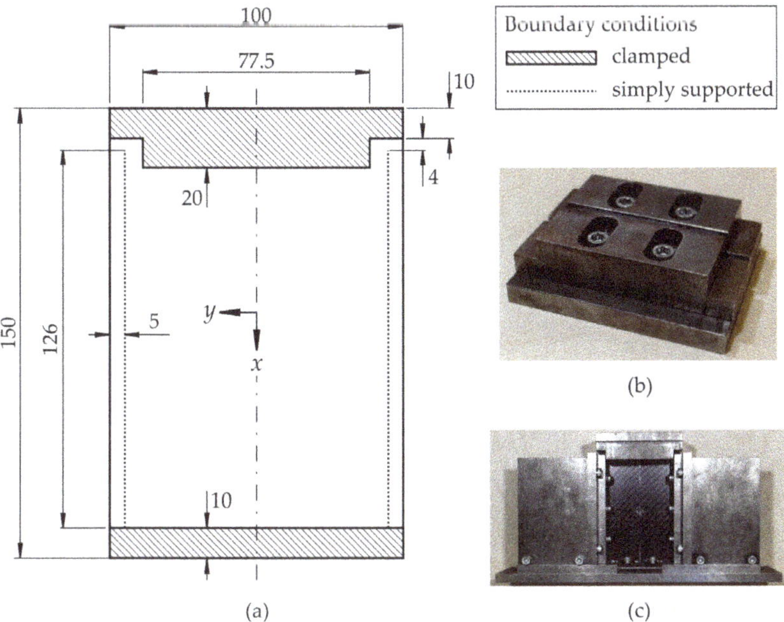

Figure 2. (**a**) Schematic fixture of composite plate (nominal dimension 100 mm × 150 mm) within modified CAI test device, (**b**) top part of modified CAI test device and (**c**) schematic fixture after modification with integrated sample.

In general, it has to be mentioned that this type of fixture is accompanied by a reduction of the free specimen area and therefore by an increase of the specimen stiffness which must be taken into account when comparing failure loads between the two different testing devices.

3. Specimens and Test Campaign

Specimens composed of CFRP are used. It is a unidirectional (UD) as well as a twill weave epoxy prepreg (SGL Technologies GmbH, Meitingen, Germany) which is processed by compression moulding through Clip Carbono (ClipCarbono.com, As Pontes, Spain) to plane plates with thickness of 2.2 mm. Both outer layers are composed of twill carbon weave (CW200-0/TW2/2 with 200 g/m^2 according to [21]). Carbon UD-layers (CU150-0/SO with 150 g/m^2 according to [21]) are positioned between them with 0° and 90° direction where the 0°-direction coincides with the x-axis according to the Figures 1a and 2a. The layer set-up is symmetrical [twill weave, 0°, 90°, 0°, 90°, 0°, 90°]$_s$. The matrix system is epoxy FT1021 according to [22]. The fiber volume content is estimated to 55% for twill weave layers and to 58% for UD layers.

Specimens are investigated in an undamaged condition because it is intended to check whether the modifications of the CAI-modified device lead to failure within the free range of the specimens and not at the supports of the loaded edges where it is expected for the CAI-standard. Five specimens are tested each with the CAI-standard respective CAI-modified device. Sample geometry for CAI-standard testing agrees with the one according to [1], except for the width. In CAI testing with the CAI-modified device, the sample geometry is 98 mm × 150 mm and satisfied the tolerances specified by the AITM standard. The width is reduced by two millimeters in order to integrate the samples into the CAI-modified device without taking the risk of touching the lateral guides during testing as their distance amounts to 100 mm. Otherwise, an additional sample loading could result due to a prevented transversal displacement.

Measurements using the CAI-standard device are carried out with a universal material testing device (Shimadzu Autograph AG-X series with maximum 100 kN, Shimadzu Europe GmbH, Duisburg, Germany). The compression is displacement controlled with a rate of 0.5 mm/min. The displacement data is obtained from the optical strain system Shimadzu Digital Video Extensometer TRViewX (Shimadzu Europe GmbH, Duisburg, Germany). The data recording rate amounts to 1 Hz. The shortening of the samples is measured between two lines having a distance of 50 mm each to the plate center respective middle of the measuring range.

The CAI-modified testing is performed with a servo-hydraulic material testing device (Schenck Hydropuls PSA, PZV 1865, with maximum 100 kN, Schenck AG, Darmstadt, Germany) equipped with Instron measuring instrumentation (Instron Deutschland GmbH Calibration Service, Darmstadt, Germany). The compression is load controlled with a rate of 0.25 kN/s leading to an averaged displacement rate of about 0.5 mm/min. Displacements are measured with the stereo pair camera digital imaging correlation (DIC) system ARAMIS (5 M sensor configuration, 2448 × 2050 pixel, Gom Gesellschaft für optische Messtechnik mbH, Braunschweig, Germany). Displacements are detected on the sample surface as far as both cameras have free view on the sample. The data recording rate is 1 Hz.

4. Test Results

The fracture of the specimens tested with the CAI-standard device is illustrated in Figure 3. A photograph of the samples is shown in Appendix A, Figure A1. Concerning sample numbering, it has to be mentioned that the specimens are taken from a batch production of several panels. This batch production is the basis for several different investigations concerning CAI testing, among others, for impacted coupons, too. Therefore, the samples used here have a non-sequential numbering. This is also valid for the CAI-modified testing below.

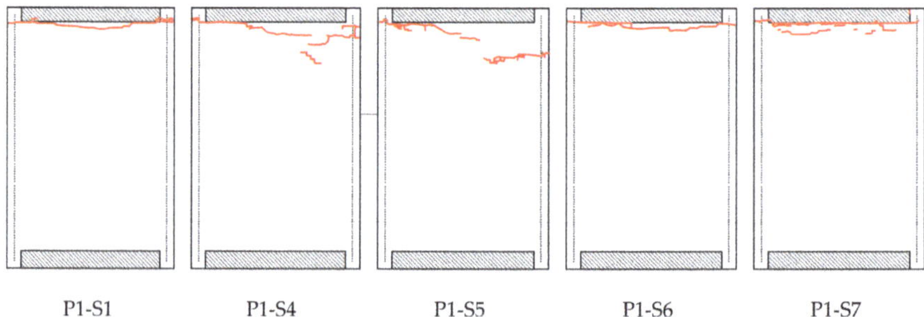

Figure 3. Observable fractures on the surface of the specimen front side tested with CAI-standard indicated by red lines (boundary conditions are described in Figure 1a) for the specimens P1-S1, P1-S4, P1-S5, P1-S6 and P1-S7.

The red lines in Figure 3 indicate the fracture which can be observed on the surface of the specimens, i.e., on the side where the displacements are optically measured (called front side below). Although the red lines are not continuously connected from the left to the right side for the specimens P1-S4 and P1-S5, all specimens show fracture running through the whole specimen width. It is mainly a compression-shear fracture mode. Fracture occurs at the free edge of the CAI device and completely runs along the loaded edge at the clamped support for the specimens P1-S1, P1-S4, P1-S6 and P1-S7. Only specimen P1-S5 exhibits a fracture line which does not completely run along the support. It is a mixed fracture at the support and the free coupon area. The latter specimen also shows the highest failure stress for the CAI-standard testing (see Table 1). In order to cancel out slight deviations concerning the loaded specimen area, the failure stress σ_F is used as failure load. It is computed by the measured failure load divided by the specimen area.

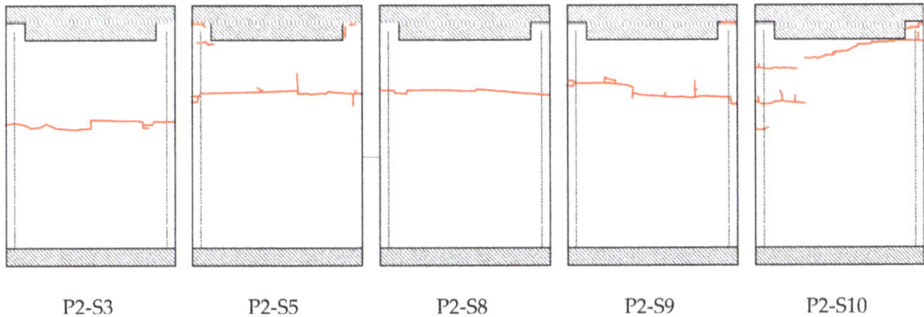

Figure 4. Observable failures on the surface of the specimen front side tested with CAI-modified (boundary conditions are described in Figure 2a) for the specimens P2-S3, P2-S5, P2-S8, P2-S9 and P2-S10.

Table 1. Failure loads and place of fracture (also indicated in Figures 3 and 4) of different specimens for devices CAI-standard as well as CAI-modified with mean failure stress and standard deviation.

Denotation	CAI-Standard Failure Stress (Mpa)	Place of Fracture	Denotation	CAI-Modified Failure Stress (Mpa)	Place of Fracture
P1-S1	239.3	at support	P2-S3	276.9	free plate
P1-S4	242.7	at support	P2-S5	266.9	free plate
P1-S5	253.4	mixed	P2-S8	263.6	free plate
P1-S6	234.9	at support	P2-S9	279.8	free plate
P1-S7	240.4	at support	P2-S10	255.6	mixed
Average:	242.2		**Average:**	268.6	
Deviation:	6.9		**Deviation:**	9.9	

The fracture of the specimens tested with the CAI-modified device is illustrated in Figure 4. A photograph of the samples is shown in Appendix A, Figure A2. Again, the fracture on the surface of the front side is indicated by red lines. All specimens show fracture that runs along the whole width. The specimens typically exhibit compression-shear fracture modes that are combined with some delamination. The main failure modes are matrix cracks, delaminations and fiber breakage. A typical shear failure mode is observed, with interlaminar cracking in the middle part of the specimen. It can be noticed that some samples also exhibit a kink zone and fracture, as can be seen in Figure 5. The plastic deformation of the matrix followed by microbuckling evolves to kink zones, which provoke the fiber buckling, and the formation of two planes of fracture.

This is also valid for specimen P2-S10 although the red lines are not completely connected on the surface. Fracture that is limited to the free range of the plate occurs for the specimens P2-S3, P2-S5, P2-S8 and P2-S9. Partially, fracture can be observed at the clamped support (for specimens P2-S5 and P2-S9) but these fractures are isolated and are not linked to the fracture line running through the free range of the plates. Contrary to that, specimen P2-S10 exhibits a mixed fracture, i.e., the fracture line runs at the clamped support into the free range of the plate. Its failure stress is lower than the ones for the specimens with fracture in the free range only (see Table 1). But the failure stress is in the range of specimen P1-S5 (CAI-standard testing) where the fracture line also runs from the support into the free range of the plate.

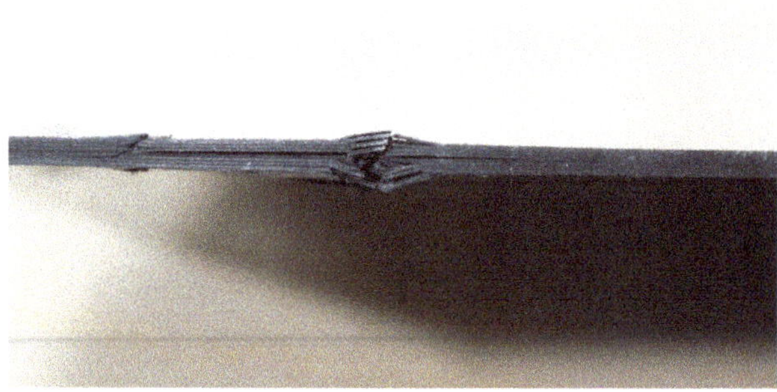

Figure 5. Illustrative kink zone and fracture, specimen P2-S10.

The mean values of failure stress σ_F as well as the corresponding standard deviation s are computed based on the data according to Table 1. Results are illustrated in Figure 6. They amount for the CAI-standard testing device to

$$\sigma_F = 242.2 \text{ MPa}, s = 6.9 \text{ MPa}$$

and for the CAI-modified testing device to

$$\sigma_F = 268.6 \text{ MPa}, s = 9.9 \text{ MPa}.$$

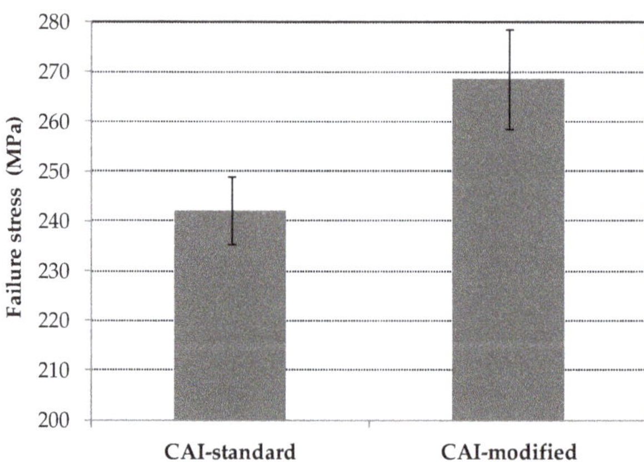

Figure 6. Failure stresses in CAI testing as well as the standard deviations for testing devices CAI-standard as well as CAI-modified.

The mean stress versus the shortening of the samples is shown in Figure 7. For the CAI-standard device, the shortening is directly measured according to Section 2 whereas the shortening concerning results of the CAI-modified device is computed based on DIC data related to the definition of the shortening of CAI-standard measurements. For the sake of simplicity, results of the CAI-modified respective CAI-standard device are illustrated with black respective colored lines.

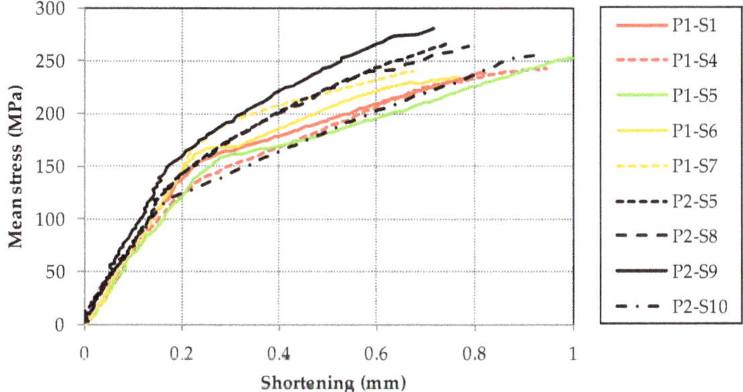

Figure 7. Mean stress versus shortening of specimens (for P2-S3 no digital imaging correlation data is available and the results for CAI-standard device are shown up to the maximum load).

In principle, all samples exhibit a linear trend between mean stress and shortening at the beginning of the measurement. According to DIC data of the CAI-modified testing, a global buckle starts to occur typically at 100 MPa which is then fully established of about 140 MPa in the plate center. From there on, the shape of the buckle remains constant. Only the absolute value of the out-of-plane displacement increases till failure (cp. out-of-plane displacements of measurements with the CAI-modified device in Appendix B in the Figures A3–A5). The failure of all samples occurs suddenly.

5. Discussion of Results

The mean failure stress is significantly increased with the CAI-modified device compared to the standard one. The increase amounts to about 10% which is greater than the resulting standard deviations. The weakening effect of the free edge is consequently significant for CAI testing as the strength loss for low impact energies is in the order of magnitude of the increase. For quasi-isotropic thin-walled CFRP laminates with different layer set-ups, e.g., the decrease in CAI strength amounts to about 26 to 34% for impact energies of 0.5 J according to [23]. In [12] respective [14] a strength reduction of 5% (impact energy of 5.5 J) respective 7% (impact energy of 7.5 J) for quasi-isotropic thin-walled CFRP composites is reported. Due to that, the evaluation of the influence of damages due to very weak impact energy scenarios on the specimen strength should be estimated more precisely with the CAI-modified device. However, it has to be mentioned that the free measuring area between the supports is reduced, too, from 91 by 132 mm^2 to 90 by 120 mm^2. This comes along with a stiffening effect of the specimen. As a consequence, the bending loading of the plate is lowered compared to the CAI-standard test (if the same external load and equal geometric imperfections are assumed). Therefore, a thin-walled specimen investigated with the CAI-modified device will probably exhibit a higher failure load so that the increase of the mean failure stress is expected to be partly caused by the reduced free measuring area. However, this is probably of minor influence as similar quasi-isotropic CFRP laminates according to [12] exhibit a CAI strength difference of about 7% if the free measuring area respective lateral length is reduced to one half (compared to only 90% for the device discussed above). Moreover, results for similar fracture modes (according to P1-S5 and P2-S10) obtained from both devices exhibit failure stresses in the same stress range (cp. Figures 3 and 4 as well as Table 1). Nevertheless, in order to facilitate a comparison where the influence of a reduced free measuring range is not existent, the measuring area should be increased to the size that agrees with typical standards.

The standard deviation of the mean failure stress is increased in CAI-modified testing compared to CAI-standard from 2.8% to 3.7% of the mean value. However, this increase is mainly caused by an undesirable failure mode in CAI-modified testing where the rupture runs partially along the loaded upper support (cp. sample P2-S10 in Figure 4). If the sample P2-S10 is excluded, we obtain a standard deviation of 2.9% (for a mean failure stress of 271.8 MPa). This deviation is in the range of CAI-standard tests which should not be exceeded by a modified device. If CAI testing campaigns are carried out using the proposed CAI-modified device, the validation procedures of the selected samples should consequently conform to typical standards where unacceptable failure modes are discarded. For example, failure modes are typically not accepted if failure does not occur in the central free measuring area (between 20% to 80% scale of the free lateral measuring length).

The fractures lines of the specimens tested with the CAI-modified device are no longer positioned at the bearings but within the free measuring area also indicating that the weakening effect of the free edge no longer is relevant. But it has to be outlined that these lines are not positioned in or near the center of the free measuring area although their occurrence might be expected there due to the theoretical maximum bending loading. They occur in the upper part of the specimens. But no rupture is observed in the lower part. This phenomenon is unclear, in particular, as fracture modes of undamaged plates concerning CAI testing are not reported in open literature. However, there is a high probability that this is at least partly influenced by the device. The methodology followed is easily extensible to other combinations of fabrics or matrices, so its study is proposed for future work [24].

6. Conclusions

Due to the modification of the CAI test device at the loaded edge, the place of fracture of thin-walled CFRP samples no longer occurs at the clamped support only as it can be observed for standard CAI testing devices. The fracture lines usually run through the free range of the specimens. Furthermore, the failure stress is clearly increased indicating that the free edge of standard CAI devices weakens thin-walled samples in CAI testing. The intended aim of the research work is therefore achieved by the CAI-modified test device. The modified CAI testing device enables a more accurate measurement of CAI strengths of undamaged specimens. The modifications can be easily transformed to other CAI testing equipment as these usually work with similar loading edges.

Author Contributions: Conceptualization, Methodology, Data Analysis, Visualization by M.L.; Experimental Investigations, Writing-Original Draft Preparation by M.L. and J.A.G.-M.

Funding: This research was partially supported by the Spanish government under grant number DPI2013-44903-R-AR through funding the specimens as well as the CAI-standard device. The Article Processing Charge is also funded under that grant number.

Acknowledgments: The authors gratefully acknowledge the partial funding by the Spanish government under grant number DPI2013-44903-R-AR. The authors also gratefully acknowledge the support by Hamburg University of Applied Sciences (HAW Hamburg, Germany) through a sabbatical term during the winter term 2017/2018 that allowed a two month research stay of researcher Linke at Universitat Politécnica de Valencia (UPV, Spain) in 2017. The financial support for travel and accommodation at UPV by Instituto de Diseño para la Producción y Fabricación Automatizada is highly welcome. The authors are grateful for the active support by Herbert Theilen, Thomas Sterr (both HAW Hamburg) and David Busquets-Mataix (UPV). In particular, the authors would like to thank Herbert Theilen (Laboratory of Lightweight Design) for the fruitful discussions about CAI testing devices as well as for the assistance while carrying out the experimental investigations with the CAI-modified device at HAW Hamburg. The support by Thomas Sterr (Laboratory of Lightweight Design) for realization the modifications of the CAI-modified device as well as by David Busquets-Mataix (Materials Research Institute) for performing the experimental investigations with the CAI-standard device at UPV is very much appreciated.

Conflicts of Interest: The authors declare no conflict of interest.

Appendix A

The damaged front sides of the specimens tested with the CAI-standard respective CAI-modified device are shown in Figure A1 respective Figure A2.

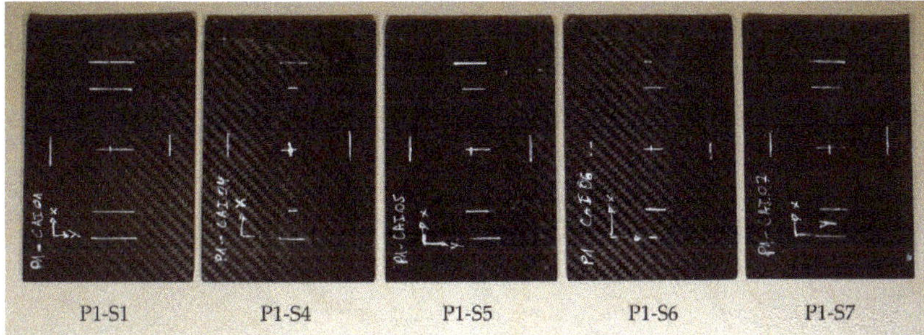

Figure A1. Front side of tested specimens P1-S1, P1-S4, P1-S5, P1-S6 and P1-S7 using CAI-standard device.

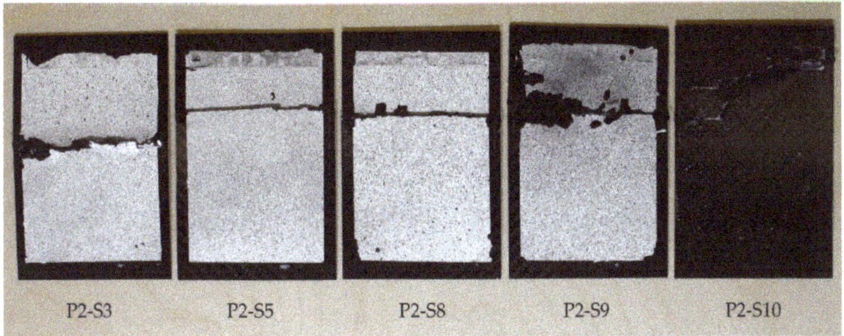

Figure A2. Front side of tested specimens P2-S3, P2-S5, P2-S8, P2-S9 and P2-S10 using CAI-modified device.

Appendix B

The normalized out-of-plane displacement on the symmetrical line according to Figure 2a is shown for specimens investigated with the CAI-modified device in Figure A3.

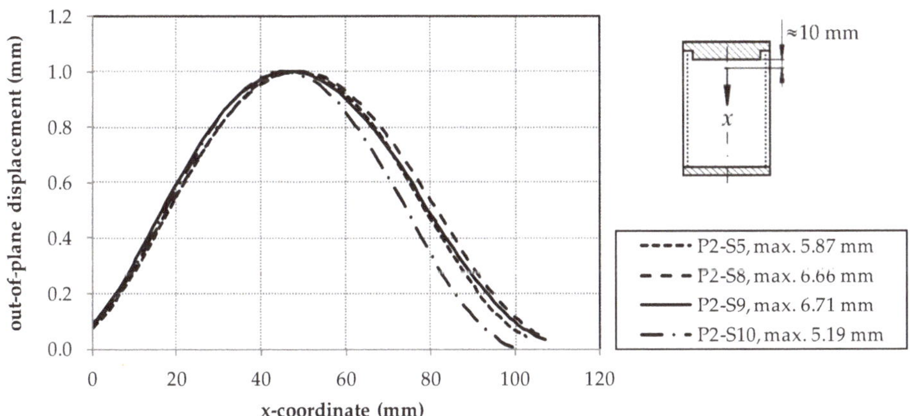

Figure A3. Normalized as well as maximum out-of-plane displacements of specimens investigated with CAI-modified device on the symmetrical line in global x-direction (cp. Figure 2a) starting near top bearing (for P2-S3 no digital imaging correlation data is available).

The out-of-plane displacement patterns of the specimens P2-S5, P2-S8, P2-S9 and P2-S10 shortly before sample failure occurs are illustrated in the Figures A4 and A5. Only the surface area is shown which is in view of both cameras of the DIC measuring device. The gaps are caused by screws and shadows of the surrounding.

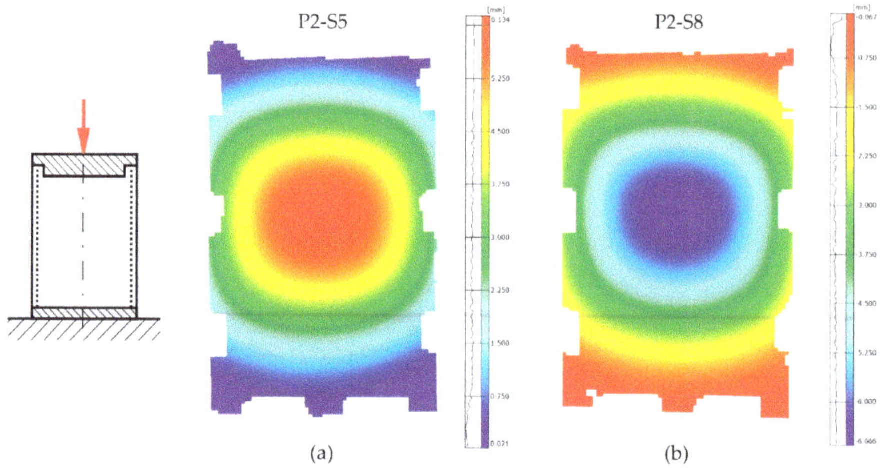

Figure A4. Qualitative illustration of out-of-plane displacement of specimen (**a**) P2-S5 and (**b**) P2-S8 measured with CAI-modified device.

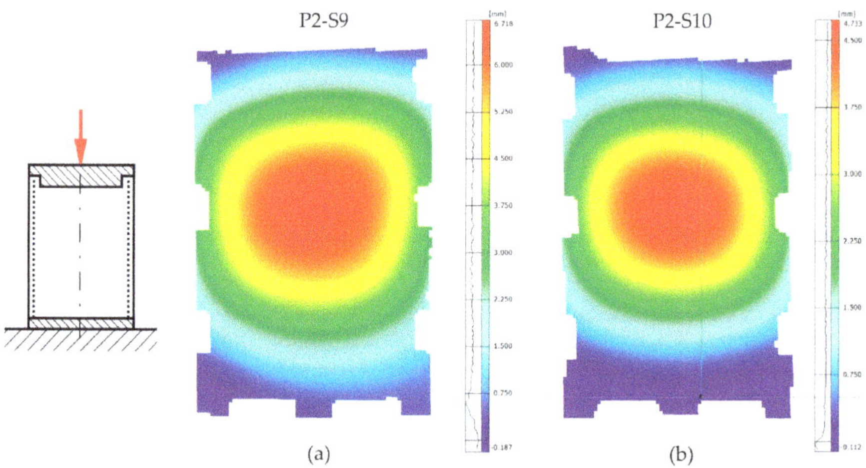

Figure A5. Qualitative illustration of out-of-plane displacement of specimen (**a**) P2-S9 and (**b**) P2-S10 measured with CAI-modified device.

References

1. *Airbus Test Method: Fibre Reinforced Plastics—Determination of Compression Strength after Impact*; AITM1-0010; Airbus S.A.S.: Blagnac, France, 2005.
2. *ASTM Test Method: Standard Test Method for Compressive Residual Strength Properties of Damaged Polymer Matrix Composite Plates*; ASTM D7137/D7137M; American Society for Testing and Materials (ASTM): West Conshohocken, PA, USA, 2017.
3. *Boeing Advanced Composite Compression Tests*; Boeing Specification Support Standard BSS 7260; The Boeing Company: Seattle, WA, USA, 1988.

4. *European Standard: Aerospace Series—Fibre Reinforced Plastics—Test Method—Determination of the Compression Strength after Impact*; German and English Version EN 6038; Beuth Verlag GmbH: Berlin, Germany, 2016.
5. *International Standard ISO 18352 Carbon-Fibre-Reinforced Plastics—Determination of Compression after Impact Properties at a Specified Impact-Energy Level*; Beuth Verlag GmbH: Berlin, Germany, 2009.
6. *SACMA Recommended Test Method for Compression after Impact Properties of Oriented Fiber-Resin Composites SRM 2R-94*; Suppliers of Advanced Composite Materials Association: Arlington, VA, USA, 1994.
7. *International Standard ISO 14126: Fibre-Reinforced Plastic Composites—Determination of Compressive Properties in the In-Plane Direction*; Beuth Verlag GmbH: Berlin, Germany, 2000.
8. Hautier, M.; Lévêque, D.; Huchette, C.; Olivier, P. Investigation of composite repair method by liquid resin infiltration. *Plast. Rubber Compos.* **2010**, *39*, 200–207. [CrossRef]
9. Linke, M.; Nesslinger, S.; García-Manrique, J.A. Re-Infiltration Repairs of Composites in Civil Aviation. In Proceedings of the 7th International Conference on Mechanics and Materials in Design, Albufeira, Portugal, 11–15 June 2017; Silva Gomes, J.F., Meguid, S.A., Eds.; INEGI/FEUP: Porto, Portugal, 2017.
10. Sánchez-Sáez, S.; Barbero, E.; Zaera, R.; Navarro, C. Compression after impact of thin composite laminates. *Compos. Sci. Technol.* **2005**, *65*, 1911–1919. [CrossRef]
11. Aymerich, F.; Priolo, P. Characterization of fracture modes in stitched and unstitched cross-ply laminates subjected to low-velocity impact and compression after impact loading. *Int. J. Impact Eng.* **2008**, *35*, 591–608. [CrossRef]
12. Ghelli, D.; Minak, G. Low velocity impact and compression after impact tests on thin carbon/epoxy laminates. *Compos. Part B* **2011**, *42*, 2067–2079. [CrossRef]
13. Khondker, O.A.; Herszberg, I.; Hamada, H. Measurements and prediction of the compression-after-impact strength of glass knitted textile composites. *Compos. Part A* **2004**, *35*, 145–157. [CrossRef]
14. Mendes, P.A.A.E.; Donadon, M.V. Numerical prediction of compression after impact behavior of woven composite laminates. *Compos. Struct.* **2014**, *113*, 476–491. [CrossRef]
15. Nettles, A.T.; Hodge, A.J. Compression-after-impact testing of thin composite materials. In Proceedings of the 23rd International SAMPE Technical Conference, Kiamesha Lake, NY, USA, 21–24 October 1991.
16. Rhead, A.T.; Butler, R.; Liu, W.; Kim, B.C.; Hallett, S.R. Compression after Impact Strength of a Buckling Resistant, Tow Steered Panel. In Proceedings of the 19th International Conference on Composite Materials, Montreal, QC, Canada, 28 July–2 August 2013; Hoa, S.V., Hubert, P., Eds.; Electronic Publishing Bytepress.com Inc.: Montreal, QC, Canada, 2013.
17. Rhead, A.T.; Hua, S.; Butler, R. Damage resistance and damage tolerance of hybrid carbon-glass laminates. *Compos. Part A* **2015**, *76*, 224–232. [CrossRef]
18. Remacha, M.; Sánchez-Sáez, S.; López-Romano, B.; Barbero, E. A new device for determining the compression after impact strength in thin laminates. *Compos. Struct.* **2015**, *127*, 99–107. [CrossRef]
19. Lopresto, V.; Papa, I.; Langella, A. Residual strength evaluation after impact tests in extreme temperature conditions. New equipment for CAI tests. *Compos. Part B* **2017**, *127*, 44–52. [CrossRef]
20. Abdulhamid, H.; Bouvet, C.; Michel, L.; Aboissiere, J.; Minot, C. Experimental study of compression after impact of asymmetrically tapered composite laminate. *Compos. Struct.* **2016**, *149*, 292–303. [CrossRef]
21. SGL Carbon Company/Products. Available online: http://www.sglgroup.com/cms/de/Produkte/Produktgruppen/Verbundwerkstoffmaterialien/Prepregs/index.html?__locale=de (accessed on 28 July 2018).
22. AMT Composites/Products. Available online: https://www.amtcomposites.co.za/products/prepreg-reinforcements/glass/epoxy-prepreg-system-ft1021 (accessed on 28 July 2018).
23. Nettles, A.T.; Sabo, S. Compression after impact strength of thin laminates with various percentages of 0° plies. *J. Compos. Mater.* **2014**, *48*, 345–354. [CrossRef]
24. Barile, C. Innovative mechanical characterization of CFRP by using acoustic emission technique. *Eng. Fract. Mech.* **2018**. [CrossRef]

© 2018 by the authors. Licensee MDPI, Basel, Switzerland. This article is an open access article distributed under the terms and conditions of the Creative Commons Attribution (CC BY) license (http://creativecommons.org/licenses/by/4.0/).

Article

Optimization of Hole-Flanging by Single Point Incremental Forming in Two Stages

Domingo Morales-Palma *, Marcos Borrego, Andrés J. Martínez-Donaire, Gabriel Centeno and Carpóforo Vallellano

Department of Mechanical Engineering and Manufacturing, Universidad de Sevilla,
Camino de los Descubrimientos s/n, 41092 Sevilla, Spain; mborrego@us.es (M.B.); ajmd@us.es (A.J.M.-D.); gaceba@us.es (G.C.); carpofor@us.es (C.V.)

* Correspondence: dmpalma@us.es; Tel.: +34-954-481355

Received: 20 September 2018; Accepted: 17 October 2018; Published: 18 October 2018

Abstract: Single point incremental forming (SPIF) has been demonstrated to accomplish current trends and requirements in industry. Recent studies have applied this technology to hole-flanging by performing different forming strategies using one or multiple stages. In this work, an optimization procedure is proposed to balance fabrication time and thickness distribution along the produced flange in a two-stage variant. A detailed analytical, numerical and experimental investigation is carried out to provide, evaluate and corroborate the optimal strategy. The methodology begins by analysing the single-stage process to understand the deformation and failure mechanisms. Accordingly, a parametric two-stage SPIF strategy is proposed and evaluated by an explicit Finite Element Analysis to find the optimal parameters. The study is focused on AA7075-O sheets with different pre-cut hole diameters and considering a variety of forming tool radii. The study exposes the relevant role of the tool radius in finding the optimal hole-flanging process by the proposed two-stage SPIF.

Keywords: sheet metal forming; flanging; hole-flanging; incremental sheet forming; Single Point Incremental Forming (SPIF); finite element analysis; thickness distribution

---

1. Introduction

Single point incremental forming (SPIF) is the simplest process variety within the incremental sheet forming (ISF) technologies. These technologies, with their germ in the late 1960's of last century (as reported in [1]) and with a strong development since the 2000's [2], have entirely proved to accomplish the current trends and requirements in industry including rapid, flexible, economic as well as environmental friendly manufacturing.

Besides, ISF processes applied to sheet metals present an additional advantage compared to conventional processes, consisting of an enhancement of formability [3,4] well above the Forming Limit Curve (FLC) for necking, allowing to deform the sheet until the onset of ductile fracture represented in the material Forming Limit Diagram (FLD) by means of the Fracture Forming Limit (FFL) in the case of in-plane tension corresponding to mode I of fracture mechanics. This benefit of ISF processes in terms of formability has permitted the application of SPIF (and other ISF varieties) to the manufacturing of geometrically complex and demanding components, such as automotive [5], aerospace [6] or even biomedical devices [7], as well as the production of specific shapes into an industrial part, as it is the case of flanges manufactured by incremental forming [8].

Regarding the production of hole flanges by means of incremental forming processes, the process consists of a sheet blank with an initial pre-cut hole which is progressively deformed by the forming tool, usually carried out in a number of steps or stages. Indeed, in this usual case of multi-stage hole-flanging by SPIF, the definition of the strategy is of paramount importance in order to optimize formability and/or other process outputs. Different multi-stage strategies have been proposed since 2010 in a series of research works such as [9,10]. Nevertheless, all the strategies presented so far including multiple stages present an obvious and not easily resolvable drawback: they are very time consuming.

With the aim of minimizing the process time attained by multi-stage strategies in single point incremental forming, the authors has recently proposed the use of a single-stage process for the manufacturing of hole flanges by SPIF [11]. This analysis of the single-stage variant has demonstrated its capability for the production of functional hole flanges with a remarkable reduction of the process duration. However, the results obtained in terms of thickness distribution along the resulting flange have to be definitely improved.

In this scientific and technological framework, the authors have recently proposed in two preliminary numerical investigations [12,13] the adoption of an alternative two-stage variant for the design of the hole-flanging process by SPIF. Indeed, this two-stage variant can be seen as a balance between the very time consuming of multi-stage strategies and the poor results in terms of thickness distribution provided by single-stage SPIF, allowing to produce hole flanges with homogeneous thickness in a relative short process time.

In this context, this paper presents a detailed numerical and experimental investigation with the overall objective of providing an optimal two-stage strategy in terms of thickness distribution for the hole-flanging process by SPIF. With this purpose, the deformation mechanism along the hole-flanging process by SPIF is analytically evaluated under the assumption of membrane stresses. This evaluation allows understanding the factors affecting the thickness distribution in the single-stage variant and permitted to propose and evaluate different strategies for the tool trajectory corresponding to the first stage. After selecting the most adequate strategy, the geometrical parameters defining the first stage are numerically optimized by launching an established Finite Element (FE) numerical procedure with Abaqus®/Explicit. The FE model makes use of a series of scripts in Python® for automating the tool path definitions and the whole optimization process. Finally, the numerical results after the first and second stages are compared with the corresponding experimental thickness distributions for the case of the optimal hole-flange.

This research work has improved the current state of the art in the hole-flanging process by SPIF with the following contributions: (i) proposing and corroborating experimentally the two-stage variant of the hole-flanging process by SPIF as a balance compared to the very time consuming multi-stage strategies proposed so far and the single-stage process recently proposed by the authors; (ii) presenting the most adequate parametric two-stage strategy that was schematically defined after identifying analytically the deformation mechanism during the single-stage variant; and (iii) establishing an experimentally validated and fully automated numerical procedure that permits an optimization of the process parameters for obtaining a satisfactory thickness distribution along the flange after only two forming stages.

2. Methodology

The proposed methodology consists in analysing the process deformation and the flange thickness of the single-stage SPIF process with the aim of establishing a new SPIF strategy to homogenise the thickness profile along the flange.

2.1. Experimental Background: Hole-Flanging by SPIF in a Single Stage

In an above cited previous work [11], the authors developed an experimental study consisting on the analysis of the physical mechanisms involved in the single-stage hole-flanging process by SPIF, aiming to evaluate the mechanical conditions upon which sheet failure takes place. With this purpose, the proposed single-stage strategy was utilized to perform a series of experimental tests on AA7075-O metal sheets of initial thickness $t_0 = 1.6$ mm using different forming tools. As a result, the maximum flange that can be successfully achieved was manufactured and analysed.

The experimental tests were carried out on a 3-axis milling CNC machine VMC-200 (EMCO, Lingen, Germany). The experimental setup and a schema of the helical tool trajectory is shown in Figure 1a. A blank holder and a backing plate with a 100-mm diameter circular hole were used to fix the sheet blanks to the machine table. Specimens were electro etched with a circle grid pattern at the outer sheet surface to analyse strains. Three different hemispherical tool radii were used ($R = 6$, 8 and 10 mm) to analyse the bending effect (measured by the sheet thickness to tool radius ratio, t/R) on formability. The feed rate was set to 1000 mm/min and 2 spindle speed were used, 0 rpm (locked tool) and 1000 rpm clockwise. A step-down of 0.2 mm/rev was used for the helical trajectories. A special lubricant for metal forming applications, Iloform TDN81 (Castrol, Liverpool, UK) was used to minimise friction. The final flange height, thickness profile along the flanges and surface roughness were analysed on the final parts.

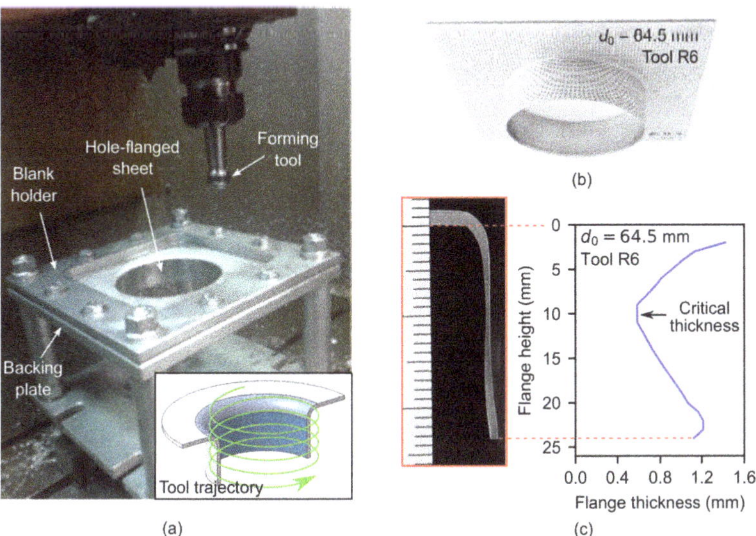

Figure 1. Hole-flanging by Single point incremental forming (SPIF) in a single stage of an AA7075-O sheet blank of 1.6-mm thickness: (**a**) experimental setup; (**b**) hole-flanged sheet part; and (**c**) thickness distribution along the flange.

Sheet blanks with different pre-cut hole diameter d_0 ranged from 55 to 82 mm were progressively deformed until producing hole flanges with a final inner diameter of $d_f = 95.8$ mm. Figure 1b shows a sample of successful hole-flanged sheet. The limiting values of d_0 to define the limit forming ratio ($LFR = d_0/d_f$) and obtain successful specimens were 64.5, 61.0 and 57.5 mm for tool radii of 6, 8 and 10 mm, respectively.

Some successful and failed specimens were cut for measuring the thickness in the profile view and, in case of fractured specimens, to validate the previous measurements along the crack. Figure 1c shows the thickness profile of a successful hole-flanged sheet. As can be observed, there are 3 distinguishable zones along the flange, from top to bottom: (1) a zone near the flat undeformed sheet, whose thickness is progressively decreasing; (2) an intermediate or critical zone, where the thickness is the smallest and the sheet failure tends to occur; and (3) the edge zone, where thickness is progressively increasing.

2.2. Analysis of the SPIF Process Deformation

In previous work, the stress/strain states accomplished in sheet metal deformed by SPIF were evaluated by means of membrane analysis [14–16]. The analytical model provides insight to the fundamentals behind the enhanced material formability of the SPIF process. The membrane analysis in axisymmetric components, illustrated in Figure 2a, was used to explain aspects of the process deformation and formability. As a result, it was pointed out that the meridional stress component (σ_ϕ) is mainly responsible of sheet thinning and opening of cracks (mode I of fracture mechanics) at the transition zone between the inclined wall and the corner radius of the sheet, in contact with the forming tool. These conclusions seems to be also valid for hole-flanging operations, represented schematically in Figure 2b. Indeed, the authors observed in failed specimens by single-stage SPIF that fracture initiates at the wall-tool interface limit [11], as can be seen in Figure 2c. Note that the onset of the crack coincided with the greatest thickness reduction or critical thickness.

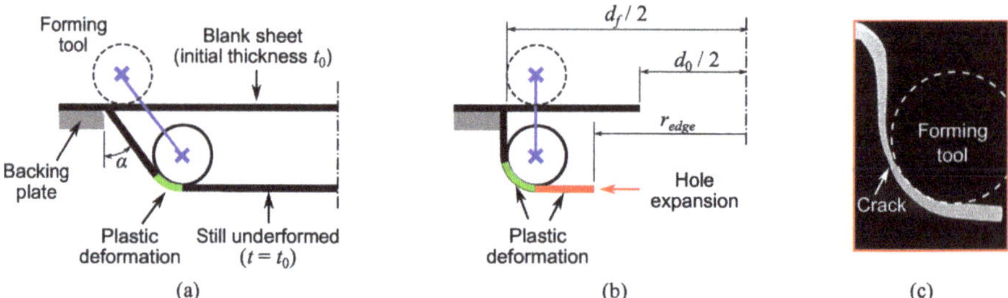

Figure 2. Schema of axisymmetric part deformation by SPIF of (**a**) inclined-wall part and (**b**) hole-flanged part; and (**c**) fractography of the failure zone for an interrupted hole-flanging test [11].

The membrane analysis for SPIF of inclined-wall parts shows that the meridional stress σ_ϕ increases and sheet thickness t decreases along the sheet-tool interface, from the formed wall to the still undeformed zone [15]. This qualitative description of the thickness distribution agrees very well with experimental observations in hole-flanging by single-stage SPIF, as depicted in Figure 2c. The membrane theory also predicts well the influence of the bending ratio t/R on the material formability of hole-flanging components [11].

On the other hand, hole-flanging has some particularities with respect to the conventional testing geometries in SPIF (usually cones and pyramids) that were previously analysed by the membrane theory. The main difference of both deformation processes illustrated in Figure 2a,b relies on the wall angle α, as well as in the circular hole of the sheet blank. In a SPIF process with a low or moderate α value, the plastic deformation is localised in the small zone in contact with the tool. Thus, the sheet thickness of the inclined-wall part can be estimated by the sine law, $t = t_0 \sin\alpha$, while the still flat material has not been plastically deformed yet and remains its initial thickness t_0. On the other hand, SPIF of components with severe α values up to 90° increase considerably the meridional stress distribution along the wall. In this situation, the plastic deformation extends beyond the area of contact with the forming tool, as the one that produces the hole expansion in hole-flanging (see schema in Figure 2b).

The hole expansion leads a plastic deformation of material that is not in contact with the forming tool. The resistance of this material zone during its radial and circumferential stretching and its effect on the forming tool action is of great importance to understand the deformation process.

Figure 3 shows a forming tool acting on a sheet during hole-flanging by SPIF in a single stage. The following analysis is based on the one developed by Silva et al. [15] for axisymmetric inclined-wall parts. The analysis focuses on a shell element located in the material being radially and circumferentially stretched due to the hole expansion and in the influence zone of the forming tool. Because of axial symmetry, principal stresses are assumed to be the circumferential (σ_θ), meridional (σ_ϕ), and thickness (σ_t) stresses.

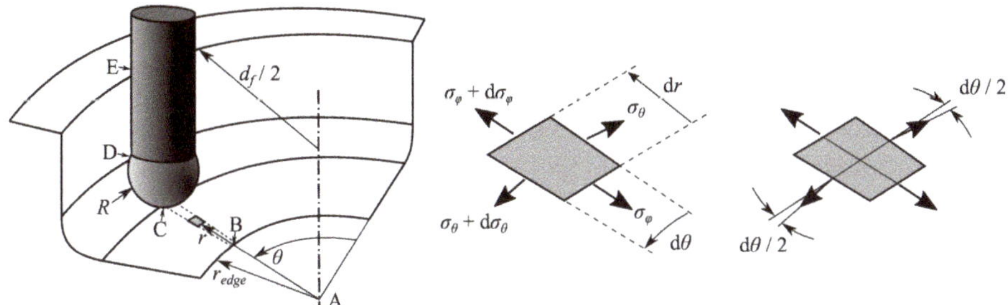

Figure 3. Membrane analysis of hole-flanging by SPIF in a single stage: schematic representation of a shell element located in the still flat zone under circumferential and radial stretching and details showing the acting stresses.

In Figure 3, the tool is positioned along the sheet section A-B-C-D-E and the shell element is in the flat zone adjacent to the hole edge, between points B and C. The shell element is located at coordinates (r, θ) and has a thickness $t < t_0$ due to the radial and circumferential expansion. The stresses acting on the shell element are also shown. Resolving the force equilibrium along the meridional direction one obtains

$$\sigma_\phi\, t\, r\, \mathrm{d}\theta + \sigma_\theta\, t\, \mathrm{d}r + (\sigma_\theta + \mathrm{d}\sigma_\theta)\, t\, \mathrm{d}r\, \sin(\mathrm{d}\theta/2) = (\sigma_\phi + \mathrm{d}\sigma_\phi)\, t\, (r + \mathrm{d}r)\mathrm{d}\theta \tag{1}$$

After neglecting higher order terms and simplifying the above equation by taking into account that $\sin(\mathrm{d}\theta/2) \approx \mathrm{d}\theta/2$, Equation (1) results in

$$\sigma_\theta\, \mathrm{d}r = \sigma_\phi\, \mathrm{d}r + r\, \mathrm{d}\sigma_\phi \quad \rightarrow \quad \frac{\mathrm{d}\sigma_\phi}{\mathrm{d}r} = \frac{\sigma_\theta - \sigma_\phi}{r}. \tag{2}$$

The principal stresses acting on the shell element can be identified as

$$\sigma_1 = \sigma_\theta > 0 \quad , \quad \sigma_2 = \sigma_\phi > 0 \quad , \quad \sigma_3 = 0 \tag{3}$$

According to the Tresca yield criterion, the term $\sigma_1 - \sigma_3$ should be equal to the yield stress σ_Y. Inserting in Equation (3) one obtains $\sigma_Y = \sigma_\theta$. Substituting in Equation (2) and integrating, the meridional stress in the shell element is found as

$$\sigma_\phi = \sigma_Y \left(1 - \frac{r_B}{r}\right) \tag{4}$$

where the boundary condition $\sigma_\phi = 0$ at the hole edge ($r = r_B$) was used to find the integration constant. Equation (4) allows to evaluate the meridional stress $\sigma_{\phi,C}$ in the transition zone with the corner radius of the sheet in contact with the tool ($r = r_C = d_f/2 - R$).

According to Silva et al. [15], the reason why thinning occurs in the corner radius C-D has to do with σ_ϕ acting in that zone. Indeed, thinning in C-D tends to balance the increment of meridional stress so that $t \cdot \sigma_\phi$ remains constant. According to this expression, thinning is especially relevant at point D. Besides, given that thickness at point C decreases as the SPIF process evolves and the hole expands, the sheet thickness in point D is getting more and more thin, in agreement with the experimental observations [11].

Therefore, the sheet thinning at point D (critical thickness along the flange) could be diminished during the SPIF process by controlling the resistance of the material being radially and circumferentially stretched. According to Equation (4), this could be achieved by reducing the yield stress or by increasing the hole edge radius r_B. This later solution can be achieved by performing a previous SPIF stage to modify the part geometry in the zone B-C.

2.3. Proposal for an Improved SPIF Strategy

Figure 4 presents 3 multi-stage SPIF strategies for hole-flanging proposed by Cui and Gao [9] to analyse the thickness distribution along the sheet flange of AA6010 sheets. Strategies consist of producing one or more simple intermediate part shapes, and the final hole-flanged part, by programming linear sections. Experimental tests were performed using a number of stages ranged from 2 to 5. Different combinations of increasing values for wall angle and part diameter were used. In a later work, Bambach et al. [10] used similar strategies to evaluate and exploit the flexibility of the SPIF technology to perform hole-flanging of mild steel sheets. Both experimental works concluded that flange thickness distribution obtained with the strategy illustrated in Figure 4a was the most uniform among the forming strategies considered. However, this strategy produced actually a lower thinning in the initial part of the flange compared to the lower and more homogeneous thickness as approaching to the edge.

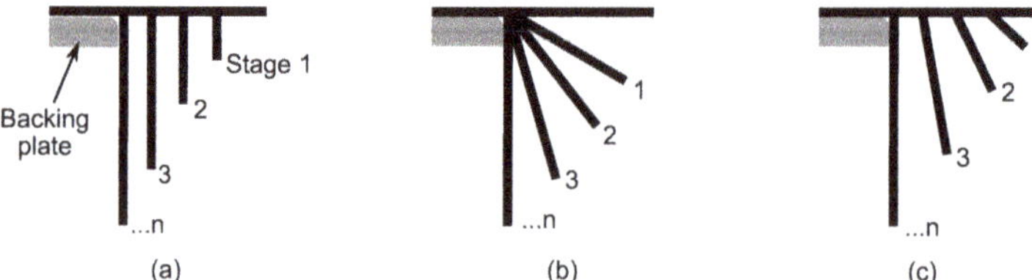

Figure 4. Schematic representation of hole-flanged sheet sections produced by SPIF using different multi-stage strategies producing successive (**a**) concentric cylinders; (**b**) conical frustums with increasing angle; and (**c**) conical frustums with increasing angle and major diameter.

Consequently, the first strategy proposed to perform hole-flanging by SPIF in two stages consists in programming two concentric helical paths. Figure 5a represents the tool path along the theoretical straight part section that defines the first stage. The distance *W* defines the hole diameter of the intermediate sheet part. As exposed in Section 4.3, this two-stage strategy produces a final thickness distribution of the flange according to the observations of the studies mentioned above [9,10]. Thus, an alternative strategy is needed to further reduce the thickness at the initial part of the flange.

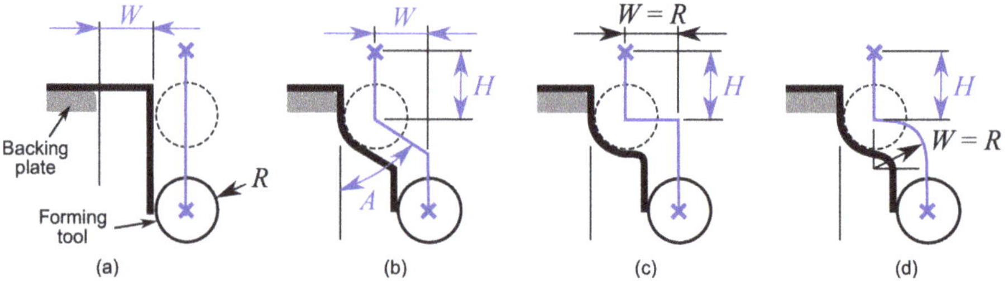

Figure 5. Schema of the first stage in hole-flanging by two-stage SPIF: (**a**) straight section due to Cui and Gao [9]; (**b**) 3-parameter section analysed in [12]; (**c**) prototype based on the three-parameter section; and (**d**) proposed section.

Regarding the analysis of the single-stage SPIF process presented in the previous section, the sheet thickness distribution could be better homogenised by modifying the part geometry adjacent to the hole edge (zone B-C in Figure 3). This can be done by performing a previous hole-flanging operation localised in that zone to increase the hole diameter, as illustrated in Figure 5b–d. The main difference with the previous strategy is that the operations start as the single-stage SPIF process and the tool trajectories are later modified to reduce the pressure with the forming tool in the critical zone of the sheet.

The SPIF strategy shown in Figure 5b was analysed in a recent numerical study [12]. An automated procedure was performed to simulate the flange thickness using Abaqus/Explicit. A series of 18 sets of consistent values of W, A and H were chosen to perform the study. It was concluded that the most homogeneous flange thickness was obtained using the highest angle value ($A = 80°$ in the study) and a distance W equals to the tool radius R.

In the present work a SPIF strategy is proposed by taking into account the conclusions of the previous study. Thus, the angle and width parameters are fixed to $A = 90°$ and $W = R$, as represented in Figure 5c. It should be noted that the horizontal transition movement in the case of $A = 90°$ requires an appropriate trajectory to avoid unexpected collisions due to the material springback. For instance, a spiral path could be used to smooth the sheet surface. Another option is to use a retract rapid movement upwards and inwards the hole and start the second drop from above. The optimum value of the height H is assumed to be dependent on the material and process geometry.

The above strategy can be further improved by programming a faster circular transition movement instead of the horizontal one, as represented in Figure 5d. Indeed, simulations performed with both options have shown almost identical thickness distributions. Therefore, in this work only the results obtained with strategies (a) and (d) shown in Figure 5 are analysed and compared.

3. Numerical Model

In order to evaluate the proposed strategy, numerical simulations have been performed using the commercial finite element code Abaqus/Explicit. The FE model has been parametrised to reproduce the experimental tests with different tool radii R and pre-cut hole diameters d_0 of the sheet blank developed in previous work [11].

Figure 6 shows the FE model. The sheet blank was modelled as a deformable circular ring of 2D-shell elements with five integration points through the thickness. All degrees of freedom of the nodes located at the outer edge of the blank were fixed to reproduce the blank holding. The backing plate and the forming tool were modelled as discrete rigid surfaces. The former was modelled as a circular ring of 100-mm inner diameter and the latter as a sphere. The material was 7075-O aluminium alloy sheet of 1.6 mm thickness. The mechanical properties of the material are summarized in Table 1. The von Mises yield criterion and the Hollomon type law were used to describe the plasticity behaviour and hardening, respectively.

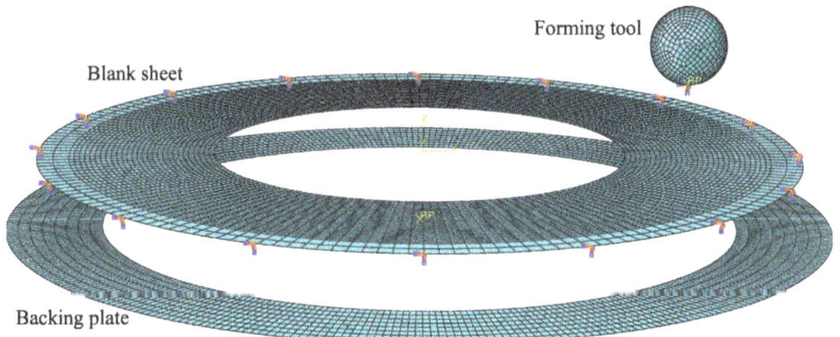

Figure 6. FE model including the meshed pre-cut sheet, backing plate and forming tool.

Table 1. Mechanical properties for AA7075-O sheets.

E (GPa)	ν	YS (MPa)	K (MPa)	n
65.7	0.3	109.7	314	0.13

It should be noticed that the assumption of material isotropy aims to reduce the complexity of the FE model and accelerate the simulations. In previous work [11], the values of Lankford's anisotropy coefficients obtained experimentally (r_0, r_{45}, r_{90}) did not differ strongly from an isotropic behaviour. On the other hand, some recent research works [17,18] have confirmed that the use of an anisotropic yield criterion is not a determinant factor for predicting the final geometry of axisymmetric conical testing parts deformed by SPIF.

The sheet blank was meshed with 180 nodes around the circumference and equispaced every 0.8 mm approximately along the radial direction. The total number of elements ranged from 7380 to 8100 for the maximum and minimum values of d_0 used in simulations: 64.5 and 57.5 mm, respectively. A *surface-to-surface* contact algorithm with the finite sliding formulation was used. A friction coefficient of 0.1 was assumed.

The tool trajectories were generated by programming Python scripts to locate precisely the tool tip coordinates. A simple helix was used to simulate the hole-flanging process by SPIF in a single stage. As can be observed in Figure 7a, the tool tip trajectory for the stage 1 of the 2 proposed process reproduces the sheet geometry of the intermediate part (see Figure 5d). A *retract and approach* transition movement was modelled to link both forming stages. Note in Figure 7b that the approach movement consists in a helical path with increasing diameter to take into account the springback of the metal sheet. In stage 2, the forming tool continues from the position it was separated from the wall at the end of the first stage.

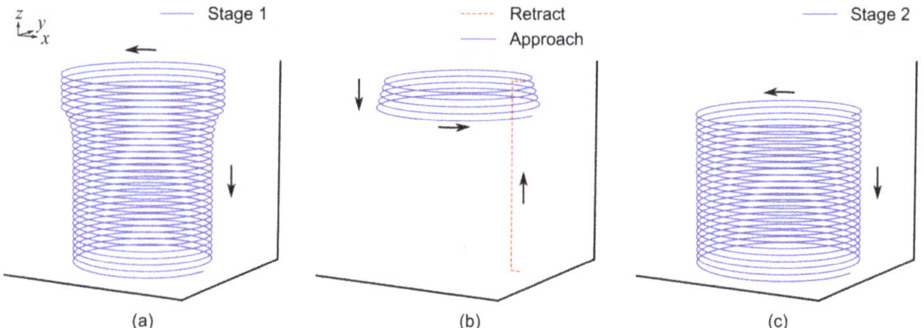

Figure 7. Schema of the tool tip trajectory for the proposed hole-flanging by SPIF in two stages: (**a**) first stage; (**b**) transition movement and (**c**) second stage.

According to the experimental tests in [11], the tool step-down was set to 0.2 mm/rev. As in previous work [12], mass scaling technique was used to speed up te computing time. Every single simulation required around 6–8 h in a 64-bit PC with an Intel Core i5 processor and 8 GB of RAM.

4. Results and Discussion

4.1. Validation of the FE Model

As said in Section 2.1, the limiting values of pre-cut hole diameter 64.5, 61.0 and 57.5 mm were experimentally determined in a previous work to obtain successful hole-flanged specimens by single-stage SPIF using tool radii of 6, 8 and 10 mm, respectively [11]. These 3 experiments have been simulated to validate the prediction capabilities of the FE model. Figure 8a presents the sheet thickness distribution on 3D views. The section *A-B-C* along the flange represents the zone near the flat undeformed sheet (point *A*), the critical zone (point *B*) and the flange edge (point *C*). As can be seen, a higher thickness reduction in the intermediate zone along the flange is shown, in well agreement with the experimental results.

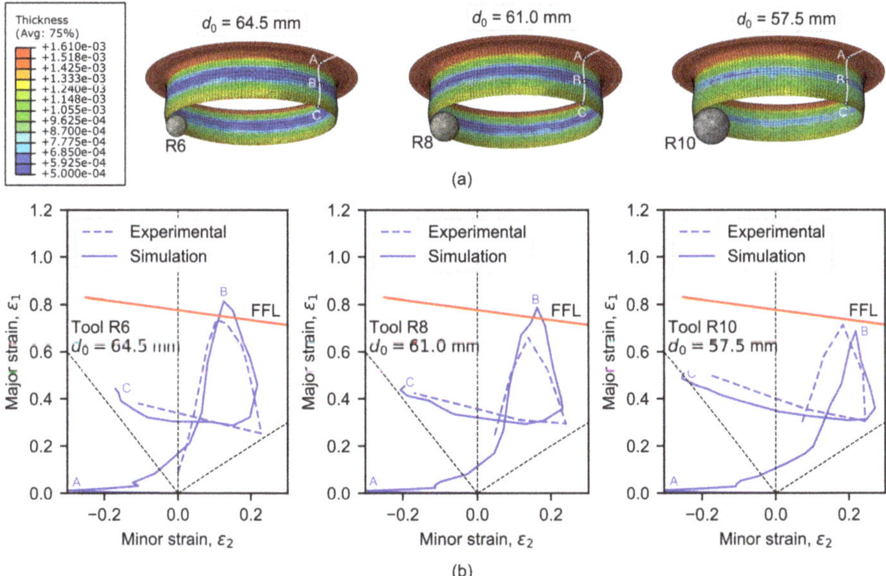

Figure 8. Simulations of hole-flanging processes by single-stage SPIF and comparison with experiments performed in [11]: (**a**) thickness distribution on 3D views and (**b**) FLD showing strain distributions along the flange.

Figure 8b depicts the principal strains measured at the outer sheet surface along the section *A-B-C* of the flange in a FLD for experiments vs. simulations. The FFL curve which was determined in previous work [11] is also represented. As can be seen, the numerical strain evolutions reproduces successfully the experimental trend, describing a loop, beginning from the top of the flange, reaching a maximum value around the critical zone and ending at the flange edge. The thickness distribution for the experimental test using the 6-mm tool radius is compared with that obtained by the FE model in Figure 9. In general, the thickness predictions of the FE model are quite fair.

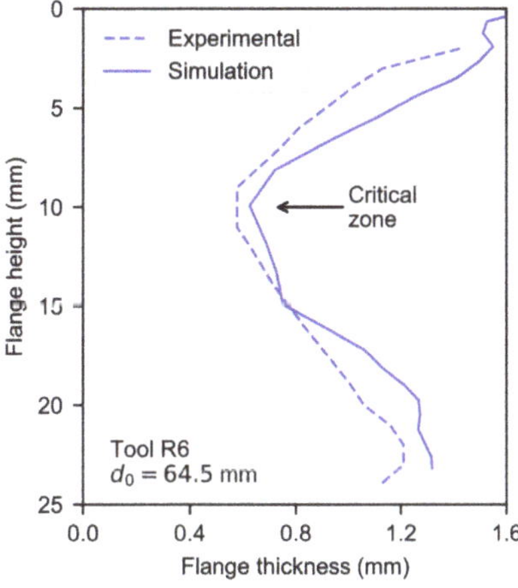

Figure 9. Thickness distribution along the flange of a hole-flanging process by single-stage SPIF: simulation and experiment [11].

4.2. Analysis of Flange Deformation by Single-Stage SPIF

Figure 10a shows the thickness profile evolution and the tool position for the simulated single-stage SPIF processes. Four simulation states are depicted: (0) the initial state; (1) a step-down equal to the tool radius; (2) the tool position when the minimum thickness is reached; and (3) the final state. A red cross indicates the critical position, that is, where the maximum thinning will occur in the flange.

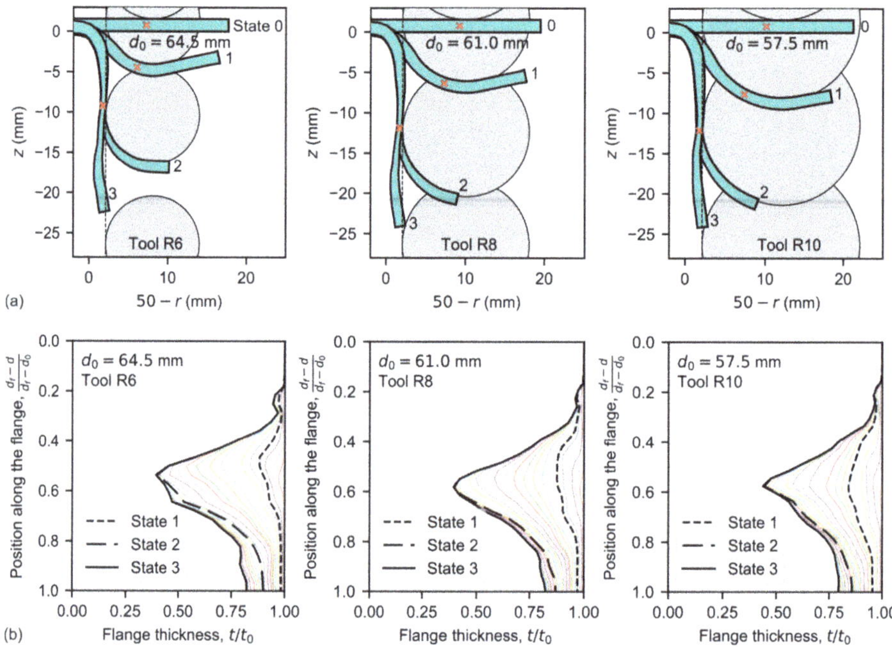

Figure 10. Simulations of hole-flanging processes by single-stage SPIF: (**a**) flange profile and tool position; and (**b**) flange thinning.

At the beginning of the SPIF process, forming tool sinks into the sheet and the contact area increases. In states 1 to 2 (see Figure 10a), the above exposed analytical framework for SPIF can be used to describe the process deformation. Indeed, the most pronounced thinning occurs in the formed flange closest to the contact with the tool for all the simulations, as can be observed in state 2 of Figure 10a. At the end of the process, the material from the critical thickness point to the hole edge seems at first glance to bend without thinning. According to the analytical framework, the reason is that the edge has expanded almost to the end and, thus, this material zone offers much less resistance in the meridional direction. Notice that the length of this region seems to be independent of the tool radius.

The flange thinning evolution obtained from simulations above is presented in Figure 10b. The successive thickness curves have been evaluated every 1 mm of step-down increment. As can be seen, the minimum thickness position moves upwards along the flange while the tool is sinking into the sheet (until state 1). Then, it moves down from state 1 until reach the critical thickness at state 2 according to the analytical framework presented above. Notice that final *bending* of the edge zone (states 2 to 3) produces a slight progressive thinning localised in that zone due to the hole expansion.

4.3. Analysis of the Flange Deformation in the Two-Stage SPIF Process

Figure 11 depicts the thickness distributions of the flange from FE simulations using a 64.5-mm pre-cut hole diameter and a 6-mm tool radius. Thickness curves correspond to the straight section and proposed strategies by two-stage SPIF illustrated in Figure 5a,d, respectively. The former has been evaluated by setting $W = 6, 7$ and 8 mm. According to the conclusions of Cui and Gao [9], it is expected that these parameter values are closed to the optimal value. On the other hand, a simple choice $H = R = 6$ mm based on experimental observations has been used for selecting the parameter value of the proposed two-stage strategy. Besides, $H = 7$ and 8 mm have also been evaluated for comparison. The thickness profile for the hole-flanging process by single-stage SPIF is also represented.

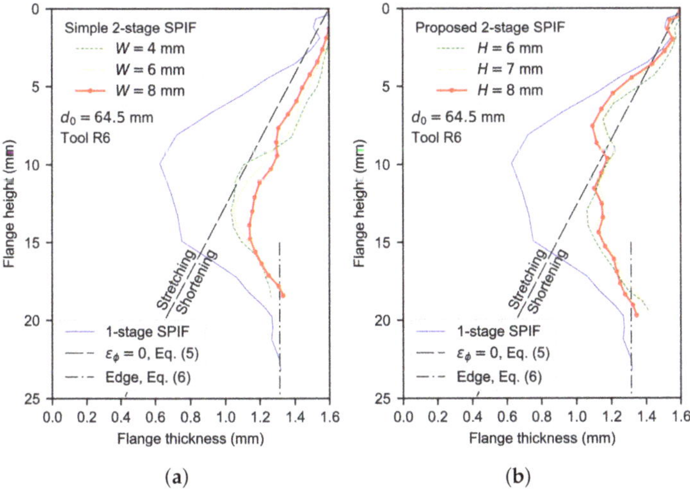

Figure 11. Flange thickness distribution by two-stage SPIF: (**a**) straight section and (**b**) proposed strategies.

As can be seen, the two-stage SPIF processes decrease the thickness reduction significantly, about a 40–50% in average, compared to the single-stage process. The straight section strategy (see Figure 11a) yields a similar shape of the thickness profile to the single-stage process: a progressive thinning along the top of the flange, a minimum thickness in the midway (around a flange height of 13–15 mm) and a progressive thickening in the edge zone. This description agrees well with experimental observations by Cui and Gao [9] for AA6010 sheets.

The proposed two-stage SPIF process (see Figure 11b) produces a more pronounced flange thinning in the top zone of the flange compared to the two-stage SPIF process using the straight section. This is attributed to the edge resistance to be circumferentially and radially expanded. Such difference can be further analysed by evaluating flange stretching in meridional direction, ε_ϕ. Thus, the thickness distribution of a hypothetical hole-flanging operation where the material is not strained in this direction can be found as

$$\varepsilon_\theta : \varepsilon_t : \varepsilon_\phi = 1 : -1 : 0 \quad \rightarrow \quad t(r) = t_0 \cdot \frac{2r}{d_f} \tag{5}$$

where r is the radial coordinate. This linear function is also depicted in Figure 11 to distinguish stretched and shortened material along the flange. As can be observed, thickness distributions of the two-stage SPIF using the straight section indicate a shortened flange. Instead, stretching occurs in both single-stage SPIF and the proposed two-stage SPIF in the top zone of the flange.

Figure 11b reveals that material stretching produced by the proposed two-stage SPIF led to a more uniform thickness in the intermediate part of the flange than the straight section strategy, around 1.1–1.2 mm thickness. The higher the H value, the higher thinning in the top zone (around 7-mm flange height) and the lower thinning in the bottom zone (around 13-mm flange height). The optimum H value that seems to homogenise flange thickness is 8 mm. Nevertheless, given that thickness differences are less than 10%, the simplest choice of $H = R$ has been assumed.

It should be noticed that edge thinning in hole-flanging processes is controlled by pure tension conditions and, therefore, it can not be modified by any forming tool. In this situation, the final thickness at the hole edge can be predicted as

$$\varepsilon_{\theta,edge} : \varepsilon_{t,edge} : \varepsilon_{\phi,edge} = 1 : -0.5 : -0.5 \quad \rightarrow \quad t_{edge} = t_0 \cdot \sqrt{\frac{d_0}{d_f}} \tag{6}$$

This expression is represented in Figure 11a,b as vertical lines. As can be seen, it reproduces very well the edge thickness predicted by SPIF simulations.

The sheet profile evolutions and tool positions by the proposed two-stage SPIF process in both stages 1 and 2 are shown in Figure 12a,b, respectively. In stage 1, four simulation states are represented: (0) the initial state; (1) a tool step-down increment of $H = R$; (2) an intermediate tool position during the transition circular movement; and (3) the final state. In stage 2, three states are depicted: (0) the initial state whose sheet profile was obtained in stage 1; (1) an intermediate tool position; and (2) the final state. A red cross indicates again the position where maximum thinning occurs. Notice that this position is closer to the edge than those obtained for single-stage SPIF.

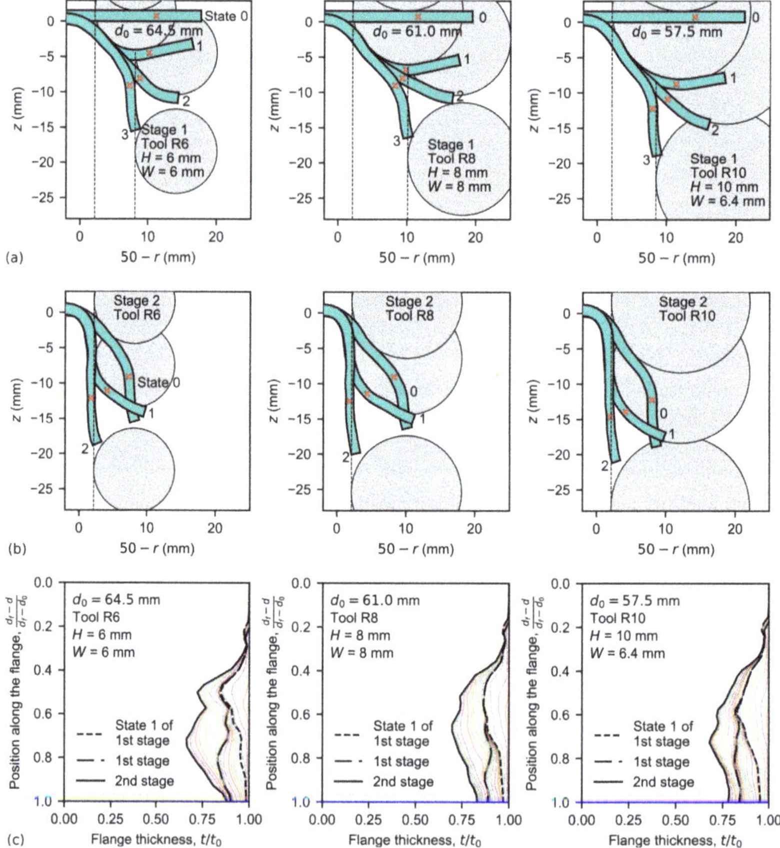

Figure 12. Simulations of hole-flanging by the proposed two-stage SPIF process: flange profile and tool position in (**a**) stage 1 and (**b**) stage 2; and (**c**) flange thinning.

Figure 12a reveals the deformation mechanism inherent to the proposed two-stage SPIF process: the material resistance is being diminished by reducing the hole-flanged diameter to $d_f - 2W$. Note that the flange section being bent in states 1–3 is similar for the three simulations. It should be remarked that simulation using a 10-mm tool radius and $W = R = 10$ mm bent a very small zone of the edge and this led to a poorer thickness distribution. Instead, $W = 6.4$ mm was set by locating the tool tip position just on the hole edge in state 1.

As deduced from Figure 12b, hole-flanging evolves in stage 2 by two consecutive reversal bending processes at states 1 and 2, respectively, to increase the inner diameter of the sheet part.

The thickness evolution is represented in Figure 12c. Thinning in state 1 of the first stage matches the one obtained in same state of the single-stage process, as expected. From states 1 to 3 of this stage, slight thinning occurs due to the edge bending process. In general, stage 2 distributes the thickness homogeneously along the flange.

4.4. Experimental Study

Hole-flanging by two-stage SPIF experimental tests was performed for evaluating the numerical analysis. The experiments were carried out using the same SPIF setup of the previous work for hole-flanging by single-stage SPIF [11]. A 6-mm tool radius was used to deform a 7075-O aluminium alloy of 1.6-mm thickness and 64.5-mm initial hole diameter. The parameter process was set to $W = R = 6$ mm. The manufactured parts were cut along the flange and the resulting thickness profiles were measured by using an optical microscope. The test was performed twice leading to a similar thickness distribution, exhibiting differences lower than 2%.

Figure 13 compares the numerically predicted thickness profiles for both single- and two-stage SPIF operations with their experimental counterpart. As can be seen, the simulation of the two-stage process corresponds fairly to the thickness profile except at the edge zone. This slight divergence may be attributed to that the forming tool pressed up the hole edge during the last turns of its helical trajectory, producing a kind of burr at the edge which is highlighted with a red arrow in Figure 13. A small burr was also observed in specimens obtained by single-stage SPIF.

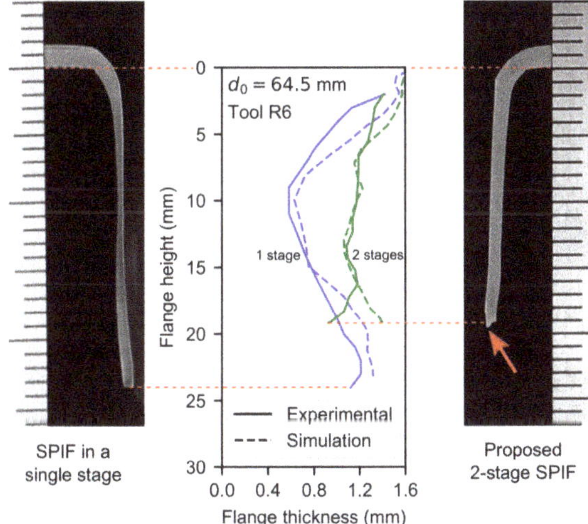

Figure 13. Views of hole-flange sections obtained by single-stage and two-stage SPIF experimental tests and comparison with numerical simulations.

The maximum numerical deviations in terms of thickness with respect to the corresponding experimental results (see Figure 13) are about 10% in the critical zone for the single-stage SPIF process, whereas in the two-stage variant they do not exceed 5%. These results point out the ability of the proposed methodology to homogenize the thickness of the flange with a two-stage SPIF operation.

Likewise, according to the numerical analysis presented here, significant improvements are also expected for the other two specimens analysed above ($d_0 = 61.0$ and 57.5 mm using tool radii $R = 8$ and 10 mm, respectively). An experimental campaign is planned to assess the influence of tool radius on this methodology and will be discussed in a further publication.

5. Conclusions

An optimization procedure using two stages has been proposed, improving the thickness homogenisation along the flange as well as fabrication times in hole-flanging by SPIF. The methodology begins by analysing the single-stage process to understand the material deformation mechanism and failure. Accordingly, a customised two-stage SPIF strategy has been proposed and evaluated by FE simulations. Different tool radii and hole diameters in AA7075-O sheets have been considered.

The understanding of flange deformation during the single-stage variant has allowed defining a new improved two-stage SPIF strategy and reducing the process design times compared to trial-and-error methods. In this sense, the explicit analysis carried out in this work helped to perform FEA with reasonable computing times. The proposed two-stage SPIF process has also been corroborated experimentally.

The optimized two-stage strategy for hole-flanging by SPIF has significantly improved the thickness distribution and fabrication time compared to other multi-stage strategies found in the scientific literature. The forming tool radius R has played an important role to determine the intermediate part geometry that has been defined by parameters height H and width W. The higher the tool radius, the higher the parameter values to obtain the best thickness distribution. Indeed, good results in terms of homogeneous thickness have been obtained by using the simple rule $H = W = R$.

Author Contributions: D.M.-P. and C.V. developed the analytical framework for hole-flanging by SPIF; M.B., A.J.M.-D. and D.M.-P. performed the simulations; M.B., G.C. and D.M.-P. designed the experimental plan and conducted the experimental tests; all the authors contributed in the interpretation of the results obtained; the paper was written by D.M.-P.

Funding: This research was funded by the Spanish Government grant number DPI2015-64047-R.

Conflicts of Interest: The authors declare no conflict of interest.

Abbreviations

The following abbreviations are used in this manuscript:

SPIF	Single Point Incremental Forming
ISF	Incremental Sheet Forming
FLC	Forming Limit Curve (by necking)
FLD	Forming Limit Diagram
FFL	Fracture Forming Limit
FE	Finite Element
FEA	Finite Element Analysis

References

1. Emmens, W.; Sebastiani, G.; van den Boogaard, A. The technology of Incremental Sheet Forming—A brief review of the history. *J. Mater. Process. Technol.* **2010**, *210*, 981–997. [CrossRef]
2. Jeswiet, J.; Micari, F.; Hirt, G.; Bramley, A.; Duflou, J.; Allwood, J. Asymmetric Single Point Incremental Forming of Sheet Metal. *CIRP Ann.* **2005**, *54*, 88–114. [CrossRef]
3. Soeiro, J.; Silva, C.; Silva, M.; Martins, P. Revisiting the formability limits by fracture in sheet metal forming. *J. Mater. Process. Technol.* **2015**, *217*, 184–192. [CrossRef]
4. Centeno, G.; Martínez-Donaire, A.J.; Bagudanch, I.; Morales-Palma, D.; Garcia-Romeu, M.L.; Vallellano, C. Revisiting Formability and Failure of AISI304 Sheets in SPIF: Experimental Approach and Numerical Validation. *Metals* **2017**, *7*, 531. [CrossRef]

5. Zha, G.; Xu, J.; Shi, X.; Zhou, X.; Lu, C. Forming Process of Automotive Body Panel based on Incremental Forming Technology. *Metall. Min. Ind.* **2015**, *12*, 350–357.
6. Hussain, G.; Khan, H.R.; Gao, L.; Hayat, N. Guidelines for Tool-Size Selection for Single-Point Incremental Forming of an Aerospace Alloy. *Mater. Manuf. Process.* **2013**, *28*, 324–329. [CrossRef]
7. Centeno, G.; Morales-Palma, D.; Gonzalez-Perez-Somarriba, B.; Bagudanch, I.; Egea-Guerrero, J.J.; Gonzalez-Perez, L.M.; García-Romeu, M.L.; Vallellano, C. A functional methodology on the manufacturing of customized polymeric cranial prostheses from CAT using SPIF. *Rapid Prototyp. J.* **2017**, *23*, 771–780. [CrossRef]
8. Silva, M.; Bay, N.; Martins, P. 2-Hole-flanging by single point incremental forming. In *Materials Forming and Machining*; Davim, J.P., Ed.; Mechanical Engineering Series; Woodhead Publishing Reviews: Cambridge, UK, 2016; pp. 25–50.
9. Cui, Z.; Gao, L. Studies on hole-flanging process using multistage incremental forming. *CIRP J. Manuf. Sci. Technol.* **2010**, *2*, 124–128. [CrossRef]
10. Bambach, M.; Voswinckel, H.; Hirt, G. A New Process Design for Performing Hole-flanging Operations by Incremental Sheet Forming. *Procedia Eng.* **2014**, *81*, 2305–2310. [CrossRef]
11. Borrego, M.; Morales-Palma, D.; Martínez-Donaire, A.; Centeno, G.; Vallellano, C. Experimental study of hole-flanging by single-stage incremental sheet forming. *J. Mater. Process. Technol.* **2016**, *237*, 320–330. [CrossRef]
12. Morales-Palma, D.; Borrego, M.; Martínez-Donaire, A.J.; López-Fernández, J.A.; Centeno, G.; Vallellano, C. Numerical study on the thickness homogenization in hole-flanging by single-point incremental forming. *J. Phys. Conf. Ser.* **2018**, *1063*, 012183. [CrossRef]
13. Morales-Palma, D.; Borrego, M.; Martínez-Donaire, A.; Centeno, G.; Vallellano, C. Preliminary investigation on homogenization of the thickness distribution in hole-flanging by SPIF. *Procedia Manuf.* **2017**, *13*, 124–131. [CrossRef]
14. Martins, P.; Bay, N.; Skjoedt, M.; Silva, M. Theory of single point incremental forming. *CIRP Ann.* **2008**, *57*, 247–252. [CrossRef]
15. Silva, M.; Skjoedt, M.; Martins, P.; Bay, N. Revisiting the fundamentals of single point incremental forming by means of membrane analysis. *Int. J. Mach. Tools Manuf.* **2008**, *48*, 73–83. [CrossRef]
16. Silva, M.; Skjoedt, M.; Atkins, A.; Bay, N.; Martins, P. Single—Point incremental forming and formability—Failure diagrams. *J. Strain Anal. Eng. Des.* **2008**, *43*, 15–35. [CrossRef]
17. Flores, P.; Duchêne, L.; Bouffioux, C.; Lelotte, T.; Henrard, C.; Pernin, N.; Bael, A.V.; He, S.; Duflou, J.; Habraken, A. Model identification and FE simulations: Effect of different yield loci and hardening laws in sheet forming. *Int. J. Plast.* **2007**, *23*, 420–449. [CrossRef]
18. Eyckens, P.; Belkassem, B.; Henrard, C.; Gu, J.; Sol, H.; Habraken, A.M.; Duflou, J.R.; Van Bael, A.; Van Houtte, P. Strain evolution in the single point incremental forming process: digital image correlation measurement and finite element prediction. *Int. J. Mater. Form.* **2011**, *4*, 55–71. [CrossRef]

© 2018 by the authors. Licensee MDPI, Basel, Switzerland. This article is an open access article distributed under the terms and conditions of the Creative Commons Attribution (CC BY) license (http://creativecommons.org/licenses/by/4.0/).

 *materials* 

Article

Replicability of Ultrasonic Molding for Processing Thin-Wall Polystyrene Plates with a Microchannel

I. Ferrer [1,*], M. Vives-Mestres [2], A. Manresa [1] and M. L. Garcia-Romeu [1]

1 Department of Mechanical Engineering & Industrial Construction, University of Girona, 17004 Girona, Spain; manresa.ariadna@gmail.com (A.M.); mluisa.gromeu@udg.edu (M.L.G.-R.)
2 Department of Computer Science, Applied Mathematics & Statistics, University of Girona, 17004 Girona, Spain; marina.vives@imae.udg.edu
* Correspondence: ines.iferrer@udg.edu; Tel.:+34-972-418-265

Received: 6 July 2018; Accepted: 26 July 2018; Published: 30 July 2018

Abstract: Ultrasonic molding is a new technology for processing small and micro polymeric components with reasonable cost and energy savings when small and medium batch sizes are required. However, when microcomponents are manufactured, the replicability of different micro features has to be guaranteed. The aim is to investigate the capability of ultrasonic molding technology for processing thin-wall plates of polystyrene with a microchannel, analyzing the filling behavior, the optical transparency, and the dimensional accuracy of the thin plate. The replicability of the manufactured microchannel is studied according to dimension and shape. The results reveal that plunger velocity influences transparency and filling cavity, whereas the vibration amplitude has less effect in both cases. The thickness deviation achieved on the final part is below 7% and the replication of the microchannel is better in depth than width, obtaining an average deviation of 4% and 11%, respectively. This replication also depends on the orientation of the microchannels and the distance from the injection gate. The replicability and repeatability for processing thin-wall plates with microchannel in polystyrene polymer are proved in this paper.

Keywords: ultrasonic molding; microchannel; thin-wall plate; replicability; polystyrene; medical devices

1. Introduction

In recent years, the use of polymer materials in industry has increased notably, with sectors such as micro-electromechanical systems (MEMS), aeronautics, and biomedicine slightly influenced by this trend. Some reasons for this strong tendency include the reduced weight of the component, the increment of the complexity of 3D geometries, the trend toward miniaturization of parts, the increment of functionality of products, the high performance requirement of the materials, and the emergence of many new polymers with very different characteristics [1,2].

Miniaturization of mechanical systems and devices leads to the development of microcomponents (such as microgears, microfluidic devices, and micro-optical lenses), small parts with microfeatures of high aspect ratio, or parts having microfeatures pattered on the surface. When processing these miniature polymeric components, injection and microinjection molding are the commonly used manufacturing techniques because of their high productivity, cost-effectiveness, high replicability and repeatability, high capability for producing an extended variety of polymeric components, tight tolerances, and complex shapes [1,3].

However, the replicability of thermoplastic components at the micro scale is quite complex, and there are several parameters that influence the process, such as injection speed, injection pressure, melting temperature, mold temperature, and others [4]. Actually, there are strong interactions among process parameters, process variations, mold design and configuration, part shrinkage, and internal material stress that reduce process repeatability and replicability [4]. Consequently, new technologies are being researched, such as compression molding [4,5], vacuum injection molding [4], ultrasonic injection molding [6,7], ultra-high-speed injection molding [8,9], and 3D printing [10].

Focusing on these molding manufacturing processes, in recent years, several researchers have studied the influence of process parameters to obtain microgeometries with high aspect ratio using different polymeric materials. Attia and Alcock [11] evaluated the weight of five products with microscaled features of polymethylmethacrylate (PMMA) manufactured by microinjection molding and found that the more relevant parameter to increase product weight was the holding pressure, but they also found that when complex geometries are molded, injection speed and mold temperature also become significant. Chen et al. [12] replicated a microfluidic chip by microinjection molding using a plate of 80 mm length, 40 mm width, and 0.2 mm thickness with microchannels of 100 μm width and 30 μm depth. They tested different processing parameters for manufacturing four polymer resins, cyclic olefin copolymer (COC), polystyrene (PS), polycarbonate (PC), and polymethylmethacrylate (PMMA), and found that, in general, the microchannels manufactured in COC and PS achieved the required accuracy, with higher dimensional accuracy when COC resin was used. Moreover, they found that microchannel width and depth improved when higher temperatures, for both molding and melting, higher injection speeds, and higher packing pressures were used. Lucchetta et al. [13] replicated a circular microfilter of PS that included squared ribs of 150 × 150 μm features separated by 50 μm. They focused on analyzing the influence of the mold material with different thermal diffusivity and mold temperature on part filling, without measuring in detail the microfeature dimensions achieved. As a result, ceramic mold material seems to replicate better than metallic due to its capability to delay polymer skin solidification when the cavity is being filled. Masato et al. [14] studied the replication of micro holes (4 μm in diameter and 5 μm in depth) regarding the thickness variation of the molded part of PS material. Their results revealed that part thickness and mold temperature influence replication. The replication accuracy improved for higher mold temperature and thinner part thickness, because the pressure of the melt inside the mold increased and the polymer could reach further. Mahmoodi et al. [2] investigated the filling behavior and dimensional accuracy of a thin part (6 × 6 mm square, 200 μm thick) using microinjection molding and adding chemical blowing agents and wood fibers. Injection pressure and mold temperature were modified in both experiments and the numerical model obtained consistent results. The higher the mold temperature and pressure were, the better the filling ratio achieved. Their results also revealed that the in-flow shrinkage percentage became slightly higher than the cross-flow shrinkage percentage. Sortino et al. [4] evaluated the replicability of conventional injection molding, injection compression molding (ICM), and vacuum injection molding (VIM) when manufacturing micro-optical devices with prism patterns using PMMA. They found the holding pressure to be the most influential parameter in the injection molding process, then the injection pressure, and injection speed the least influential. In this regard, the ICM process resulted in the most precise and best feature replication, and VIM the worst. Sato et al. [6] studied the application of ultrasonic injection molding (UIM) when replicating optical lens surfaces. What makes that process different from USM is that ultrasonic energy is applied to the melted material during the mold cavity filling. They varied the oscillation time and obtained the best results in replication when the ultrasonic energy was applied at end of the injection process. This energy locally increases the polymer temperature and causes an oscillatory polymer flow that improves replicability and avoids shrinkage during the packaging and cooling stages.

Ultrasonic molding is a new manufacturing technology to obtain micropolymeric parts. In ultrasonic molding, the raw material is melted by the direct application of ultrasonic energy. In this process, first the pellets are located in plasticization chamber of the mold (Figure 1), then the sonotrode moves into the mold, vibrates, and transmits the vibrational energy to the material. Then, while the sonotrode is vibrating and the material is being melted, the plunger moves, pushing the melted material into the mold cavity. When the plunger reaches the final position, the mold cavity is filled and it starts the cooling and packing stages. After applying the compaction forces and once the material solidifies, the sonotrode returns to the initial position and the mold can be opened to extract the final part.

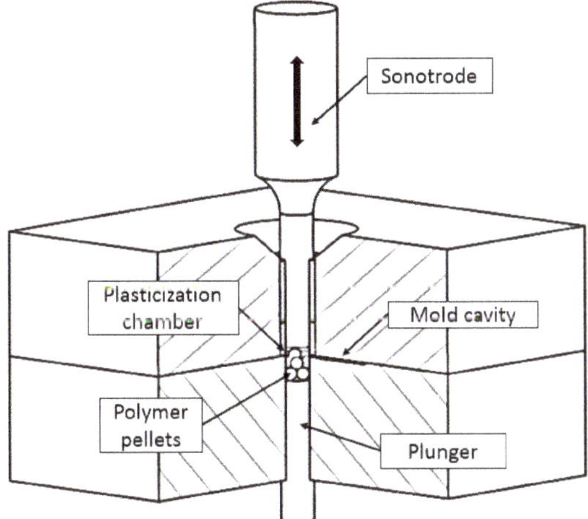

Figure 1. Ultrasonic molding process schema.

Each molding cycle only requires the amount of material to fill the mold cavity and the plasticizer chamber, thus no extra material is melted into the barrel before the process, as in conventional or microinjection molding. This represents an important material and energy savings for small-batch production and high-cost materials (i.e., implantable grade or reinforced materials). Moreover, the process is quite flexible for testing different materials and geometries, because all the material placed into the mold is removed after each cycle. The vibrational energy applied directly to the material produces high heating rates, which allows fast polymer plasticization [15]. Thus, the material is kept at higher temperatures during shorter periods of time, typically less than 5 s, reducing the residence time in comparison to conventional injection molding, and therefore material degradation can be avoided [7].

The ultrasonic vibration is applied by means of the acoustic unit, which is composed of the generator, the transducer, the booster, and the sonotrode. The transducer converts the electrical energy from the generator into vibrations, which can be amplified or reduced by the booster, and the sonotrode is the final element that transmits the vibrational amplitude to the material. The sonotrode is a mechanical element that has to be designed to vibrate in a longitudinal mode at the operating frequency of the acoustic unit, and with the required tolerances to fulfill the clearance requirements with the plasticization chamber dimensions.

There are a few published studies regarding this novel technology dealing with the influence of the processing conditions on manufacturing components. Michaeli et al. [16] can be considered the pioneers of this technology. Through several research works, they proved that ultrasonic energy is capable of melting small amounts of thermoplastic polymer for molded parts. Moreover, they verified the integrity of the polymer after being processed in two ways: internal homogeneity and physical properties, such as weight [15–18]. They processed PC [15], polyoxymethylene (POM), and polypropylene (PP) [19], and in all of the cases simple geometries were used, such as cylinders or plates. Polymer plasticization occurs mainly by two mechanisms, internal material heating and friction between pellets [19]. In this regard, the melt temperature varies as a function of different processing parameters, such as vibration amplitude and vibration frequency, the pressure applied to the material, or the material properties of the pellets [18]. In recent studies, the influence of processing conditions on the fabrication of different polymeric components has been analyzed. Sacristán et al. [20] studied the influence of vibration amplitude and pressure on polymer properties. They manufactured mini tensile test specimens (1.5 × 0.1 × 0.1 cm^3) of poly (lactic acid) (PLA) and noted that the cavity filling improved when high amplitude and lower pressure were applied. However, polymer degradation and material homogeneity had the opposite behavior. When higher levels of vibration amplitude and pressure were used, the polymer degraded, whereas using the lowest levels achieved better polymer homogeneity. Later, Planellas et al. [21] used the same geometry to test the feasibility of ultrasonic molding for processing polylactide (PLA) and polybutylene succinate (PBS). Although both polymers were manufactured without molecular degradation, optimization of the processing conditions would be required to obtain better results on the final part. The introduction of clay silicate particles in the polymers did not influence PLA properties but affected PBS degradation. In both cases, structures with high orientation of clay nanosheets were oriented in the direction of polymer flow.

Actually, the literature reveals that there is strong interest in studying the influence of processing conditions on several precision molding processes of microfeature replicability in recent years. However, so far, there are only a few studies about the manufacturability of microfeatures using ultrasonic molding technology. Consequently, the aim in this research is to investigate the process capability of ultrasonic molding for processing thin-plate components with a microchannel of high aspect ratio using PS polymer. Specimens were manufactured under different conditions of vibration amplitudes and injection velocities to evaluate the filling behavior, the optical transparency, and the dimensional accuracy in the plate and the microchannel. The replicability of the manufactured microchannel was studied from the point of view of dimension and shape. The material was used for its good replicability and optical properties, which are required features when manufacturing microfluidic components.

2. Materials and Methods

2.1. Part Geometry, Materials, and Mold

The test geometry used in this experiment was a thin-walled rectangular specimen of 15 mm × 8 mm × 0.55 mm with a direct fan gate (Figure 2a). The plate contained a microchannel of 150 μm depth and 80 μm width and had both parallel and perpendicular zones to flow direction at different distances from the material entry.

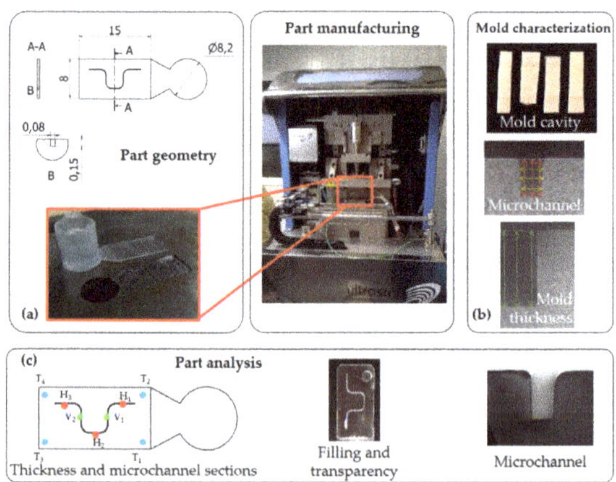

Figure 2. Method and experimental setup: (**a**) part geometry; (**b**) mold characterization; and (**c**) part analysis.

Transparent STYRON™ 678E PS resin was used for the experiment. This material is characterized by its optical clarity, transparency, biocompatibility, low water absorption, high flow ability, and high impact strength compared to other polymers such as silicon or glass.

An experimental mold of steel with the cavity on the lower half was used to replicate the geometry and obtain the final product. It included an ejector to facilitate the expulsion of the part, making a circular mark in final part (Figure 2c). It was manufactured using a Deckel-Maho 64 V Linear and the plasticization chamber was allocated at the mold upper half by a cavity of Ø 8.2 mm. The mold was characterized after being manufactured to guarantee repeatable measurements of the microchannels (Figure 2b). A silicon pattern was created to measure it. The thickness was measured in each corner of the specimen (T_1 Mold, T_2 Mold, T_3 Mold, and T_4 Mold) by a digital micrometer (Micromar 40 EWV). The microchannel dimensions were measured in 5 sections (Figure 2c), 3 in the inflow direction (H_1 Mold, H_2 Mold, and H_3 Mold) and 2 in the cross-flow direction (V_1 Mold and V_2 Mold) by a Leica DMR-XA optical microscope (Meyer Instruments Inc., Houston, TX, USA). Three measurements were taken of each section and thickness (see mean results in Table 1). The results reveal that the thickness of the manufactured mold was 14.3% bigger than the theoretical one, whereas the microchannel dimensions were 6.6% smaller in depth and 10% in width.

Table 1. Mold characterization results.

		Cavity Thickness				Microchannel			
						Depth		Width	
		Mean (μm)	Error (%)			Mean (μm)	Error (%)	Mean (μm)	Error (%)
Corner	$T_{1\,Mold}$	566	13	Microchannel Section	$H_{1\,Mold}$	137	8.7	73	7
	$T_{2\,Mold}$	576	15.2		$V_{1\,Mold}$	151	0.7	71	9
	$T_{3\,Mold}$	580	16		$H_{2\,Mold}$	139	7.3	70	10
	$T_{4\,Mold}$	565	13		$V_{2\,Mold}$	136	9.3	70	10
					$H_{3\,Mold}$	139	7.3	76	4
	Avg.		14.35		**Avg.**		6.67		8

According to the mold dimensions and considering the extra amount of material in the plasticizer chamber, around 300 mg of material was provided in each experiment.

2.2. Part Analysis

The specimens were manufactured using a Sonorus 1G Ultrasound Micro Molding Machine. The responses analyzed were part filling, transparency, thickness, and microchannel formation in 5 sections of the microchannel path (Figure 2c).

- Part filling was assessed by a ratio that compares the obtained part length with the expected theoretical length. Despite only complete parts being valid, corresponding to 100% filling, the influence of the processing parameters on the percentage of filling was also evaluated.
- Part transparency was evaluated visually and categorized in a qualitative scale divided into 7 levels and transmittance values (Table 2). Levels 1–4 represent blurred parts classified according to the percentage of cloudy zones, and levels 5–7 are transparent parts ordered regarding the presence of defects. Blurred parts could be caused by the presence of nonmelted material, weld lines, or polymer degradation, whereas defects include bubbles, sink marks, or burrs. Moreover, transmittance values according to this qualitative scale were measured by a UV-2401PC Shimadzu spectrometer using wavelengths between 400 and 600 nm. Six specimens were taken to measure each category and the mean was calculated.
- Part thickness was measured in the same corners as mold thickness (Figure 2c), although in this case the variables are referred to as T_1 Part, T_2 Part, T_3 Part, and T_4 Part. A Micromar 40 EWV digital micrometer was used to measure them.
- Microchannel formation was assessed by dimensions (width and depth) and shape. Five sections of the microchannel path were analyzed in each specimen and 3 specimens were selected for each

combination of process parameter, on the condition of complete parts. The selected specimens were cut in each section (in-flow and cross-flow direction), embedded into 40 mm of Axson RSF816 epoxy resin, prepolished by Struers Knuth Rotor-3 equipment and finished by a Buehler Ecomet sander polishing with alumina powder of 1, 3, and 9.5 microns. The observation of the microchannel was done using a Leica DMR-XA optical microscope with 5×, 10×, and 20× Plan-Apochromat lenses. The microchannel dimensions were measured by Fiji software and 3 measures of depth and width were taken for each specimen.

Table 2. Part transparency classification.

	Level	Qualitative Description	Transmittance Value (%)	Transparency Level (Examples)
Blurred	1	76–100% of cloudy zones	<85	Level 1 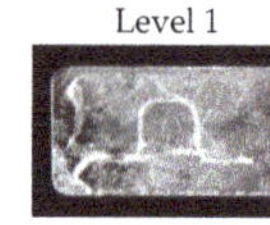
	2	51–75% of cloudy zones		
	3	26–50% of cloudy zones	87.366	Level 3 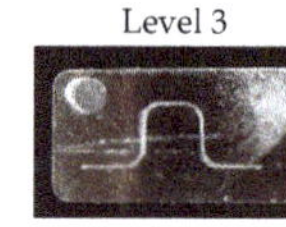
	4	1–25% of cloudy zones		
Transparent	5	Transparent with some defects	93.436	Level 6 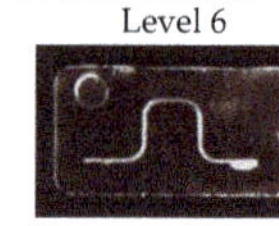
	6	Transparent with isolated defects	94.043	
	7	Transparent without defects	98.647	

2.3. Experimental Design and Procedure

In this research, a sequential design of experiments was performed. That is, start with simple designs and few experiments (Phase 1), analyze them, and, if necessary, proceed with further experiments to conclude with the relationship between the process parameters and the part responses (Phase 2). This methodology is known as response surface analysis [22] and it allows selection of the best manufacturing conditions to achieve the part requirements [23]. In this research, two process parameters were varied: plunger velocity (V) and amplitude (A) (also referred as factors of interest).

However, before starting the sequential design of experiments, a preliminary screening was carried out to limit the parameter values to guarantee a filling cavity and a level of transparency acceptable to the polymer application. Velocity was varied from 1 to 6 mm/s and amplitude from 10 to 90 μm. The results revealed that the filling cavity improved when velocity decreased, whereas amplitude did not have much effect. Otherwise, visual transparency improved when velocity increased and when low amplitudes were used. Consequently, the process parameters in the design of experiments (DOE) were limited to amplitude range 14.06–70.31 μm and velocity 1–4 m/s.

Phase 1 started deploying a two-level factorial design (2^k, where k = 2) with a central point (2^k + CP). Figure 3 shows the 5 experimental runs represented by ○. For each combination of process parameters 13 parts were manufactured; the first 3 were discarded to stabilize the process and the other 10 were analyzed. Each experimental run was replicated 3 times in completely randomized order. The central point runs allow for checking the trend in the middle point of the process parameters values. When this trend is a curve, more experiments should be done to estimate the quadratic effect, which was done in Phase 2. Phase 2 went on with a 3-level factorial design (3^k), which was an extension of the previous design (2^k + CP). Here, 4 new runs were performed (◆ in Figure 3) following the same pattern of 10 replicates and 3 repeats.

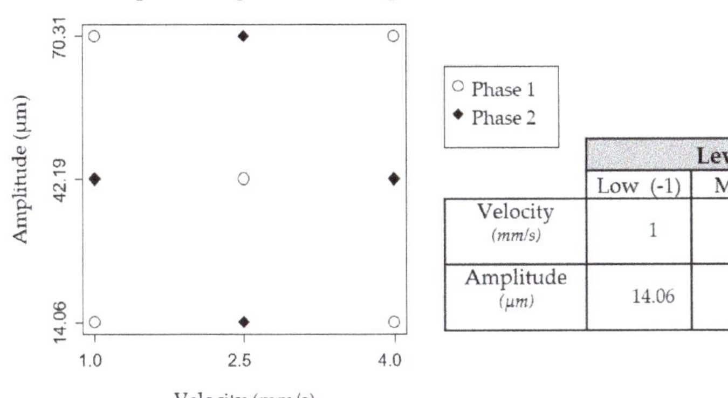

	Levels of the factors		
	Low (-1)	Middle (0)	High (+1)
Velocity (mm/s)	1	2.5	4
Amplitude (µm)	14.06	42.19	70.31

Figure 3. Sequential experimental design: Phase 1 (2^k with a central point) and Phase 2 (3^k); k = 2.

A set of process parameters was kept constant during both experimental phases. Mold temperature was set according to a raw material datasheet for injection molding applications (50 °C), cooling time was 10 s, with a packing force of 5000 N. Material dosage was approximately 0.3 g and vibration frequency was 30,000 Hz. Regarding the injection strategy selected, the plunger pushes the materials against the sonotrode, which is fixed in a position close to the mold cavity beginning. Its vibration melts the material and the plunger force pushes the melting material into the mold cavity.

The full model was estimated in each phase and for each response, including first-order and interactions in both phases and the quadratic terms in Phase 2. Then, the nonsignificant terms (one at each step) were sequentially removed based on the tests of individual regression coefficients. Each model was analyzed in terms of the fit statistics: R^2-adjusted and root mean square error (RMSE). The model shows a good fit if the R^2 statistic is high and RMSE is low. For each model, the test for significance of regression (*p*-value associated with the model in an analysis of variance, ANOVA) was observed. Analyzing the *p*-value resulting from ANOVA, *p*-value < α (α is the significance level) indicates that the regression is significant. The last step before validating the model is to check adequacy. It allows examination of whether the fitted model offers an adequate approximation to the real system and corroboration that the least squares regression assumptions are met, that is, that residuals are normal independent and identically distributed N (0, σ2). The first is checked by plotting observed values on the model response surface and the second by looking at different model diagnostic graphs. In the present paper, the residual vs. predicted values graph was used to check for independency and homoscedasticity and the normal probability plot was used to check normality. Conclusions on graphical analysis can be confirmed with tests such as the Shapiro–Wilk normality test or the test on correlation coefficient.

Finally, the retained model is plotted by a response surface to enhance interpretability and identify the process conditions where the responses become optimal. R software [24] was used to carry out the statistical analysis, and the significance level in all cases is $\alpha = 0.1$.

3. Results and Discussion

Table 3 presents the average results for the percentage of completed specimens, the transparency achieved, and the average thickness for each combination of processing parameters. There is no thickness dimension when any completed part was achieved.

Table 3. Summary of the experimental results; k = 2.

	Velocity (mm/s)	Amplitude (μm)	Filling (%)	Transparency (Qualitative Scale)	Thickness Dimensions (Standard Deviation) (μm)			
					T_{1Part}	T_{2Part}	T_{3Part}	T_{4Part}
Phase 1: 2^K experiment + central point	1	14.06	100	5.3	590 (±13)	583 (±9)	582 (±4)	586 (±7)
			100	5.4	583 (±8)	582 (±6)	579 (±10)	575 (±6)
			100	5.7	585 (±11)	578 (±12)	581 (±10)	586 (±13)
		70.31	100	6	590 (±5)	586 (±4)	585 (±7)	586 (±5)
			100	5.9	587 (±6)	583 (±12)	578 (±7)	579 (±10)
			100	5.8	582 (±12)	578 (±10)	577 (±6)	579 (±9)
	4	14.06	46	7	–	–	–	–
			80.50	5.50	566 (±3)	565 (±7)	571 (±12)	561 (±8)
			83.33	5.67	582 (±10)	569 (±1)	576 (±6)	574 (±12)
		70.31	62	7	–	–	–	–
			65.50	6.4	–	–	–	–
			70	6.2	585 (±2)	562 (±3)	571 (±5)	563 (±8)
	2.5	42.19	87.50	6.6	582 (±3)	573 (±1)	576 (±4)	577 (±6)
			85	6.7	569 (±7)	561 (±9)	577 (±9)	571 (±8)
			81	6.7	567 (±8)	562 (±11)	565 (±11)	566 (±4)
Phase 2: 3^K experiment	1	42.19	100	6	576 (±7)	577 (±6)	570 (±9)	571 (±9)
			100	5.7	592 (±9)	594 (±4)	603 (±9)	586 (±8)
			100	6	583 (±12)	592 (±7)	590 (±13)	581 (±8)
	4	42.19	45	7	–	–	–	–
			42.50	7	–	–	–	–
			41	6.7	–	–	–	–
	2.5	70.31	82	6.6	565 (±5)	565 (±3)	564 (±4)	555 (±13)
			90	5.78	–	–	–	–
			82	6.75	–	–	–	–
		14.06	98	6	572 (±17)	577 (±9)	575 (±10)	575 (±7)
			94	6	–	–	–	–
			99	5	578 (±9)	577 (±16)	572 (±14)	576 (±16)

3.1. Transparency

First, the transparency results of the Phase 1 2^k + CP experimental runs are analyzed, although initially the central point is excluded. Figure 4a shows the main effects plot of the response mean for each level of the process parameters. It reveals that transparency is better when both velocity and amplitude levels are higher. Besides, the Pareto chart of the standardized effects (Figure 4b) reflects those factors, which contribute significantly to explain transparency. Really, both velocity and amplitude affect transparency, although amplitude is a near-significant factor. On the contrary, the interaction between the two is not significant, so it is excluded from the model and re-estimated. In the final model, the resulting regression is significant, and both velocity and amplitude become significant parameters. The residuals of that model are normal (Shapiro–Wilk normality test *p*-value = 0.7574) and homoscedastic, and the model is able to explain almost 50% of the variability of the response (R^2adj = 0.49) and has RMSE = 0.45.

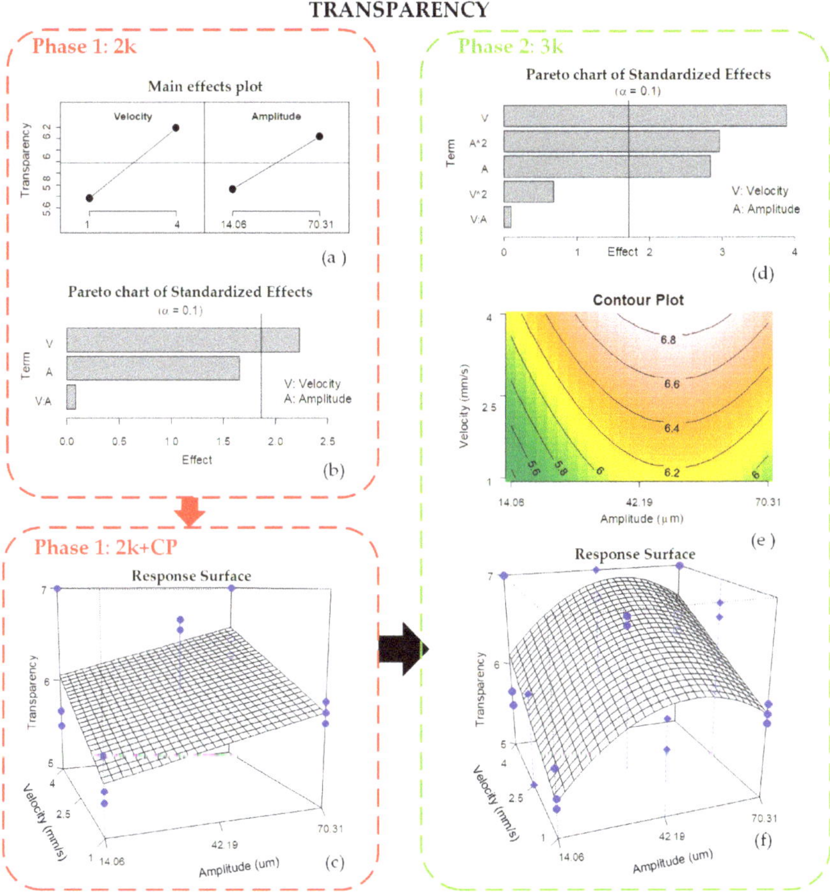

Figure 4. Analysis of transparency results: (**a**) main effects plot 2^k; (**b**) Pareto chart of effects 2^k; (**c**) response surface 2^k + CP; (**d**) response surface 3^k; (**e**) Pareto chart of standardized effects 3^k; and (**f**) contour plot of final 3^k model; k = 2.

Figure 4c shows the response surface of Phase 1 including the experimental results of the central point experiments (2^k + CP). Three replicates per experimental combination resulting from the mean of 10 repeats are represented by dots. In this graph, the responses at the central point are far from the surface, which means a lack of linearity between low and high factor levels and the possibility of a curve trend. Indeed, the test on the existence of a curve effect was significant, therefore more experiments were run to estimate this quadratic effect (as introduced in Section 2.3). Consequently, four new experiments with three replicates were performed and analyzed (Figure 3; Phase 2, 3^k).

Taking all the experiments from Phase 2, the factors influencing transparency are velocity (V) and amplitude (A and A2) (Figure 4d). The model is re-estimated after suppressing the nonsignificant terms and the resulting response surface is represented in Figure 4f. In this figure, the response surface is curved along the amplitude factor and is quite close to the experimental results (●), which is reflected by lower RMSE of 0.39. This model is able to explain 60% of the variability of transparency (R2adj = 0.6), which is a better result than the previous one (Phase 1, 2^k). Residuals are normally distributed (Shapiro–Wilk normality test *p*-value = 0.7672) and homoscedastic. The contour plot of the retained model is shown in Figure 4e. This process map shows the manufacturing conditions in which part transparency becomes better. It is at high velocity of 4 mm/s and amplitude around 42.19 μm.

High plunger velocity fills the mold cavity quickly, reducing the contact time of melting into the mold cavity and dropping the melt temperature, the same as the injection speed in microinjection molding [10]. Otherwise, low velocity increases the contact time of the polymer with ultrasonic energy, increasing the melting temperature and cooling time. As a result, degradation, internal cracking, and bubbles appear in the final part, notably reducing the transparency.

Analyzing the transparency variability of the three replicates (Figure 5a), it is observed that the dispersion is really low. Maximum variability of SD (0.6) is achieved. For low velocity values, transparency repeatability is better, whereas across amplitude, the standard deviation is not greatly affected.

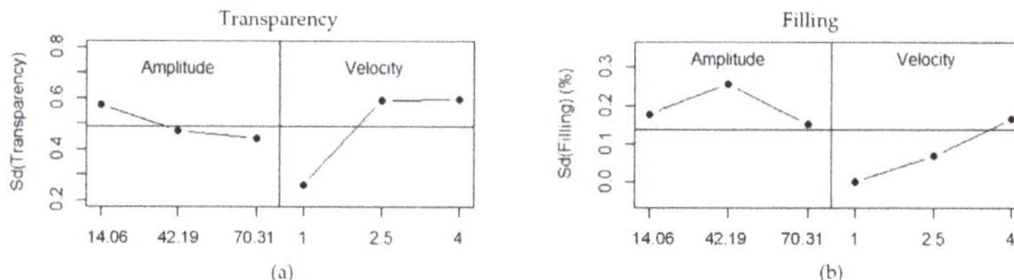

Figure 5. Standard deviation analysis: (**a**) transparency; and (**b**) part filling.

3.2. *Filling*

As shown in Table 3, most fillings are above 50% and even between 80% and 100%, therefore highly biased to the highest levels. Although it was expected, due to the range of process parameters adjusted during the preliminary screening, this response variable had to be transformed to accomplish the regression assumptions. A logit transformation was applied. In our case, we decided to logit transform the variable logit (Filling) = log(Filling/(100 − Filling) and adjust the model to this new response variable. When filling is 100%, its logit cannot be computed, and after some trials, we decided to replace those cases by 10.

The filling ratio analysis was done directly, taking the results from Phase 2, because the quadratic response surface estimated to explain transparency was shown to be appropriate, and measuring it is not expensive or time-consuming.

The main effects plot for filling is shown in Figure 6a and the Pareto chart of standardized effects in Figure 6b. Here, velocity is the factor that most contributes to the filling ratio. When velocity is low

(1 mm/s), filling is highest (100%). On the other hand, changes in amplitude do not greatly change the filling; however, as the Pareto diagram shows, this small change becomes significant.

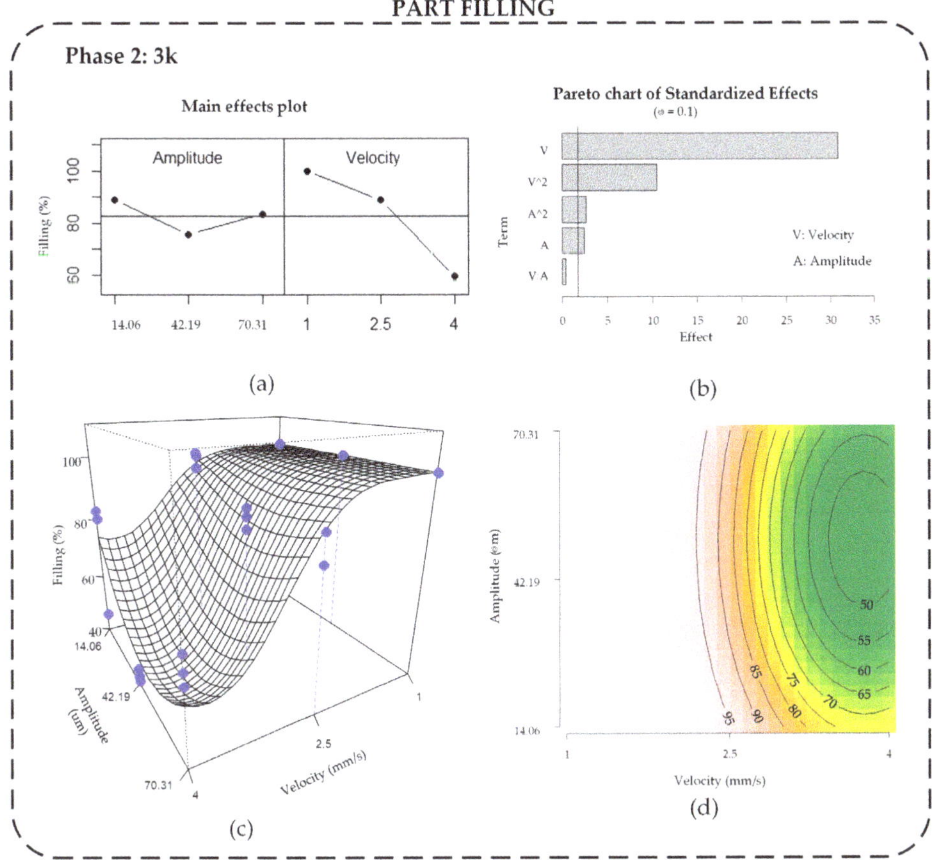

Figure 6. Analysis of cavity filling results: (**a**) main effects plot, 3^k; (**b**) Pareto chart of standardized effects, 3^k; (**c**) response surface, 3^k; and (**d**) contour plot of final 3^k model, k = 2.

The model that better explains part filling includes velocity and amplitude factors, with both linear and quadratic terms, and without interaction effect. It explains 97.7% of the data variability (R2adj = 0.977), with normally distributed residuals (Shapiro–Wilk normality test *p*-value = 0.6944) and homoscedasticity.

Figure 6c illustrates the response surface including the experimental results (●), and Figure 6d shows the contour plot of the same model. In both plots, the responses are represented in original scale. It can be seen how the logit transformation enables a response surface limited to 100%, which could not be attained using the original filling variable. Maximum filling is attained at low velocity levels for any amplitude values, getting mostly completed parts. Actually, when velocity is lower, melting temperature increases due to more ultrasonic energy being applied, and complete filling is guaranteed [7,25]. The polymer temperature drops more slowly and it has more time to fill the mold cavity properly. At high velocity, the influence of amplitude is higher, due to the filling reaching the minimum value at the middle value, and improves when it gets away, although the majority are below 70% of filling. During the experiment, when both amplitude and velocity were at the highest values, the parts became too breakable due to the high level of energy that the polymer received, and often broke or seemed to be degraded.

The standard deviation of the filling is also very low, because less than 0.3% of the variability is achieved (Figure 5b). It can be seen that an increment of velocity is associated with an increment of the standard deviation of filling, which means that the experiments replicate less. Indeed, at a velocity of 1 mm/s, the standard deviation is 0 because all three replicates have the same 100% filling result. Amplitude does not greatly affect the standard deviation.

Comparing filling and transparency response, we observe that is not possible to find the manufacturing conditions that maximize both responses. High velocity and medium amplitude values improve the transparency but worsen the filling cavity. Thus, depending on the application, the manufacturing conditions should be adapted.

3.3. Part Thickness

Part thickness was only measured in completely filled parts (Table 3 shows average results) and the thickness difference (D) was calculated to evaluate this response. The thickness difference (D) is the result for comparing part thickness measured in each corner of the parts (T_i Part, i = 1 ... 4) regarding the real mold thickness measured in the same corresponding corner (T_1 Mold, i = 1 ... 4). Thus $D_i = T_i$ Part—T_i Mold, i = 1 ... 4, resulting in four variables for each part. On the one hand, a positive value of D_i indicates that the part is thicker than the real mold thickness. On the other hand, a negative value of D_i indicates that the piece is thinner than the real mold thickness.

The objective here is to analyze those four variables and relate them to the manufacturing factors (amplitude and velocity). Nevertheless, as a four-dimensional space cannot be plotted, they were represented in a lower-dimensional space by means of principal component analysis (PCA). PCA is a statistical technique used to reduce the number of variables to represent the dimensions that have more variability in the results. In this case, the four variables (Di, i = 1 ... 4) are represented in a biplot that retains almost 83% of the variability of the dataset (Figure 7). Each number plotted in this graph is a vector that contains the four differences (Di) measured in each complete part (see numerical examples in Figure 7). In Figure 7, the first two principal components (PC1 and PC2) are represented.

The first component (PC1) in the horizontal axis is interpreted as the sum of all thickness differences, that is, PC1 $\approx \Sigma$Di. Thus, parts that are on the right-hand side of the graph have positive ΣDi values, meaning that, in general, the part thickness is thicker than the mold (see values of experiments 7 and 77 in Figure 7). On the contrary, the parts that are on the left of the graph are narrower than the mold (see experiments 56 and 79). This component (PC1) is the most important one, because it explains 70.4% of variability.

Actually, the expected result would be that most of the pieces were equal to or thinner than the mold thickness, due to the shrinkage effect that commonly happens in injection molding processes, such as microinjection and conventional injection molding, although, as is known, amorphous polymers are less affected by this. However, the experimental results in this research determined by ultrasonic molding technology reveal that around 50% of the obtained parts had an opposite trend to what was expected and seemed to be affected by a releasement phenomenon. The main hypothesis to explain this is that the polymer was still too warm when the mold opened and the pressure released. Thus, the polymer relieved its stored internal energy (both thermal and ultrasonic) in the easier direction, lightly expanding the parts, similar to the swelling behavior in extrusion processes. Possibly the cooling time used in the experiment was not enough to guarantee final cooling of the polymer, although the parts seemed complete when the mold opened. This cooling time was taken from the microinjection molding process; however, these results reveal that the melting temperature in ultrasonic molding becomes higher than in microinjection molding.

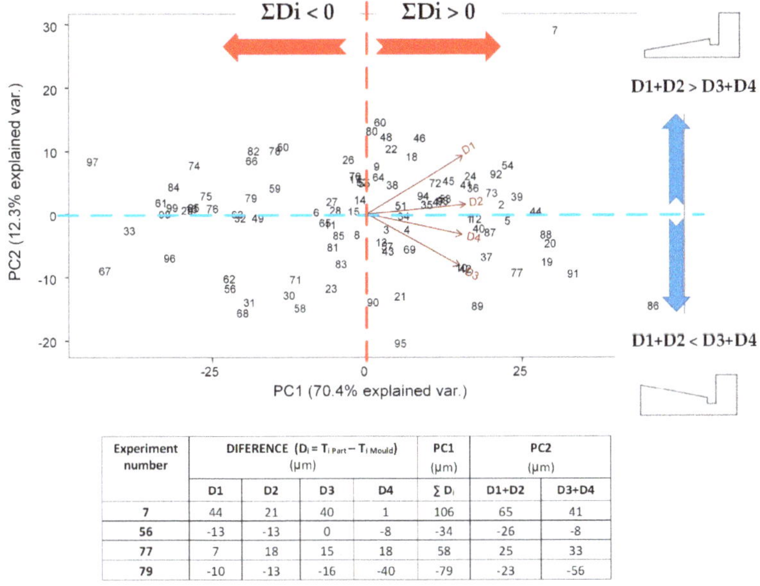

Experiment number	DIFERENCE ($D_i = T_{i\,Part} - T_{i\,Mould}$) (µm)				PC1 (µm)	PC2 (µm)	
	D1	D2	D3	D4	ΣD_i	D1+D2	D3+D4
7	44	21	40	1	106	65	41
56	-13	-13	0	-8	-34	-26	-8
77	7	18	15	18	58	25	33
79	-10	-13	-16	-40	-79	-23	-56

Figure 7. Biplot resulting from principal component analysis of Di, where $D_i = T_i$ Part—T_i Mold, i = 1 . . . 4 measured at four different part points.

Despite this effect, the average of PC1 $\approx \Sigma D_i$ is around 50 µm, which represents less than 7% of thickness dimensional error, which is acceptable for many applications.

The second component (PC2) in the vertical direction axis is interpreted as D1 + D2 against D3 + D4. It lets us compare the thickness of the corners nearest to the injection entry those farthest. When the parts have positive values of PC2 (upper side of the graph), this means that they have larger values of D1 and D2 than D3 and D4, meaning that the part is thicker on the side of the injection flow (see experiments 7 and 79). On the other hand, the parts on the lower side of the graph have lower D1 and D2 values and higher D3 and D4 values, which means that the part is thicker at the farthest side of the injection entry (see experiments 77 and 56). Note that variables D1 and D2 are grouped together, as they are closely associated or correlated from a statistical point of view, meaning that parts with high (low) values of D1 also have high (low) values of D2. The same applies to D3 and D4.

Once the characterization of part thickness was explained, it was interesting to relate those characteristics with velocity and amplitude factors. Actually, the first principal component (PC1 = ΣDi) showed more significant differences between factor levels, therefore it was analyzed. Taking PC1 results, a graphical analysis by multiple box plots (Figure 8a,b) and numerical ANOVA (Figure 8c,d) were done, comparing the sum of all difference thicknesses (ΣDi) for the different factor levels.

Regarding velocity, when it is low (1 mm/s), the parts are, on average, thicker than the mold, whereas when velocity is at a medium or high level, the parts are, on average, thinner than the mold. Considering amplitude, the parts are, on average, thicker than the mold. As with transparency and the part filling response, amplitude does not really influence part thickness, but velocity does. ANOVA enables us to test the hypothesis that the means of the three levels are equal. In the case of amplitude, the averages are equals, and, for velocity, the medium and highest level are equal.

When the process is slow (velocity 1 mm/s), the ultrasonic time is longer and the material achieves higher temperatures. Consequently, when the mold is opened, the part temperature is too hot and the polymer releases its internal energy, leading to slightly increased thickness.

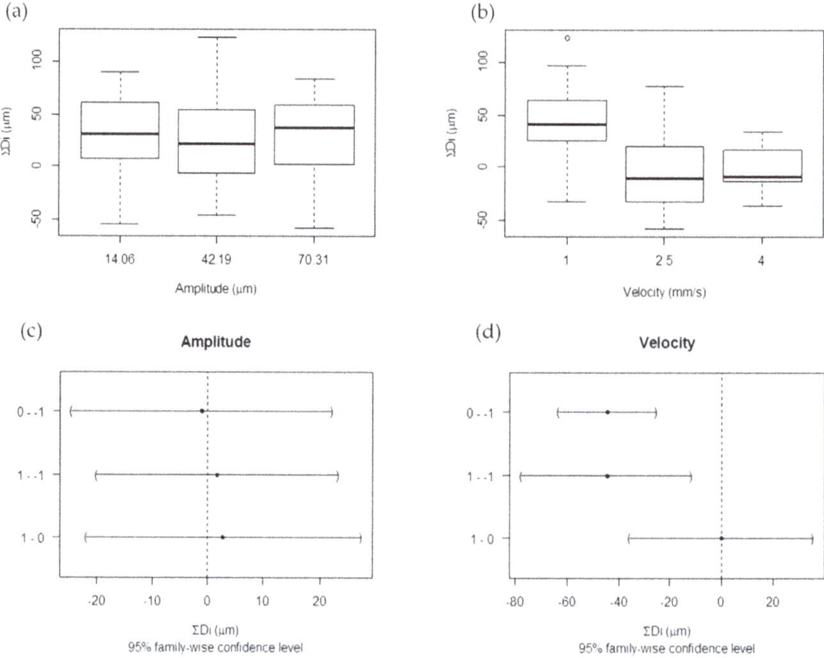

Figure 8. Multiple box plots: (**a**) amplitude and (**b**) velocity and, ANOVA analysis: (**c**) amplitude and (**d**) velocity for first principal component (PC1 = ΣDi).

3.4. Microchannel Results

The microchannels were studied in three sections in the in-flow direction (H_1, H_2, and H_3) and two in the cross-flow direction (V_1 and V_2). Three completed parts for each combination of process parameters were selected, resulting in 15 sections for each combination of process parameters. Three measures of depth and width of each microchannel were taken in each section, similar to the mold characterization shown in Figure 2b. Altogether, 630 measures were collected, and the average values were used to evaluate these results.

Three issues of the microchannel results are assessed. The first is the influence of the process parameters on the width and depth of the microchannel independently of section location. The second is an examination of these dimensions according the section location (H_1, H_2, H_3, V_1, and V_2). Finally, the microchannel shape is observed and discussed in detail.

The average microchannel dimension is around 139 μm in depth and 81 μm in width, resulting in 1 μm smaller than the mold depth (140 μm), 9 μm bigger than the mold width (72 μm), and a low variability, as shown in Figure 9. It is an error of 1% and 12%, respectively. Thus, polystyrene has good replicability when microfeatures are manufactured by ultrasonic molding technology. According to Chakraborty et al. [10], the replicability is better in depth than in width.

ANOVA reveals that the dimensions of the microchannel are not affected by velocity, independently of section location. However, amplitude does slightly affect the width dimension. In contrast, the section location (H_1, H_2, H_3, V_1, and V_2) is the factor that contributes most to microchannel replicability (Figure 10). This figure plots the averages of microchannel dimensions, specifying section location (by the shape of the label), amplitude (by color), and velocity (by the color intensity). It reveals that the depth results are nearer the mold dimensions and less dispersed than the width results, just as in [26] with molded PMMA microchannels, where they obtained good replication in depth but poor in width. On average, the replication accuracy for depth is less than 4% and for width is less than 11%, the same as achieved by Chakraborty et al. [10]. Focusing on in-flow sections, the H_1 dimensions (▲) are very close to the target values for both depth and width. Basically, proximity to the injection entry supports this result [14]. The H_2 results (–) are the worst, since both depth and width are very far away from the target values. The H_3 dimensions (■) are between H_1 and H_2 results. Analyzing the cross-flow sections, V_1 (♦) and V_2 (●), the results are grouped more around the target depth than the target width. Without considering H_2 results, it is observed that the results of cross-flow sections are more dispersed than in-flow ones. The deviated results of H_2 can be due to a lack of homogeneity in the mold temperature and the proximity of this section to the hottest mold zone. For this reason, the polymer of the H_2 section relieved more stored internal energy, causing results to be away from the target values.

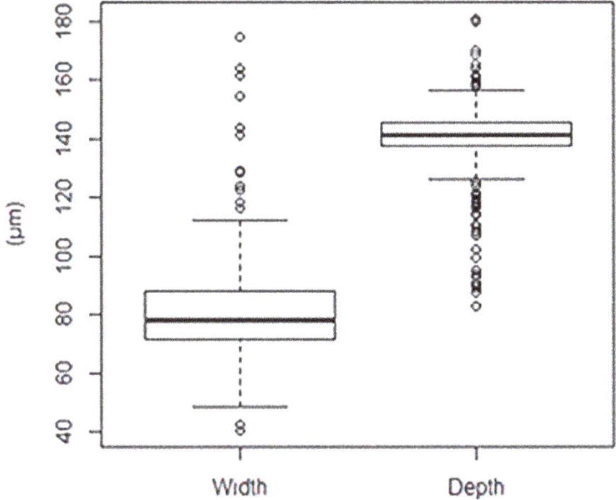

Figure 9. Multiple box plot of the microchannel.

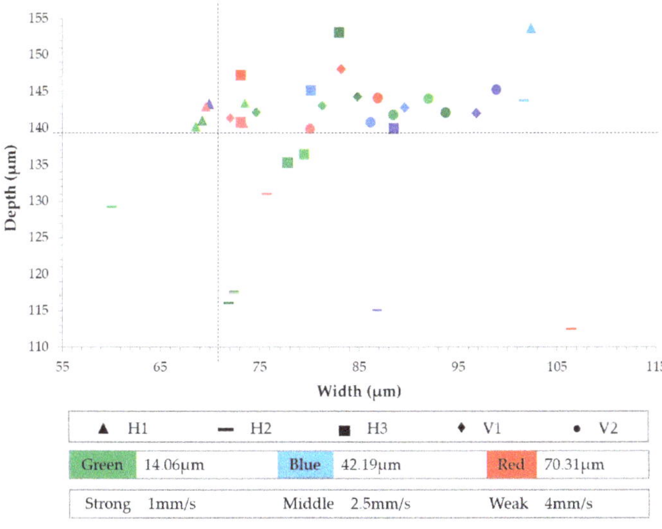

Figure 10. Microchannel dimensions.

Finally, the microchannel shapes are presented in Figure 11. Two sets of specimens of different combinations of process parameters are shown for each microchannel section and three replicates for each combination. It can be seen that there are no differences in microchannel shape between the replicates of each combination, even when different sets of different combinations of process parameters are compared. This reinforces that velocity and amplitude are not significant to microchannel shape.

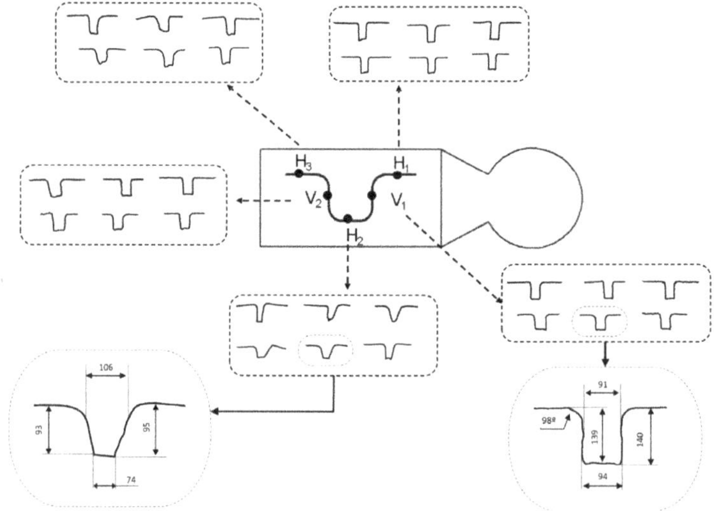

Figure 11. Microchannel shape analysis.

The shape of sections H_1, V_1, and V_2 are similar and mostly well defined in both the walls and the bottom. On average, the difference of microchannel width between the surface and the top is around 11 μm, the difference in depth is 7 μm, and the radius of the top is around 98°. In contrast, the worst resolution is observed for section H_2. In that case, deviations are around 45 μm in width and 15 μm in depth. Here, the shape is not well defined and the walls are not parallel to each other. In the case of H_3, the shape of the microchannel is between those of H_1 and H_2.

4. Conclusions

In this research, thin-wall plates with polystyrene microchannels were successfully manufactured by ultrasonic molding technology. A rigorous statistical procedure was followed using a large set of experimental data that were analyzed to demonstrate the repeatability and replicability of this technology.

The results reveal that plunger velocity affects both transparency and filling cavity. Higher velocity improves transparency but decreases cavity filling, so intermediate values are required to guarantee both requirements. Overlapping the corresponding response surfaces, the optimal values can be selected. The vibration amplitude has slightly less influence in both cases. Degradation symptoms appear when the material receives too much ultrasonic energy, coming from high levels of amplitude and low plunger velocity.

The average deviation of part thickness is less than 7%, but the research shows that cooling time becomes a relevant factor to assure this property. The principal component analysis (PCA) method becomes very useful to evaluate the thickness data and analyze them regarding the mold dimensions.

Replication accuracy of the microchannel is better in depth than in width, obtaining an average deviation of 4% and 11%, respectively. This replication also depends on the orientation of the microchannels, so that the results of cross-flow sections are little more dispersed than in-flow ones. The distance from the injection entrance also has an influence on these results.

In further research, a more complex microchannel geometry should be tested to guarantee the feasibility of this technology for processing microfluidic devices. In that sense, other polymers should be studied, like cyclic olefin copolymer (COC) and polymethyl methacrylate (PMMA). Regarding the process parameters, the influence of cooling time and compaction force should be studied. S

Author Contributions: F.I. and M.A. conceived and designed the experiments. M.A. performed the USM experiments and measurements. F.I. and G.-R.M. analyzed the data and directed the research. V.-M.M. carried out the statistical analysis. F.I. and V.-M.M. wrote the manuscript. The manuscript was finalized with contributions from all authors, and all authors approve the final manuscript.

Funding: This research received funding from the University of Girona (MPCUdG2016/036), the Spanish Ministry of Education (DPI2016-77156-R), and the Catalan Agency for Management of University and Research Grants (2017-SGR-0385).

Conflicts of Interest: The authors declare no conflict of interest.

References

1. Ciofu, C.; Mindru, D.T. Injection and Micro Injection of Polymeric Plastics Materials: A Review. *Int. J. Mod. Manuf. Technol.* **2013**, *3604*, 49–68.
2. Mahmoodi, M.; Park, S.S.; Rizvi, G. Feasibility study of thin microinjection molded components. *Polym. Eng. Sci.* **2012**, *52*, 180–190. [CrossRef]
3. Martínez, R.; Trotta, G.; Volpe, A.; Bernava, G.; Basile, V.; Paturzo, M.; Ferraro, P.; Ancona, A.; Fassi, I.; Osellame, R. Rapid Prototyping of Plastic Lab-on-a-Chip by Femtosecond Laser Micromachining and Removable Insert Microinjection Molding. *Micromachines* **2017**, *8*, 328. [CrossRef]
4. Sortino, M.; Totis, G.; Kuljanic, E. Comparison of Injection Molding Technologies for the Production of Micro-optical Devices. *Procedia Eng.* **2014**, *69*, 1296–1305. [CrossRef]
5. Hong, S.; Hwang, J.; Kang, J.; Yoon, K. Comparison of injection molding and injection/compression molding for the replication of microstructure. *Korea-Aust. Rheol. J.* **2015**, *27*, 309–317. [CrossRef]
6. Sato, A.; Ito, H.; Koyama, K. Study of application of ultrasonic wave to injection molding. *Polym. Eng. Sci.* **2009**, *49*, 768–773. [CrossRef]
7. Grabalosa, J.; Ferrer, I.; Elías-Zúñiga, A.; Ciurana, J. Influence of processing conditions on manufacturing polyamide parts by ultrasonic molding. *Mater. Des.* **2016**, *98*, 20–30. [CrossRef]
8. Jiang, K.; Yu, F.; Su, R.; Yang, J.; Zhou, T.; Gao, J.; Deng, H.; Wang, K.; Zhang, Q.; Chen, F.; et al. High speed injection molding of high density polyethylene. Effects of injection speed on structure and properties. *Chin. J. Polym. Sci.* **2011**, *29*, 456–464. [CrossRef]
9. Chen, S.; Liao, W.; Yeh, J.; Chien, R. Rheological behavior of PS polymer melt under ultra high speed injection molding. *Polym. Test.* **2012**, *31*, 864–869. [CrossRef]
10. Chakraborty, A.; Xiang, M.; Luo, C. Fabrication of Super-Hydrophobic Microchannels via Strain-Recovery Deformations of Polystyrene and Oxygen Reactive Ion Etch. *Materials* **2013**, *6*, 3610–3623. [CrossRef] [PubMed]
11. Attia, U.M.; Alcock, J.R. An evaluation of process-parameter and part-geometry effects on the quality of filling in micro-injection moulding. *Microsyst. Technol.* **2009**, *15*, 1861–1872. [CrossRef]
12. Chen, C.-S.; Chen, S.-C.; Liao, W.-H.; Chien, R.-D.; Lin, S.-H. Micro injection molding of a micro-fluidic platform. *Int. Commun. Heat Mass Transf.* **2010**, *37*, 1290–1294. [CrossRef]
13. Lucchetta, G.; Ferraris, E.; Tristo, G.; Reynaerts, D. Influence of mould thermal properties on the replication of micro parts via injection moulding. *Procedia CIRP* **2012**, *2*, 113–117. [CrossRef]
14. Masato, D.; Sorgato, M.; Lucchetta, G. Analysis of the influence of part thickness on the replication of micro-structured surfaces by injection molding. *Mater. Des.* **2016**, *95*, 219–224. [CrossRef]
15. Michaeli, W.; Kamps, T. Heating and plasticizing thermoplastics with ultrasound for micro injection molding. In Proceedings of the Annual Technical Conference—ANTEC 2010, Orlando, FL, USA, 16–20 May 2010; pp. 323–327.
16. Michaeli, W.; Spennemann, A.; Gärtner, R. New plastification concepts for micro injection moulding. *Microsyst. Technol.* **2002**, *8*, 55–57. [CrossRef]

17. Michaeli, W.; Opfermann, D. Ultrasonic plasticising for micro injection moulding. In *4M 2006—Second International Conference on Multi-Material Micro Manufacture*; Elsevier Science: New York, NY, USA, 2006; pp. 345–348.
18. Jiang, B.; Hu, J.; Li, J.; Liu, X. Ultrasonic plastification speed of polymer and its influencing factors. *J Cent. South Univ.* **2012**, *19*, 380–383. [CrossRef]
19. Michaeli, W.; Kamps, T.; Hopmann, C. Manufacturing of polymer micro parts by ultrasonic plasticization and direct injection. *Microsyst. Technol.* **2011**, *17*, 243–249. [CrossRef]
20. Sacristán, M.; Plantá, X.; Morell, M.; Puiggalí, J. Effects of ultrasonic vibration on the micro-molding processing of polylactide. *Ultrason. Sonochem.* **2014**, *21*, 376–386. [CrossRef] [PubMed]
21. Planellas, M.; Sacristán, M.; Rey, L.; Olmo, C.; Aymamí, J.; Casas, M.T.; del Valle, L.J.; Franco, L.; Puiggali, J. Micro-molding with ultrasonic vibration energy: New method to disperse nanoclays in polymer matrices. *Ultrason. Sonochem.* **2014**, *21*, 1557–1569. [CrossRef] [PubMed]
22. Myers, R.H.; Montgomery, D.C.; Anderson-Cook, C.M. *Response Surface Methodology: Process and Product Optimization Using Designed Experiments*; John Wiley & Sons: Hoboken, NJ, USA, 2016.
23. Liu, Y.-T.; Zhang, L. An investigation into the aspheric ultraprecision machining using the response surface methodology. *Precis. Eng.* **2016**, *44*, 203–210. [CrossRef]
24. Team, R. *R: A Language and Environment for Statistical Computing*; R Foundation for Statistical Computing: Vienna, Austria, 2013.
25. Negre, P.; Grabalosa, J.; Ferrer, I.; Ciurana, J.; Elias-Zuñiga, A.; Rivillas, F. Study of the Ultrasonic Molding Process Parameters for Manufacturing Polypropylene Parts. *Procedia Eng.* **2015**, *132*, 7–14. [CrossRef]
26. Chien, R.D. Micromolding of biochip devices designed with microchannels. *Sens. Actuators A Phys.* **2006**, *128*, 238–247. [CrossRef]

© 2018 by the authors. Licensee MDPI, Basel, Switzerland. This article is an open access article distributed under the terms and conditions of the Creative Commons Attribution (CC BY) license (http://creativecommons.org/licenses/by/4.0/).

Article

Effect of Temperature Distribution in Ultrasonically Welded Joints of Copper Wire and Sheet Used for Electrical Contacts

Jeyaraj Pradeep Kumar

Department of Production Engineering, PSG College of Technology, Coimbatore 641 004, Tamil Nadu, India; jp.psgtech@gmail.com; Tel.: +91-944-387-2965

Received: 31 May 2018; Accepted: 12 June 2018; Published: 14 June 2018

Abstract: The temperature distribution occurring at the interface while joining a simple electrical contact comprising of a copper wire and a copper sheet using ultrasonic metal welding was analyzed using finite element method. Heat flux due to plastic deformation and friction was calculated and provided as input load for simulation of temperature distribution. The results of temperature obtained from simulation are found to be in good agreement with the results of temperature from experiments measured using thermocouple. Special focus was given to how the heat generated at the wire–sheet interface affect the strength of the joint in tension. With the knowledge of heat generated at the interface while welding, it is possible to control the strength of the joint and produce defect free joints. Based on the results from finite element analysis and experiments, it is observed that the influence of heat developed due to friction and plastic deformation of metallic specimens has a significant effect on the progress of welding and strength of the joint.

Keywords: ultrasonic metal welding; finite element analysis; temperature distribution; strength of the joint in tension; heat flux; friction; plastic deformation; thermocouple

1. Introduction

Recent technological developments in producing electrical contacts are evolving at a faster rate, resulting in consistent functioning of wide variety of customer durable products. Due to advent of component miniaturization and improvement in manufacturing light weight portable products, there exists a significant demand for superior quality electrical contacts. The electrical and electronic components used in these products are subjected to severe operating conditions during the service life of the products. Ultrasonic metal welding (USMW) has received significant attention in the past few years, and has become a more reliable and suitable process for producing quality electrical contacts to be used in these products. USMW is an environmentally friendly and rapid solid state joining process in which similar or dissimilar metals are joined by the application of ultrasonic vibrations and adequate pressure at the interface. The high frequency relative motion between the parts to be welded result in progressive shearing and plastic deformation between surface asperities. This rubbing action disperses oxides, contaminants and form a sound metallurgical bond between the two metallic parts.

Ding et al. [1] performed thermo-structural numerical analysis on ultrasonic wire bonding to study the effect of bonding parameters on temperature developed at the interface. The authors presented the temperature distribution in wire and the bond pad and reported that the bond force had more effect on development of flash temperature at the contact zone between the interacting asperities and marginal effect on bulk temperature rise developed along the frictionally heated surface. The flash temperature at the asperities between wire and pad on the microscopic scale was found to play a significant role in the formation of solid state bonds. During welding, the temperature at the contact interface increased rapidly at the initial stages of bond formation followed by gradual increase until the end of the vibrations.

Sooriyamoorthy et al. [2] presented a study on temperature distribution at the interface and stress distribution in sonotrode and the welded joint during ultrasonic welding of aluminum sheets. A finite element model was developed for prediction of temperature developed during the process by considering various parameters such as material thickness, clamping force, weld time and coefficient of friction. It was observed that the temperature at the interface increases with increase in weld time and decreases with increase in clamping force. The temperature developed at the interface was measured using thermocouple and the results from experiments were found to be in good agreement with the results from finite element analysis.

Zhao et al. [3] developed thin film thermocouple for measurement of temperature in ultrasonic welding of battery tabs. The dynamic temperature rise during welding was measured using thermocouple with high repeatability. The experimental trials for measurement of temperature were carried out based of design of experiments. The temperature measured by the thermocouple was compared with the temperature measured by an infrared thermal imager and were found to be in good agreement.

Shin et al. [4] conducted an experimental parametric study on welding of aluminum alloy sheets using ultrasonic spot welding. The process parameters such as vibrational amplitude and weld time were considered in this work. During welding, maximum temperature was observed around the tip of the sonotrode using infrared thermal imager. Significant heat generation occurred at the interface due to frictional vibration and large plastic deformation. The maximum temperature developed for varying vibrational amplitude and weld time was approximately 500 °C and this temperature was about 80% of the melting temperature of the work material. The phenomenon of part marking and part sticking was noticed while welding at higher temperatures.

Elangovan [5] investigated theoretically and experimentally on temperature distribution in ultrasonic welding of copper specimens. Crank Nicolson method was adopted to characterize the unsteady heat transfer scenario at the interface and to determine the temperature developed at each node for varying time steps. It was reported that the temperature at the interface increases with weld time and decreases with increase in thickness of the sheet. It was also inferred that the maximum temperature was obtained in spot welding when compared with seam welding.

Lee and Cai [6] performed 2-D finite element simulations to understand the effect of four different knurl designs provided in the tip of the sonotrode on strength of the joint. The results from simulation revealed that the sharp knurl design tip severely deform the metallic specimens resulting in reduction of strength of the joint.

Jedrasiak et al. [7] described a finite element model for prediction of temperature during ultrasonic welding of aluminum–aluminum, aluminum–magnesium and aluminum–steel work materials. The temperature data obtained were used for estimation of growth of intermetallic phase between aluminum-magnesium welds. The K-type thermocouple was embedded in the tip of the sonotrode and the aluminum sheets. The predictive capability of the finite element model was evaluated by comparing results from analysis with the data obtained using thermocouple.

Chen and Zhang [8] constructed a 3-D finite element model for analyzing the temperature distribution while welding dissimilar automotive alloys. The heat generated due to interface friction was in linear relationship with weld time. The heat generated due to material plastic deformation was found to increase rapidly during the initial phase of the welding and gradually decreases due to decrease in plastic deformation over time. Temperature contours and deformation in shape were obtained using results from simulation. The maximum temperature was located at the interface and the temperature decreases towards the other interfaces such as sonotrode/upper specimen and anvil/lower specimen. High temperature was observed in the upper specimen when compared with the lower specimen.

De Vries [9] presented the mechanics and mechanism of ultrasonic metal welding of aluminum. The temperature developed at the interface for varying welding conditions were measured using an infrared thermal imager. The temperature at the interface was found to vary 40–80% of the melting temperature of the work material. An empirical relation was developed for calculation of heat flux due to deformation of work material and friction between the parts to be welded.

A comprehensive understanding on formation of joint in ultrasonic metal welding due to heat developed at the interface is still in dispute [10–12]. Based on literature survey, research pertaining to correlation of temperature generated at the interface during ultrasonic welding of electrical contacts with strength of the welded joints seems to be not reported. This work was carried out to fill this gap. The primary purposes of this work were to analyze the amount of heat developed during ultrasonic metal welding of an electrical contact comprising of copper wire (100 mm length and 1.2 mm diameter) and copper sheet (100 mm length × 25 mm width × 0.2 mm thickness) and to evaluate the correlation between the heat generated at the interface and strength of the electrical contact joint in tension using finite element analysis and experiments. Finite element analysis was performed using commercially available ABAQUS 6.12 software.

2. Experimental Details

The experiments were carried out based on Taguchi's L9 orthogonal array [13,14] using a conventional lateral drive ultrasonic metal welding machine (National Indosonic, Bangalore, India) (2.5 kW, 20 kHz). Before welding, the weld samples were cleaned with acetone to remove any surface impurity as it may affect the strength of the joint. Based on literature survey, the controllable factors considered in this work for carrying out simulation and experimental trials were the clamping force, amplitude of vibration of the sonotrode and weld time (Table 1). Each factor was varied at three levels. The quality characteristic response variables were the temperature at the interface and the strength of the joint in tension.

Table 1. Factors and levels.

Factors	Units	Designation	Level 1	Level 2	Level 3
Clamping force	N	A	795	995	1195
Amplitude of vibration of sonotrode	μm	B	30	42.5	57
Weld time	second	C	2	2.5	3

A 10 kN tensile testing machine (Hitech, Coimbatore, India) was used to measure the strength of the joint in tension. The schematic representation of the joint and tensile loading condition is shown in Figures 1 and 2.

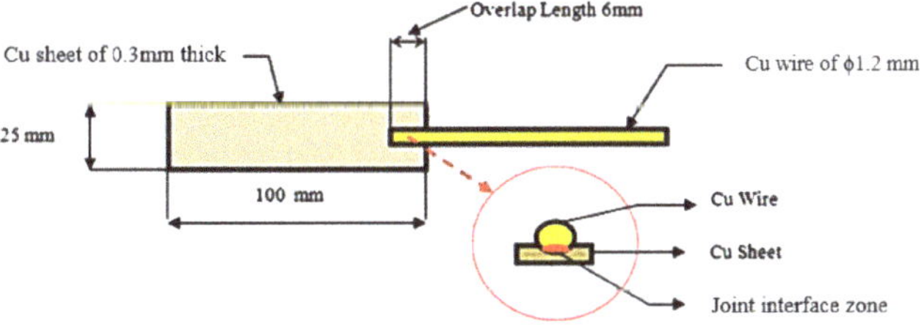

Figure 1. Schematic representation of the joint.

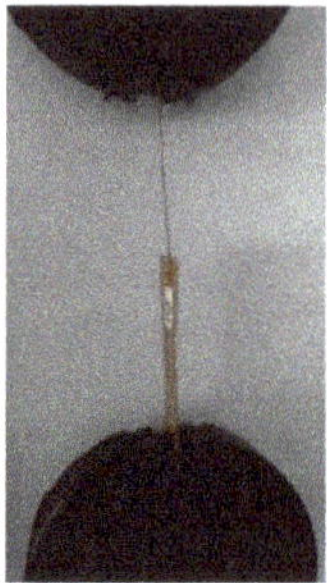

Figure 2. Tensile loading.

A suitable instrumentation consisting of K-Type thermocouple (Chromel/Alumel) and data acquisition system (DAQ) interfaced with LabVIEW software (2017, National Instruments, Pune, India) was used to obtain real-time temperature data developed at the interface. Figure 3 depicts the flow diagram of data transfer in data acquisition system.

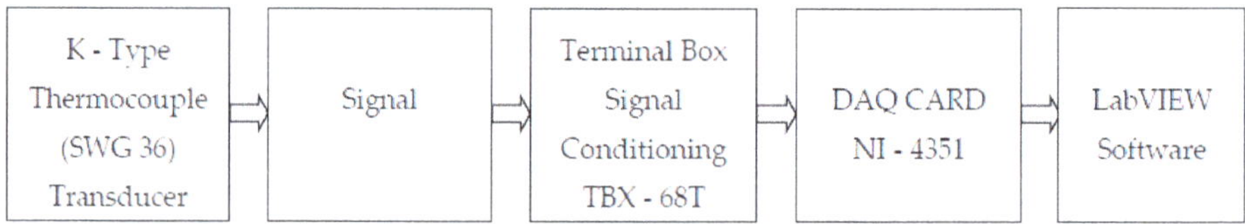

Figure 3. Flow diagram for data transfer in a DAQ system: Temperature.

The junction of K-Type thermocouple was made with chromel (positive leg: 90% nickel, 10% chromium) and alumel (negative leg: 95% nickel, 2% aluminum, 2% manganese and 1% silicon) wires of SWG 36 (0.193 mm diameter) twisted together for a length of 10 mm. One end of the thermocouple was placed transverse to the longitudinal axis at the interface, as shown in Figure 4, and the other end was connected to the terminal block of the data acquisition system. A 10 Hz notch filter setting was used to optimize measurement frequency and minimize noise level. This DAQ system was interfaced with LabVIEW software to obtain the real-time temperature data developed at the interface. The electrical signal generated from thermocouple was converted to temperature values and displayed in graphical form by LabVIEW software. A custom designed stepped sonotrode with similar circular cross section and gradual change in cross section using a tapered profile at the middle of the sonotrode, as shown in Figure 5, was used to carry out the experiments. The stepped sonotrode has a rectangular tip of 20 mm × 5 mm with serration depth of 0.2 mm, wherein the ultrasonic vibrations and clamping pressure are transmitted to the interface effectively, thereby improving the strength of the joint.

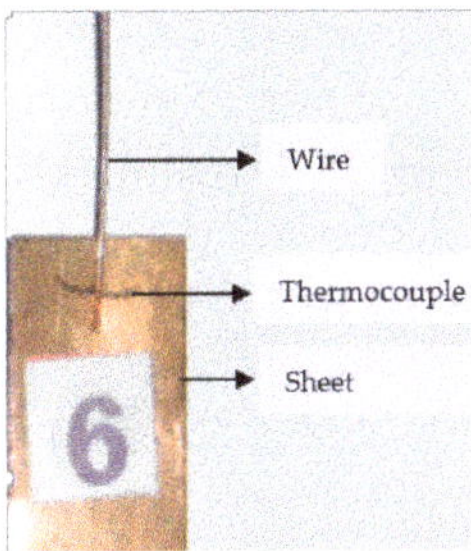

Figure 4. Thermocouple at the interface.

Figure 5. Sonotrode.

3. Finite Element Analysis

3.1. CAD Model

A CAD model of the joint made of copper wire and copper sheet with an overlap of 6 mm was developed using ABAQUS 6.12 software (Dassault Systems, Vélizy-Villacoublay, France) for performing thermal analysis, as shown in Figure 6.

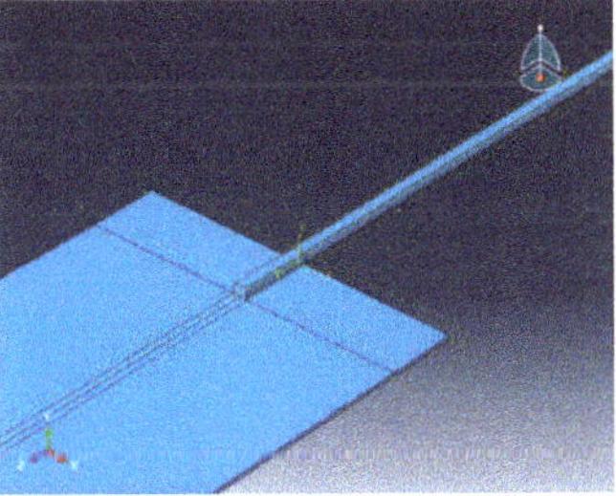

Figure 6. CAD model of the joint.

3.2. Material Properties

The material properties considered for thermal analysis are tabulated in Table 2. The properties of copper specimens are taken from ASM Hand Book [15].

Table 2. Material properties of copper.

Properties	Value
Young's Modulus (GPa)	115
Poisson ratio	0.3
Density (kg/m^3)	8940
Thermal conductivity (W/m °C)	391
Specific heat (J/Kg °C)	385
Thermal expansion co-efficient(°C^{-1})	1.66×10^{-5}

3.3. Element Selection and Meshing of CAD Model

The specimens were modeled and the element type was selected based on the type of analysis using ABAQUS 6.12 element reference guide [16]. ABAQUS C3D8RT was used for meshing the model. This kind of element is a thermomechanical coupled brick element and has eight-node trilinear displacement and one degree of freedom for temperature with reduced integration scheme and hourglass control. The meshing technique adopted in this study is free meshing which allows the finite element analysis software to generate high quality mesh. A total of 10,400 elements were used for meshing the sheet and 5678 elements were used for meshing the wire based on convergence test and considering the computational intensity and time. The finite element model of the specimen is shown in Figure 7.

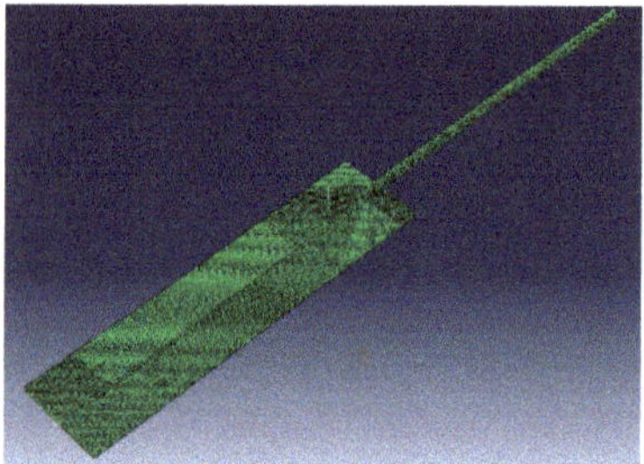

Figure 7. Finite element model of joint.

3.4. Assumptions for Thermal Analysis

The following assumptions were made in simulation of temperature distribution

i. Unsteady state is considered for thermal analysis.
ii. Full contact is established with no air gap between the specimens.
iii. Room temperature is 30 °C.
iv. The area in which the friction is effective is assumed to be the area of deformation
v. Surfaces exposed to air are set under free convection.

Surface–surface contact was established between wire and sheet. Free convection was applied with overall heat transfer coefficient of 5 W/m^2·°C [17]. The thermal contact conductance was assumed as 393 W/m^2 °C [18,19]. The transient analysis was selected with a time step of 0.5 s.

3.5. Measurement of Area of Deformation

In USMW, the upper specimen and the lower specimen are pressed and rubbed against each other to create a solid-state joint. The part of specimens under the tip of sonotrode deform due to static clamping pressure and swaying of specimen surfaces against each other due to ultrasonic vibrations result in formation of joint between the weld specimens. The area of deformation (A_{DZ})

play a significant role in calculation of heat flux required for simulation and it is estimated as the rectangular area bordering the interface as shown in Figure 8. The area of deformation for each joint is measured for all the trials of experiments using Tool maker's microscope (Mitutoya South Asia, NewDelhi, India) [20].

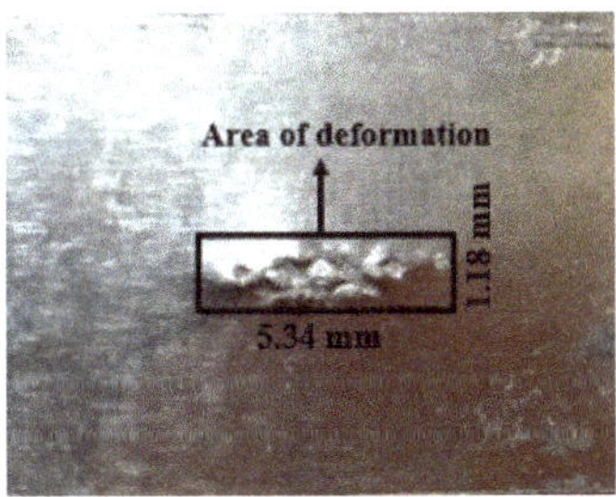

Figure 8. Area of deformation.

3.6. Calculation of Heat Flux

The heat generation due to deformation at the interface is the power dissipated over the weld area. Power at the weld area depends on weld force which is a function of temperature dependent yield strength and the clamping force, as shown in Equation (1).

$$Q_w = \frac{P}{A_w} = \frac{F_w \times V_{avg}}{A_w} \tag{1}$$

where Q_w is the heat flux due to deformation in W/m^2, P is the power in W, A_w is the weld area in m^2, F_w. is the weld force in N, V_{avg} is the average sonotrode velocity and equals $4 \times \varepsilon_0 \times f_w$, ε_0 is the amplitude of vibration of the sonotrode, and f_w is the welding frequency. The weld force (F_w) is given by Equation (2).

$$F_w = \sqrt{\left(\frac{Y_T}{2}\right)^2 - \left(\frac{F_N/A_{DZ}}{2}\right)^2} \times A_{DZ} \tag{2}$$

where Y_T is the average temperature dependent yield strength in N/m^2, F_N is the clamping force in N, and A_{DZ} is the area of deformation in m^2. By substituting Equation (2) into Equation (1), the heat flux due to deformation is obtained as shown in Equation (3).

$$Q_w = \frac{\sqrt{\left(\frac{Y_T}{2}\right)^2 - \left(\frac{F_N/A_{DZ}}{2}\right)^2} \times A_{DZ} \times 4 \times \varepsilon_0 \times f_w}{A_w} \tag{3}$$

At the end of the welding cycle, the area in which the friction is effective is assumed to be the area of deformation. Thus, A_{DZ} is the A_W. Applying this condition, the heat flux due to deformation can be calculated, as shown in Equation (4).

$$Q_w = \sqrt{\left(\frac{Y_T}{2}\right)^2 - \left(\frac{F_N/A_{DZ}}{2}\right)^2} \times 4 \times \varepsilon_0 \times f_w \tag{4}$$

The temperature dependent yield strength of copper is determined by using third order polynomial equation using the data points available for copper material [15]. The fitted polynomial curve along with equation is shown in Figure 9. The temperature dependent yield strength of copper is calculated using Equation (5).

$$Y_T = \frac{\left[\int_0^{800}\left(2 \times 10^{-6}T^3 - 0.002T^2 - 0.164T + 419.2\right) \times 10^6 dT\right]}{\Delta T} \quad (5)$$

where ΔT is the 800 °C. Therefore, $Y_T = \frac{\left[\int_0^{800}(2\times10^{-6}T^3 - 0.002T^2 - 0.164T + 419.2)\times10^6 dT\right]}{800} = 182.933 \times 10^6\ N/m^2$.

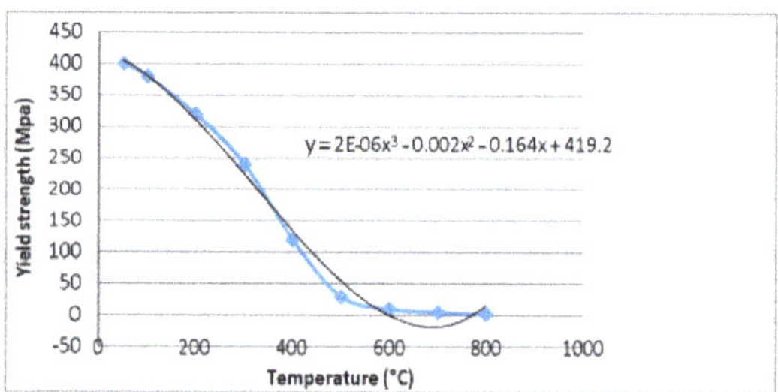

Figure 9. Temperature dependent yield strength of copper.

The heat flux due to deformation while joining a copper wire of diameter (1.2 mm) and a copper sheet of thickness (0.2 mm) with clamping pressure of 795 N, amplitude of vibration of the sonotrode of 30 μm, weld time of 2 s and area of deformation of 6.3×10^{-6} m^2, thus obtained as discussed in Section 3.5, is given by

$$Q_w = \sqrt{\left(\frac{182.933 \times 10^6}{2}\right)^2 - \left(\frac{795/6.3 \times 10^{-6}}{2}\right)^2} \times 4 \times 30 \times 10^{-6} \times 20000 = 158.92 \times 10^6\ W/m^2$$

Heat flux due to friction (Q_{FR}) is calculated using Equation (6).

$$Q_{FR} = \frac{\mu \times F_N \times 4 \times \varepsilon_0 \times f_w}{A_{DZ}} \quad (6)$$

where μ is the coefficient of friction. Therefore, $Q_{FR} = \frac{0.3\times795\times4\times30\times10^{-6}\times20000}{6.3\times10^{-6}} = 90.85 \times 10^6\ W/m^2$

Total heat flux is calculated by adding both heat flux due to deformation and heat flux due to friction. Total heat Flux = $Q_W + Q_{FR} = 158.92 \times 10^6\ W/m^2 + 90.85 \times 10^6\ W/m^2 = 249.77 \times 10^6\ W/m^2$.

4. Simulation and Experimental Trials

The simulation and experimental trials were carried out based on parameters provided in Taguchi's L9 orthogonal array. The heat flux as a function of participating process parameters was calculated and given as input for each simulation trial. The results from simulation and experiments are shown in Table 3.

Table 3. Results of temperature from simulation and experiments.

Trial No.	Clamping Force (N)	Amplitude of Vibration of Sonotrode (μm)	Weld Time (s)	Temperature from Simulation (°C)	Temperature from Experiments (°C)			Strength of the Joint in Tension * (N)
					Trial 1	Trial 2	Average	
1	795	30	2	79.26	79.97	79.73	79.85	187.721
2	995	30	2.5	110.9	110.06	110.25	110.16	210.107
3	1195	30	3	137.2	137.38	137.01	137.20	224.946
4	795	42.5	2	83.27	83.25	83.39	83.32	193.548
5	995	42.5	2.5	117.8	117.55	117.73	117.64	213.342
6	1195	42.5	3	139.9	139.01	139.38	139.20	227.621
7	795	57	2	99.16	99.74	99.37	99.56	202.369
8	995	57	2.5	119.5	119.60	119.00	119.30	217.638
9	1195	57	3	141.2	141.13	141.87	141.50	231.432

* Average of two experimental trials.

5. Results and Discussions

Simulation Trial 1 was carried out with clamping force 795 N, amplitude of vibration of sonotrode 30 μm and weld time 2 s. The process parameters in this trial were set at lower level. The heat flux calculated for this combination of process parameters is 249.77×10^6 W/m^2. The temperature obtained from simulation is 79.26 °C. The average temperature obtained from experiments for the same combination of process parameter using thermocouple is 79.85 °C. The strength of the joint in tension obtained is 187.721 N. As all the process parameter values are set at the lower levels, the temperature developed at the interface and the strength of the joint obtained in this trial is minimum when compared with all the other trials. The results from simulation and experiments are shown in Figure 10.

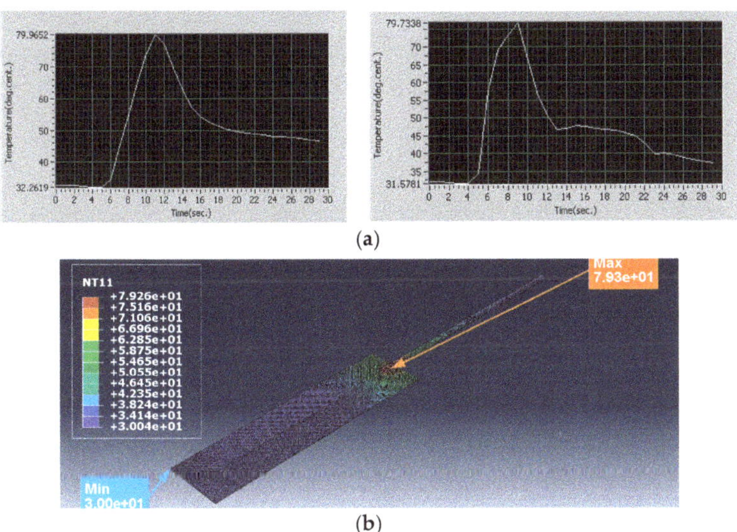

Figure 10. Trial 1, comparison of temperatures (°C): (**a**) results from experiments (Trial 1); and (**b**) result from simulation (Trial 1): Minimum temperature 30.04 °C, Maximum temperature 79.26 °C.

Simulation Trial 5 was carried out with clamping force 995 N, amplitude of vibration of sonotrode 42.5 μm and weld time 2.5 s. The process parameters in this trial were set at medium level. The heat flux calculated for this combination of process parameters is 355.73 × 10^6 W/m^2. The temperature obtained from simulation is 117.8 °C. The average temperature obtained from experiments for the same combination of process parameter is 117.64 °C. The strength of the joint obtained is 213.342 N. The results from simulation and experiments are shown in Figure 11.

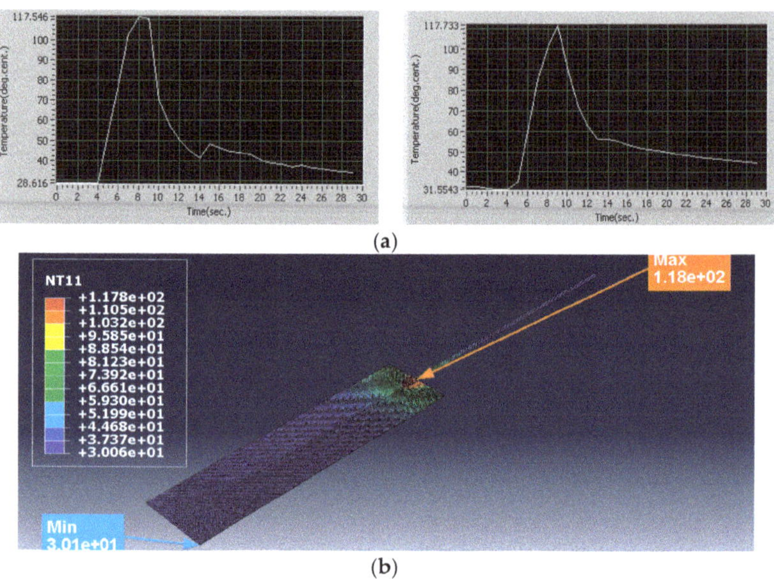

Figure 11. Trial 5, comparison of temperatures (°C): (**a**) results from experiments (Trial 5); and (**b**) result from simulation (Trial 5): Minimum temperature 30.06 °C, Maximum temperature 117.8 °C.

Simulation Trial 9 was carried out with clamping force 1195 N, amplitude of vibration of sonotrode 57 μm and weld time 3 s. The process parameters in this trial were set at higher level. The heat flux calculated for this combination of process parameters is 483.5 × 10^6 W/m^2. The temperature obtained from simulation is 141.2 °C. The average temperature obtained from experiments for the same combination of process parameter is 141.50 °C. The strength of the joint obtained is 231.432 N. As all the process parameter values were at higher levels, the temperature developed at the interface and the strength of the joint obtained in this trial is maximum when compared with all the other trials. The results from simulation and experiments are shown in Figure 12.

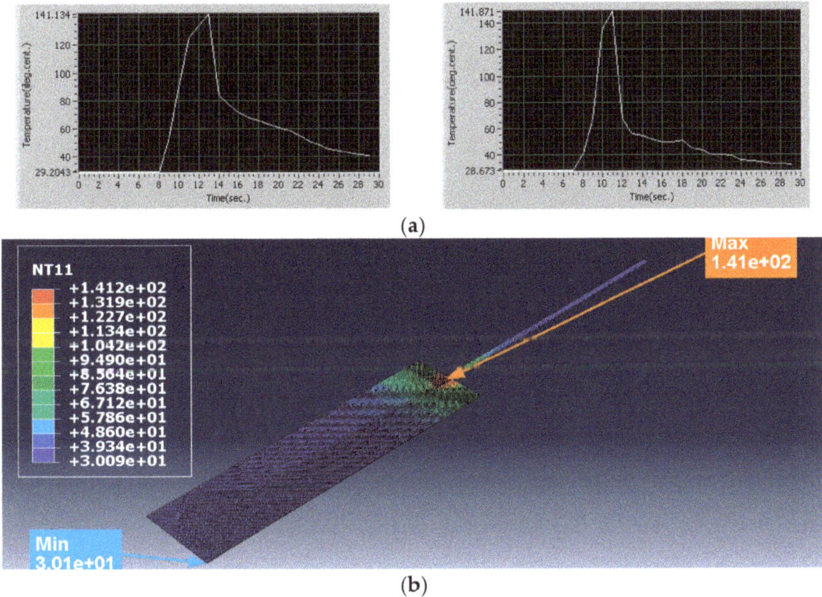

Figure 12. Trial 9, comparison of temperatures (°C): (**a**) results from experiments (Trial 9); and (**b**) result from simulation (Trial 9): Minimum temperature 30.09 °C, Maximum temperature 141.12 °C.

In Simulation Trials 1, 4 and 7, the clamping force and the weld time were set constant at 795 N and 2 s, respectively, whereas the amplitude of vibration of the sonotrode was varied at 30 μm, 42.5 μm and 57 μm. The temperature generated at the interface in these trials are 79.26 °C, 83.27 °C and 99.16 °C, respectively, as shown in Table 3. Based on the results, it is observed that the temperature at the interface increases with increase in amplitude. This is in agreement with the observations from the experiments. Increase in amplitude of vibration of the sonotrode results in increased sliding action between the specimens leading to plastic deformation and rise in temperature. Similar trend was observed with varying amplitude of vibration of the sonotrode in Simulation Trials 2, 5 and 8 where the clamping force and weld time were set constant at medium level (995 N, 2.5 s) and in Simulation Trials 3, 6 and 9 where the clamping force and weld time were set constant at higher level (1195 N, 3 s).

Simulation Trials 1,5 and 9 correspond to lower level (795 N, 30 μm and 2 s), medium level (995 N, 42.5 μm, 2.5 s) and higher level (1195 N, 57 μm and 3 s) of process parameters, respectively. The temperature generated at the interface in these trials are 79.26 °C, 117.8 °C and 141.2 °C, respectively, as shown in Figures 10–12. The strength of the joints obtained in these trials are 187.721 N, 213.342 N and 231.432 N, respectively, as shown in Table 3. Based on the results, it is observed that higher level parameters produce maximum strength of the joint in tension of 231.432 N when compared with lower level and medium level combinations of process parameters.

The results from simulation were compared with results from experiments, as shown in Figure 13. The temperature obtained from simulation is found to be in good agreement with the temperature from experiments measured using thermocouple. Therefore, the developed finite element analysis model is found useful to predict the temperature. It is also observed in Figure 13 that the strength of the joint under tensile loading correlate well with the temperature. The co-efficient of correlation between temperature and strength of the joint in tension is determined as 0.99. Hence, it is observed the temperature at the interface has significant effect on progress of the welding and strength of the joint.

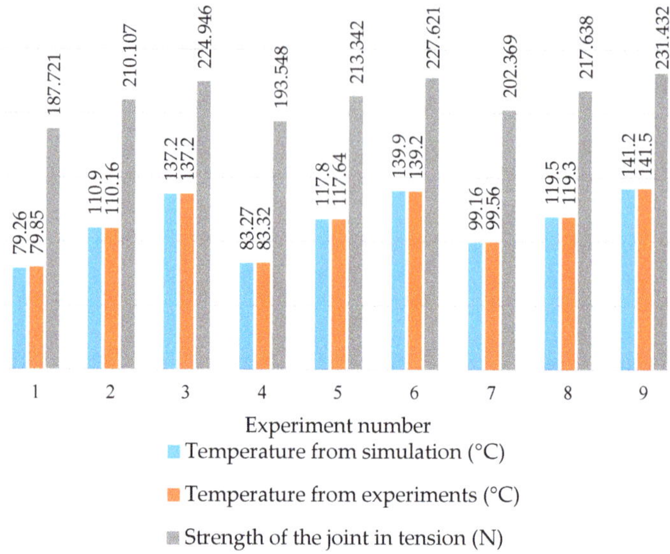

Figure 13. Comparison of temperature and strength of the joint under tensile loading.

6. Conclusions

Temperature distribution while joining metallic wire and a sheet made of copper was analyzed using finite element analysis. The results of the analysis reveal the following salient observations.

i. The results from simulation and experiments conducted based on Taguchi's L9 orthogonal array reveal that the maximum temperature developed during welding is less than the melting point of the work material, validating that the USMW is a solid state welding process.

ii. It is observed from the analysis that the influence of heat generated due to deformation and friction is significant in the process of formation of joint. The results of temperature from simulation are found to be in good agreement with results of temperature from experiments measured using thermocouple. Thus, the developed finite element model is validated.

iii. The results of temperature developed at the interface are compared with results of strength of the joint under tensile loading. It is inferred that the strength of the joint correlate well with the temperature developed at the interface indicating that the temperature at the interface has significant effect on strength of the joint. It is observed that the strength of the joint depends on the variations of heat generated during welding under different process parametric conditions.

Author Contributions: The author exerted best effort to ensure that the paper becomes qualified for publishing.

Funding: This research was funded by University Grants Commission, New Delhi, under Major Research Project Scheme (F.No.42-876/2013(SR)).

Acknowledgments: The author expresses his sincere and heartfelt thanks to his Ph.D. guide Dr.K. Prakasan, Professor and Head, Department of Production Engineering, the Management and the Principal of PSG College of Technology, Coimbatore for extending the essential support and infrastructure to carry out this work.

Conflicts of Interest: The author declares no conflict of interest.

References

1. Yong, D.; Kim, J.-K. Numerical analysis of ultrasonic wire bonding: Part 2. Effects of bonding parameters on temperature rise. *Microelectron. Reliabil.* **2008**, *48*, 149–157.

2. Elangovan, S.; Ponnayya, S.; Henry, J.; Kalakkath, P. Experimental studies on optimization of process parameters and finite element analysis of temperature and stress distribution on joining of Al–Al and Al–Al_2O_3 using ultrasonic welding. *Int. J. Adv. Manuf. Technol.* **2011**, *55*, 631–640.
3. Jingzhou, Z.; Li, H.; Choi, H.; Cai, W.; Abell, J.A.; Li, X. Insertable thin film thermocouples for in situ transient temperature monitoring in ultrasonic metal welding of battery tabs. *J. Manuf. Process.* **2013**, *15*, 136–140.
4. Hyung-Seop, S.; de Leon, M. Parametric study in similar ultrasonic spot welding of A5052-H32 alloy sheets. *J. Mater. Process. Technol.* **2015**, *224*, 222–232.
5. Elangovan, S. Experimental and theoretical investigations on temperature distribution at the joint interface for copper joints using ultrasonic welding. *Manuf. Rev.* **2014**, *1*, 18. [CrossRef]
6. Dongkyun, L.; Cai, W. The effect of horn knurl geometry on battery tab ultrasonic welding quality: 2D finite element simulations. *J. Manuf. Process.* **2017**, *28*, 428–441.
7. Jedrasiak, P.; Shercliff, H.R.; Chen, Y.C.; Wang, L.; Prangnell, P.; Robson, J. Modeling of the thermal field in dissimilar alloy ultrasonic welding. *J. Mater. Eng. Perform.* **2015**, *24*, 799–807. [CrossRef]
8. Chen, K.K.; Zhang, Y.S. Numerical analysis of temperature distribution during ultrasonic welding process for dissimilar automotive alloys. *Sci. Technol. Weld. Join.* **2015**, *20*, 522–531. [CrossRef]
9. De Vries, E. Mechanics and Mechanisms of Ultrasonic Metal Welding. Ph.D. Thesis, The Ohio State University, Columbus, OH, USA, 2004.
10. Elangovan, S.; Semeer, S.; Prakasan, K. Temperature and stress distribution in ultrasonic metal welding—An FEA-based study. *J. Mater. Process. Technol.* **2009**, *209*, 1143–1150. [CrossRef]
11. Gallego-Juárez, J.A.; Graff, K.F. (Eds.) *Power Ultrasonics: Applications of High-Intensity Ultrasound*; Elsevier: New York, NY, USA, 2014.
12. Kumar, S.; Wu, C.S.; Padhy, G.K.; Ding, W. Application of ultrasonic vibrations in welding and metal processing: A status review. *J. Manuf. Process.* **2017**, *26*, 295–322. [CrossRef]
13. Mathews, P.G. *Design of Experiments with MINITAB*; ASQ Quality Press: Milwaukee, WI, USA, 2005.
14. Montgomery, D.C. *Design and Analysis of Experiments*; John Wiley & Sons: Hoboken, NJ, USA, 2017.
15. ASM International Handbook Committee. Properties and selection: nonferrous alloys and special-purpose materials. *ASM Int.* **1992**, *2*, 1143–1144.
16. Dassault Systems. *Abaqus 6.12 Analysis User's Manual and Element Reference Guide*; Dassault Systems: Vélizy-Villacoublay, France, 2013.
17. Chen, K.; Zhang, Y. Thermal-mechanical analysis of ultrasonic spot welding considering acoustic softening effect. *Procedia Eng.* **2014**, *81*, 2117–2122. [CrossRef]
18. Jeng, Y.-R.; Chen, J.-T.; Cheng, C.-Y. Theoretical and experimental study of a thermal contact conductance model for elastic, elastoplastic and plastic deformation of rough surfaces. *Tribol. Lett.* **2003**, *14*, 251–259. [CrossRef]
19. Kim, W.; Argento, A.; Grima, A.; Scholl, D.; Ward, S. Thermo-mechanical analysis of frictional heating in ultrasonic spot welding of aluminium plates. *Proc. Inst. Mech. Eng. Part B J. Eng. Manuf.* **2011**, *225*, 1093–1103. [CrossRef]
20. Zhou, B.; Thouless, M.D.; Ward, S.M. Determining mode-I cohesive parameters for nugget fracture in ultrasonic spot welds. *Int. J. Fract.* **2005**, *136*, 309–326. [CrossRef]

© 2018 by the author. Licensee MDPI, Basel, Switzerland. This article is an open access article distributed under the terms and conditions of the Creative Commons Attribution (CC BY) license (http://creativecommons.org/licenses/by/4.0/).

Article

Methodology to Reduce Distortion Using a Hybrid Thermal Welding Process

Javier Souto [1,*], Enrique Ares [2], Paulino Alegre [2] and Jorge Cerqueiro [2]

1 Calculus and Simulation Unit at AIMEN Technology Center, C/Relva 27A Torneiros, 36410 Porriño-Pontevedra, Spain

2 Escuela de Ingeniería Industrial, Campus Universitario, University of Vigo, 36310 Vigo-Pontevedra, Spain; enrares@uvigo.es (E.A.); alegre@uvigo.es (P.A.); jcerquei@uvigo.es (J.C.)

* Correspondence: javier.soutogrela@outlook.com; Tel.: +34-677-311-791

Received: 6 July 2018; Accepted: 3 September 2018; Published: 7 September 2018

Abstract: Welding is a thermal process which results in high strain and stress values in the material and leads to its change of phase. This might cause significant distortions in the welded structure, which often becomes a relevant design and manufacturing issue. This study deals with a line-heating-based forming process that is applied at the moment of the welding operation, with the final objective of minimizing distortion. A FEM (finite element method) based on a thermo-elastoplastic approach is used here. The computational method is first calibrated in three stages: heatline forming calibration, flame heat source calibration, and the weld process. The final model presented in this work simulates a hybrid process called htTTT (high-temperature thermal transient tensioning) which was optimized over large T welds to minimize the final distortion of the components.

Keywords: welding distortion; FEM; LSND (low stress no distortion); htTTT; model validation

1. Introduction

Welding is a joining process which generates changes in the material state at high temperatures. Welding technology is widely used in many different industries, such as shipbuilding or the nuclear or automotive industries. Welding has interesting properties with respect to other joining technologies: its continuity in the final assembly, speed and versatility in the design and fabrication stages, etc. However, this technology in its traditional form has disadvantages because the high temperature involved generates thermal expansion, shrinkage, and microstructural transformations. For that reason, the welded joints in structures need constant supervision by highly qualified workers.

The prediction and control of distortion is particularly important for the design and manufacturing of stiffened welded structures [1]. The extensive use of fusion-welding processes in the manufacturing of this kind of structures generates high temperature in the parts, resulting in high distortion in the final assemblies. This undesirable effect leads to direct costs due to the reworking that is needed to assure the compliance of the final product. Thus, numerical simulation becomes an essential tool for the forecasting and control of distortion in order to avoid high reprocessing costs in the fabrication stage.

The distortion phenomenon has been studied to develop new ways of minimizing its effects. A reduction of the distortion can be achieved through different strategies based on treatments applied either previous to, throughout, or after the weld process. These techniques can be mechanical [2], thermal [3], electromagnetic [4] or use other welding processes [5,6], and can be based on minimizing net stress in the assembly or inducing additional stresses after the welding process, such as in straightening.

Avoiding the distortion phenomenon by using predictive methods is an actual challenge that is addressed in this work. The thermal transient tensioning (TTT) process was studied for the case of a typical shipbuilding stiffener. Parametric analysis for different torch positions and heat power densities was performed. The result will point to a new constructive technique which aims to reduce cost and construction time.

The transient thermal tensioning process was patented by Dull, R.M. [7] (Edison Welding Institute, Columbus, OH, USA). This process is one of the more promising techniques used to mitigate the distortion produced by welding processes. In this method, different heat sources are located in different areas over the plates. These sources are transient, thus implying the presence of moving and localized heat sources, which produce both temperature gradients and thermal stresses. The substantial idea of TTT is to control the stresses in the assembly during the welding by combining the auxiliary heat sources with the welding process heat sources.

2. Method Details

In this work, a modified TTT procedure was developed and applied to a large T-joint. High-temperature TTT (htTTT) was applied over a typical stiffener used in the shipbuilding industry, but the conclusions are relevant to other industries where these kinds of assemblies are used.

The T-joint (P1) and the asymmetrical T-joint (P2) are assemblies of two plates, each 3 m long (Figure 1). The difference between both assemblies lies in the position of the web over the flange. Long T-joint stiffeners are sometimes welded in several bead depositions using specific sequences [8]. Those sequences can minimize the total distortion, but the process is slower and complex, and therefore other strategies to avoid the distortions were studied.

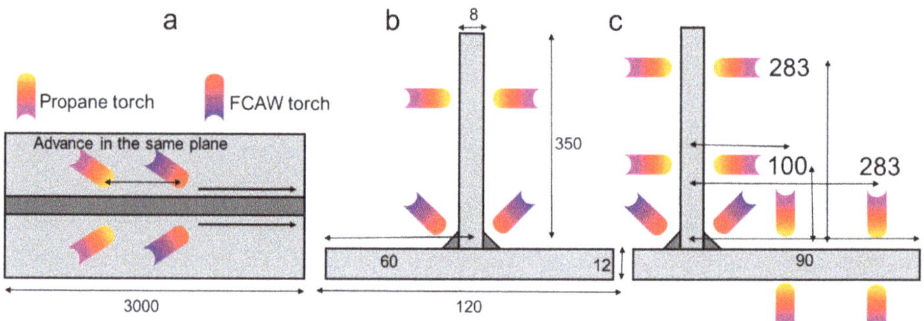

Figure 1. Schematic of the manufacturing process with the size of the T-joint. (dimensions in mm). (**a**) Distance between process torches; (**b**) P1 geometry; (**c**) P2 geometry.

The reference assemblies (P1 and P2) are welded by using double-sided symmetric flux cored arc welding (FCAW) welding. The weld was performed in a single pass from start to end, and therefore the total time for the weld process was short. The TTT process was applied to P1 and P2 (Figure 1); the use of the transient heat-forming torches together with the welding process is specific to the TTT process. The technique implemented in this case involved using two symmetrical propane torches over the FCAW torches aimed towards the web of the structure, as shown in Figure 1. Two torches were used to avoid asymmetrical thermal loads, so that, at the end of the process, no transversal deflection was observed.

The TTT developed for this case was a stressing method; it operates as with heatline forming, using a high thermal power to introduce strains/stresses into the plates. This operation is different from the low stress no distortion method [9], where the minimization of the final residual stresses implies lower distortions.

The case described here shows a high-temperature TTT process, where the residual compressive stresses work against the weld distortion. The reduction of distortion was controlled by acting on the flame intensity and the position of the flame torches. The distortion produced by both the weld and the flame heating will depend on the geometry and the thermal power of the heat sources; therefore, to obtain the optimal final non-distortion configuration, it will be necessary to calibrate the position and the power of both sources.

The clamping strategy for the TTT process studied in this research involves minimum clamping, implying that the assembly rests on a surface without any geometrical restriction. The flange and the web of the T assembly must be correctly positioned and joined by short welds; this will prevent separation or misalignment of the plates.

The distance of the torches to the plates and the power of each single torch should be carefully controlled to ensure that the power acting on the plates is symmetrical; also, the heat power applied on the plates must be controlled because a low power of the flame torches will produce low distortion correction, while a high power could yield higher final distortions. This effect is investigated in detail in the following sections.

The advantage of TTT comes from being an in-process method; therefore, from the viewpoint of manufacturing efficiency and cost, this process is more desirable than other alternative processes such as straightening. The process studied in this research was applied together with the weld at the same time and same speed as the weld torch; therefore, the total time of manufacturing was the one-pass welding time.

3. Methodology

The welding/heat forming simulation problem has a complex nature. The process is highly nonlinear and it couples together thermal, mechanical and metallurgical fields: thermal loads induce changes in the mechanical fields, while high temperatures and high cooling rates can lead to phase change and metallurgical transformations which make the material properties dependent on the temperature and metallurgical phase proportion. The geometry was meshed using 132.642 hexahedron elements. No rigid clamping was used; instead, the T-joint was simply supported. The constant convection film coefficient (5 W/m^2) was defined for the exterior surfaces of the assembly. The model was built using the ESI.Syweld software (2014, ESI Group, Paris, France).

However, it is not possible to simulate all of the physics involved in the welding process in a realistic way, as detailed in the study by Lindgren [10]. At present, the FEM simulations for welding often use a Lagrangian mesh [1–8,11]. In this work, the heat source was modeled with the Goldak [12] approximation (Equation (1)) for the weld heat input, and the propane torch source was modeled by 3D conical approximation. The thermo-metallurgical-elastoplastic model approximation (2) has showed great accuracy for weld simulation as well as line forming [13], but in the case of big assemblies, the computation cost increases significantly.

$$P = \int Q(x,y,z,t) = Q_f \cdot e^{-\frac{[x+v(\tau-t)]^2}{a_f^2}} \cdot e^{-\frac{y^2}{b^2}} \cdot e^{-\frac{z^2}{c^2}} + Q_r \cdot e^{-\frac{[x+v(\tau-t)]^2}{a_r^2}} \cdot e^{-\frac{y^2}{b^2}} \cdot e^{-\frac{z^2}{c^2}} \quad (1)$$

$$\overset{\bullet}{\varepsilon} = \overset{\bullet e}{\varepsilon} + \overset{\bullet p}{\varepsilon} + \overset{\bullet tp}{\varepsilon} + \overset{\bullet th}{\varepsilon} \quad (2)$$

In the equations above, Q is the power of the source, Q_f and Q_r are the energy in the front and rear zones of the source, respectively, a, b and c are the dimensions of the elliptical shape of the source, r is the radius of the Gaussian source over the plate, v is the speed of the source and x, y and z are the position coordinates of the source over the plate. The weld beads were added during the process using the rebirth elements method [14]. The present study is performed taking the strains in Equation (2) into account: the strains due to the elastic and plastic domain ($\overset{\bullet e}{\varepsilon}$, $\overset{\bullet p}{\varepsilon}$), thermal strains ($\overset{\bullet tp}{\varepsilon}$) and strains produced by the phase transformations ($\overset{\bullet th}{\varepsilon}$). The welding process can be a cyclic process (multi-pass welding) and therefore the plasticity was calculated using a linear kinematic hardening approximation for the elastic-plastic constitutive model [15]. The material (DH-36) was characterized at the AIMEN Technology Centre facilities (Porriño-Pontevedra, Spain). Its properties—specific heat, thermal conductivity, young modulus, elastic limit and stress–strain curves—were temperature-dependent (0–800 °C), and values higher than 800 °C were extrapolated from those. These values were completed using other ones from references [16].

The htTTT methodology was simulated using a highly non-linear model with large-displacement and large-strain behaviors. Also, the materials were modeled with non-linear properties, and therefore the thermo-metallurgical-mechanical model used to compute the weld process was executed with the updated Lagrangian formulation and Newton–Raphson method [13].

First, the three initial experimental procedures were carried out (Figure 2) and characterized and validated by FEM, as described below. Small laboratory tests were used in the first and second stage. For the final stage, some representative structures in the shipbuilding industry were welded. Naval grade steel (DH-36) was used in every experiment. Shipbuilding process parameters (current, voltage and torch speed) were used depending on the thickness, weld process and materials. The experiments were performed at the AIMEN Technology Centre facilities. The results from the experimental tests were used to validate the numerical models [17].

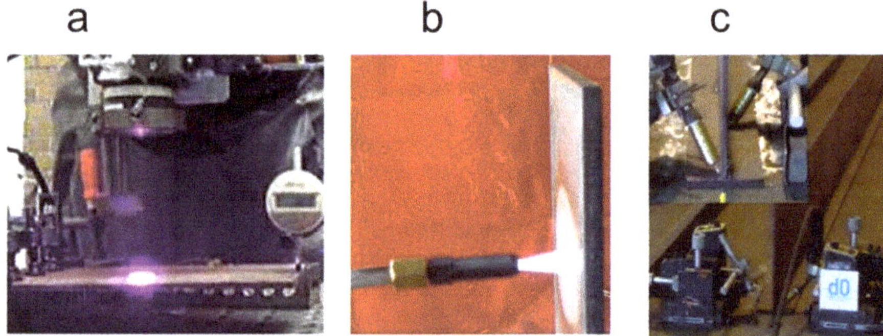

Figure 2. Experimental procedures, (**a**) heatline laser forming process; (**b**) propane heating process; (**c**) flux cored arc welding (FCAW) welding process.

1. The heatline forming process using a heatline laser diode was applied over small plates (300 × 200 mm) with two thickness values (6 and 8 mm) under different power density configurations. The assembly chosen was a cantilever-like structure. The laser scanning was performed in the middle line of the plate;
2. The propane gas static heating process characterization was performed, using a heating propane torch over a vertical plate. The heating of the heatprint and the cooling the plate were analyzed;
3. Weld process characterization over a large T-joint using a FCAW (flux cored arc welding) process. The tests were performed with continuum and double-sided welds. Every sample used was three meters long. The thickness values chosen were 8 mm for the web and 12 mm for the flange.

Every experiment was recorded with a thermal camera to analyze the evolution of the temperature field; also, the dimensions of every sample were measured before and after the thermal process, as was the flatness of each component. The plates were carefully positioned and measured to ensure the correct initial state, because any misalignment or deviation would generate inconsistencies in the distortion. In this work, the maximum vertical deflection was chosen as a reference to verify the accuracy of the FEM model chosen. The experimental results were used to validate the FEM model and to generate the htTTT numerical model [17].

The numerical model was based on the mathematical description from [14–21]. The model of the htTTT process was developed incorporating the flame torches to the weld process. In this work, the four-heat-sources approach is shown with two welding torches plus two propane torches. For the htTTT comparison, the welding process used was the same for every case, but the heatline torches' powers and positions were varied in the study. Seven different positions–heights (mm) for the flame torches over the web of the assembly (Figure 3) were studied.

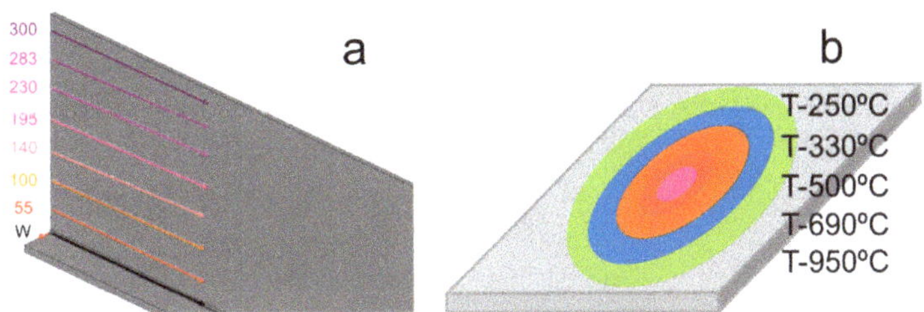

Figure 3. (**a**) Vertical parametrization in mm; (**b**) Temperature objective parametrization.

Five different power density values (Figure 3) were applied in this study, while only the net power was changed in the definition of the source model. It is not possible to measure the net power experimentally because the experimental propane setup is carried out by mixing the volume and the pressure of oxygen and propane, so the torches were identified by their corresponding maximum temperature produced on the plate (Figure 3). The maximum temperature is the resulting combined effect of the power source and the velocity of the torch and the thickness of plate.

The power, velocity, and therefore temperature was almost constant throughout the whole process. The trajectories of the two pairs of torches—welding and TTT—were parallel and their displacements synchronized, and all four torches therefore advanced with the same direction and at the same speed.

4. Results

A de-coupled thermo-mechanical method was used to predict the TTT behavior and for the calculations for the different torch positions over the web. Figure 4 shows the maximum vertical deflection for the parametrization of the TTT torch position and heat power.

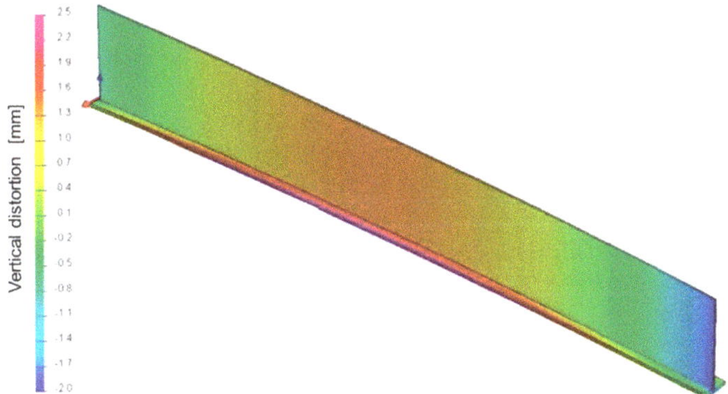

Figure 4. Vertical deflection of P1, the reference T-joint. Scale of vertical distortion (mm).

The Figure 5 shows the numerical simulation results of the htTTT compared against the experimental results of the welded P1. The black dashed line represents the distortion previously calculated and validated against the experimental results (corresponding to the P1 welding process alone). The colored squares (datapoints) show the distortion obtained by the numerical simulation model results for each htTTT parameterization. In the graph, the y-axis indicates the representative distortion value while the x-axis shows the position of the TTT torches over the length of the web.

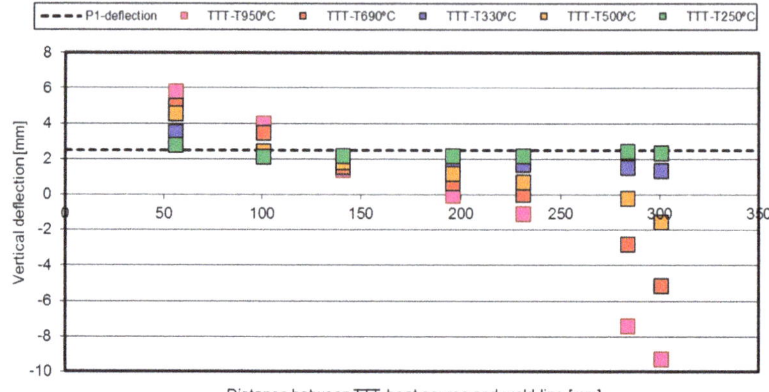

Figure 5. Vertical deflection comparison against torch temperature and torch position parametrization.

For instance, the red dots indicate that both the HL-55 and HL-100 cases show a higher final distortion value when the TTT torch is too close to the joint. When the TTT torch is 140 mm up the weld joint, an improvement in the final distortion is noticed. For HL-230, the distortion after the process is almost zero, while above 230 mm over the joint base, the final distortion obtained is negative, meaning that it is produced in the direction opposite to the natural one caused by the welding process.

Similar trends were observed for the blue, orange, red and pink curves. Higher distortion values were obtained with the TTT torches positioned closer to the weld torches, while those values decreased as the position of TTT torches was further away from the weld line. The behavior of the distortion is amplified for higher temperature values, so the slopes of the curves are higher. For the lower TTT temperature studied here—green dots—the performance is small, with the points laying very close to the black line. Surprisingly, the optimal distortion behavior for the green points happened for the HL-100 case due to the thermal coupling because of the reduction in the thermal gradients around the weld zone.

It is possible to extract from the dots' tendencies, shown in Figure 5, the behavior of the structure for each case, and to calculate the optimal torch position to obtain null distortion: 198 mm for TTT-950 °C, 231 mm for TTT-690 °C and 283 mm for TTT-500 °C. The cases with source power TTT-330 °C and TTT-250 °C do not reach the zero distortion value for the P1 geometry.

It can be seen that the TTT action may reduce or increase the final distortion. Therefore, there should be an inflexion point between both behaviors, and that point is called the 'neutral zone'. As shown in these figures, the position of that neutral zone depends on the heat power and the geometry of the plates, and the dependency is strongly non-linear.

The residual stresses were extracted from the FEM postprocessing. The stresses were measured on the surface of the plates located in the middle of the assembly (Figure 6). The black line shows a reference solution representing the action of the welding process alone. The stresses over the bead are similar in every case, with the major differences appearing in the place upon which the propane torch acts. Increasing the source heat power results in higher stress values and a wider affected zone. Moreover, it produces higher compressive stresses along the plate. This is a clear difference that can be seen between the conventional welding and the TTT process. The welding and TTT processes together induce high tensile stress in the position where the propane torches are working. Thus, two main zones with peak stress values can be seen to appear for the welding-plus-TTT process.

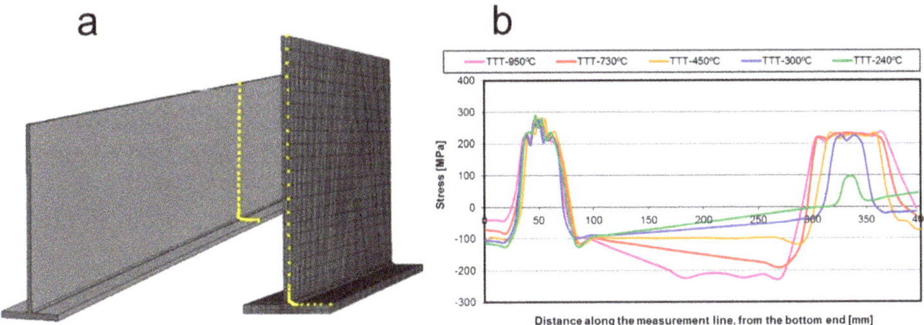

Figure 6. (**a**) Measurement points for the residual stresses; (**b**) Residual stresses over the measurement line for thermal transient tensioning (TTT)-HL283.

The green line shows a small stress peak at the propane torch position due to the low-power source applied in that case. It can be seen that, except for the torch position, the longitudinal stress curve is parallel to the welding stress curve, the final distortion being the same in both cases. With the high-power source, lower peak stress values are reached for the bead position, but that difference is almost negligible, and so the distortion reduction is performed by the introduction of the compressive stresses.

The reduction in the final distortion was caused by the compressive stresses from the auxiliary thermal sources. Comparing the results along the weld line (Figure 7b), the longitudinal stress shows the same values and trends. Therefore, the TTT applied does not affect directly the action of the weld pass.

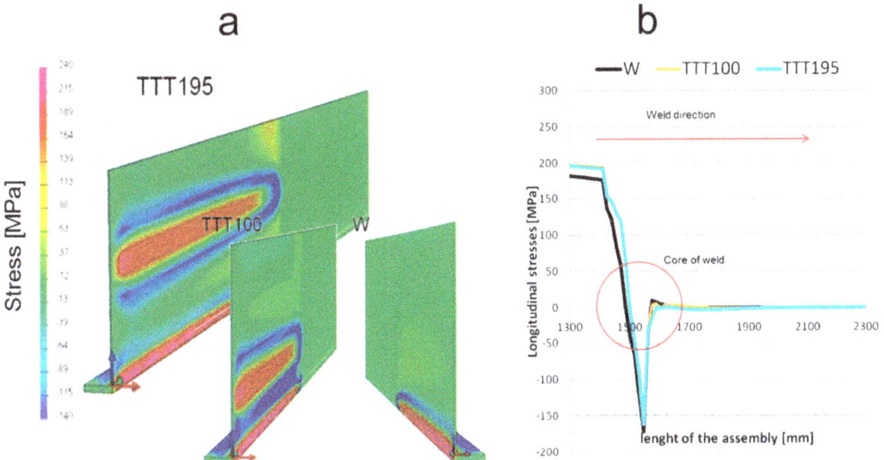

Figure 7. (**a**) Stresses distribution during the high-temperature TTT (htTTT) process (MPa); (**b**) stress distribution during the process.

The residual stress values for different TTT processes (Figure 8a) show the slope of the residual stresses between the maximum peaks (a direct effect of the heat sources). The slope of these curves defines the direction of the final distortion. That final distortion is dependent on the intensity of the heat sources and the position of these sources with respect to the vertical center of mass. Figure 8b shows a simplified schematic of the theoretical residual stresses for the ideal htTTT process, helping to understand Figure 8a. Three possible slopes in the graphs are possible, and a horizontal curve is aimed for during the htTTT process.

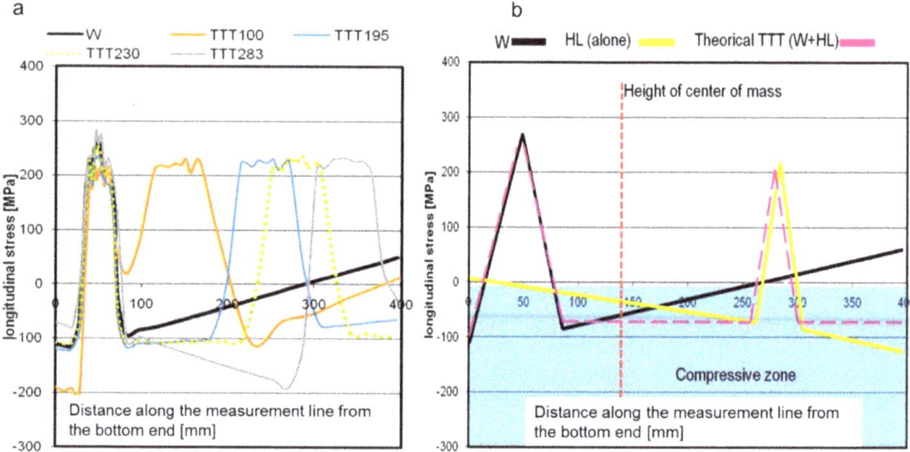

Figure 8. (**a**) Longitudinal residual stresses for different locations of the additional heat source (Figure 3a); (**b**) schematic superposition of residual stresses from FCAW welding and an additional heat source.

The compressive stresses produced by htTTT determine the final distortion. Therefore, when the compressive stresses work below the center of mass, the structure becomes concave, while if the compressive stresses work over the center of mass, the final shape of the structure will be convex (Figure 9). The htTTT process makes it possible to find an optimal configuration of the propane torches that allows reaching the equilibrium of the compressive stresses, finally leading to zero out-of-plane distortion.

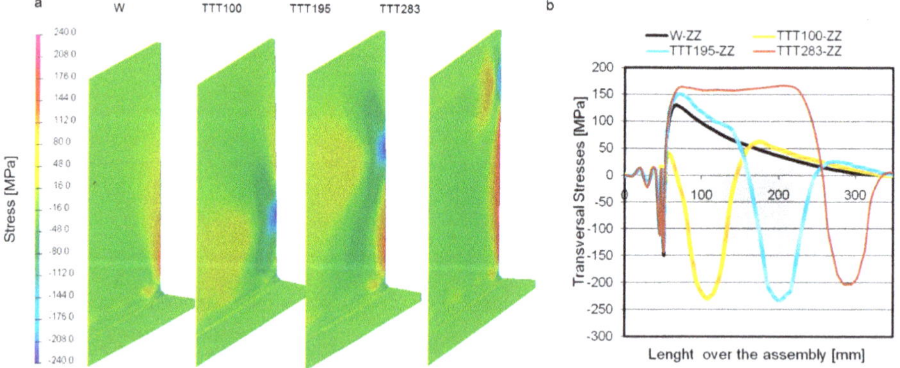

Figure 9. (**a**) Residual stresses for several htTTT tests; (**b**) stress comparison along the measurement line (Figure 3a).

The study of the residual stresses helps us to understand the evolution of the distortion caused by the action of the htTTT. The stresses induced by the thermal loads and the edge effects contribute to producing the distortion of the assembly.

Finally, the htTTT process was applied over non-symmetrical T-joints (P2) (Figure 10) and the same behavior was observed. For a non-symmetrical geometry, transversal distortion appears during a double-sided welding process. The vertical distortion of the T-joint reaches 3.52 mm while the transversal distortion reaches 8.74 mm (Figure 10b). Using htTTT at 690 °C on the web and wing, the values of the distortion change (Figure 10b).

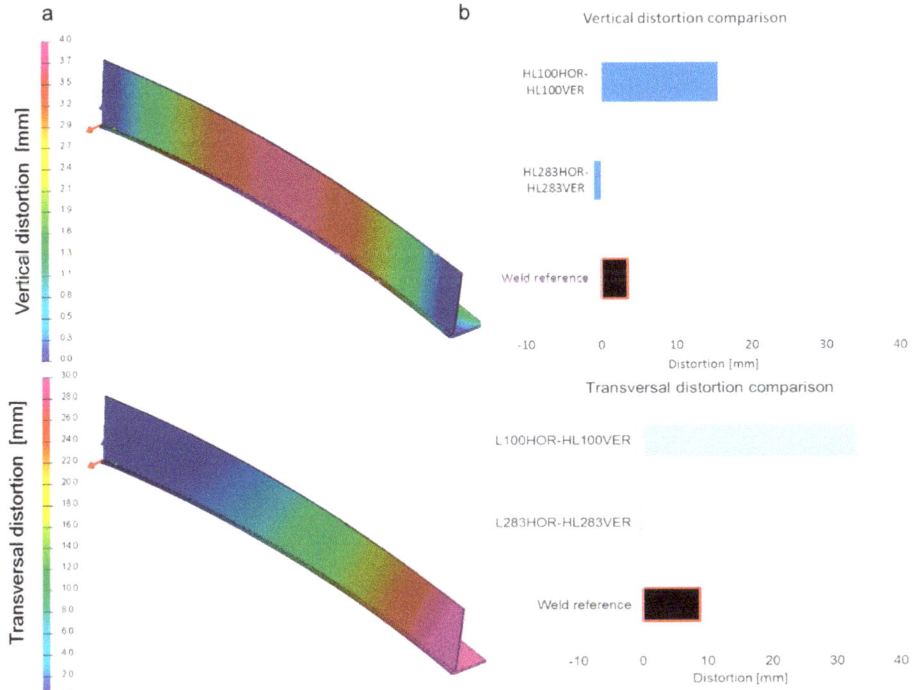

Figure 10. (**a**) P2 distortion results on the reference non-symmetrical weld (mm); (**b**) results of the distortion with htTTT.

When the auxiliary torches are located near the weld torches, the distortion increases as in the P1 case, while with an adequate positioning of the auxiliary htTTT torches, it is possible to reduce the distortion values. Therefore, the htTTT can be used to reduce the distortion during the welding process in both cases.

5. Discussion

htTTT has proven itself to be an efficient strategy to reduce distortions from welding processes. It causes the reduction in the distortion by adding new residual stresses, and therefore is a truly different process from the LSND (low stress no distortion) methodology, because the LSND aims to minimize residual stresses and the htTTT adds stresses into the assembly to minimize the distortion. htTTT was performed over two different T-joint geometries, and the results and conclusions obtained were similar. The temperature and position of the propane torches were studied, as both parameters are necessary to optimize the htTTT process for a given geometry of the component.

The htTTT does not affect the residual stresses on the bead zone if the auxiliary torches are far enough, and therefore the temperature over the bead remains almost constant, although future studies will be focused on their thermal behavior. If the temperatures are more uniform over the plates, the distortion at the final of the process is minimized. Moreover, other techniques such as preheating or slow cooling generate a temperature coupling that results in a more uniform distribution of temperature in the welded plates. The idea behind the htTTT is to minimize the weld distortion with compensative distortion applied in a different location.

6. Conclusions

In this work, htTTT was developed, studied and parametrized. The position and the heat power of the torches over a big symmetric T-joint were studied. Extreme reductions of the out-of-plane distortion in T-joints were obtained by applying htTTT. Therefore, high-temperature transient thermal tensioning can be used to fully eliminate the vertical distortion in T-joints.

The application of high-temperature transient thermal tensioning should be done far away from the bead and always over the neutral line. The higher power density source used for the TTT torches produces a distortion, and this phenomenon can be used to reduce the total distortion of the assembly. It has been found that the optimal htTTT configuration for minimizing the distortion of P1 T-joints involves applying the auxiliary propane torches at a distance of 330 mm from the flange with a 418 °C surface temperature.

It has been shown that htTTT does not have any relevant impact on the residual stresses around the molten weld zone when the htTTT is performed far away from the focus of the weld. The improvement of the distortion behavior in the assembly is a result of the generation of new compressive stresses by the auxiliary torches. These new stresses produce compensation forces during the weld process, leading to a lower distortion at the end of the cooling. The process can be used with an asymmetrical setup of the stiffeners.

The use of higher temperatures (950 °C) results in higher distortion reductions, but they also cause metallurgical transformations, and therefore are not useful in the case of the material used in this study.

Author Contributions: Conceptualization, J.S.; Methodology, J.S. and E.A.; Validation, J.S.; Formal Analysis, J.S., E.A., P.A., J.C.; Investigation, J.S.; Writing-Original Draft Preparation, J.S.; Writing-Review & Editing, J.S., E.A., P.A. and J.C.; Supervision, J.S. and E.A.; Project Administration, J.S.

Funding: This research received no external funding.

Acknowledgments: This work had been developed by AIMEN Technology Centre in collaboration with a significative number of companies from industries such as shipbuilding and boiler-making, in the search for the optimal design of new structures. The study was developed within the R&D project: 'Shipbuilding manufacturing with zero distortion (D0)' funded by CDTI and co-funded by FEDER.

Conflicts of Interest: The authors declare no conflicts of interest

References

1. Deng, D.; Zhou, Y.; Bi, T.; Liu, X. Experimental and numerical investigations of welding distortion induced by CO_2 gas arc welding in thin-plate bead-on joints. *Mater. Des. (1980–2015)* **2013**, *52*, 720–729. [CrossRef]
2. Cozzolino, L.D.; Coules, H.E.; Colegrove, P.A. Modelling distortion reduction on pre- and post-weld rolled gas metal arc welded plates. In Proceedings of the International Workshop on Thermal Forming and Welding Distortion, Bremen, Germany, 6–7 April 2011.
3. Pazooki, A.M.A.; Hermans, M.J.M.; Richardson, I.M. Reduction of welding distortion in DP600 steel by manipulation of temperature distribution during welding. In Proceedings of the International Workshop on Thermal Forming and Welding Distortion, Bremen, Germany, 6–7 April 2011.
4. Jahn, A.; Kratzsch, M.; Brener, B. Induction assisted laser beam welding of HSLA steel sheets. In Proceedings of the International Scientific Colloquium Modelling for Electromagnetics Processing, Hannover, Germany, 27–29 October 2008.
5. Bakir, N.; Artinov, A.; Gumenyuk, A.; Bachmann, M.; Rethmeier, M. Numerical Simulation on the Origin of Solidification Cracking in Laser Welded Thick-Walled Structures. *Metals* **2018**, *8*, 406. [CrossRef]
6. D'Urso, G.; Giardini, C. Thermo-Mechanical Characterization of Friction Stir Spot Welded AA7050 Sheets by Means of Experimental and FEM Analyses. *Materials* **2016**, *9*, 689. [CrossRef] [PubMed]
7. Dull, R.M.; Dydo, J.R.; Russell, J.J.; Shaghvi, J. Method of Reducing Distortion by Transient Thermal Tensioning. US Patent 6,861,617, 1 March 2005.
8. Souto Grela, J.; Blanco Viana, E.B.; Martinez, D.; Piñeiro, E. Numerical simulation in welding process: Optimizing structures with sequence and inertial study. *Matér. Tech.* **2012**, *100*, 317–326. [CrossRef]

9. Chen, S. Low Stress Non-Distortion Welding. Ph.D. Thesis, University of Wollongong, Wollongong, NSW, Australia, 2013.
10. Lindgren, L.E. Finite element modelling and simulation of welding, part 1: Increased complexity. *J. Therm. Stress.* **2001**, *24*, 141–192. [CrossRef]
11. Anca, A.; Cardona, A.; Risso, J.; Fachinotti, V. Finite element modeling of welding processes. *Appl. Math. Model.* **2011**, *35*, 688–707. [CrossRef]
12. Goldak, J.; Chakravarti, A.; Bibby, M. A new finite element model for welding heat sources. *Metall. Trans. B* **1984**, *15*, 299–305. [CrossRef]
13. Smith, M.C.; Smith, A.C. NeT bead-on-plate round robin: Comparison of residual stress predictions and measurements. *Int. J. Press. Vessels Pip.* **2009**, *86*, 79–95. [CrossRef]
14. Hong, J.K.; Tsai, C.L.; Dong, P. Assessment of Numerical Procedures for Residual Stress Analysis of Multipass Welds. *Weld. J.* **1998**, *77*, 372s–382s.
15. Hinton, E. *NAFEMS: Introduction to Nonlinear Finite Element Analysis*; NAFEMS Birniehill East Killbridge Glasgow. Glasgow, UK, 1992.
16. Cozzolino, L.D.; Coules, H.E.; Colegrove, P.A.; Wen, S. Investigation of post-weld rolling methods to reduce residual stresses and distortion. *J. Mater. Proc. Technol.* **2017**, *247*, 243–256. [CrossRef]
17. Souto, J.; Ares, E.; Alegre, P. Procedure in Reduction of Distortion in Welding Process by High Temperature Thermal Transient Tensioning. *Procedia Eng.* **2015**, *132*, 732–739. [CrossRef]
18. Leblond, J.B.; Devaux, J. A new kinetic model for anisothermal metallurgical transformations in steels including effect of austenite grain size. *Acta Metall.* **1984**, *32*, 137–146. [CrossRef]
19. Lindgren, L.E. Finite element modelling and simulation of welding. Part 2. Improved material modelling. *J. Therm. Stress.* **2001**, *24*, 195–231. [CrossRef]
20. Lindgren, L.E. Finite element modeling and simulation of welding. Part 3. Efficiency and integration. *J. Therm. Stress.* **2001**, *24*, 305–334. [CrossRef]
21. Ueda, Y.; Yamakawa, T. Analysis of thermal elastic-plastic stress and strain during welding by finite element method. *JWRI* **1971**, *2*, 90–100.

© 2018 by the authors. Licensee MDPI, Basel, Switzerland. This article is an open access article distributed under the terms and conditions of the Creative Commons Attribution (CC BY) license (http://creativecommons.org/licenses/by/4.0/).

Article

Unexpected Event Prediction in Wire Electrical Discharge Machining Using Deep Learning Techniques

Jose A. Sanchez [1,*], Aintzane Conde [2], Ander Arriandiaga [3], Jun Wang [4] and Soraya Plaza [5]

1 Aeronautics Advanced Manufacturing Center, CFAA (UPV/EHU), Bizkaia Technology Park, Building 202, 48170 Zamudio, Spain
2 Machine-Tool Institute (IMH), Azkue Auzoa 1 48, 20870 Elgoibar, Spain; aintzane@imh.eus
3 iCub Facility, Istituto Italiano di Tecnologia Via Morego, 30, 16163 Genova, Italy; ander.arriandiaga@gmail.com
4 Faculty of Mechanical Engineering, Tianjin University of Science & Technology (TUST), Dongjiang Rd, Hexi Qu, Tianjin 300222, China; jwang003@ikasle.ehu.eus
5 Faculty of Engineering of Bilbao, UPV/EHU, Plaza Torres Quevedo 1, 48013 Bilbao, Spain; soraya.plaza@ehu.eus
* Correspondence: joseantonio.sanchez@ehu.eus; Tel.: +34-94-6014068

Received: 13 June 2018; Accepted: 26 June 2018; Published: 28 June 2018

Abstract: Theoretical models of manufacturing processes provide a valuable insight into physical phenomena but their application to practical industrial situations is sometimes difficult. In the context of Industry 4.0, artificial intelligence techniques can provide efficient solutions to actual manufacturing problems when big data are available. Within the field of artificial intelligence, the use of deep learning is growing exponentially in solving many problems related to information and communication technologies (ICTs) but it still remains scarce or even rare in the field of manufacturing. In this work, deep learning is used to efficiently predict unexpected events in wire electrical discharge machining (WEDM), an advanced machining process largely used for aerospace components. The occurrence of an unexpected event, namely the change of thickness of the machined part, can be effectively predicted by recognizing hidden patterns from process signals. Based on WEDM experiments, different deep learning architectures were tested. By using a combination of a convolutional layer with gated recurrent units, thickness variation in the machined component could be predicted in 97.4% of cases, at least 2 mm in advance, which is extremely fast, acting before the process has degraded. New possibilities of deep learning for high-performance machine tools must be examined in the near future.

Keywords: WEDM; deep learning; deep neural networks; Industry 4.0

1. Introduction

Machine tools and in general, manufacturing industries, have traditionally been characterized by relying on empirical approaches when it comes to process optimization. Due to the large number of phenomena and variables involved in each operation, the practical application of theoretical models is difficult. In fact, although theoretical models are very interesting for understanding the underlying physical phenomena, they usually exhibit important limitations for industrial practice.

This fact is particularly evident in the case of manufacturing components for high-added value sectors, such as aircraft manufacturing. The aerospace industry has experienced an exponential increase in recent years. It is expected that by 2032 there will be double the total number of aircraft worldwide [1]. This trend has generated great investment by manufacturing companies in order to adapt their products to the increasing tolerance and accuracy requirements that are demanded by this sector. Unconventional machining processes have gained acceptance and within them, much attention has been directed towards wire electrical discharge machining (WEDM). This technology allows for the processing of difficult-to-cut, extremely hard materials with very tight tolerances and with impressive surface finish [2,3]. Nonetheless, trial and error approaches are still required for process optimization due to the above-mentioned limited accuracy of theoretical models [4]. In this context, artificial intelligence (AI) and more specifically, deep learning (DL) techniques appear to be an interesting approach, provided that massive amounts of data can be collected from the process.

Deep learning using Deep Neural Networks (DNNs) has achieved impressive state-of-the-art results in very difficult learning tasks such as image recognition [5], handwriting recognition [6], natural language processing [7], image description [8], and mitosis detection [9]. In contrast to shallow neural networks (SNNs), in a DNN, a series of hidden layers extract abstract features from a sequence or images [10]. Because of this, an overwhelming number of new applications are being developed in the field of information and communication technologies (ICTs), including automatic translation and voice recognition, thus increasing the interest from both academia and industry. A brief summary of the main network architectures for deep learning is presented in the following paragraphs.

For processing sequential data, recurrent neural networks (RNNs) are a common approach in many fields [10]. The earliest attempts to train RNNs were made by Rumelhart et al. using back-propagation through time [11]. Later, Elman introduced the Elman network with feedback from the output of the hidden layer to the input of said layer [12]. However, these training methods and architectures do not deal properly with long-term time dependencies due to vanishing and exploding gradients [13]. Thus, in order to solve the vanishing gradients problem, in 1997 Hochreiter & Schmidhuber introduced the long short-term memory networks (LSTMs) [14]. Unlike the classic RNN, an LSTM uses gates to decide whether or not to keep the existing memory [15]. Thus, an LSTM unit is able to keep an important feature over a long distance and therefore, deal with long-term time dependencies. Although variations of the LSTM architecture have been proposed, probably the most commonly used is the gated recurrent unit (GRU) that replaces the input, forget, and output gates by an update gate and a reset gate, reducing the number of gates from 3 to 2 [16].

To extract features from sequences of data, convolutional neural networks (CNNs) exhibit outstanding performance. CNNs are feed-forward neural networks that combine three ideas: local receptive fields, subsampling, and shared weights [17]. The local receptive fields and subsampling ideas were already in the neocognitron neural network model proposed by Fukushima [18]. By using CNNs with local receptive fields, neurons can extract features from images (2D structures), sequences, or time series (1D structures). From a convolutional layer, multiple futures can be extracted using several future maps. Furthermore, by combining these features in the subsequent layers, CNNs are capable of detecting higher-order features. Generally, each convolutional layer is followed by a subsampling layer that reduces the resolution of the feature map to reduce the sensitivity of the output to distortions [19]. Unlike the neocognitron, a CNN is trained with the back-propagation technique. Thus, weight sharing reduces the number of free parameters, improving generalization.

When looking at industrial applications outside the cope of ICTs, DNNs have been traditionally used in fault diagnosis for various sectors. For instance, Yin et al. [20] presented a novel method for fault diagnosis in high-speed railways, which is currently based on manual operation. In particular, the authors proposed an automated diagnosis network in order to detect failures in vehicle-on-board equipment. The results show that a deep belief network outperforms other trained networks and improves the accuracy of fault diagnosis by up to 95%. A further illustration can be found in the selection of different techniques for improving fault diagnosis in rolling bearings [21]. Following a thorough study of different AI techniques, the authors concluded that rule based method could become an extremely versatile tool in the fault diagnosis of rotating machinery.

Efficient training of deep neural networks is only possible if a massive number of labeled data is available to apply back-propagation training algorithms and this is not always possible in manufacturing environments. Though, some interesting approaches can be found in scientific literature.

Most published research is devoted to optimization of process parameters in advanced machining techniques (such as laser cutting, electro-chemical machining, ion beam micro-milling, and grinding) using SNNs [22–25]. Only a very limited number of studies have examined the use of DL in machining. In a very interesting recent study, Wang developed a DL based approach to material removal rate prediction in polishing technologies [26]. A pattern recognition, identification, and process control system has also been developed by Gunter to achieve an intelligent laser-welding machine [27].

To the knowledge of the authors, none of the published research focusing on the WEDM process has addressed process modeling using deep learning techniques. Selection of optimum process parameters has been a classic application of SNNs [28], which have also been used [29] to extract information about degradation of the cutting conditions during WEDM. In a more recent study, Conde et al. [30] proposed using a variant of SNNs to predict the accuracy of components machined by WEDM. By combining the predictions of the network with the simulated annealing optimization technique, wire paths of variable radii can be designed so that radial deviations due to wire deformations can be minimized. The results revealed that the average deviation between network predictions and actual components is below 6μm, which falls within the current limits of process accuracy. The search and recognition of behavioral patterns of voltage and current signals in the WEDM process has been studied by Caggiano et al. [31,32], who presented a SNN that effectively correlates voltage and current signals with the defects and marks originated on the machined component during the WEDM process. In all cases, the success of the network exceeded 81%.

At this point, it must be highlighted that during the WEDM process, extremely large amounts of data can be collected using high-frequency voltage and current probes. Data from every single discharge during the process can be collected and patterns that contain useful information about the actual machining process, no matter process conditions, can be examined. Taking into account the efficiency shown by DNNs in tasks related to pattern recognition, in this work an original contribution to advance unexpected event prediction during practical WEDM operations using deep learning techniques is presented. The occurrence of an unexpected event, namely a change of the thickness of the machined part, can be predicted in advance by recognizing hidden patterns from process signals. Raw data are directly obtained from the machining process carried out in industrial conditions by using voltage and current probes and a high-frequency oscilloscope. Various DNN architectures have been studied and it was found that the combination of a convolutional layer with gated recurrent units achieved the best performance. By adopting this approach, thickness variation can be predicted in 97.4% of cases, at least 2 mm in advance, which is fast enough as to act before the WEDM process is degraded. New possibilities for applying DNNs in the field of advanced manufacturing and high-performance machine tools must therefore be examined in the future.

2. Materials and Methods

2.1. Instrumentation and Measured Variables

The WEDM erosion mechanism is based on the generation of discrete short discharges between two electrically conductive electrodes (wire and workpiece) in a fluid dielectric medium. In a recent paper, Almacinha et al. [33] proposed the feasible possibility that in the sinking electrical discharge machining (EDM) process, with hydrocarbon oil as a dielectric, multiple discharges occur during pulses with long on-time. However, in the case of ire WEDM, the short duration of the pulses (1.2 μs in commercial machines as the one used in the experiments) and the use of deionized water as a typical dielectric medium, the possibility of occurrence of multiple discharges during one pulse has not yet been proven. Therefore, in this work, the common hypothesis in WEDM [2] that workpiece material removal mechanism is attributed to the consecutive occurrence of discharges was used. As each single discharge generated a crater of a few micrometers on the workpiece, the combined contribution of millions of these resulted in the removal of part material, thus drawing the shape of the part [34].

Figure 1 shows voltage evolution during a single discharge, as collected in an actual WEDM operation. A discharge can be divided into three parts. Before a discharge occurred (Phase 1, Figure 1), an off-time (voltage zero) was programmed, during which the gap between electrodes was cooled down and dielectric flow tried to remove the debris resulting from the previous discharge. In Phase 2 (Figure 1) the isolating capacity of the dielectric (deionized water in this case) was locally broken by the application of a voltage (commonly known as open-circuit voltage, see Table 1) between the wire and workpiece. The open-circuit voltage was applied by the machine generator. Then, ionization started (voltage signal was constant and current was zero amperes). This period, known as ionization time, was not controlled by the machine generator but by the local conditions of the dielectric. In other words, ionization time was not a machine parameter: for each single discharge, it depended on the specific electrical conductivity conditions of the dielectric. If flushing was effective and the gap was clean, ionization time was long. On the contrary, if flushing was difficult and debris was present in the gap, ionization time was short or even zero. Finally Phase 3 (Figure 1), when the electrical local conductivity of the dielectric between wire and workpiece was high enough (ionization ended), voltage dropped and current flowed during the on-time, resulting in part material removal due to the generated heat. For the experiments carried out, discharge current during on-time was 5A (see Table 1). This general pattern was reproduced during the process, although it is difficult to model phenomena such as the presence of debris between the electrodes, since differences were introduced between the discharges, affecting process performance.

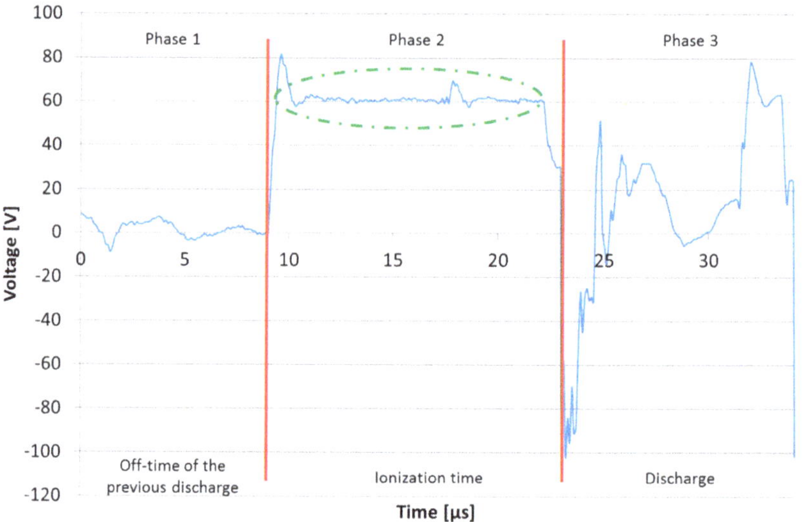

Figure 1. Voltage signal evolution during a single discharge in an industrial wire electrical discharged machining (WEDM) operation.

Table 1. Electrical parameters as selected by machine table look-up.

WEDM Parameters	Settings
Height [mm]	100
Off-time [µs]	9.0
On-time [µs]	1.2
Current intensity [A]	5.0
Open-circuit voltage [V]	60.0
Initial dielectric pressure [bar]	17.0
Wire tension [kg]	1.2

Each single discharge contained valuable information about process performance. As shown in the previous section, other authors have recently attempted to correlate process signals with final part quality. However, advanced pattern recognition may be far more efficient in evaluating process performance. The fact that large amounts of information can be collected during the process (with sampling rates as high as 10.0 MS/s, as explained below) opens the possibility of training DNNs that have already proven their excellence in other fields (see Section 1).

Voltage sequences were acquired during WEDM using a high-frequency oscilloscope Tektronix DPO5034B (Tektronix UK Ltd., Berkshire, UK) and a voltage probe Tektronix TMDP0200 (Tektronix UK Ltd., Berkshire, UK) connected to the WEDM machine. Figure 2 shows an example of voltage signal measurement. The sampling rate was 10.0MS/s, with a resolution of 100 ns. This is required to record any possible event during ionization of the discharge channel. A significant number of consecutive discharges must be recorded, because, as shown in Figure 2, there was a certain degree of variability between them. Therefore, a signal length of 200 μs was chosen. Measuring range for the voltage probe was set at −120 V to +120 V. The reason for this was that, although open-circuit voltage was set at 60 V (see Table 1), some random voltage peaks appeared and they would also be recorded. Also, commercial WEDM machines implement the feature of ensuring zero average voltage to avoid parasite currents that may affect process performance. This is why wire polarity changed as shown in Figure 2.

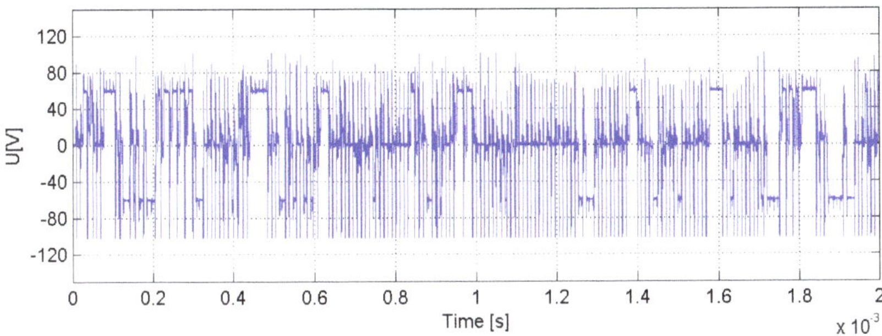

Figure 2. Example of voltage data sequence.

2.2. Experimental Methodology: WEDM Tests

In order to simulate the conditions of a degraded WEDM operation, a typical industrial situation in which process parameters cannot be controlled in advance was reproduced in our experiments. During an industrial operation, WEDM process parameters are set by machine-table look-up. Those parameters apply to a given combination of factors including part material, part thickness, and machining time. The parameters are available in the machine, and have been obtained through controlled experiments by the machine manufacturer. The machine user therefore finds in the WEDM machine the optimum combination of parameters for his/her application.

However, during WEDM cutting, operation conditions may vary. A typical example is an unexpected variation in part thickness. Parameters can be changed on-line once thickness change has happened [35] but anticipation of thickness change before it occurs has not been addressed. In fact, this feature is not present in commercial machines.

Controlled experiments were designed and carried out involving variation of part thickness during a WEDM operation. Part material for the samples was AISI D2 (ISO 160CrMoV12) 62 HRc tool steel, quenched and tempered. Stepped sample parts were prepared for the experiments, in which the WEDM cut faced a sudden thickness variation from 100 mm to 80 mm. Figure 3 shows a scheme of the process during the cutting of the test part.

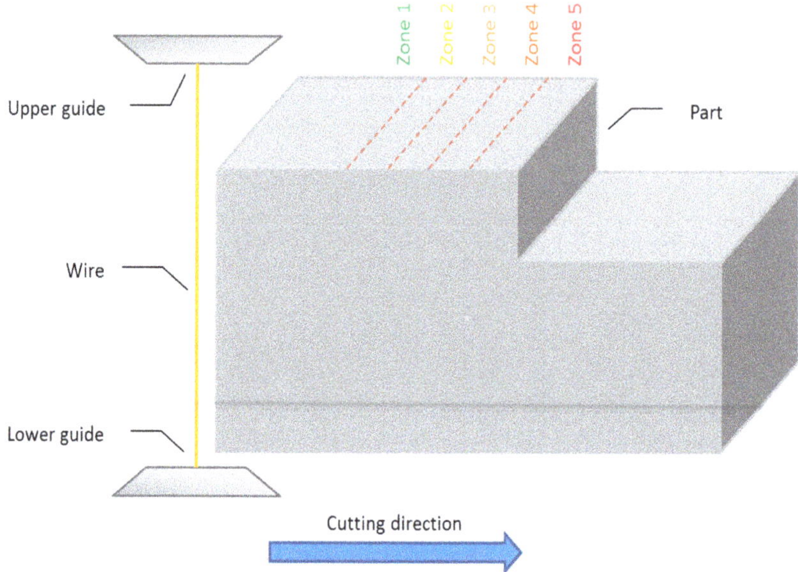

Figure 3. Scheme of the process with the sample geometry.

The experiments were conducted under industrial conditions using commercial WEDM machinery (ONA AX3 WEDM machine, ONA Electroerosión S.A., Durango, Spain). The wire used was an uncoated brass wire (CuZn37) of 0.25 mm diameter, with ultimate strength of 900 N/mm^2 and 1% elongation. As explained previously, WEDM electrical parameters (listed in Table 1) were selected by machine table look-up, and they correspond to roughing conditions.

When the wire approached the point of thickness change, the cut began to degrade because dielectric pressure was lost. This resulted in the occurrence of different voltage patterns in the discharges with respect to those occurring when the cut was performed under optimum conditions. To the best of our knowledge, there is not yet an industrial system that can detect such a pattern change. In order to conduct a systematic analysis, voltage sequences were collected at different distances from the point of thickness change. To do so, five different zones were defined: 5 mm away (Zone 1) from the point of thickness change, 4 mm away (Zone 2), 3 mm away (Zone 3), 2 mm away (Zone 4), and 1 mm away (Zone 5). The closer the wire to Zone 5 (in other words, to the point of thickness change), the more degraded the cut will be.

Hence, five different zones of 1 mm length were established in order to adequately describe the cutting process degradation. The recording length was set to 0.8 mm, so that the oscilloscope could be reset during the remaining time. This allowed for a total of 567 sequences of 2 ms to be recorded, with a resolution of 100 ns and a sample rate of 10.0 MS/s. Thus, mean values of 140 discharges per sequence were recorded. Furthermore, this process was repeated 16 times in order to accumulate an appropriate number of tests.

These collected data were used to generate three different datasets. The first was used to study different DNN architectures to classify the voltage sequences of each zone. The second dataset was used to check the performance of the DNN to classify the sequences for Zones 1, 3, and 5. Finally, the third dataset was used to check the performance of the DNN for the slightly less ambitious task of classifying the sequences of the first and fifth zones. The Z_all dataset was balanced, so that there are an equal number of sequences for each class. For the other two cases, all of the available data were used in order to avoid significantly reducing the training dataset. However, the number of samples for each class was similar and there was little difference between the zones, as can be seen in Table 2.

Table 2. Dataset used.

Denotation	Zones	Sequences
Z_all	1, 2, 3, 4, 5	2835 (5 × 567)
Z_135	1, 3, 5	2088 (688 + 677 + 723)
Z_15	1, 5	1411 (688 + 723)

2.3. Deep Learning Architectures Tested

The efficiency of different DNN architectures was evaluated by dividing the Z_all dataset into training (70%), validation (15%), and testing (15%) categories. The training, validation, and testing datasets were balanced, i.e., there was an equal number of sequences from each zone. All models were trained up to 100 epochs with categorical cross-entropy loss function and Adam optimizer [36]. Once the model was trained, the performance of the model was measured using the testing dataset, evaluating precision, recall, and F1 score.

The DNN models evaluated were a CNN, an RNN, and a bidirectional RNN and CNN combined with RNN. For the RNN, the GRU was used since it has recently been shown [34] that GRU slightly outperformed vanilla LSTM on almost most tasks and because the GRU is faster to train. The model studied is the following:

- CNN: the first layer was a convolutional layer with 50 filters of 10 × 1 dimension (10 × 1 × 50). The signals were of one dimension and, therefore, a 1D convolutional layer was used. The following layer was a stacked convolutional layer composed of 100 layers of 10 × 1 dimensions. After these two layers, a max pooling layer was used to down sample the input by two. Moreover, a dropout of 0.2 was also used to avoid overfitting. A further two convolutional layers of 150 filters with a smaller dimension (5 × 1) were then used. In this case a max-pooling layer was also used to down sample by two and dropout. The fully connected layer of 150 neurons with ReLu and dropout was used. Finally, a softmax activation function was used in the last layer. In all convolutional layers, a ReLu is used.
- GRU: in the case of the model with GRUs, the model had three layers of 50, 50, and 25 units with dropout of 0.2 intercalated between the layers to avoid overfitting. Finally, the last layer was fully connected with a softmax activation function.
- Bidirectional GRU: the model with bidirectional GRU layers was quite similar to the GRU model but with fewer units in each layer. Thus, the model was composed of three bidirectional GRU (BiGru) of 10, 50, and 25, with 0.2 dropout intercalated and softmax activation function in the last fully connected layer, as in the GRU model.
- Convolutional layer + GRU: the last model had a convolutional first layer to extract features from the signals, followed by two layers of GRU units. Hence, the first convolutional layer had 100 filters of 10 × 1 dimension followed by a max pooling of 2 for down sampling with a dropout and ReLu. Two GRU layers of 150 and 50 with dropout were then stacked. Finally, as in the case of both the GRU and BiGRU models, a fully connected layer with softmax activation function was used.

3. Results and Discussion

The results of Table 3 clearly show that the architecture combining a convolutional layer and GRU network outperformed other models for all the metrics analyzed. Therefore, for the other datasets (z_135 and z_15) this model was used to analyze the performance of the model and the complexity of the dataset. Due to the reduced dataset for classifying between Zones 1 and 5 and the fact that selecting the best model was not among the aims of the current study, the dataset was divided into training (70%) and testing (30%).

Table 3. Model results for Z_all dataset.

Model	Precision	Recall	F1 Score
CNN	0.5806	0.5765	0.5785
GRU	0.6969	0.5788	0.6324
BiGRU	0.6968	0.6706	0.6835
CGRU	0.7260	0.7106	0.7182
Model	Precision	Recall	F1 Score

The results achieved with fewer zones (Table 4) were much higher than those yielded from all the zones. Moreover, the F1 Score for Z_135 dataset was 0.9169 and for Z_15 was 1. These results were outstanding, as they highlight the ability of the CGRU (Convolutional Gated Recurrent Unit) network to classify voltage sequences with high accuracy. The models with GRU units clearly outperformed those with CNN. This appeared logical because current research has indicated that GRU units deal accurately with sequences. In fact, CNN without any gate unit cannot satisfactorily classify WEDM spark sequences with a F1 score lower than 60%. Therefore, for classifying WEDM spark sequences, it is highly recommended to use DNNs with GRU units.

Table 4. CGRU model results for Z_135 and Z_15 datasets.

	Precision	Recall	F1 Score
Z_135	0.9361	0.9361	0.9361
Z_15	1	1	1

Focusing on models with GRU units, it is interesting to see that, in terms of precision, the model with bidirectional GRU achieved almost the same result as the GRU model but outperformed the GRU in terms of recall and F1 Score. This is interesting because the BiGRU model has less GRU units in the input layer (10 in the BiGRU and 50 in the GRU). However, the results were not sufficiently clear to draw the conclusion that for sequence classification, a BiGRU model would outperform the GRU model in all cases.

Similarly, analyzing the results from Table 4, it appears that adding a convolutional layer in the input of a GRU model helped to classify WEDM spark sequences. Thus, the first convolutional layer helped to extract features from spark sequences and then GRU units modeled these new sequences generated by the convolutional layer. Therefore, the results show that a CGRU model works accurately when classifying WEDM sequences with high precision (0.7260). Moreover, Table 4 shows that this model classified almost perfectly when dealing with less complicated datasets. Indeed, the model was capable of achieving 100% precision for classifying sequences of Zones 1 and 5.

In contrast, from the process point of view, these results can be analyzed as follows. As the wire got closer to the thickness variation point, the behavior of the signal varied. This can be observed in the confusion matrixes of Figure 4. As explained in Section 2.2, Zone 1 (Z1) was the one that described a stable process, while Zone 5 (Z5) was the one closest to the point at which the thickness change occurred.

Figure 4 displays the confusion matrix considering the available data for the five zones (Z_all dataset). It can be observed that there was no interference between Zones 1 and 2 and Zones 4 and 5 or vice versa, or between Zones 3 and 5. From this result it can be stated that, when the cutting process came to be unstable, this was always successfully detected in advance by the neural network. For the experiment carried out, since the average feed for part thickness 100 mm was 1.4 mm/min, this means that there were at least 1.4 min to act before thickness variation occurred. Clearly, this time decreased when part thickness was smaller because of the higher feed, but in any case it would be tens of seconds (about 30 s when part thickness is 50 mm), which is more than enough time to make corrective actions. Moreover, the misclassification between zones was less than 3% among all consecutive zones throughout the degradation process, excluding Zones 1 and 2 (in any case, less than 10%), which was the starting point of the cutting degradation process.

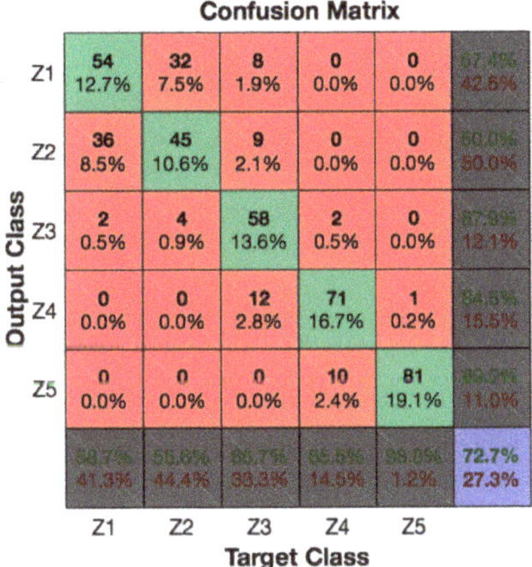

Figure 4. Confusion matrix for Z_all datasets with the CGRU model.

Figure 5 shows the confusion matrix results when only Zones 1, 3, and 5 were used, that is, an stable cutting zone, an intermediate degraded zone and finally, the one in which the thickness variation occurred. Again, the first and most obvious consideration was that there is no confusion between Zones 1 and 5, which means that a stable cut was clearly distinguished from the nearest region to the thickness variation point. The mix between Zones 1 and 3 occurred in less than 4% of the cases; however, it is more likely that a degraded cut is confused for a stable cut than vice versa. Finally, it is worth noting that there were 1% of cases in which Zone 5 was classified as Zone 3. In conclusion, even if in 4% of the cases a thickness variation of 4 mm could not be predicted in advance, this change can always be predicted 2 mm earlier than it occurs.

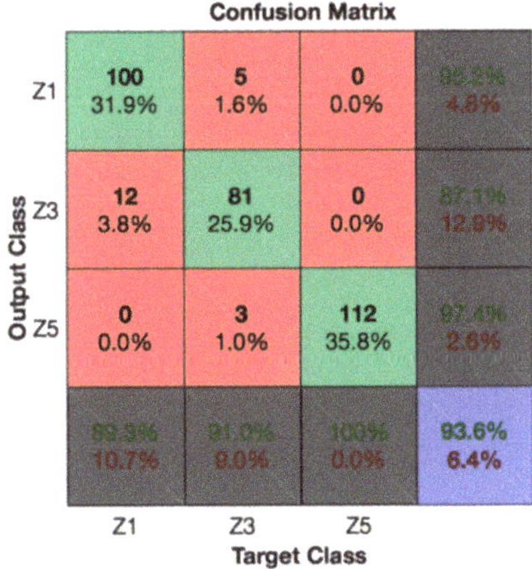

Figure 5. Confusion Matrix for Z_135 dataset with CGRU model.

To conclude, to the extent that the regions have been arbitrarily chosen, a variation in the behavioral pattern between a stable cut and a degraded cut was clearly shown in both cases (Z_all dataset and Z_135 dataset). Thus, the DNN enabled rapid action to be taken during the WEDM process.

Finally, Goodfellow et al. [10] stated that in supervised deep learning, to achieve acceptable results, more than 5000 labelled examples per category are needed and with at least 10 million labelled examples it is possible to match or exceed human performance. However, in this study, only 567 labelled examples for each category were available, considerably less than the recommended 5000. Although from a process point of view the results are excellent, one might consider that there is much room for improvement, because with more labelled examples, deeper ANNs can be used to reach higher precision results. However, as stated in the introduction, due to the difficulties in collecting large amounts of data in machine-tool workshops, in many cases it is not feasible to use DNNs. Therefore, if improvements in deep learning are to be exploited to their full potential for use in pattern recognition, then the companies and researchers involved in manufacturing and data collection must play a role in ensuring that new strategies are put in place.

4. Conclusions

The aim of this study was to evaluate the possibility of predicting an unexpected event during an industrial WEDM machining process by using DNNs to recognize hidden patterns from process raw voltage signals. A precision, recall, and F1 score comparison for different DNN models and datasets was provided. The results clearly showed that a model with a first convolutional layer with two GRU layers outperformed the other models. Moreover, this model achieved outstanding performance for the other datasets, with a precision of around 100%. From the process point of view, confusion matrixes showed that thickness variation can be predicted, at least 2 mm in advance, which allows sufficient time to act on machining parameters. New possibilities for applying DNNs in the field of advanced manufacturing and high-performance machine tools must be examined in the future. In particular, given the difficulty in collecting large quantities of labeled examples from machining processes, new strategies will need to be developed to resolve this problem. When large amounts of labeled data become available, the possibility of extensively applying DNNs in manufacturing will become a reality.

Author Contributions: Conceptualization, J.A.S. and A.A.; Methodology, A.C.; Software, A.A.; Validation, S.P. and J.W.; Formal Analysis, J.A.S. and A.A.; Investigation, A.C.; Resources, J.A.S.; Data Curation, A.C. and S.P.; Writing-Original Draft Preparation, A.A.; Writing-Review & Editing, J.A.S.; Visualization, A.A.; Supervision, J.A.S.; Project Administration, J.A.S.; Funding Acquisition, J.A.S.

Funding: The authors gratefully acknowledge the funding support received from the Spanish Ministry of Economy and Competitiveness and the FEDER operation program for funding the project "Scientific models and machine-tool advanced sensing techniques for efficient machining of precision components of Low Pressure Turbines" (DPI2017-82239-P) and UPV/EHU (UFI 11/29). The authors would also like to thank Euskampus and ONA-EDM for their support in this project.

Conflicts of Interest: The authors declare no conflict of interest. The funders had no role in the design of the study; in the collection, analyses, or interpretation of data; in the writing of the manuscript, and in the decision to publish the results.

References

1. Steinhardt, E. Future Aero Engine: Impact on Design and Production Technology. In Proceedings of the International Conference on Turbomachinery Manufacturing, Aachen, Germany, 20–21 February 2013.
2. Ho, K.H.; Newman, S.T.; Rahimifard, S.; Allen, R.D. State of the art in wire electrical discharge machining (WEDM). *Int. J. Mach. Tools Manuf.* **2004**, *44*, 1247–1259. [CrossRef]
3. Klocke, F.; Welling, D.; Klink, A.; Veselovac, D.; Nöthe, T.; Perez, R. Evaluation of Advanced Wire-EDM Capabilities for the Manufacture of Fir Tree Slots in Inconel 718. *Proced. CIRP* **2014**, *14*, 430–435. [CrossRef]
4. Klocke, F.; Welling, D.; Klinka, A.; Perez, R. Quality Assessment through in-process Monitoring of Wire-EDM for Fir Tree Slot Production. *Proced. CIRP* **2014**, *24*, 97–102. [CrossRef]
5. Krizhevsky, A.; Sutskever, I.; Hinton, G.E. ImageNet Classification with Deep Convolutional Neural Networks. In Proceedings of the International Conference on Neural Information Processing Systems, Lake Tahoe, NV, USA, 3–6 December 2012; Volume 25, pp. 1106–1114.

6. Ciresan, D.C.; Schmidhuber, J. Multi-Column Deep Neural Networks for Offline Handwritten Chinese Character Classification. *arXiv* **2013**, arXiv:1309.0261.
7. Sutskever, I.; Vinyals, O.; Quoc, V.L. Sequence to sequence learning with neural networks. In Proceedings of the 27th International Conference on Neural Information Processing Systems, Montreal, QC, Canada, 8–13 December 2014; pp. 3104–3112.
8. Vinyals, O.; Toshev, A.; Bengio, S.; Erhan, D. Show and tell: A Neural Image Caption Generator. In Proceedings of the IEEE Conference on Computer Vision and Pattern Recognition (CVPR), Boston, MA, USA, 7–12 June 2015; pp. 3156–3164.
9. Ciresan, D.C.; Giusti, A.; Gambardella, L.M.; Schmidhuber, J. Mitosis detection in breast cancer histology images with deep neural networks. In Proceedings of the International Conference on Medical Image Computing and Computer-Assisted Intervention, Nagoya, Japan, 22–26 September 2013; pp. 411–418.
10. Goodfellow, I.; Bengio, Y.; Courville, A. *Deep Learning*; MIT Press: Cambridge, MA, USA, 2016; Available online: http://www.deeplearningbook.org (accessed on 27 June 2018).
11. Rumelhart, D.E.; Hinton, G.E.; Williams, R.J. *Learning Internal Representations by Error Propagation*; Technical Report, DTIC Document; MIT Press: Cambridge, MA, USA, 1985.
12. Elman, J. Finding Structure in Time. *Cogn. Sci.* **1990**, *14*, 179–211. [CrossRef]
13. Bengio, Y.; Simard, P.; Frasconi, P. Learning long-term dependencies with gradient descent is difficult. *IEEE Trans. Neural Netw.* **1994**, *5*, 157–166. [CrossRef] [PubMed]
14. Hochreiter, S.; Schmidhuber, J. Long short-term memory. *Neural Comput.* **1997**, *9*, 1735–1780. [CrossRef] [PubMed]
15. Chung, J.; Gulcehre, C.; Cho, K.H.; Bengio, Y. Empirical Evaluation of Gated Recurrent Neural Networks on Sequence Modeling. In Proceedings of the NIPS Deep Learning Workshop, Montreal, QC, Canada, 11 December 2014.
16. Cho, K.; Merrienboer, B.; Bahdanau, D.; Bengio, Y. On the properties of neural machine translation: Encoder-decoder approaches. *arXiv* **2014**, arXiv:1409.1259.
17. LeCun, Y.; Bottou, L.; Bengio, Y.; Haffner, P. Gradient-based learning applied to document recognition. *Proc. IEEE* **1998**, *86*, 2278–2324. [CrossRef]
18. Fukushima, K. Neocognitron: A Self-organizing Neural Network Model for a Mechanism of Patter Recognition Unaffected by Shift Position. *Biol. Cybern.* **1980**, *36*, 193–202. [CrossRef] [PubMed]
19. LeCun, Y. *Generalization and Network Design Strategies*; Connectionism in Perspective; Elsevier: Zurich, Switzerland, 1989.
20. Yin, J.; Zhao, W. Fault diagnosis network design for vehicle on-board equipments of high speed railway: A deep learning approach. *Eng. Appl. Artif. Intell.* **2016**, *56*, 250–259. [CrossRef]
21. Dou, D.; Zhou, S. Comparison of four direct classification methods for intelligent fault diagnosis of rotating machinery. *Appl. Soft Comput.* **2016**, *46*, 459–468. [CrossRef]
22. Samanta, S.; Chakraborty, S. Parametric optimization of some non-traditional machining processes using artificial bee colony algorithm. *Eng. Appl. Artif. Intell.* **2011**, *24*, 946–957. [CrossRef]
23. Rao, R.V.; Kalyankar, V.D. Parameter optimization of modern machining processes using teaching–learning-based optimization algorithm. *Eng. Appl. Artif. Intell.* **2013**, *26*, 524–531.
24. Rao, R.V.; Rai, D.P.; Balic, J. A multi-objective algorithm for optimization of modern machining processes. *Eng. Appl. Artif. Intell.* **2017**, *61*, 103–125. [CrossRef]
25. Arriandiaga, A.; Portillo, E.; Sánchez, J.A.; Cabanes, I.; Zubizarreta, A. Downsizing training data with weighted FCM for predicting the evolution of specific grinding energy with RNNs. *Appl. Soft Comput.* **2017**, *61*, 211–221. [CrossRef]
26. Wang, P.; Gao, R.X.; Yan, R. A deep learning-based approach to material removal rate prediction in polishing. *CIRP Ann.* **2017**, *66*, 429–432. [CrossRef]
27. Günther, J.; Pilarski, P.M.; Helfrich, G.; Shen, H.; Diepold, K. Intelligent laser welding through representation, prediction, and control learning: An architecture with deep neural networks and reinforcement learning. *Mechatronics* **2016**, *34*, 1–11. [CrossRef]
28. Mukherjee, R.; Chakraborty, S.; Samanta, S. Selection of wire electrical discharge machining process parameters using non-traditional optimization algorithms. *Appl. Soft Comput.* **2012**, *12*, 2506–2516. [CrossRef]
29. Portillo, E.; Marcos, M.; Cabanes, I.; Zubizarreta, A. Recurrent ANN for monitoring degraded behaviors in a range of workpiece thicknesses. *Eng. Appl. Artif. Intell.* **2009**, *22*, 1270–1283. [CrossRef]

30. Conde, A.; Arriandiaga, A.; Sanchez, J.A.; Portillo, E.; Plaza, S.; Cabanes, I. High-accuracy wire electrical discharge machining using artificial neural networks and optimization techniques. *Robot. Comput. Integr. Manuf.* **2018**, *49*, 24–38. [CrossRef]
31. Caggiano, A.; Teti, R.; Perez, R.; Xirouchakis, E.D.M. Wire Monitoring for Zero-Defect Manufacturing based on Advanced Sensor Signal Processing. *Proced. CIRP* **2015**, *33*, 315–320. [CrossRef]
32. Caggiano, A.; Perez, R.; Segreto, T.; Teti, R.; Xirouchakis, P. Advanced Sensor Signal Feature Extraction and Pattern Recognition for Wire EDM Process Monitoring. *Proced. CIRP* **2016**, *42*, 34–39. [CrossRef]
33. Almacinha, J.A.; Mendes-Lopes, A.; Rosa, P.; Duarte-Marafona, J. How Hydrogen Dielectric Strength Forces the Work Voltage in the Electric Discharge Machining. *Micromachines* **2018**, *9*, 240. [CrossRef]
34. Cabanes, I.; Portillo, E.; Marcos, M.; Sánchez, J.A. An industrial application for on-line detection of instability and wire breakage in wire EDM. *J. Mater. Process. Technol.* **2008**, *195*, 101–109. [CrossRef]
35. Liao, Y.S.; Chuang, T.J.; Yu, Y.P. On-line Workpiece Height Estimation and its Application in Servo Feed Control of WEDM Process. Procedia CIRP, 6 (2013) 226-231.Kingma, D.P.; Ba, J.L. Adam: A method for stochastic optimization. *arXiv* **2015**, arXiv:1412.6980.
36. Jozefowicz, R.; Zaremba, W.; Sutskever, I. An Empirical Exploration of Recurrent Network Architectures. In Proceedings of the International Conference on Machine Learning, Lille, France, 7–9 July 2015.

© 2018 by the authors. Licensee MDPI, Basel, Switzerland. This article is an open access article distributed under the terms and conditions of the Creative Commons Attribution (CC BY) license (http://creativecommons.org/licenses/by/4.0/).

Article

Combination of Laser Material Deposition and Laser Surface Processes for the Holistic Manufacture of Inconel 718 Components

Jon Iñaki Arrizubieta *, Magdalena Cortina, Jose Exequiel Ruiz and Aitzol Lamikiz

Department of Mechanical Engineering, University of the Basque Country, Plaza Torres Quevedo 1, Bilbao 48013, Spain; magdalena.cortina@ehu.eus (M.C.); joseexequiel.ruiz@ehu.eus (J.E.R.); aitzol.lamikiz@ehu.eus (A.L.)

* Correspondence: joninaki.arrizubieta@ehu.eus; Tel.: +34-946-013-932

Received: 27 June 2018; Accepted: 18 July 2018; Published: 20 July 2018

Abstract: The present work proposes a novel manufacturing technique based on the combination of Laser Metal Deposition, Laser Beam Machining, and laser polishing processes for the complete manufacturing of complex parts. Therefore, the complete process is based on the application of a laser heat source both for the building of the preform shape of the part by additive manufacturing and for the finishing operations. Their combination enables the manufacture of near-net-shape parts and afterwards removes the excess material via laser machining, which has proved to be capable of eliminating the waviness resulting from the additive process. Besides, surface quality is improved via laser polishing so that the roughness of the final part is reduced. Therefore, conventional machining operations are eliminated, which results in a much cleaner process. To validate the capability of this new approach, the dimensional accuracy and surface quality as well as the microstructure of the resulting parts are evaluated. The process has been validated on an Inconel 718 test part, where a previously additively built-up part has been finished by means of laser machining and laser polishing.

Keywords: laser; additive manufacturing; laser beam machining; laser polishing; waviness; roughness; Inconel 718

1. Introduction

Laser Material Processing is an alternative to many traditional manufacturing processes, such as arc welding, electrochemical machining, hand polishing, electron beam welding, etc. Laser Material Processing's main characteristic is the use of a high-power laser as a heat source, which results in a very high concentration of the energy density that reduces the Heat Affected Zone (HAZ) and thermally induced distortions [1].

One of the laser-based processes that is experiencing a continuous growth is the Laser Metal Deposition (LMD). This additive manufacturing (AM) technique consists on generating a melt pool on the surface of the substrate, while wire or powder shaped filler material is added simultaneously [2]. Besides, LMD enables to obtain near-net-shape parts, which reduces the amount of wasted material [3,4]. Regarding environmental impact considerations, if material reductions as high as 50% with respect to the initial part are required during the manufacturing process, AM becomes environmentally friendlier compared with machining and forging [5]. In the same way, the aeronautical industry uses the buy-to-fly ratio as an efficiency factor, since it relates the weight of the part that really flights with the weight of the initial part stock. Laser Material Deposition can reduce the buy-to-fly ratio below 1.5:1, comparable to laser welding processes [6]. However, LMD manufactured parts do not meet the final surface roughness and dimensional requirements, and a finishing operation is always required [7]. Usually, conventional machining is applied for the finishing of the parts.

Another laser-based process that has found a niche in the market is the Laser Beam Machining (LBM), where the laser beam is directly applied for melting and vaporizing unwanted material from the substrate surface [8]. As the LBM is a laser-based process, no cutting tools are required, and materials can be machined regardless their hardness [9]. In addition, LBM process applies a laser beam (usually smaller than 75 μm beam diameter) directly for removing surface material. Hence, this process is especially suitable for the machining of small details on hard materials [10]. Moreover, high aspect-ratio grooves and holes can also be achieved [11] and almost no HAZ is generated when nano or femto pulse-duration lasers are used [12]. Nevertheless, as Dubey et al. stated, LBM process is not fully developed, and it is still waiting to its industrial use [8].

LBM does not always provide the desired surface quality and a finishing operation is therefore required. To this end, highly skilled operators using abrasive tools have traditionally performed finishing operations manually. For instance, Peng et al., proposed the Abrasive Flow Machining for removing the falling effect and the powder adhesion generated during AM [13].

An alternative to reduce the surface roughness of previously manufactured parts, which has caught the interest of many researchers, is the laser polishing (LP) [14–16]. In LP, peaks of the surface roughness are melted, and the material is redistributed in the valleys due to the surface tension and the gravity [17]. Therefore, when laser-polishing material is not removed, nor the final shape of the part is modified, but material is relocated while melted. To improve the understanding of the effect of LP on additively manufactured parts, Marimuthu et al., studied the influence of the melt pool dynamics on the resulting surface topology and roughness [18].

Other authors have studied experimentally the improvement of the surface quality when AM and LP are combined. For example, Zhihao et al., studied the surface roughness reduction of additively built-up parts using LP [19]. They concluded that LP improves the surface roughness of Inconel 718 Selective Laser Melting manufactured parts. On the other hand, Ma et al., also studied the improvement of the surface roughness of additively manufactured Ti alloys [17]. Nevertheless, the reference surface on which authors applied LP was a W-EDM cut surface and not the wavy surface characteristic of AM.

Up to now, the roughness and excess material resulting from the AM process is eliminated mechanically via milling or other abrasive processes, such as grinding. In this direction, the current trend of modern industry is to combine additive and subtractive technologies within the same machine [20]. However, laser-based processes are not always easily combined with other manufacturing techniques. For instance, the combination of LMD with milling or turning may result problematic, especially when cutting fluids are used. The problems arisen can be classified in two groups. On the one hand, the handling and filtering of the moisture generated when the powder particles and the cutting fluid are mixed results problematic. On the other hand, pore phenomena do appear if the surface is not properly cleaned before the LMD process [21].

Consequently, if LMD, LBM and LP processes are combined, the machining operation could be eliminated from the production chain, which leads to a much cleaner and environmentally friendlier manufacture. Moreover, the use of coolants, tooling, etc. is eliminated, which simplifies the management of the generated residues during the manufacturing process.

To demonstrate the validity of this statement, a novel manufacturing procedure, fully based on laser, which combines LMD, LBM and LP technologies is developed, where Laser Beam Machining is employed for removing the overstock and waviness generated by Laser Material Deposition. Finally, LP is used for reducing the roughness resulting from the LBM process. Topographies of the attained surfaces are obtained for each operation and roughness values as well as the microstructure are analyzed to evaluate the surface quality.

2. Materials and Methods

The proposed process involves very different laser operations. On the one hand, LMD is usually carried out with Continuous Wave lasers, while LBM and LP are usually performed with pulsed lasers. On the other hand, laser beam diameters for LMD are usually between 100 μm and 1 mm, while LBM and LP processes used to be carried out with much smaller laser beams (usually below 75 μm). Therefore, two different machines have been used to perform the proposed procedure. Firstly, the Kondia Aktinos 500 laser center (Kondia, Elgoibar, Spain) coupled with a 1 kW Rofin FL010 fiber laser (ROFIN-SINAR Laser GmbH, Bergkirchen, Germany) has been employed for the LMD tests. The LMD head includes a 200 mm focal length lens that concentrates the laser beam in a 0.75 mm diameter spot, values provided by the laser supplier. Powder material is supplied using a Sulzer Metco Twin 10 C powder feeder (Oerlikon Metco, Pfäffikon, Switzerland) and focused by an in house designed coaxial nozzle, denominated as EHUCoax-2015 (UPV/EHU, Bilbao, Spain) [22]. Argon has been used as protective and carrier gas. Then, a Trumpf TruMark Station 5000 (Trumpf, Ditzingen, Germany) is used for the LBM and LP operations [23]. This marking station has a fiber laser with a Q-switch pulse technology that concentrates a 50 W laser power in 7–500 ns duration pulses. A 2D galvanometric scanner (Trumpf, Ditzingen, Germany) controls the laser beam position and focuses it at a 212 mm focal distance and a 45 μm diameter; these values are supplied by Trumpf (Ditzingen, Germany).

The material used for the tests is Inconel 718 superalloy, which is supplied by Oerlikon Metco (Pfäffikon, Switzerland) under the name MetcoClad 718. The chemical composition of the powder material is shown in Table 1 and, as it can be observed, it is similar to that of Inconel 718. Powder is supplied with a particle size between 44 and 90 microns in diameter and the spherical shape of the particles is ensured as they are manufactured via Argon-gas atomization.

Table 1. Chemical composition (wt. %) of the MetcoClad 718 [24].

Cr	Mo	Nb	Fe	Ti	Si	Mn	C	B	Ni
19	3	5	18	1	0.2	0.08	0.05	0.005	Bal.

Before manufacturing a final test part, three types of tests are performed:

(1) First, a preliminary test (Test Part 1) for evaluating the capability of LBM for machining LMD manufactured Inconel 718 parts is performed. For this purpose, a 3 mm thickness layer is deposited by means of LMD. Afterwards, the surface of the deposited material is grinded to ensure a flat reference surface. On this surface, different LBM parameters are tested, and, in each case, the reached depth and the resulting surface quality are evaluated. Based on the obtained results, the maximum effective depth at which the laser could remove material is defined.

(2) Secondly, following the same procedure and based on the results obtained in Test 1, the capability of LP for improving the roughness resulting from LBM is evaluated. Based on these results, the optimum LP parameters are defined. Besides, the recast layer generated by LP is measured.

(3) Finally, the capability of LBM for eliminating the surface waviness resulting from LMD is evaluated. In this case, no intermediate grinding operation is performed.

Process parameters for LMD of MetcoClad 718 were obtained in a previous work [21] and they are detailed in Table 2. In Figure 1 a cross section of a single clad is shown, where the dimensions and dilution can be observed. The sample is etched using Kalling's 2 reagent to reveal the microstructure originated during the cooling stage. Generated clads have 2 mm width and a constant 0.8 mm height is obtained with each layer.

Table 2. LMD process parameters for the MetcoClad 718 [21].

Process Parameter	Value
Continuous wave laser power (W)	570
Scan velocity (mm·min^{-1})	525
Track offset (mm)	1.036
Overlap between tracks (%)	26
Powder mass flow (g·min^{-1})	8.78
Powder preheating temperature (°C)	60
Protective gas flow rate (L·min^{-1})	14

Figure 1. (**a**) Cross section of a single clad; (**b**) Detail of the microstructure.

For the first test, material is deposited over an AISI 1045 substrate. This substrate has no influence on the subsequent LBM operations since they are performed only in the LMD zone. Nevertheless, for the final tests, Inconel 718 substrate is used. Figure 2 shows the substrate with the deposited area after the grinding operation.

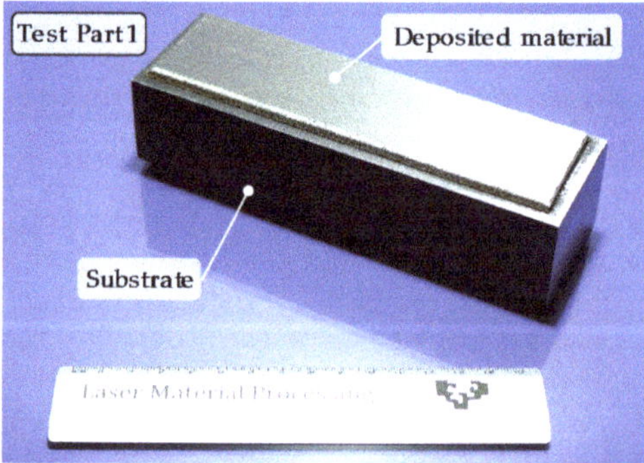

Figure 2. Test part 1 after the LMD and grinding processes.

To determine the best conditions for LBM, a parameter scanning is performed over the grinded flat surface in Test Part 1. Obtained results are shown in Figure 3, whereas the employed parameters in these tests are shown in Table A1 (see Appendix A).

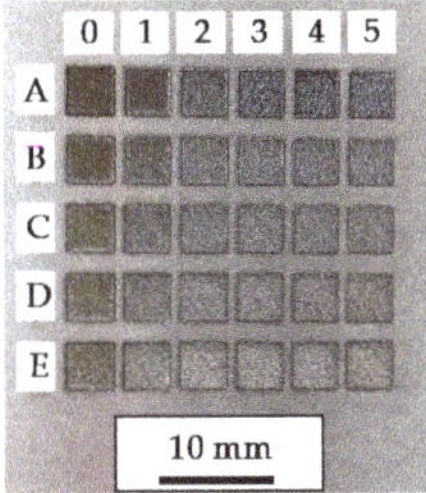

Figure 3. Parameter tests for obtaining the best LBM conditions (Test Part 1).

Likewise, with a view to determining the best LP conditions, a parameter scanning has been performed over the LBM surface resulted from applying the optimum process conditions determined previously, see Figure 4. Test codes for the LP tests are named with lower case letters to avoid misunderstandings with the LBM test naming. The parameters of these tests are showed in Table A2 (see Appendix B).

Figure 4. Parameter tests for obtaining the best LP conditions (Test Part 2).

Once Tests 1 and 2 are carried out and the optimum parameters for LBM and LP are defined for an LMD manufactured Inconel 718 part, Test 3 is performed. Test Part 3, which is also used for the manufacture of the Final Test Part, is manufactured layer-by-layer via LMD and the result is a 50 mm high wall with a 4 mm thickness and 60° inclination, see Figure 5. Please note that in this test, no grinding operation is executed and surface waviness resulting from the LMD process is eliminated exclusively via LBM.

Figure 5. Part manufactured via LMD for Test Part 3.

3. Results and Discussion

3.1. Material Removal via LBM

The aim of the LBM operation is to remove as much material as possible from the substrate. Therefore, to determine the optimum parameters for the LBM operation, the depth reached in each case is measured based on the average surface profiles obtained by means of a Leica DCM 3D confocal microscope. The depth reached in each case after a single repetition is detailed in Table 3, whereas the process conditions employed in each test are detailed in Table A1 (see Appendix A).

Table 3. Depth reached, in microns, after a single repetition for the different LBM process parameters.

Test Code	A	B	C	D	E
0	52.1	10.6	4.0	5.5	6.6
1	57.4	24.9	8.2	7.2	7.9
2	53.5	16.9	8.3	6.5	10.9
3	54.2	27.1	7.5	7.4	9.2
4	59.3	31.7	8.8	8.1	8.7
5	57.3	24.5	8.2	10.5	12.5

Process parameters corresponding to the test A4 are considered the best in terms of penetration and low recast layer; therefore, these parameters are employed for the following LBM operations, see Tables 4 and 5.

Table 4. LBM parameters for the MetcoClad 718.

Process Parameter	LBM
Mean pulse power (W)	6720
Velocity (mm·s^{-1})	800
Pulse frequency (Hz)	372,000
Pulse duration (ns)	20
Defocusing (mm)	0

Table 5. Hatching parameter values for LBM.

Process Parameter	LBM
Line spacing (mm)	0.05
Number of hatchings (-)	20
Angle increment (°)	17

Once the process parameters are determined, LBM is performed on the surface of the Test Part 1, with the laser beam focused on its surface and without changing the focal position between the consecutive repetitions. After every 10 repetitions, the mark generated on the surface of the substrate is analyzed by means of a Leica DCM 3D confocal microscope. In Figure 6, the topographies of two different marks are shown.

As the number of repetitions is increased, the depth increment is lower, and after 100 repetitions, it is noticed that the laser is not capable of removing material anymore. Therefore, the LBM process is concluded to be capable of removing material until a 1.6 mm maximum distance from the focal plane position (fpp), see Figure 7. It must be highlighted that the laser beam is focused on the original grinded surface of the substrate and its position remains unchanged as the number of repetitions is increased.

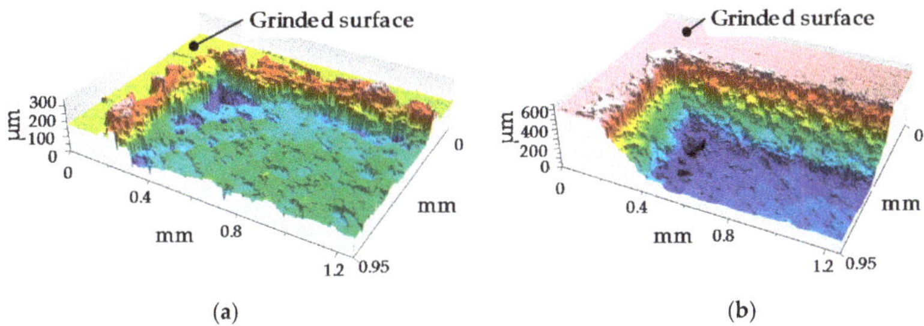

Figure 6. Topographies of the mark after (**a**) One repetition; (**b**) 10 repetitions.

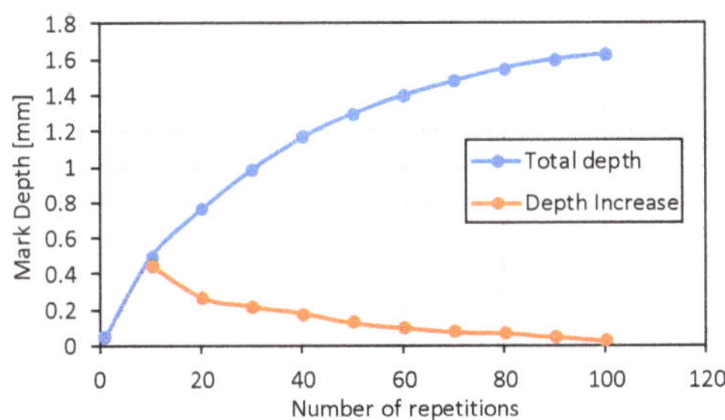

Figure 7. Depth increase as the number of repetitions is increased.

Besides, the time required for processing a 1 cm^2 area until a 0.08 mm depth is of 507.31 s, which results in a 8.3×10^{-3} $g \cdot min^{-1}$ material removal rate in the LBM process.

3.2. Roughness Reduction via LP

In Test 2, the roughness resulting from Test 1 is reduced via LP. With the aim of determining the optimum LP parameters, the resulting Ra roughness is measured in every polished square shown in Figure 4. Roughness measurements are performed according to the standard ISO 4287 and using a Leica DCM 3D confocal microscope. The resulting Ra values in microns are detailed in Table 6, whereas the process conditions employed in each test are shown in Table A2 (see Appendix B).

Table 6. Ra value, in microns, after the different LP tests.

Test Code	0	1	2	3	4
a	1.12	0.98	0.79	0.87	1.54
b	0.95	0.90	0.67	0.77	1.42
c	0.91	0.89	0.68	0.75	1.38
d	1.15	1.03	0.83	0.72	1.61
e	0.98	0.93	0.72	0.72	1.45
f	0.93	0.92	0.75	0.62	1.41
g	1.34	1.18	0.95	1.04	1.89
h	1.14	1.04	0.80	0.89	1.70
i	1.09	1.05	0.82	0.90	1.74
j	1.36	1.24	0.95	0.86	2.03
k	1.18	1.10	0.86	0.85	1.74
l	1.10	1.10	0.90	0.74	1.78

After the results analysis, it is concluded that for the same process parameters (laser power, frequency, hatching, defocus, etc.) an increase of the laser scan velocity results directly in higher surface roughness. Besides, it is also noticed that the surface roughness improves as the laser frequency is increased, but 200 kHz becomes a limit value, after which Ra value increases.

Regarding the number of hatches used for polishing, the resulting roughness value is improved as the number of repetitions is increased and a lower Ra value is obtained in all cases with 10 repetitions rather than with 5. However, when the number of repetitions is further increased, until 20, there is no considerable roughness reduction, whereas the required time for the process is doubled. Hence, it is decided that 10 repetitions is the optimum parameter tested.

Process parameters corresponding to the test b2 provided the lowest roughness value, and therefore, these parameters are employed for the following LP operations, see Tables 7 and 8. Notice that due to the 4 mm defocusing, the laser spot becomes approximately 120 microns in diameter at the working plane.

Table 7. LP parameters for the MetcoClad 718.

Process Parameter	LP
Mean pulse power (W)	621
Velocity (mm·s^{-1})	100
Pulse frequency (Hz)	175,000
Pulse duration (ns)	460
Defocusing (mm)	4

Table 8. Hatching parameter values for LP.

Process Parameter	LP
Line spacing (mm)	0.02
Number of hatchings (-)	10
Angle increment (°)	36

The idea of combining LMD and LBM processes arises as a methodology aiming to remove the surface waviness that LMD generates, and thus, obtain a flat surface. To that end, the laser is defocused 1 mm above the desired final surface. Therefore, the laser eliminates all material until a distance of 1.6 mm from the focal plane position, see Figure 8, and the process does only affect material located in this concrete region. However, as the resulting surface quality from the LBM process has a high roughness value, a polishing stage is afterwards performed.

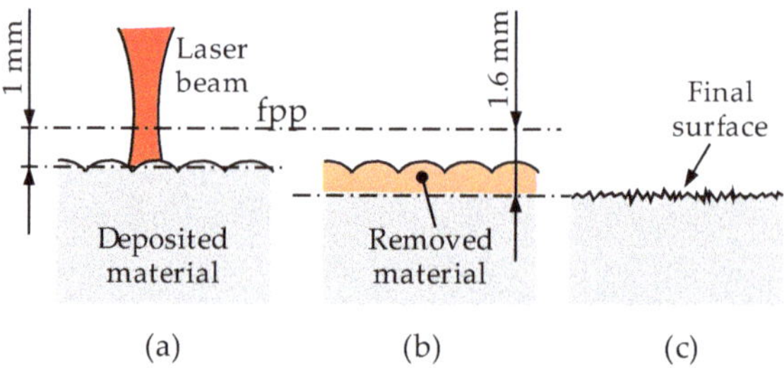

Figure 8. Followed strategy in LBM for obtaining a flat surface from the waved LMD surface. (**a**) LMD manufactured part; (**b**) Material removal via LBM; (**c**) Resulting flat surface after LBM.

A 3D view of the surfaces attained after the different laser-based processes are shown in Figures 9–11. In the three figures, the same height axis scale is used to make results visually comparable. In the case of the LMD surface, roughness is measured perpendicularly to the LMD direction, because LMD is a directional process and so is the resulting surface pattern. On the contrary, in LBM and LP the hatching direction is changed in every repetition to avoid any directional pattern on the surface, and therefore, roughness is independent from the measured direction.

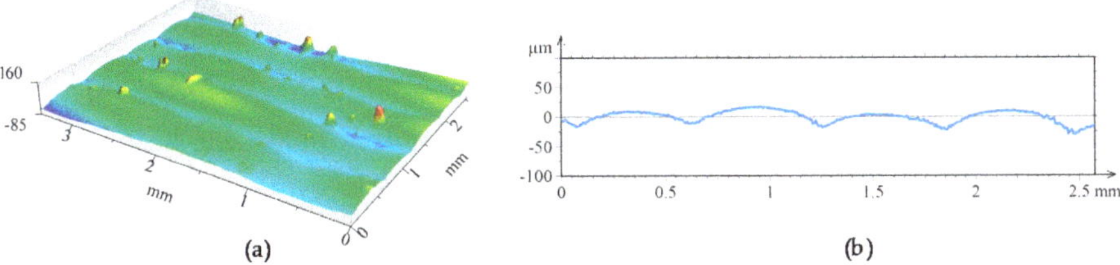

Figure 9. (**a**) Topography and (**b**) surface profile LMD.

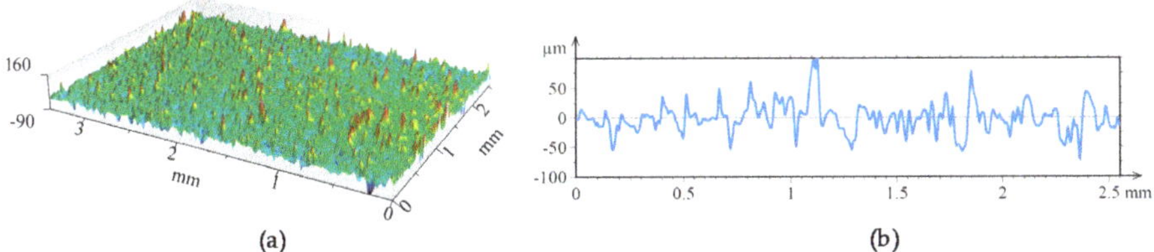

Figure 10. (**a**) Topography and (**b**) surface profile LMD + LBM.

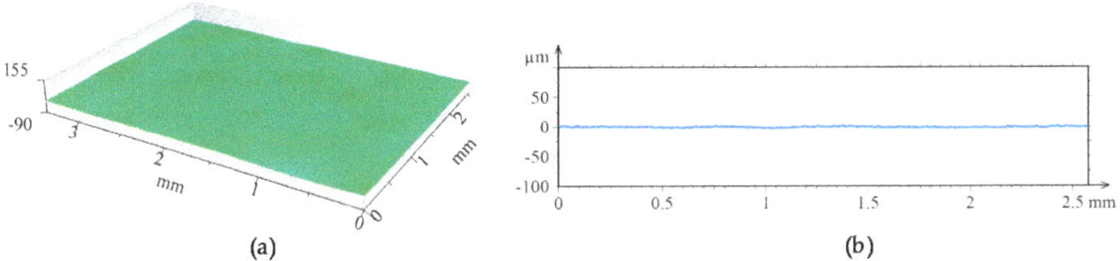

Figure 11. (**a**) Topography and (**b**) surface profile LMD + LBM + LP.

In a second step, to compare numeric roughness values, the roughness of each surface is obtained. For this purpose, the arithmetic mean deviation of the surface roughness (Ra) of five different profiles is measured in each surface and the average value is calculated. Measurements are performed according to the standard ISO 4287. As it is shown in Table 9, the Ra value is higher after the LBM process, than that after LMD. However, LBM provides a waviness-free surface, but the roughness needs to be reduced with the subsequent polishing stage.

Table 9. Arithmetic Mean Deviation of the Roughness Profile (Ra) in microns, according to ISO 4287. 0.25 mm Gaussian filter applied.

Measurement	LMD	LBM	LM
1	1.56	20.45	0.53
2	2.38	20.24	0.66
3	1.32	20.21	0.57
4	1.85	24.80	0.71
5	2.01	16.17	0.56
Average Ra	1.82	20.37	0.61

3.3. Influence of the LP on Material Microstructure

LP is proved to be capable of modifying the surface roughness. However, it also affects the microstructure of the material and generates a recast layer that may modify the mechanical properties of the final part. To evaluate the influence of the LP on the microstructure, both LBM and LBM + LP surfaces have been cross-sectioned, polished, and etched using Kalling's 2 reagent. Notice that the polished surface shown in Figure 12b is the same LBM surface shown in Figure 12a that has been later subjected to LP. The thickness of the recast layer due to the polishing is of 22 μm, which is a circumstance to be considered depending on the final application of the part.

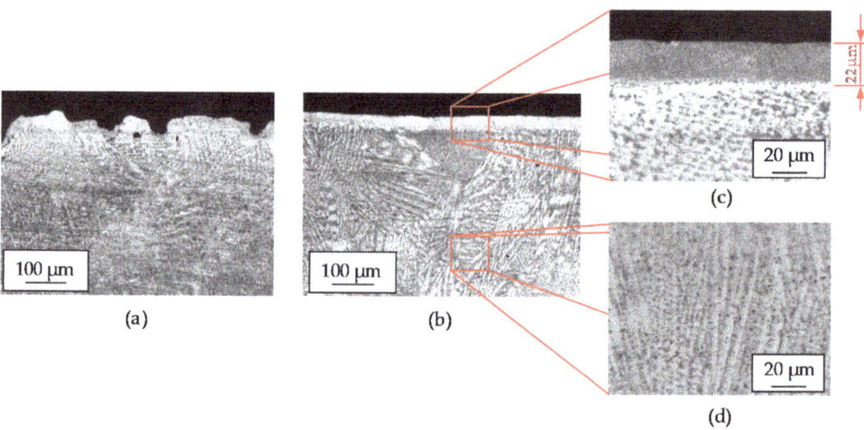

Figure 12. (**a**) Cross section of the LBM manufactured surface; (**b**) Cross section of the LBM + LP manufactured surface; (**c**) Detail of the recast layer generated on the surface due to polishing; (**d**) Internal microstructure developed as a result of the LMD process.

3.4. Final Test Part

To demonstrate the potential of combining LMD, LBM and LP processes, a final test part is manufactured, Figure 13. First, starting from an Inconel 718 substrate, an oblique wall is built using MetcoClad 718 filler material with the same strategy and conditions used in the previous tests. Afterwards, the wavy surface resulting from the LMD is processed via LBM up to a 0.5 mm depth. Finally, the desired regions are laser polished.

Figure 13. Final shape of the manufactured final test part.

4. Conclusions

In the present work, a full laser-based manufacturing technique is proposed. According to the attained results, the following conclusions are drawn:

(1) The LBM process is proved capable of eliminating the waviness generated in the LMD process and enables to obtain a flat surface.
(2) Surface quality resulting from LBM may not comply with the desired requirements. However, high surface quality (N5–N6 roughness grade) is obtained after the LP stage.
(3) LP generates a recast layer with a thickness of 22 μm. Depending on the final application of the part, this circumstance must be considered, because it might be detrimental to the mechanical properties. Further investigations must be performed to determine the influence of this recast layer.
(4) In LBM a maximum material removal rate of 8.3×10^{-3} g·min^{-1} is obtained. Therefore, LBM is proved to be slow when compared with the machining processes. Consequently, the combination of LMD + LBM is only advantageous when difficult-to-cut materials are processed, or high-resolution detail operations are required.
(5) The LBM process is capable of manufacturing small details that may not be possible to attain with other traditional machining processes, such as milling.

Author Contributions: Conceptualization and methodology, J.I.A.; LMD experiments, M.C.; engraving experiments, J.I.A.; polishing experiments, J.E.R.; Analysis of the obtained data, J.I.A., M.C. and J.E.R.; Writing-Original Draft Preparation, J.I.A.; Supervision, A.L.

Acknowledgments: Special thanks are addressed to the Industry and Competitiveness Spanish Ministry for the support on the DPI2016-79889-R INTEGRADDI project and PARADDISE project H2020-IND-CE-2016-17/H2020-FOF-2016 of the European Union's Horizon 2020 research and innovation program.

Conflicts of Interest: The authors declare no conflict of interest.

Appendix A

Process parameters for LBM tests carried out to determine the optimal parameters are detailed in Table A1. In all tests, the hatching parameters are kept constant according to the values detailed in Table 4 (line spacing of 0.05 mm and 20 hatching with an angle increment of 17°). Test codes for the LBM tests are named with upper case letters.

Table A1. LBM process parameters.

Test	Velocity [mm·s^{-1}]	Defocusing (mm)	Pulse Frequency [Hz]	Pulse Duration [ns]
A0	800	0	144,000	55
A1	800	0	201,000	37
A2	800	0	258,000	27
A3	800	0	315,000	23
A4	800	0	372,000	20
A5	800	0	429,000	17
B0	1100	0	144,000	55
B1	1100	0	201,000	37
B2	1100	0	258,000	27
B3	1100	0	315,000	23
B4	1100	0	372,000	20
B5	1100	0	429,000	17
C0	1400	0	144,000	55
C1	1400	0	201,000	37

Table A1. *Cont.*

Test	Velocity [mm·s^{-1}]	Defocusing (mm)	Pulse Frequency [Hz]	Pulse Duration [ns]
C2	1400	0	258,000	27
C3	1400	0	315,000	23
C4	1400	0	372,000	20
C5	1400	0	429,000	17
D0	1700	0	144,000	55
D1	1700	0	201,000	37
D2	1700	0	258,000	27
D3	1700	0	315,000	23
D4	1700	0	372,000	20
D5	1700	0	429,000	17
E0	2000	0	144,000	55
E1	2000	0	201,000	37
E2	2000	0	258,000	27
E3	2000	0	315,000	23
E4	2000	0	372,000	20
E5	2000	0	429,000	17

Appendix B

Process parameters for LP are shown in the following Table A2. In all tests, the line spacing is kept constant with a value of 0.02 mm. The angle increment between the hatchings is defined to sweep a total angle of 360° with the defined number of hatches. Test codes for the LP tests are named with lower case letters.

Table A2. LP process parameters.

Test	Velocity [mm·s^{-1}]	Defocusing [mm]	Number of Hatches	Angle Increment between Hatchings [°]	Pulse Frequency [Hz]
a0	100	4	5	72	125,000
a1	100	4	5	72	150,000
a2	100	4	5	72	175,000
a3	100	4	5	72	200,000
a4	100	4	5	72	225,000
b0	100	4	10	36	125,000
b1	100	4	10	36	150,000
b2	100	4	10	36	175,000
b3	100	4	10	36	200,000
b4	100	4	10	36	225,000
c0	100	4	20	18	125,000
c1	100	4	20	18	150,000
c2	100	4	20	18	175,000
c3	100	4	20	18	200,000
c4	100	4	20	18	225,000
d0	100	5	5	72	125,000
d1	100	5	5	72	150,000
d2	100	5	5	72	175,000
d3	100	5	5	72	200,000
d4	100	5	5	72	225,000
e0	100	5	10	36	125,000
e1	100	5	10	36	150,000
e2	100	5	10	36	175,000
e3	100	5	10	36	200,000
e4	100	5	10	36	225,000
f0	100	5	20	18	125,000

Table A2. *Cont.*

Test	Velocity [mm·s^{-1}]	Defocusing [mm]	Number of Hatches	Angle Increment between Hatchings [°]	Pulse Frequency [Hz]
f1	100	5	20	18	150,000
f2	100	5	20	18	175,000
f3	100	5	20	18	200,000
f4	100	5	20	18	225,000
g0	200	4	5	72	125,000
g1	200	4	5	72	150,000
g2	200	4	5	72	175,000
g3	200	4	5	72	200,000
g4	200	4	5	72	225,000
h0	200	4	10	36	125,000
h1	200	4	10	36	150,000
h2	200	4	10	36	175,000
h3	200	4	10	36	200,000
h4	200	4	10	36	225,000
i0	200	4	20	18	125,000
i1	200	4	20	18	150,000
i2	200	4	20	18	175,000
i3	200	4	20	18	200,000
i4	200	4	20	18	225,000
j0	200	5	5	72	125,000
j1	200	5	5	72	150,000
j2	200	5	5	72	175,000
j3	200	5	5	72	200,000
j4	200	5	5	72	225,000
k0	200	5	10	36	125,000
k1	200	5	10	36	150,000
k2	200	5	10	36	175,000
l3	200	5	10	36	200,000
l4	200	5	10	36	225,000
l0	200	5	20	18	125,000
l1	200	5	20	18	150,000
l2	200	5	20	18	175,000
l3	200	5	20	18	200,000
l4	200	5	20	18	225,000

References

1. Steen, W.M.; Mazumder, J. *Laser Material Processing*, 4th ed.; Springer: London, UK, 2010; ISBN 978-1-84996-062-5.
2. Cortina, M.; Arrizubieta, J.I.; Calleja, A.; Ukar, E.; Alberdi, A. Case study to illustrate the potential of conformal cooling channels for hot stamping dies manufactured using hybrid process of Laser Metal Deposition (LMD) and milling. *Metals* **2018**, *8*, 102. [CrossRef]
3. Klocke, F.; Klink, A.; Veselovac, D.; Aspinwall, D.K.; Soo, S.L.; Schmidt, M.; Schilp, J.; Levy, G.; Kruth, J.P. Turbomachinery component manufacture by application of electrochemical, electro-physical and photonic processes. *CIRP Ann.-Manuf. Technol.* **2014**, *63*, 703–726. [CrossRef]
4. Ashby, M.F. *Materials and the Environment: Eco-Informed Material Choice*; Butterworth-Heinemann: Oxford, UK, 2012; ISBN 9780123859716.
5. Ingarao, G.; Priarone, P.C.; Deng, Y.; Paraskevas, D. Environmental modelling of aluminium based components manufacturing routes: Additive manufacturing versus machining versus forming. *J. Clean. Prod.* **2018**, *176*, 261–275. [CrossRef]
6. Caiazzo, F.; Alfieri, V.; Corrado, G.; Argenio, P.; Barbieri, G.; Acerra, F.; Innaro, V. Laser Beam Welding of a Ti–6Al–4V Support Flange for Buy-to-Fly Reduction. *Metals* **2017**, *7*, 183. [CrossRef]
7. Gasser, A. Chapter 11.6: Laser Metal Deposition. In *Tailored Light 2, Laser Application Technology*; Poprawe, R., Ed.; Springer: Berlin/Heidelberg, Germany, 2011; pp. 216–224, ISBN 978-3-642-01236-5.

8. Dubey, A.K.; Yadava, V. Laser beam machining—A review. *Int. J. Mach. Tools Manuf.* **2018**, *48*, 609–628. [CrossRef]
9. Borse, S.C.; Kadam, M.S. Experimental Study in Micromilling of Inconel 718 by Fiber Laser Machining. *Procedia Manuf.* **2018**, *20*, 213–218. [CrossRef]
10. Yilbas, B.S. Laser Machining Processes. In *Reference Module in Materials Science and Materials Engineering, from Comprehensive Materials Finishing*; Elsevier: New York, NY, USA, 2017; Volume 1, pp. 344–363.
11. Lee, S.W.; Shin, H.S.; Chu, C.N. Fabrication of micro-pin array with high aspect ratio on stainless steel using nanosecond laser beam machining. *Appl. Surf. Sci.* **2013**, *264*, 653–663. [CrossRef]
12. Kononenko, T.V.; Freitag, C.; Sovyk, D.N.; Lukhter, A.B.; Skvortsov, K.V.; Konov, V.I. Influence of pulse repetition rate on percussion drilling of Ti-based alloy by picosecond laser pulses. *Opt. Lasers Eng.* **2018**, *103*, 65–70. [CrossRef]
13. Peng, C.; Fu, Y.; Wei, H.; Li, S.; Wang, X.; Gao, H. Study on improvement of surface roughness and induced residual stress for Additively Manufactured metal parts by Abrasive Flow Machining. *Procedia CIRP* **2018**, *71*, 386–389. [CrossRef]
14. Willenborg, E. Chapter 11.3: Polishing with Laser Radiation. In *Tailored Light 2, Laser Application Technology*; Poprawe, R., Ed.; Springer: Berlin/Heidelberg, Germany, 2011; pp. 196–202, ISBN 978-3-642-01236-5.
15. Alfieri, V.; Argenio, P.; Caiazzo, F.; Sergi, V. Reduction of Surface Roughness by Means of Laser Processing over Additive Manufacturing Metal Parts. *Materials* **2017**, *10*, 30. [CrossRef] [PubMed]
16. Dewey, M.P.; Ulutan, D. Development of Laser Polishing as an Auxiliary Post-Process to Improve Surface Quality in Fused Deposition Modeling Parts. In Proceedings of the ASME 12th International Manufacturing Science and Engineering Conference, Los Angeles, CA, USA, 4–8 June 2017; Volume 2. [CrossRef]
17. Ma, C.P.; Guan, Y.C.; Zhou, W. Laser polishing of additive manufactured Ti alloys. *Opt. Lasers Eng.* **2017**, *93*, 171–177. [CrossRef]
18. Marimuthu, S.; Triantaphyllou, A.; Antar, M.; Wimpenny, D.; Morton, H.; Beard, M. Laser polishing of selective laser melted components. *Int. J. Mach. Tools Manuf.* **2015**, *95*, 97–104. [CrossRef]
19. Zhihao, F.; Libin, L.; Longfei, C.; Yingchun, G. Laser Polishing of additive Manufactured Superalloy. *Procedia CIRP* **2018**, *71*, 150–154. [CrossRef]
20. Flynn, J.M.; Shokrani, A.; Newman, S.T.; Dhokia, V. Hybrid additive and subtractive machine tools-Research and industrial developments. *Int. J. Mach. Tools Manuf.* **2016**, *101*, 79–101. [CrossRef]
21. Cortina, M.; Arrizubieta, J.I.; Ukar, E.; Lamikiz, A. Analysis of the Influence of the Use of Cutting Fluid in Hybrid Processes of Machining and Laser Metal Deposition (LMD). *Coatings* **2018**, *8*, 61. [CrossRef]
22. Arrizubieta, J.I.; Tabernero, I.; Ruiz, J.E.; Lamikiz, A.; Martínez, S.; Ukar, E. Continuous coaxial nozzle design for LMD based on numerical simulation. *Phys. Procedia* **2014**, *56*, 429–438. [CrossRef]
23. TruMark Station 5000. Available online: https://www.trumpf.com/en_INT/products/machines-systems/marking-systems/trumark-station-5000/ (accessed on 11 June 2018).
24. Oerlikon Metco. *MetcoClad 718 Material Product Data Sheet*; Oerlikon Metco: Pfäffikon, Switzerland, 2018. Available online: https://www.oerlikon.com/ecomaXL/files/metco/oerlikon_DSMW-0002.6_NiSuperalloys_LaserCladding.pdf&download=1 (accessed on 11 June 2018).

© 2018 by the authors. Licensee MDPI, Basel, Switzerland. This article is an open access article distributed under the terms and conditions of the Creative Commons Attribution (CC BY) license (http://creativecommons.org/licenses/by/4.0/).

Article

Single-Shot near Edge X-ray Fine Structure (NEXAFS) Spectroscopy Using a Laboratory Laser-Plasma Light Source

Przemysław Wachulak [1,*], Martin Duda [2], Tomasz Fok [1], Andrzej Bartnik [1], Zhanshan Wang [3], Qiushi Huang [3], Antoni Sarzyński [1], Alexandr Jancarek [2] and Henryk Fiedorowicz [1]

1 Institute of Optoelectronics, Military University of Technology, 2 Urbanowicza Str., 00-908 Warsaw, Poland; tfok@wat.edu.pl (T.F.); abartnik@wat.edu.pl (A.B.); asarzynski@wat.edu.pl (A.S.); hfiedorowicz@wat.edu.pl (H.F.)
2 Faculty of Nuclear Sciences and Physical Engineering, Czech Technical University in Prague, Břehová 7, 115 19 Praha 1 Prague, Czech Republic; dudamar1@fjfi.cvut.cz (M.D.); alexandr.jancarek@fjfi.cvut.cz (A.J.)
3 Key Laboratory of Microstructured Materials of MOE, School of Physics Science and Engineering, Tongji University, Shanghai 200092, China; wangzs@tongji.edu.cn (Z.W.); huangqs@tongji.edu.cn (Q.H.)
* Correspondence: wachulak@gmail.com; Tel.: +48-261-839-540

Received: 27 June 2018; Accepted: 25 July 2018; Published: 28 July 2018

Abstract: We present a proof of principle experiment on single-shot near edge soft X-ray fine structure (NEXAFS) spectroscopy with the use of a laboratory laser-plasma light source. The source is based on a plasma created as a result of the interaction of a nanosecond laser pulse with a double stream gas puff target. The laser-plasma source was optimized for efficient soft X-ray (SXR) emission from the krypton/helium target in the wavelength range from 2 nm to 5 nm. This emission was used to acquire simultaneously emission and absorption spectra of soft X-ray light from the source and from the investigated sample using a grazing incidence grating spectrometer. NEXAFS measurements in a transmission mode revealed the spectral features near the carbon K-α absorption edge of thin polyethylene terephthalate (PET) film and L-ascorbic acid in a single-shot. From these features, the composition of the PET sample was successfully obtained. The NEXAFS spectrum of the L-ascorbic acid obtained in a single-shot exposure was also compared to the spectrum obtained a multi-shot exposure and to numerical simulations showing good agreement. In the paper, the detailed information about the source, the spectroscopy system, the absorption spectra measurements and the results of the studies are presented and discussed.

Keywords: NEXAFS; soft X-rays SXR; SXR absorption spectroscopy

1. Introduction

X-ray absorption fine structure (XAFS) spectroscopy allows for the chemical environment of the sample to be investigated by studying X-ray absorption of the sample in the vicinity and above the core level binding energy of the considered atom. A modulation in the X-ray absorption spectrum of an atom carries information about its physical and chemical states. A near-edge X-ray absorption fine structure (NEXAFS) spectroscopy is typically performed in the energy range from a few eV below the absorption edge of the investigated atom to, typically, 20–30 eV above the absorption edge. It is a well-established technique for the characterization of chemical and environmental compounds [1], including organic materials (composed of carbon, oxygen, and nitrogen) that exhibit absorption edges in the soft X-ray (SXR) spectral region, $\lambda = 0.1–10$ nm wavelength. The SXR NEXAFS yields information about elemental composition through the observation of the spectral features in the vicinity of the X-ray absorption edge [2], for studies of the intermolecular bond structure of polymers [3] and saccharides [4], or for obtaining polymer fingerprint of the material [5]. NEXAFS was also used to study the functions of low dimensional nanostructures [6], to investigate liquids [7] and nanomaterials [8], or probing electronic and chemical states of battery materials [7,8].

NEXAFS spectroscopic investigations are usually performed using synchrotron radiation (SR) facilities, which provide radiation of very high brilliance and intensity over a broad wavelength/energy range in the X-ray spectral region. However, for SR NEXAFS spectrum acquisition a wavelength//energy scanning approach is used, which is time-consuming and does not allow for time-resolved studies. A concise overview of this can be found in the article, in which motivation for an easy accessed complimentary NEXAFS technique based on a compact X-ray source was pointed out [9]. Several NEXAFS systems, operating in the SXR range, have been developed using laboratory laser plasma light sources driven with picosecond and nanosecond lasers [10–13]. These systems were used in the studies of various materials in vacuum [11–13], while the NEXAFS system based on the source driven with a picosecond laser was used in the investigations on photo induced phase transitions studies [14]. More recently, high order harmonic generation (HHG) sources were also used for soft X-ray spectroscopy [15] reaching even the "water window" spectral range (λ = 2.3–4.4 nm) [16]. Most of the studies, however, required relatively long, multi-pulse exposure to acquire a single NEXAFS spectrum [12], which may become an obstacle to investigating biological specimen or samples that change with time. An example of this is a single-gas jet laser-plasma SXR source employed for NEXAFS experiments [17], in which a very long exposure time, reaching up to ten thousand pulses [18], was necessary to reach sufficient signal to noise ratio to obtain a single NEXAFS spectrum. Such approach discards the possibility for high throughput measurements. To overcome this limitation, a single pulse (single-shot) has to be used for NEXAFS spectrum acquisition. A single-shot NEXAFS has been demonstrated recently using a laser plasma light source based on a solid target [19]. However, solid targets that are known to produce debris associated with laser ablation products, which is a highly undesirable effect. Moreover, the design of the spectrometer, including two separate off-axis zone plates, may be prone to mechanical, vibration instabilities, errors in the alignment of the sample and reference spectra for two spectra acquired separately and integration errors in the minute curvature of lines of equal energy in the spectra obtained using off-axis zone plates.

In this paper, we demonstrate a single-shot NEXAFS experiment with the use of a laser plasma light source, based on a double stream gas puff target, which injects two gasses into the laser-matter interaction region, to improve the overall photon yield from such produced plasmas [20]. The target was irradiated with modest (a few joules) energies of the laser pulses. In the gas puff target, the inner gas was chosen for a specific elemental emission, while the outer gas that surrounds the inner gas decreases the density gradient of the inner gas in the direction of the nozzle axis. This significantly increases the target density in the interaction region and allows to obtain higher extreme ultraviolet (EUV) and SXR yields at more modest pumping conditions. Moreover, the gaseous target does not have a problem with a debris production.

Thus, in this work, we demonstrate single-shot NEXAFS measurements on the thin organic samples with the laser plasma SXR source employing a double stream gas puff target. The SXR emission from krypton/helium plasma, allowed one to perform NEXAFS with a 1.3 ns exposure time. As a proof of principle, a 1 μm thick polyethylene terephthalate (PET) and L-ascorbic acid samples were used. Optical density spectra of both samples were obtained with a single SXR pulse exposure and composition of the PET sample was evaluated to confirm the applicability of laser plasma source, based on a double stream gas puff target, to NEXAFS measurements, obtaining a useful single-shot signal.

As a result, a NEXAFS system was developed, based on 10 J, 1 ns Nd:YAG laser system. In this approach, a simultaneous acquisition of reference and sample spectra was possible, through a specially designed SXR spectrometer equipped with long entrance slit. Such construction facilitates the much more accurate acquisition of the spectra, which are independent of source energy fluctuations as well as mechanical instabilities of the system. The spectral resolution of this compact system is comparable with early synchrotron-based works. In the following sections, the details about this system will be presented and discussed.

2. Experimental Setup

The experimental setup for the single SXR pulse NEXAFS system using the emission from krypton/helium plasma is depicted in Figure 1 and the photograph of the system is depicted in Figure 2.

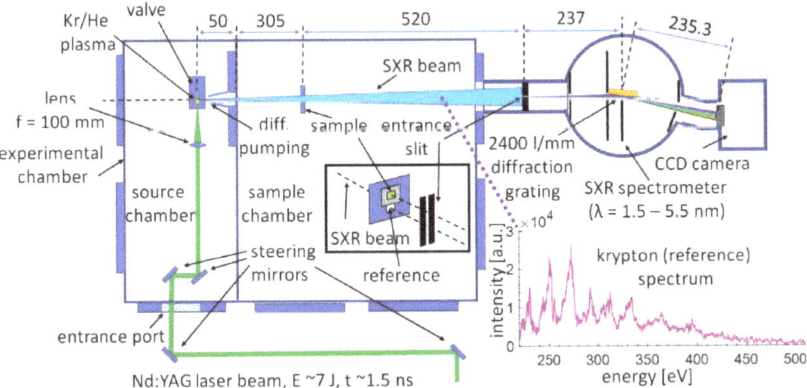

Figure 1. Optical arrangement for single soft X-ray (SXR) pulse experimental near edge soft X-ray fine structure (NEXAFS) system employing laser plasma source based on a double stream gas puff target.

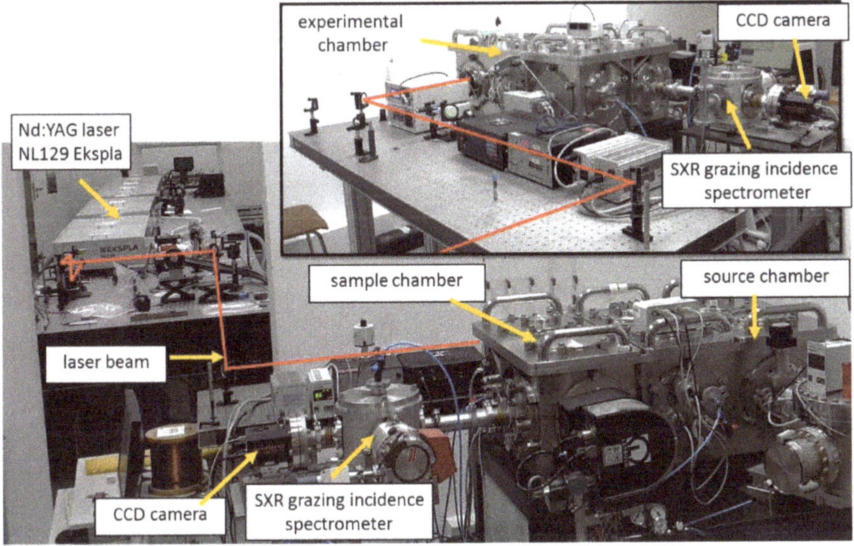

Figure 2. Photograph of the single SXR pulse NEXAFS system.

An Nd:YAG laser beam, emitted from an NL 129 laser system, maximum energy 10 J, (EKSPLA, Vilnius, Lithuania), with laser pulse energy ranging from ~2 J to ~7 J, depending on the measurements, and ~1.5 ns time duration, is focused by an f = 10 cm focal length lens onto a double stream gas puff target. The target is produced by a collinear set of two nozzles, driven independently by two electromagnetic valves.

The diameters of the nozzles are 0.4 mm for the inner nozzle and 0.7–1.5 mm for the outer, ring-shaped nozzle. The inner nozzle was pressurized with krypton gas (working gas) at an optimum backing pressure of 11 bar, while the outer nozzle was connected to helium (outer gas) pressurized to 5 bar. The double stream gas puff target approach was used to increase the gas puff target density. It is done by injecting Kr gas into a hollow stream of He gas to shape the flow of the inner gas into a vacuum through the use of the outer gas. In such case higher inner gas density and SXR yield is obtained comparing to single stream gas puff approach [21]. The valves were driven separately by a dedicated two-channel controller, which is capable of independent adjustment of the delay and opening time for each valve. Moreover, driving signals for both valves are synchronized with the laser oscillator.

Due to the interaction of the laser pulses with the gaseous target, a laser produced plasma is created. Such plasma emits radiation in the broad range of wavelengths, from soft X-rays to infrared, depending on the gas used as a target, laser beam and focusing system parameters. In this experiment, an efficient soft X-ray emission from krypton was achieved (as depicted in the inset of Figure 1, for E = 6.7 J laser pulse with 1.3 ns time duration) and employed for a single SXR pulse NEXAFS spectroscopy. The radiation from krypton plasma enters the second vacuum chamber (sample chamber, see Figure 1), where it illuminates the sample, which is being investigated, placed 355 mm from the plasma. The sample holder is designed in such a way to allow simultaneously for the SXR light to be transmitted through the sample (sample beam), but a portion of the SXR light (reference beam) also enters undisturbed the entrance slit of the spectrometer, see inset in Figure 1, located 875 mm from the plasma. Thus, in this system, in contrary to other compact systems [9] and similar to our recent measurements with the compact NEXAFS system [22,23], a simultaneous acquisition of the two spectra has been achieved. This solution has significant advantages. One of them is that the system remains unaffected by the energy fluctuations of the source, but it is also immune to mechanical instabilities of the system that may occur during a separate acquisition of sample and reference spectra. This, in turn, may lead to unpredicted spectral shifts and difficulty in calculating the optical density of the sample in the vicinity of the absorption edge. For simultaneous acquisition of two spectra the SXR beams enter the spectrometer through the elongated entrance slit, 15 mm in length, which width was 12 μm. The slit was fabricated in a 50 μm thick brass foil by repetitive ablation of the material due to interaction with a focused laser beam with a metal sheet.

The NEXAFS spectra are obtained using a home-made spectrograph with a grazing incidence diffraction grating from Hitachi High Technologies America Inc., Baltimore, MD, USA, having 2400 lines per mm, wavelength range from 1 nm to 5 nm, and a back-illuminated CCD camera (GE 20482048, greateyes GmbH, Berlin, Germany), placed downstream the diffraction grating, in configuration reported in [24]. The camera has a chip with 2052 × 2046 pixels, each 13 × 13 μm^2 in size. During the experiments the chip was cooled down to −40 °C to reduce its internal noise and the background.

For the efficient emission in the SXR region used in the NEXAFS experiment, especially to be able to record high signal to noise ratio reference and sample spectra in a single SXR pulse, the krypton/helium target laser plasma source was properly optimized. The optimization concerned the gas puff target delays in respect to the synchronization pulse from the laser oscillator, arriving 1 ms before the laser pulse, yielded delay times for krypton equal to 500 μs and krypton valve opening time of 800 μs, while the same parameters for helium were 500 μs and 700 μs, respectively. More about synchronization and timing can be found in [22,25]. To avoid reabsorption of SXR radiation generated from laser-plasma, the laser focus was located not in the center of the nozzle but shifted in direction of the optical system by 0.5 mm. Moreover, the working and outer gas pressures were also optimized for maximum SXR photon yield from krypton/helium laser-plasma resulting in optimum pressure values of 11 bar for krypton and helium pressure of 5 bar. If the pressure of the gas is being increased, the density of the target increases as well, and so does the photon yield. After the optimum value of the pressure is reached, further increasing the backing pressure causes the target density to decrease. This is due to the fact that high backing pressure in the valve reservoir prohibits the valve to fully open, because of limited energy stored in the capacitor bank that is discharged through the valve coil while it operates.

The key part of the single SXR pulse NEXAFS system is a high spectral resolution grazing incidence SXR spectrometer. For its proper operation, a precise calibration of the spectrometer is essential because the spectral features near the absorption edge ought to be properly defined and distinguished with energy accuracy of a fraction of an eV. The laser plasma SXR source based on a double stream gas puff target allows one to perform such calibration easily because the change in spectral emission can be obtained by changing the inner, working gas. For that purpose, initially, three different gasses were used: argon, oxygen, and nitrogen. Those gasses have in their emission spectra single, well visible, and easy to recognize isolated emission lines [26]

Such lines are shown in Figure 3a for those three gasses and their wavelength range spans from λ = 2.1602 nm line from O^{6+} ion in O_2–based plasma, through two well-defined and most intense SXR nitrogen lines: λ = 2.489 nm and λ = 2.878 nm from N^{5+} ions, till λ = 4.873 and λ = 4.918 nm lines from Ar^{8+} ions. In this experiment, however, instead of three different gasses for spectrometer calibration we prepared and used an $Ar:N_2:O_2$ (1:1:1 by volume) gas mixture for easier, faster and more precise calibration of the spectrometer, since in one spectrum multiple transitions (spectral emission lines) from different gasses are present and could be used for the calibration process. Those lines were used to obtain the calibration curve for the spectrometer, Equation (1). To do that a parabolic function was fitted to the data with R-squared fitting equal to $0.(9)_6648$, ($1 - R^2 = 3.52{\cdot}10^{-7}$), resulting in maximum wavelength error for 2.878 nm N^{5+} line equal to 0.017% and minimum error of −0.031% for 2.489 nm line from N^{5+} ions.

$$y(x) = 6.36{\cdot}10^{-7}x^2 + 1.27{\cdot}10^{-3}x + 0.43 \quad (1)$$

where y is the wavelength in nm, while the x value defines the pixel index in the CCD camera image, horizontal axes in Figure 3a,b.

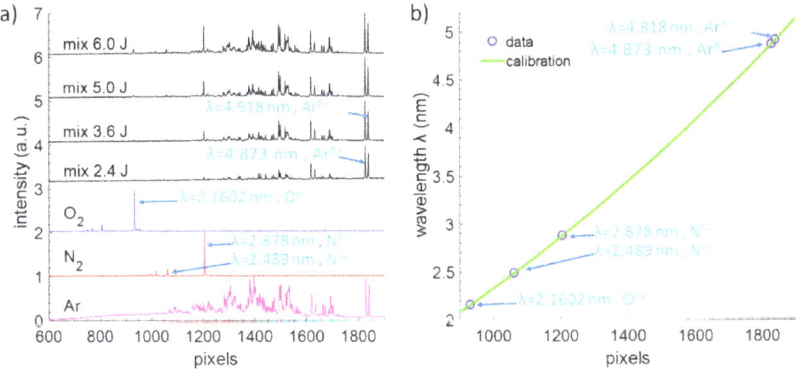

Figure 3. SXR spectrometer calibration. SXR spectra (**a**) of the pure gasses (three bottom plots) and a 1:1:1 (by volume) of Ar, N_2, and O_2 gasses for different energies of the laser pulses. A mixture (mix) is more convenient to perform calibration and allows one to see all lines of interest, later used to obtain a spectrometer calibration curve, depicted in Figure (**b**).

The resolving power of the SXR spectrometer, equipped with 12 μm entrance slit, was estimated in the vicinity of the carbon absorption edge by measuring an isolated line at 4.409 nm wavelength ($2s^22p^3$-$2s^22p^2(3P)3d$ transition from S^{9+} ions from SF_6 gas, the photon energy of 281 eV). It was observed that the line has FWHM width of 0.3 eV, thus $E/\Delta E$ was estimated to be ~ 940.

3. Experimental Results for a Single SXR Pulse near Edge X-ray Absorption Spectroscopy of PET and L-Ascorbic Acid

To demonstrate the performance of the experimental system for a single SXR pulse NEXAFS spectroscopy a proof of principle experiment for obtaining NEXAFS spectra from 1 μm thick PET foil, $(C_{10}H_8O_4)_n$, (Lebow, USA) was performed. This experiment was already performed using the compact NEXAFS system, however, for single NEXAFS spectrum acquisition typically 100 SXR pulses were required [22]. In this experiment, the foil was partially covering the elongated aperture in the sample holder to allow two beams (sample and reference beams, indicated in the inset in Figure 1) to enter the spectrometer slit. Thus, the sample $S_{sam}(E)$ and reference spectra $S_{ref}(E)$ were acquired simultaneously with one SXR pulse from krypton/helium plasma. Typical optical density *OD* NEXAFS spectra for PET foil obtained with this system are depicted in Figure 4, for various laser energies ranging from 6.3 J to 6.7 J, and were obtained with Equation (2):

$$OD(E) = -\ln\left[\frac{S_{sam}(E)}{S_{ref}(E)}\right]. \quad (2)$$

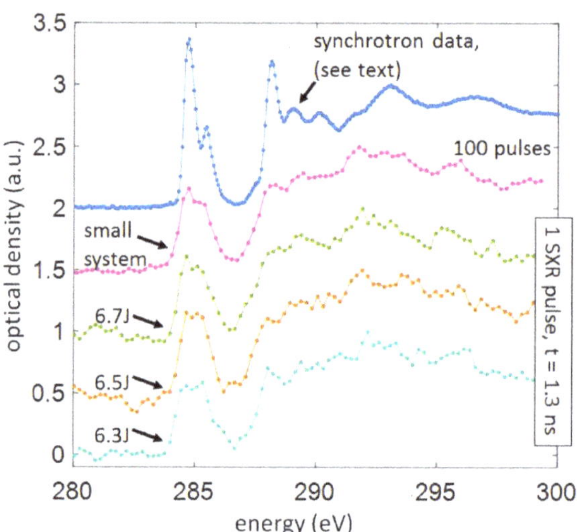

Figure 4. Polyethylene terephthalate (PET) NEXAFS spectra—optical density as a function of SXR radiation energy. The figure depicts a comparison between NEXAFS spectrum obtained using a synchrotron radiation [3], 100 SXR pulses from the compact system [22] with NEXAFS spectra of the same material and thickness, obtained with a single SXR pulse, generated by a laser interaction with a double stream gas puff target. The spectra are presented for three different laser pulse energies ranging from 6.3 J to 6.7 J and 1.3 ns laser pulse duration (FWHM).

The reference spectrum, which is the emission from Kr/He plasma, spans typically from 2.2 nm up to a detection limit of our grazing incidence spectrometer (GIS), which is 5 nm, corresponding to the energy of 225 eV to 560 eV. The sample spectrum shows clearly the carbon K-α absorption edge in the vicinity of 4.3 nm wavelength, above which the absorption of carbon is low. This results in a very well distinguishable and characteristic features in the NEXAFS spectrum, depicted in Figure 4.

The NEXAFS spectrum is typically slightly smoothed out using the Golay-Savitzky algorithm, following [9], or by averaging the separately acquired spectra. A NEXAFS spectrum of PET, presented in Figure 4, was obtained by vertical integration of 101 spectral lines using a single data set and single laser pulse, 1.3 ns in duration, without any filtering procedure or multiple spectra accumulations.

Similarly to compact system data (100 laser pulses with energies of 650 mJ and pulse duration 3 ns), reported in [22] (second-from-top plot in Figure 4), the most prominent feature of the spectrum is a $\pi^*_{C=C}$ bond from the aromatic ring in the PET structure at an energy of 284.4 eV and 285.1 eV. The other peaks in the spectrum were also identified and are listed in Table 1. The peaks were assigned based on the synchrotron data for poly(ethylene terephthalate) [27].

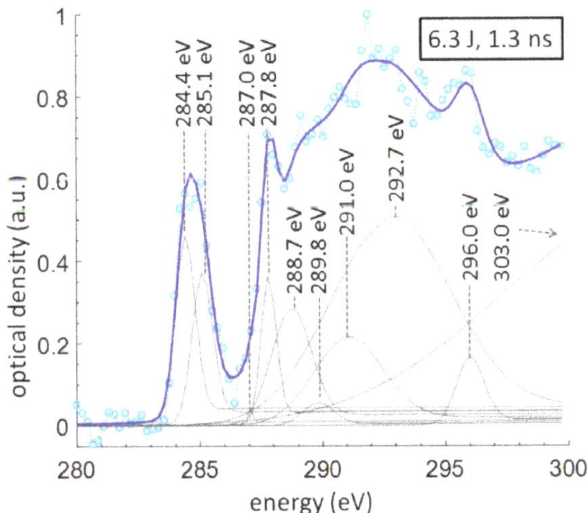

Figure 5. A NEXAFS spectrum of a 1 μm thick PET foil in the vicinity of carbon edge, obtained with one SXR pulse. Measured data points are indicated with circles, thick solid line depicts the fitting. Each contribution to the fitting is depicted by a thin solid line. A combined approach [9] was used, in which for each peak a separate 10% step function was employed. From the peaks composition of the PET sample was subsequently calculated.

Table 1. Energy positions and assignments of features depicted in Figure 5, taken from [27], in the C-1s NEXAFS spectrum of PET foil, based on synchrotron data.

Peak Energy (eV)	Main Character	1s → (Orbital)	Area (%)	Analysis (Renormalized) (%)
284.4	ring	$\pi^*_{C=C}$	10.9	29.5
285.1	ring	$\pi^*_{C=C}$	8.7	23.6
287.0	carbohydrates	σ^*_{C-H}	1.4	3.8
287.8	carbonyl (C=O)	$\pi^*_{C=O}$	6.5	17.6
288.7	ring/C=O	$\pi^*_{C=C}$, $\pi^*_{C=O}$	8.1	22.0
289.8	ring	$\pi^*_{C=C}$	1.3	3.5
291.0	—— questionable ——		8.7	-
292.7		σ^*	31.1	
296.0		σ^*	2.6	-
303.0		σ^*	20.7	-

The spectral resolution of the NEXAFS data is sufficient to successfully fit two spectral contributions in a $\pi^*_{C=C}$ bond from the aromatic ring in the PET structure at photon energies of 284.4 eV and 285.1 eV, in accordance with synchrotron data [27] and similarly to synchrotron measurements [3] (top plot in Figure 4). Based on the resolving power of the spectrometer the spectral resolution of this single-shot NEXAFS system, ~0.3 eV, is comparable with early synchrotron-based works, however, more recent data [3] report better spectral resolutions of ~0.1 eV.

For chemical composition analysis the peaks, corresponding to certain bonds in the molecular structure, as well as a step function, describing the profile of the absorption edge, are fitted, similarly to other works [9,19] and to our previous measurements with multiple SXR pulses and compact NEXAFS system [22]. The result of the fitting is depicted in Figure 5. For the step function, *arctan* was used, while for the peaks a pure Gaussian function was utilized [1]. To perform the peak fitting to the experimental NEXAFS spectral data a dedicated MATLAB software (R2018a, MathWorks, Natick, MA, USA) based on a nonlinear programming solver that searches for the minimum of an unconstrained multivariable function using the derivative-free method, was used. In the peak fitting algorithm, the only assumed parameters were the peak/step positions (energies) and the width of the step functions. All other parameters, such as all widths and amplitudes of the peak contributions (Gaussian functions) and heights of the step functions (based on the amplitudes of peak functions), were fitted automatically.

During the data processing, we found that adding 10% *arctan* local step function to each spectral contribution described by a Gaussian peak function is sufficient to perform accurate fitting to the spectral data. This method, presented in [9], was used since it is not trivial to determine the exact position of a single, *arctan* global step function [1] for the measured data. The height of each step was set to 10%, because for the PET the ratio of absorption lengths around the carbon edge is equivalent to ~10%, according to the CXRO data [28].

3.1. Chemical Composition Analysis from a Single Pulse SXR PET NEXAFS Spectrum

For the elemental composition analysis, it was assumed that the area under each Gaussian peak curve assigned to certain resonance is proportional to the frequency of occurrence of this binding form of the studied element [29,30]. This allows one to approximate the composition of the bound elements. The spectral components, listed in Table 1, were fitted to the spectrum (thin gray curves under the dotted spectrum data points). The sum of all fitted curves is depicted as a blue line and it matches well the experimental data points. The energy positions and bond assignments were based on synchrotron data [27]. The spectral components listed in Table 1 are not the only ones that contribute, however, those are the most probable transitions, according to the literature, and were used for assignment of the peaks.

For composition analysis, the sum of areas under each fitted spectral component was normalized to obtain a probability of occurrence of a particular bond. Not all spectral components from Table 1 were used for this normalization. The contribution of the σ^* bonds has been accounted for from the π^* orbitals, thus, components above 292 eV were left out of the estimation. Moreover, due to a disagreement in the assignment of the spectral component at 291 eV in the synchrotron data (Table 1 in [27], Okajima et al.) for PET, this contribution was also not accounted for. In different sources [27,31], the other spectral components are defined precisely with an energy accuracy of ~0.3–0.4 eV, while this particular spectral feature is assigned as carbonyl only in the data in [27] (Okajima et al.) while in the other reference it does not exist. Thus, since the existence and assignment of this component are not well established, we have decided to omit that contribution in the composition analysis.

The results of composition evaluation for PET, compared to theoretical values, are presented in Table 2. For more information how the assignment was performed, based on the probability of occurring bonds please see [9,22]. By a comparison to the theoretical value, calculated as a percent by weight, *w/w* %, of the composition, the error of composition analysis was evaluated using a root-mean-square deviation approach defined by Equation (3):

$$\delta = \frac{1}{N}\sqrt{\sum_{i=1}^{N}(C_{Ti} - C_{Mi})^2} \tag{3}$$

where N is the number of elements, considered in the composition analysis ($N = 3$) and i defines the index for each element {C, H, O} = {1, 2, 3}. C_{Ti} is a theoretical and C_{Mi} is a measured percentage value for each element in the molecular structure.

From the error analysis, the global error was found to be ~2% for the composition analysis based on the experimental data. By value, the lowest error was found for hydrogen, equal to 0.4%, while the highest for carbon, equal to 5.1%.

Table 2. The experimental elemental composition of the analyzed PET sample compared to the theoretical composition values. A 10% step function for each Gaussian type peak fitting curve was used as a fitting method.

Sample PET Foil 1 m Thick $(C_{10}H_8O_4)_n$	Method	Comment	Composition (%)			Global Error (%)
			C	H	O	
Theoretical	*w*/*w*%	calculated from the chemical formula	62.5	4.2	33.3	0
Experiment	10% step for each peak	NEXAFS spectrum from Figure 5	67.6	3.8	28.6	2.3

3.2. The NEXAFS Spectrum of L-Ascorbic Acid and Comparison to Numerical Simulations

Moreover, to validate the experimental NEXAFS spectra an organic sample of L-ascorbic acid, $C_6H_8O_6$, was used. The L-ascorbic acid (99.9% purity, ~10 mg) in a powder form (Stanlab Sp.J., Lublin, Poland) was dissolved in distilled water (10 mL) using an ultrasonic cleaner. A 3 μL drop was then placed on top of 75 nm thick Si_3N_4 membrane acting as a support and dried in a nitrogen atmosphere for 5 min. The NEXAFS spectrum of L-ascorbic acid was obtained with a single SXR pulse, created by a 7.6 J and 1.6 ns in duration laser pulse. Such single-shot spectrum was compared with the one obtained with the compact NEXAFS system, in which NEXAFS spectrum using multiple SXR pulses was acquired. Moreover, both spectra were also compared directly with numerical simulations based on *fdmnes* software [32]. The comparison can be seen in Figure 6, where the correspondence between spectral features in both spectra is visible. The relative values of the optical density are chosen to separate the plots vertically.

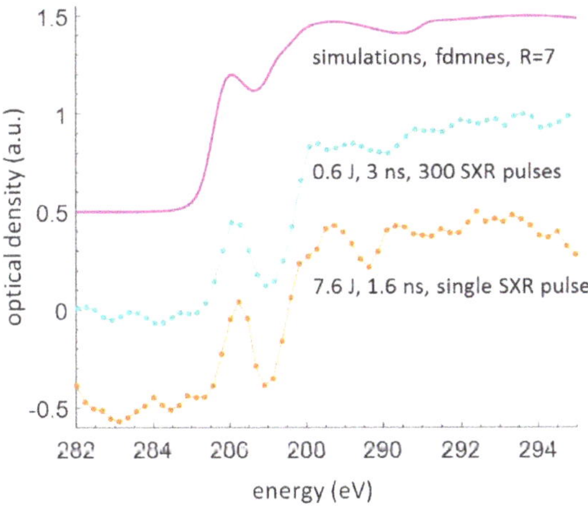

Figure 6. The relative values of the optical density of the L-ascorbic acid sample obtained near the carbon K-α absorption edge for multiple SXR pulses (compact system reported in [22]) and a single SXR pulse, compared to *fdmnes* numerical simulations (top plot).

The *fdmnes* simulations were performed with the region of interest of R = 7, and convolution parameter of 0.3 eV, based on lattice parameters and positions of atoms obtained from [33], with lattice parameters of a = 6.390(1) Å, b = 6.262(1) Å, c = 17.127(4) Å, $\alpha = \gamma = 90°$, and $\beta = 99.36°$. The multiple SXR pulse NEXAFS spectrum was obtained using 300 SXR pulses emitted from the Kr/He plasma generated by the interaction of 0.6 J energy and 3 ns duration Nd:YAG laser pulses (laser beam diameter of ~7 mm), focused by an $f = 25$ mm lens. To obtain such spectrum 300 SXR pulses were necessary. A single SXR pulse NEXAFS spectrum was obtained using radiation from Kr/He plasma as well, however, the plasma was formed by the interaction of 7.6 J energy and 1.6 ns duration laser pulse (laser beam diameter of 25 mm and 100 mm focus lens) with a double stream Kr/He gas puff target. The focusing numerical apertures were similar, NA = 0.138 for the smaller system, comparing to NA = 0.124 for higher energy system. Thus, the beam diameters were also comparable. Even though the peak power for the latter case was 4.75 GW, comparing to 200 MW for the smaller laser, roughly 24 times smaller, the much more efficient Kr plasma emission allowed to obtain a NEXAFS spectrum for the same material of similar quality with just a single SXR pulse.

4. Conclusions

In conclusion, the proof of principle experiment, showing the applicability of laser plasma source based on a double stream gas puff target to a single-shot NEXAFS, was demonstrated. The single-shot capability is important from the point of view of possible time-resolved studies. In this case, the time resolution of such studies is comparable to the duration of the single SXR pulse, 1.3 ns for PET and 1.6 ns for the L-ascorbic acid sample. This work demonstrates, that the laser-plasma source based on a double stream gas puff target can be employed for NEXAFS measurements in the laboratory environment, obtaining a useful single-shot signal. The table-top system based on this source was developed, and its application to study a polymer sample, showing its applicability to the near edge X-ray absorption fine structure spectroscopy, was demonstrated. The obtained PET spectrum is comparable in quality to the spectrum obtained with another compact desk-top system, requiring, however, 300 SXR pulses acquisition. The spectrum exhibits the expected peaks at the characteristic positions, which in turn allows performing the composition analysis, which results match the theoretical values of C, H, and O composition. Moreover, the NEXAFS spectrum of a real sample of L-ascorbic acid on top of the Si_3N_4 membrane also compares very well to multiple-shot data and to numerical simulations.

The system allows one to obtain the NEXAFS spectrum from the simultaneous acquisition of two spectra (sample and reference) with the exposure time of ~1 ns. Simultaneous acquisition of both sample and reference spectra makes possible for much more accurate data acquisition, independent of source energy fluctuations and mechanical instabilities of the system. For higher photon throughput, no thin film filters were employed. The unwanted spectral contributions were removed by the geometry of GIS and additional beam-stops inside the spectrometer housing.

The gas puff target approach allows one to change the working gas to illuminate the sample with different emission spectra, currently, the Kr gas was used with an energy range of 250–500 eV. However, other gasses can be used as well, to facilitate higher emission near other absorption edges of interest of different materials.

Also, a novelty was to use a specially prepared gas mixture for easier (quicker) and more precise calibration of the spectrometer, since in one spectrum multiple transitions (spectral emission lines) from different gasses, present in the mixture, were visible and could be accounted for during the calibration process.

Such a single-shot NEXAFS system could provide the possibility to perform test experiments on environmental, biological, and material science samples, to obtain preliminary data on novel materials and samples, without the immediate need to get beam time on the large-scale facility. Also, it may allow performing research on materials that may be too fragile or have other constraints and limitations that preclude measurements at a synchrotron source. This system may also be used for developing novel approaches to data processing with samples, which are later studied in more detail at large scale facilities. It also may, in the near future, allow for time-resolved studies and a broader spread of the NEXAFS spectroscopy to environmental, biological, and material sciences, which in turn, might benefit the development of these areas of science and technology.

Author Contributions: Conceptualization, P.W.; Methodology, P.W.; Software, P.W., and M.D.; Formal Analysis, M.D.; Investigation, P.W., M.D., and T.F.; Resources, A.B., Z.W., Q.H., A.S. A.J., and H.F.; Writing-Original Draft Preparation, P.W., and T.F.; Writing-Review & Editing, P.W.; Funding acquisition, P.W., H.F., and A.J.

Funding: This research was funded by (National Science Centre (NCN)) grant numbers (UMO-2015/17/B/ST7/03718) and (UMO-2015/19/B/ST3/00435); by (the Education, Audiovisual and Culture Executive Agency (EACEA) Erasmus Mundus Joint Doctorate Programme (EXTATIC)) grant number (2012-0033); by (European Union's Horizon 2020 research and innovation program Laserlab-Europe IV) grant number (654148); by (the Ministry of Education, Youth and Sports CR) grant number (LG15013).

Acknowledgments: We acknowledge Krystyna Jabłońska and Anna Wolska, from Institute of Physics, Polish Academy of Sciences, Warsaw, Poland, for their constructive comments during our initial experiments and related to NEXAFS spectroscopy with laser-plasma sources of SXR radiation. We would like to thank the referees for their useful comments.

Conflicts of Interest: The authors declare no conflict of interest. The funders had no role in the design of the study; in the collection, analyses, or interpretation of data; in the writing of the manuscript, and in the decision to publish the results.

References

1. Stöhr, J. *NEXAFS Spectroscopy*; Springer: Berlin, Germany, 1992; ISBN 978-3-662-02853-7.
2. Carravetta, V.; Plashkevych, O.; Ågren, H. A theoretical study of the near-edge X-ray absorption spectra of some larger amino acids. *J. Chem. Phys.* **1998**, *109*, 1456–1464. [CrossRef]
3. Dhez, O.; Ade, H.; Urquhart, S.G. Calibrated NEXAFS spectra of some common polymers. *J. Electron. Spectrosc. Relat. Phenom.* **2008**, *128*, 85–96. [CrossRef]
4. Gainar, A.; Stevens, J.S.; Jaye, C.; Fischer, D.A.; Schroeder, S.L.M. NEXAFS Sensitivity to Bond Lengths in Complex Molecular Materials: A Study of Crystalline Saccharides. *J. Phys. Chem. B* **2015**, *119*, 14373–14381. [CrossRef] [PubMed]
5. Urquhart, S.G.; Ade, H. Trends in the carbonyl core (C 1S, O 1S) → π^*c=o transition in the near-edge X-ray absorption fine structure spectra of organic molecules. *J. Phys. Chem. B* **2002**, *106*, 8531–8538. [CrossRef]
6. Guttmann, P.; Bittencourt, C. Overview of nanoscale NEXAFS performed with soft X-ray microscopes. *Beilstein J. Nanotechnol.* **2015**, *6*, 595–604. [CrossRef] [PubMed]
7. Velasco-Vélez, J.J.; Jones, T.E.; Pfeifer, V.; Dong, C.-L.; Chen, Y.-X.; Chen, C.-M.; Chen, H.-Y.; Lu, Y.-R.; Chen, J.-M.; Schlögl, R.; et al. Trends in reactivity of electrodeposited 3d transition metals on gold revealed by operando soft X-ray absorption spectroscopy during water splitting. *J. Phys. D Appl. Phys.* **2017**, *50*, 024002. [CrossRef]
8. Pfeifer, V.; Jones, T.E.; Velasco-Vélez, J.J.; Arrigo, R.; Piccinin, S.; Hävecker, M.; Knop-Gericke, A.; Schlögl, R. In situ observation of reactive oxygen species forming on oxygen-evolving iridium surfaces. *Chem. Sci.* **2017**, *8*, 2143–2149. [CrossRef] [PubMed]
9. Sedlmair, J.; Gleber, S.-C.; Peth, C.; Mann, K.; Niemeyer, J.; Thieme, J. Characterization of refractory organic substances by NEXAFS using a compact X-ray source. *J. Soils Sediments* **2012**, *12*, 24–34. [CrossRef]
10. Mallozzi, P.J.; Schwerzel, R.E.; Epstein, H.M.; Campbell, B.E. Laser-EXAFS: Fast Extended X-ray Absorption Fine Structure Spectroscopy with a Single Pulse of Laser-Produced X-rays. *Science* **1979**, *206*, 353–355. [CrossRef] [PubMed]
11. Vogt, U.; Wilhein, T.; Stiel, H.; Legall, H. High resolution X-ray absorption spectroscopy using a laser plasma radiation source. *Rev. Sci. Instrum.* **2004**, *75*, 4606–4609. [CrossRef]
12. Peth, C.; Barkusky, F.; Mann, K. Near-edge X-ray absorption fine structure measurements using a laboratory-scale XUV source. *J. Phys. D Appl. Phys.* **2008**, *41*, 105202. [CrossRef]
13. Mantouvalou, I.; Witte, K.; Grötzsch, D.; Neitzel, M.; Günther, S.; Baumann, J.; Jung, R.; Stiel, H.; Kanngießer, B.; Sandner, W. High average power, highly brilliant laser-produced plasma source for soft X-ray spectroscopy. *Rev. Sci. Instr.* **2015**, *86*, 035116. [CrossRef] [PubMed]
14. Grossmann, P.; Rajkovic, I.; Moré, R.; Norpoth, J.; Techert, S.; Jooss, C.; Mann, K. Time-resolved near-edge X-ray absorption fine structure spectroscopy on photoinduced phase transitions using a tabletop soft-X-ray spectrometer. *Rev. Sci. Instrum.* **2012**, *83*, 053110. [CrossRef] [PubMed]
15. Cousin, S.L.; Silva, F.; Teichmann, S.; Hemmer, M.; Buades, B.; Biegert, J. High-flux table-top soft X-ray source driven by sub-2-cycle, CEP stable, 1.85-μm 1-kHz pulses for carbon K-edge spectroscopy. *Opt. Lett.* **2014**, *39*, 5383–5386. [CrossRef] [PubMed]

16. Cousin, S.L.; Silva, F.; Teichmann, S.; Hemmer, M.; Biegert, J. Molecular fine structure from water-window coherent soft-X-rays. *Opt. Photonics News* **2014**, *12*, 58.
17. Barkusky, F.; Bayer, A.; Döring, S.; Flöter, B.; Grossmann, P.; Peth, C.; Reese, M.; Mann, K. Applications of compact laser-driven EUV/XUV plasma sources. *Proc. SPIE* **2009**, *7361*, 736112. [CrossRef]
18. Kühl, F.-C.; Müller, M.; Schellhorn, M.; Mann, K.; Wieneke, S.; Eusterhues, K. Near-edge X-ray absorption fine structure spectroscopy at atmospheric pressure with a table-top laser-induced soft X-ray source. *J. Vac. Sci. Technol. A Vac. Surf. Films* **2016**, *34*, 041302. [CrossRef]
19. Mantouvalou, I.; Witte, K.; Martyanov, W.; Jonas, A.; Grötzsch, D.; Streeck, C.; Löchel, H.; Rudolph, I.; Erko, A.; Stiel, H.; Kanngießer, B. Single shot near edge X-ray absorption fine structure spectroscopy in the laboratory. *Appl. Phys. Lett.* **2016**, *108*, 201106. [CrossRef]
20. Fiedorowicz, H.; Bartnik, A.; Jarocki, R.; Rakowski, R.; Szczurek, M. Enhanced X-ray emission in the 1-keV range from a laser-irradiated gas puff target produced using the double-nozzle setup. *Appl. Phys. B* **2000**, *70*, 305–308. [CrossRef]
21. Fiedorowicz, H.; Bartnik, A. X-ray laser emission from a laser-irradiated gas puff target. *Bull. Pol. Acad. Tech.* **2005**, *53*, 103–111. [CrossRef]
22. Wachulak, P.; Duda, M.; Bartnik, A.; Sarzyński, A.; Węgrzyński, L.; Nowak, M.; Jancarek, A.; Fiedorowicz, H. Compact system for near edge X-ray fine structure (NEXAFS) spectroscopy using a laser-plasma light source. *Opt. Express* **2018**, *26*, 8260–8274. [CrossRef] [PubMed]
23. Wachulak, P.; Duda, M.; Bartnik, A.; Sarzyński, A.; Węgrzyński, Ł.; Fiedorowicz, H. 2-D elemental mapping of an EUV-irradiated PET with a compact NEXAFS spectromicroscopy. *Spectrochim. Acta B* **2018**, *145*, 107–114. [CrossRef]
24. Nakano, N.; Kuroda, H.; Kita, T.; Harada, T. Development of a flat-field grazing-incidence XUV spectrometer and its application in picosecond XUV spectroscopy. *Appl. Opt.* **1984**, *23*, 2386–2392. [CrossRef] [PubMed]
25. Wachulak, P.W.; Bartnik, A.; Fiedorowicz, H.; Feigl, T.; Jarocki, R.; Kostecki, J.; Rakowski, R.; Rudawski, P.; Sawicka, M.; Szczurek, M.; et al. A compact, quasi-monochromatic laser-plasma EUV source based on a double-stream gas-puff target at 13.8 nm wavelength. *Appl. Phys. B* **2010**, *100*, 461–469. [CrossRef]
26. Kelly, R.L. Atomic and Ionic Spectrum Lines below 2000 Angstroms: Hydrogen through Krypton. *J. Phys. Chem.* **1987**, *16* (Suppl. 1), 1–649.
27. Okajima, T.; Teramoto, K.; Mitsumoto, R.; Oji, H.; Yamamoto, Y.; Mori, I.; Ishii, H.; Ouchi, Y.; Seki, K. Polarized NEXAFS spectroscopic studies of poly(butylene terephthalate), poly(ethylene terephthalate), and their model compounds. *J. Phys. Chem. A* **1998**, *102*, 7093–7099. [CrossRef]
28. Henke, B.L.; Gullikson, E.M.; Davis, J.C. X-ray interactions: Photoabsorption, scattering, transmission, and reflection at E = 50-30, 000 eV, Z = 1-92. *At. Data Nucl. Data Tables* **1993**, *54*, 181–342. [CrossRef]
29. di Stasio, S.; Braun, A. Comparative NEXAFS study on soot obtained from an ethylene/air flame, a diesel engine, and graphite. *Energy Fuels* **2006**, *20*, 187–194. [CrossRef]
30. Xia, K.; Weesner, F.; Bleam, W.F.; Bloom, P.R.; Skyllberg, U.L.; Helmke, P.A. XANES studies of oxidation states of sulfur in aquatic and soil humic substances. *Soil Sci. Soc. Am. J.* **1998**, *62*, 1240–1246. [CrossRef]
31. Lippitz, A.; Friedrich, J.F.; Unger, W.E.S.; Schertel, A.; Wöll, Ch. Surface analysis of partially crystalline and amorphous poly(ethylene terephthalate) samples by X-ray absorption spectroscopy (NEXAFS). *Polymer* **1996**, *37*, 3151–3155. [CrossRef]
32. Bunău, O.; Joly, Y. Self-consistent aspects of X-ray absorption calculations. *J. Phys. Condens. Matter* **2009**, *21*, 345501. [CrossRef] [PubMed]
33. Milanesio, M.; Bianchi, R.; Ugliengo, P.; Roetti, C.; Viterbo, D. Vitamin C at 120 K: Experimental and theoretical study of the charge density. *J. Mol. Struct. (Theochem.)* **1997**, *419*, 139–154. [CrossRef]

© 2018 by the authors. Licensee MDPI, Basel, Switzerland. This article is an open access article distributed under the terms and conditions of the Creative Commons Attribution (CC BY) license (http://creativecommons.org/licenses/by/4.0/).

Article

Experimental Study of Macro and Microgeometric Defects in Drilled Carbon Fiber Reinforced Plastics by Laser Beam Machining

Pedro F. Mayuet Ares *, Juan Manuel Vázquez Martínez, Mariano Marcos Bárcena and Antonio J. Gámez

Department of Mechanical Engineering & Industrial Design, Faculty of Engineering, University of Cadiz, Av. Universidad de Cádiz 10, E-11519 Puerto Real (Cadiz), Spain; juanmanuel.vazquez@uca.es (J.M.V.M.); mariano.marcos@uca.es (M.M.B.); antoniojuan.gamez@uca.es (A.J.G.)

* Correspondence: pedro.mayuet@uca.es, Tel.. +34-616-852-858

Received: 1 July 2018; Accepted: 13 August 2018; Published: 18 August 2018

Abstract: Plastic matrix composite materials are an excellent choice for structural applications where high strength-weight and stiffness-weight ratios are required. These materials are being increasingly used in diverse industrial sectors, particularly in aerospace. Due to the strict tolerances required, they are usually machined with drilling cycles due to the type of mounting through rivets. In this sense, laser beam drilling is presented as an alternative to conventional drilling due to the absence of tool wear, cutting forces, or vibrations during the cutting process. However, the process carries with it other problems that compromise the integrity of the material. One of these is caused by the high temperatures generated during the interaction between the laser and the material. In this work, variance analysis is used to study the influence of scanning speed and frequency on macro geometric parameters, surface quality, and defects (taper and heat affected zone). Also, in order to identify problems in the wall of the drill, stereoscopic optical microscopy (SOM) and scanning electron microscopy (SEM) techniques are used. This experimental procedure reveals the conditions that minimize deviations, defects, and damage in machining holes.

Keywords: laser beam machining; carbon fiber reinforced polymer; hole quality characteristics; geometrical parameters; roughness parameters; taper; heat affected zone

---

1. Introduction

Reinforced plastic and polymers were originally designed for improving the physicochemical properties and reducing the weight of the classic materials, mainly metallic alloys, commonly used in different industrial sectors. Particularly among them, carbon fiber reinforced polymers (CFRP) are widely applied in sectors that require lightweight materials, the ability to withstand great efforts, high stiffness, and excellent conditions for fatigue [1]. These characteristics, especially those related to their low density, make CFRP materials highly attractive for industrial applications in a great diversity of sectors, such as the aerospace or automotive sectors. However, changes in global economics and politics have led to a production standstill of CFRP, even a decrease in certain sectors, due to large cuts, mainly on the defense sector, although this effect has been diminished by the increase of consumption for commercial applications. Therefore, some producers expect an increase in demand of about 75% in the industrial sector (excluding the aerospace) by 2020, specifically in energy, transport, and civil engineering [1].

In the aeronautical industry, the use of non-metal matrix composites (NMMC) has increased sharply in the last few years as they are becoming more and more used in the building of structural elements of commercial and military airships. In these cases, organic matrix-based composite materials reinforced with high-strength fiber or particles are used [2,3]. Among them, CFRP is one of the most widely used. These composite materials are characterized by their high compositional heterogeneity and anisotropy, low thermal conductivity, matrix sensitivity at high temperatures, and the abrasive nature of the fibers used in their manufacture [4–6]. These features complicate the manufacturing processes of the airship elements based in CFRPs.

Moreover, most of the CFRP elements used in the airship's building need to be machined by drilling or contour milling. However, the aforementioned CFRP features negatively affect the machinability of these elements when traditional or conventional machining processes are used. The material heterogeneity and the abrasive behavior of carbon fiber conditions negatively affect the tool life [7–12]. For this reason, the material and geometry of the tools must be adapted to the machining conditions required by the material in order to reduce the tool wear and the thermal and mechanical defects produced during the cutting process to maintain the functional performance of the manufacturing process [13]. Moreover, tool life is one of the most relevant parameters that condition the economic performance of the process. However, currently, the optimization of the cutting tools used in the machining processes of CFRP work pieces is considered a difficult task [14]. In this sense, different non-traditional machining processes can be thought to be profitable and economical alternatives to solve the troubles previously mentioned. Therefore, alternatives to this type of machining are becoming more common, such as abrasive water jet (AWJM), electrical discharge machining (EDM), ultrasonic machining, or laser beam machining (LBM) [12,15–19].

In particular, LBM possesses several advantages. For example, it is a technology without contact and without abrasive particles, therefore eliminating contact forces, tool wear, and vibrations during cutting. In addition, laser cutting can be easily automated and performed at high speed [20–23]. On the other hand, the challenges for the laser to machining the carbon fiber are the minimization or elimination of the thermal damage due to the great difference between the mechanical and thermal properties of the material constituents and to maintain a high cutting speed [24–29]. Therefore, defects, such as the heat affected zone (HAZ), the carbonization of the fiber, the removal or loss of resin, or the delamination due to intense thermal effects, are the main obstacles for industrial applications of carbon fiber machining by laser, as seen in Figure 1.

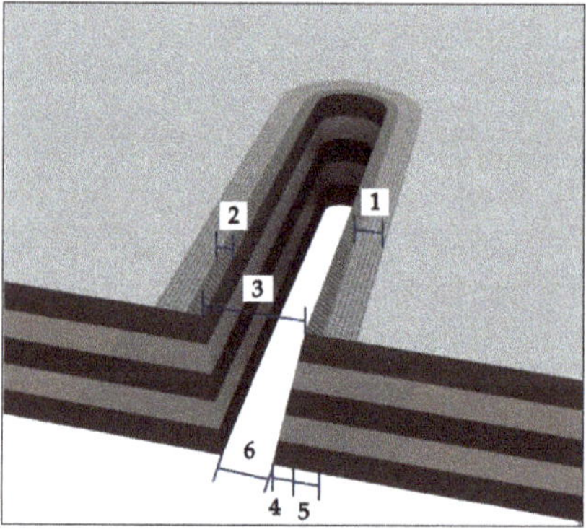

Figure 1. Typical defects on the specified surface zones of the material by laser in polymer matrix composites. At the beam entrance, 1. heat affected matrix, 2. matrix recession, and 3. kerf width. At the exit: 4. matrix recession, 5. heat affected matrix, and 6. kerf width.

In this paper, carbon fiber reinforced polymer (CFRP) is machined by a fiber laser doped with ytterbium. In order to analyze the quality of the cut, the influence of the cutting parameters has been analyzed. Specifically, a number of tests varying the scanning speed (S) and frequency (Fr) have been performed to obtain holes. The test pieces have been inspected by combining stereoscopic optical microscopy (SOM) and scanning electron microscopy (SEM) techniques. Furthermore, macro and micro-geometric parameters, such as diameter (D), cylindricity (C), straightness (St), roundness (R-In/R-Out), and roughness (Ra), have been studied. Finally, the quality of the holes has been evaluated by measuring defects such as the heat affected zone (HAZ) or the taper angle (T).

2. Materials and Methods

2.1. Experimental Procedure

Table 1 includes the main features of the CFRP pieces (80 mm × 80 mm × 5.85 mm) used in this study. Holes of 7.92 mm diameter, by combining circular trajectories using a fiber laser doped with ytterbium (Yb), have been laser drilled. This diameter is commonly used for assembling structural elements in aeronautical industry.

Table 1. CFRP pieces features.

Type of Material	Composition	Production Method	Technical Specification
Layers of unidirectional carbon fiber with epoxy resin matrix and a symmetrical stacking sequence of (0/90/45/-45/45/-45)	Intermediate module fiber (66%) and epoxy resin (34%)	Pre-preg and autoclaved at 458° ± 5° at a pressure of 0.69 MPa	AIMS-05-01-XXX

The influence of pulsed frequency, F, and scanning speed, S, on the thermal caused defects has been analyzed [2,21,22,30]. Thus, Table 2 includes the selected values in this study. The rest of parameters have remained constant during the tests, Table 3. Before them, trials with each set of parameters were carried out to determine the thickness of the material in each pass.

Table 2. Cutting parameters selected for the tests.

Parameter	Levels			
S (mm/s)	25	50	75	100
Fr (kHz)	20	40	60	

Table 3. Constant parameters during the tests.

Power (W)	Wavelength (nm)	Spot Diameter (μm)	Working Mode	Scanning Distribution	Atmosphere
10	1062	60	Pulsed	Shaded	Environment

For the realization of the drills, a strategy of marking by shading was chosen. To program it, Laser Mark software (version 2.3.0) was used. In addition, it is important to mention that each consecutive pass was performed by modifying the scanning direction by 45°. In this way, the heat damage distribution is homogenized among all the fiber directions observed in the stacking sequence.

To distinguish the most significant parameters for evaluation results, analysis of variance (ANOVA) for a 95% confidence interval was employed. Thus, the F value and the p value have been analyzed to measure the evidence against the null hypothesis. After that, contour charts for each variable studied in the experimental (D, D_{In}, D_{Out}, St, C, Ra, T and HAZ) were obtained.

2.2. Macro and Micro-Geometrical Evaluation

The diameter at the inlet and outlet of the material were evaluated using a digital profile projector Tesa Visio 300 (Hexagon, Stockholm, Sweden).

To analyze the macro and micro geometric deviations, replicas of the holes were fabricated due to the impossibility of direct measuring on the material. These replicas were made with a polymer type F80 Ra (R.G.X, Plastiform, Madrid, Spain) with the ability to guarantee stability during the measurement process for diameters greater than 4 mm. In this way, roundness was calculated at the entrance and exit of the drill (R-In/R-Out), along with straightness (St), cylindricity (C), and roughness (Ra).

A station of measurement Mahr MMQ44 Form Tester (Mahr, Göttingen, Germany) was used to measure the roundness at the entrance and exit of the drill, the cylindricity of the entire profile of the drill, and the straightness in four separate generatrices to 90°, as shown in Figure 2a. On the other hand, the evaluation of the micro-geometric defects was carried out using a Mahr Perthometer Concept PGK120 (Mahr, Göttingen, Germany) equipment to obtain four 2-D separated profiles, as seen in Figure 2b. For that, a stylus with a 2 µm tip radius and 90° tip angle was used for the measurements, where the surface finishing parameter employed to indicate the surface quality was the arithmetic mean roughness.

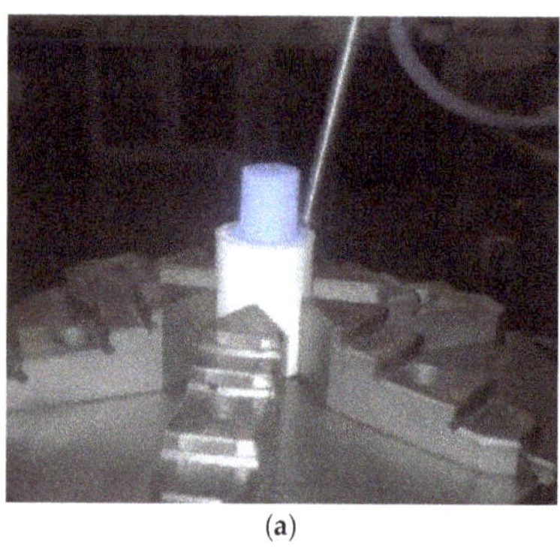

(**a**)

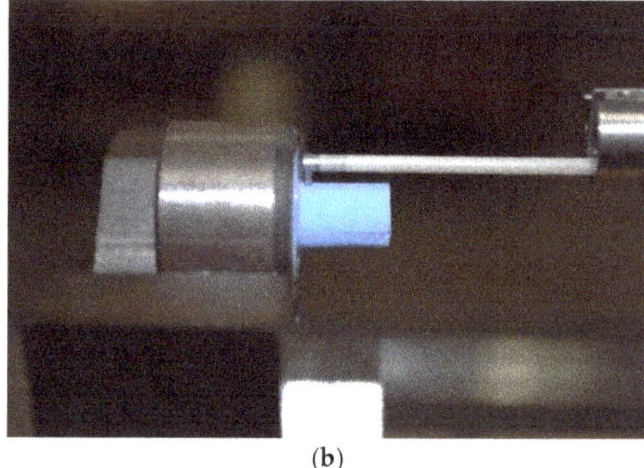

(**b**)

Figure 2. Measurements made on the replicas, (**a**) example of measurement of macro geometry; and (**b**) example of measurement of micro-geometry.

2.3. SOM/SEM Evaluation

The test pieces were inspected by microscopy techniques using Tesa Visio 300 (Hexagon, Stockholm, Sweden). Subsequently, to calculate the taper (T) with the inlet diameter (D_{In}), the outlet diameter (D_{Out}), and the thickness of the material (t), ImageJ software (1.50i) was used. Also, to obtain a thermal damage ratio on the surface of the material, the damaged diameter (D_d) was also calculated.

$$\theta = \tan^{-1}\left(\frac{D_{In} + D_{out}}{2t}\right) \quad (1)$$

$$HAZ = \frac{D_d}{D_{In}} \quad (2)$$

Finally, to complete the inspection of the defects, scanning electron microscopy Quanta 200 (SEM, Thermo Fisher Scientific, Waltham, MA, USA) was used.

3. Results and Discussion

3.1. Results of the Analysis

Tables 4 and 5 show the data obtained for each test and the ANOVA analysis, respectively.

Table 4. Results of the evaluated variables.

Test	S (mm/s)	Fr (kHz)	D (mm)	D_{In} (mm)	D_{Out} (mm)	St (mm)	C (mm)	Ra (μm)	T (rad)	HAZ
1	25	20	7.934	0.041	0.057	0.028	0.388	3.15	0.069	1.233
2	25	40	7.929	0.035	0.077	0.036	0.524	2.20	0.099	1.314
3	25	60	7.896	0.045	0.084	0.056	0.622	2.38	0.118	1.355
4	50	20	8.103	0.046	0.092	0.029	0.430	2.84	0.074	1.119
5	50	40	7.878	0.043	0.121	0.039	0.556	2.39	0.100	1.160
6	50	60	8.081	0.045	0.105	0.045	0.722	2.80	0.118	1.194
7	75	20	8.066	0.046	0.138	0.018	0.429	3.78	0.070	1.031
8	75	40	7.996	0.055	0.142	0.042	0.583	2.50	0.097	1.096
9	75	60	8.024	0.070	0.144	0.057	0.719	2.76	0.152	1.097
10	100	20	8.048	0.047	0.123	0.031	0.418	2.68	0.070	1.000 [1]
11	100	40	8.127	0.062	0.138	0.037	0.616	4.37	0.094	1.049
12	100	60	8.032	0.079	0.192	0.049	0.729	6.22	0.122	1.049

[1] Without a heat affected zone.

Table 5. ANOVA analysis of the evaluated variables.

S			Fr		
Parameter	**F-value**	***p*-value**	**Parameter**	**F-value**	***p*-value**
D	2.19	0.190	D	0.55	0.602
R-In	4.99	0.045	R-In	3.60	0.094
R-Out	12.80	0.005	R-Out	2.79	0.139
St	0.08	0.970	St	18.02	0.003
C	5.70	0.034	C	128.86	0.000
Ra	0.83	0.229	Ra	0.41	0.686
T	0.83	0.524	T	32.32	0.001
HAZ	143.19	0.000	HAZ	22.61	0.002

3.2. Analysis of Macro and Micro-Geometrical Deviations

3.2.1. Diameter Evaluation

The nominal diameter of the drilled hole was 7.92 mm. The results obtained from the diameter measurements are shown in Figure 3. In this contour chart, the closest theoretical value was obtained for low scanning speed and medium frequency level conditions. Under these conditions, 7.929 mm was the best result, obtained in test number 2. In contrast, measurement values over 8 mm in diameter were detected for S = 100 mm/s. The analysis of the results shows that a low scanning speed involves more accuracy for diameter dimensions of the drilled holes. On the other hand, the ANOVA analysis was not conclusive for rejecting the null hypothesis. However, a higher F-value was shown by S, suggesting a greater relation between this parameter with respect frequency.

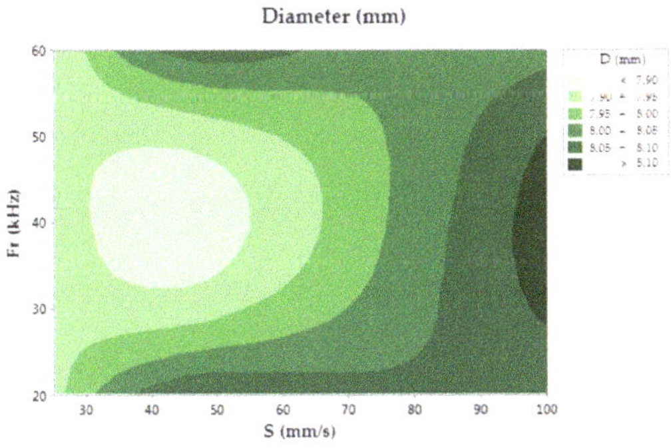

Figure 3. Contour chart of the diameter.

3.2.2. Roundness Evaluation

In the case of roundness deviations, the ANOVA analysis showed a higher influence of S, allowing for the rejection of the null hypothesis for the values obtained in the measurement of the drilled hole. Additionally, a tendency to increase the roundness deviations could be observed for high values of the studied process parameters (Fr and S), as seen in Figure 4.

Generally speaking, R-In increased Fr and S, although the minimum was found for low sweep speeds and medium frequency values. R-Out showed a similar trend, but the lowest roundness deviations were measured for the lowest frequency and scanning speed.

Also, relevant differences could be detected between the entrance and exit sections of the holes, presenting an important growth in the deviation values for the exit section with respect to the entrance, as seen in Figure 5. This behavior was specially related to the increase in the number of laser scanning stages and the focal length variation between stages.

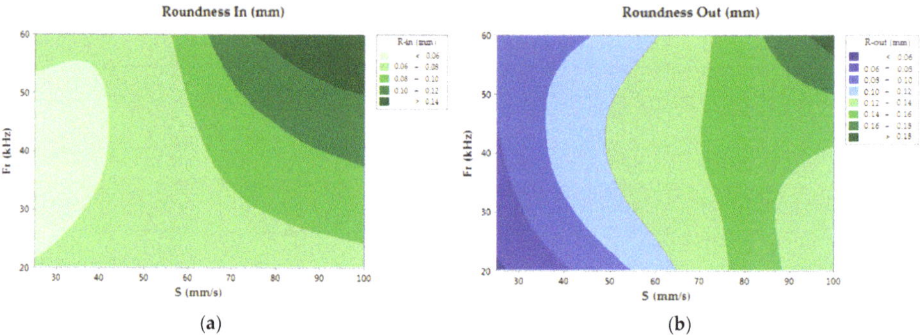

Figure 4. (**a**) Contour chart of the inlet roundness; (**b**) Contour chart of the outlet roundness.

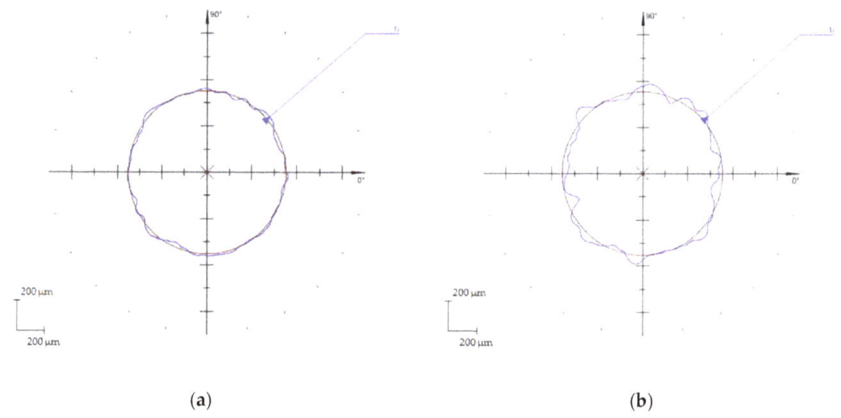

Figure 5. Roundness profile, (**a**) Hole inlet, Test 12. (S = 100 mm/s and Fr = 60 kHz); (**b**) Hole Outlet, Test 12. (S = 100 mm/s and Fr = 60 kHz).

3.2.3. Straightness Evaluation

In the case of straightness, the null hypothesis could be rejected using the frequency parameter. Regarding the scanning speed, the straightness behavior did not indicate a significant tendency, showing similar behavior for all the S values of the range studied. In contrast, an increasing trend have been followed in the straightness deviations as a function of frequency, as can be seen in Figure 6a. Under these considerations, a maximum value of 0.057 mm is reported for the Fr = 60 kHz tests. The evaluation of the measured straightness behavior allowed for the confirmation of a direct dependency with the energy of the pulse, improving the hole features for higher frequency conditions.

3.2.4. Cylindricity Evaluation

Cylindricity deviations were affected by the same considerations as the straightness parameter, showing a relevant increasing trend as a function of frequency, and obtaining a maximum value of 0.729 mm. In this aspect, a slighter influence from the scanning speed is reported, as can be seen in Figure 6b. This behavior may confirm that the Fr parameter could be used reject the null hypothesis. On the other hand, the cylindricity allowed for the calculation of a first approximation of the taper angle, being directly related as shown in Table 4. Thus, Figure 7 shows the cylindricity profiles in tests with opposite parameters showing the taper of the drills.

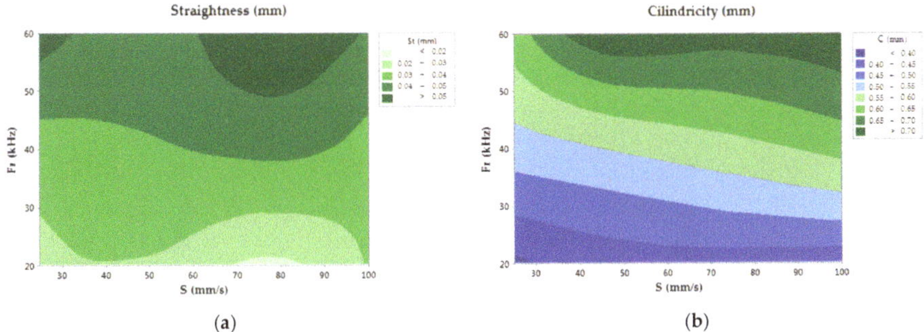

Figure 6. (**a**) Contour chart of the straightness; (**b**) Contour chart of the cylindricity.

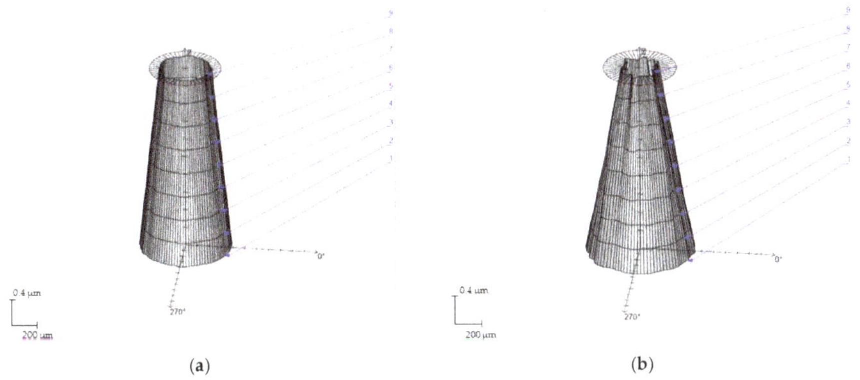

Figure 7. Cylindricity profile, (**a**) Hole inlet, Test 1. (S = 20 mm/s and Fr = 20 kHz); (**b**) Hole outlet, Test 12. (S = 100 mm/s and Fr = 60 kHz).

3.2.5. Surface Finish Evaluation

The average roughness parameter Ra was used to assess the surface finish. It was measured by constructing replicas of the drilled holes. In this evaluation, the dependency between the studied parameters could not be verified, where the F value of S shows a higher value. However, the highest Ra values were found for the highest scanning speeds and frequency conditions [29], as seen in Figure 8. Values of Ra between 2.20 µm and 6.22 µm are reported.

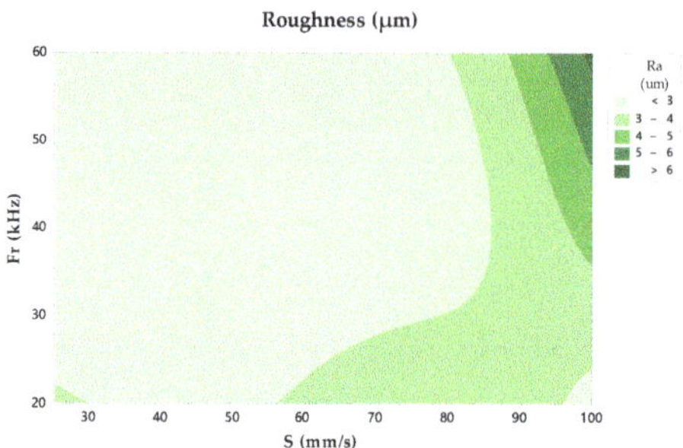

Figure 8. Contour chart of the roughness.

3.3. *Damage and Defects Analysis on the Drilled Holes*

Table 5 shows that the p-value may vary as a function of the analyzed response variable (T, HAZ). First, for taper growth, experimental data confirmed the null hypothesis for the Fr parameter, showing a direct influence in the defect formation process. On the other hand, the HAZ allowed for the rejection of the null hypothesis for the different process parameters studied.

3.3.1. Taper Angle

Results obtained from the ANOVA analysis revealed the relevance of Fr in the taper formation, as seen in Figure 9a. In this way, taper formation became critical for higher values of Fr, resulting in a maximum value of 0.152 in test number 9. This behavior may have caused an increase in the conicity of the drill hole as a function of the pulse energy of the laser beam. However, a direct dependency between S and taper defect was not detected, where similar values are reported for all of the studied speed range. Under these conditions, Fr may be considered the most influential parameter in the process of taper development [30–32], according to Figure 10.

3.3.2. Heat Affected Zone

Based on the evaluation of the HAZ, defects caused by heat damages allowed for the rejection of the null hypothesis. However, its relevant to remark that the F-value showed a higher influence of S than Fr. In Figure 10b, the taper decreased when the scanning speed increased. This was due to the fact that, with high S, the laser beam affected the composite surface for a shorter time, leading to a decrease in the HAZ as expected [33]. Additionally, in Figure 11, the defect caused by heat is shown in detail, disappearing when higher speeds were used. In addition, it can be seen how the defect appeared only at the border of the drill, which favors the extension of the zone due to the conductivity of the reinforcement in the longitudinal direction of the fiber.

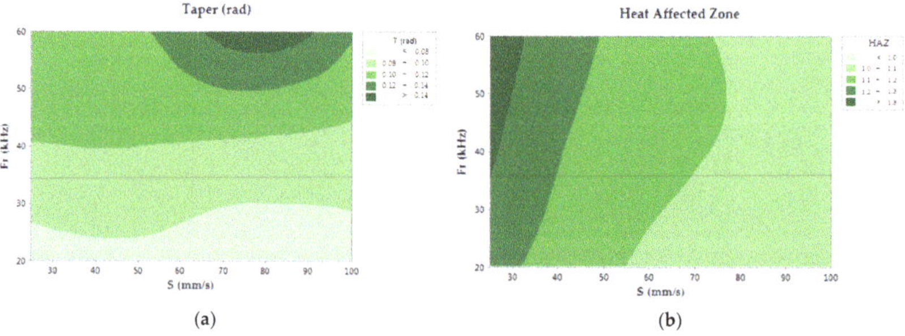

Figure 9. (**a**) Contour chart of the taper angle; (**b**) Contour chart of the heat affected zone.

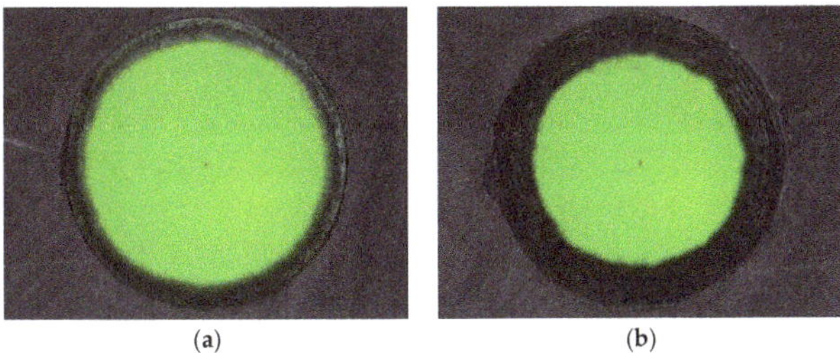

Figure 10. Effect of the frequency, (**a**) Test 7. (S = 75 mm/s and Fr = 20 kHz); (**b**) Test 9. (S = 75 mm/s and Fr = 60 kHz).

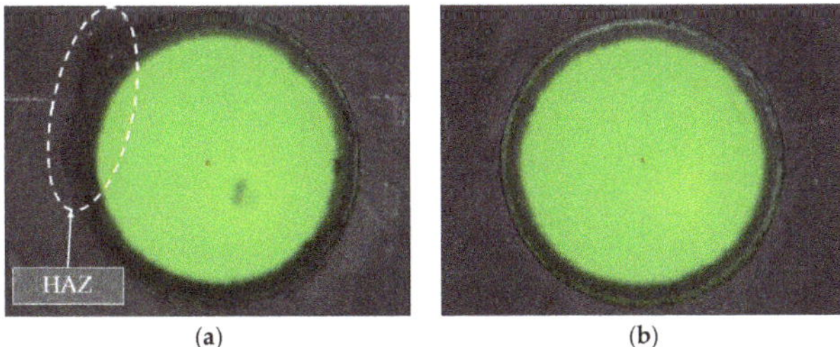

Figure 11. Effect of scanning speed, (**a**) Absence of reinforcement and matrix at the entrance of the drill, Test 1. (S = 20 mm/s and Fr = 20 kHz); (**b**) Hole without defects, Test 10. (S = 100 mm/s and Fr = 20 kHz).

3.4. SEM Analysis

The heat affected zone is considered the most important defect in the machining of CFRP. SEM analysis allows for more detailed observation distinguishing some typical defects produced in the material [34–36].

In Figure 12, it is possible to appreciate what has been described in the experimental procedure related to the number of stages necessary to eliminate material-producing defects similar to delamination [33]. In addition, charring at the drill wall has been observed, basically caused by the high temperatures reached during cutting, as seen in Figure 13a. On the other hand, it has also been possible to identify areas of material with matrix recession. This occurs when the matrix and fibers are removed at different rates owing to their different thermo-physical properties [27,37,38], as seen in Figure 13b. In addition, molten matrix deposited on the wall of the drill was also detected.

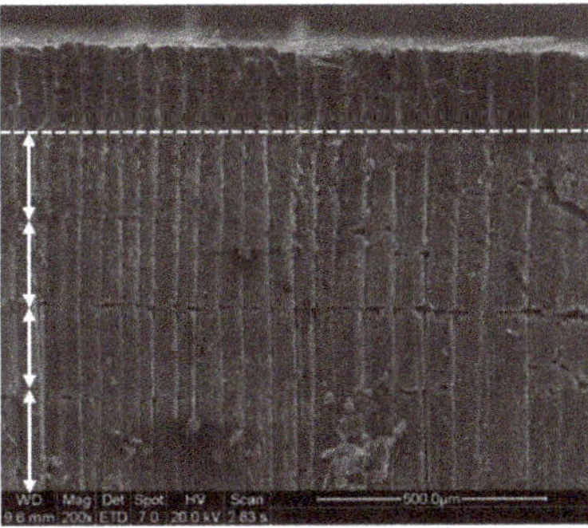

Figure 12. Clearance distance per pass in test 1 (S = 20 mm/s and Fr = 20 kHz).

(a) (b)

Figure 13. (**a**) Burned fiber debris deposited on the drill wall (charring); (**b**) Absence of resin due to the evaporation process. Detail of molten matrix adhered to the drill wall.

4. Conclusions

An experimental study was conducted to determine the influence of process parameters on hole geometry and kerf wall characteristics in laser beam drilling machining of CFRP composites. Based on this, the following conclusions can be made:

- The roundness measurements obtained at the inlet and outlet of the drill reveal that smaller dimensional deviations were obtained when selecting low scanning speeds and pulse frequencies. In addition, it was found that the deviation of the roundness at the exit is always greater (increasing in some tests up to 250%), mainly due to the influence of the focal distance.
- Pulsed frequency and scanning speed also affected straightness and cylindricity in a great depth, especially for the latter. The straightness deviation decreased with frequency and the deviation of cylindricity presented higher values relating to the taper defect, especially when the pulsed frequency was increased.
- The roughness of the drilled holes did not seem to have a significant relationship with the parameters evaluated. However, reduced scanning speeds had better surface quality values. Specifically, for S = 20 mm/s, measurements of about 2 μm were obtained. Roughness values were considered low compared to those obtained through other machining processes.
- The taper angle was closely related to the frequency and affected by the speed of scanning, where an increase in the energy of the pulse decreased the appearance of the defect. The minimum value of the angle was obtained in trial 1 with 0.069 rad. However, this drill presented high damage caused by temperature.
- This study showed the direct influence of the parameters proposed in the experiment. However, the scanning speed determined the appearance of defects in the surface of the material. In this study, no damage was recorded when selecting high speeds and low pulse frequencies (S = 100 mm/s and Fr = 20 kHz).
- Similar to the evaluation of the surface finish, the influence of the cutting parameters in the evaluation of the diameter could not be clearly stated with this test. However, the selection of reduced scanning speeds seemed to offer a higher dimensional accuracy of the hole.
- SEM analysis detected characteristic defects associated with the laser machining of composites on the hole's surface. Thus, charring localization, absence of matrix, and deposition of matrix remains on it was recorded. These defects occurred due to the high temperatures produced during the process.

Author Contributions: Conceptualization, P.F.M.A. and M.M.B.; Methodology, P.F.M.A. and J.M.V.M.; Software, P.F.M.A. and J.M.V.M.; Validation, A.J.G.; Formal Analysis, P.F.M.A. and J.M.V.M.; Investigation, P.F.M.A.; Resources, M.M.B.; Data Curation, P.F.M.A. and J.M.V.M.; Writing-Original Draft Preparation, P.F.M.A.; Writing-Review & Editing, P.F.M.A. and A.J.G.; Visualization, P.F.M.A. and J.M.V.M.; Supervision, A.J.G.; Project Administration, M.M.B.; Funding Acquisition, M.M.B.

Funding: This work has received financial support from the Spanish Government (Project DPI2015-71448-RProject DPI2011-29019) and from the Andalusian Government (PAIDI).

Acknowledgments: P.F.M.A. and J.M.V.M. thank the laboratory of corrosion and protection (LABCYP) at the University of Cadiz for allowing access to its facilities, where the tests of this work have been carried out.

Conflicts of Interest: The authors declare no conflict of interest.

References

1. Soo-Jin, P. *Carbon Fibers*, 1st ed.; Springer: New York, NY, USA, 2014.
2. Antar, M.; Chantzis, D.; Marimuthu, S.; Hayward, P. High speed EDM and laser drilling of aerospace alloys. *Procedia CIRP* **2016**, *42*, 526–531. [CrossRef]
3. Liu, D.F.; Tang, Y.J.; Cong, W.L. A review of mechanical drilling for composite laminates. *Compos. Struct.* **2012**, *94*, 1265–1279. [CrossRef]
4. Madhavan, V.; Lipczynski, G.; Lane, B.; Whitenton, E. Fiber orientation angle effects in machining of unidirectional CFRP laminated composites. *J. Manuf. Process.* **2015**, *20*, 431–442. [CrossRef]
5. Li, M.J.; Soo, S.L.; Aspinwall, D.K.; Pearson, D.; Leahy, W. Influence of lay-up configuration and feed rate on surface integrity when drilling carbon fibre reinforced plastic (CFRP). *Procedia CIRP* **2014**, *13*, 399–404. [CrossRef]
6. Feito, N.; Diáz-Álvarez, J.; Cantero, J.L.; Miguélez, M.H. Influence of special tool geometry in drilling woven CFRPs materials. *Procedia Eng.* **2015**, *132*, 632–638. [CrossRef]
7. Tsao, C.C. Investigation into the effects of drilling parameters on delamination by various step-core drills. *J. Mater. Process. Technol.* **2008**, *206*, 405–411. [CrossRef]
8. Teti, R. Machining of composite materials. *CIRP Ann. Manuf. Technol.* **2002**, *51*, 611–634. [CrossRef]
9. Abrate, S.; Walton, D.A. Machining of composite materials. Part I: Traditional methods. *Compos. Manuf.* **1992**, *3*, 75–83. [CrossRef]
10. König, W.; Wulf, C.; Grass, P.; Willerscheid, H. Machining of fibre reinforced plastics. *Ann. CIRP* **1985**, *34*, 537–548. [CrossRef]
11. Lachaud, F.; Piquet, R.; Collombet, F.; Surcin, L. Drilling of composite structures. *J. Compos. Struct.* **2001**, *52*, 511–526. [CrossRef]
12. Abrate, S.; Walton, D.A. Machining of composite materials. Part II: Non-traditional methods. *Compos. Manuf.* **1992**, *3*, 85–94. [CrossRef]
13. Krishnaraj, V.; Zitoune, R.; Davim, J.P. *Drilling of Polymer-Matrix Composites*, 1st ed.; Springer: Heidelberg, Germany, 2013.
14. Abrão, A.M.; Campos-Rubio, J.C.; Faria, P.E.; Davim, J.P. The effect of cutting tool geometry on thrust force and delamination when drilling glass fibre reinforced plastic composite. *Mater. Des.* **2008**, *29*, 508–513. [CrossRef]
15. Wang, J. A machinability study of polymer matrix composites using abrasive waterjet cutting technology. *J. Mater. Process. Technol.* **1999**, *94*, 30–35. [CrossRef]
16. Rashed, C.A.; Romoli, L.; Tantussi, F.; Fuso, F.; Burgener, M.; Cusanelli, G. Water jet guided laser as an alternative to EDM for micro-drilling of fuel injector nozzles: A comparison of machined surfaces. *J. Manuf. Process.* **2013**, *15*, 524–532. [CrossRef]
17. Phadnis, V.A.; Roy, A.; Silberschmidt, V.V. A finite element model of ultrasonically assisted drilling in carbon/epoxy composites. *Procedia CIRP* **2013**, *8*, 141–146. [CrossRef]
18. Hejjaji, A.; Singh, D.; Kubber, S.; Kalyanasundaram, D.; Gururaja, S. Machining damage in FRPs: Laser versus conventional drilling. *Compos. Part A Appl. Sci. Manuf.* **2016**, *82*, 42–52. [CrossRef]
19. Barile, C.; Casavola, C.; Pappalettere, C. The influence of stitching and unconventional fibres orientation on the tensile properties of CFRP laminates. *Compos. Part B* **2017**, *110*, 248–254. [CrossRef]
20. Dubey, A.K.; Yadava, V. Laser beam machining—A review. *Int. J. Mach. Tools Manuf.* **2008**, *48*, 609–628. [CrossRef]

21. Mishra, S.; Yadava, V. Laser beam micromachining (LBMM)–A review. *Opt. Lasers Eng.* **2015**, *73*, 89–122. [CrossRef]
22. Wang, H.J.; Lin, H.; Wang, C.Y.; Zheng, L.J.; Hu, X.Y. Laser drilling of structural ceramics—A review. *J. Eur. Ceram. Soc.* **2017**, *37*, 1157–1173. [CrossRef]
23. Arrizubieta, I.; Lamikiz, A.; Martínez, S.; Ukar, E.; Tabernero, I.; Girot, F. Internal characterization and hole formation mechanism in the laser percussion drilling process. *Int. J. Mach. Tools Manuf.* **2013**, *75*, 55–62. [CrossRef]
24. Cena, A.A.; Mathew, P. Analysis and prediction of laser cutting parameters of fibre reinforced plastics (FRP) composite materials. *Int. J. Mach. Tools Manuf.* **2002**, *42*, 105–113. [CrossRef]
25. Cenna, A.A.; Mathew, P. Evaluation of cut quality of fibre-reinforced plastics—A review. *Int. J. Mach. Tools Manuf.* **1997**, *37*, 723–736. [CrossRef]
26. Goeke, A.; Emmelmann, C. Influence of laser cutting parameters on CFRP part quality. *Phys. Procedia.* **2010**, *5*, 253–258. [CrossRef]
27. Saini, S.K.; Dubey, A.K.; Upadhyay, B.N.; Choubey, A. Study of hole characteristics in laser trepan drilling of ZTA. *Opt. Laser Technol.* **2018**, *103*, 330–339. [CrossRef]
28. Tagliaferri, F.; Leopardi, G.; Semmler, U.; Kuhl, M.; Palumbo, B. Study of the influences of laser parameters on laser assisted machining processes. *Procedia CIRP* **2013**, *8*, 170–175. [CrossRef]
29. Choudhury, I.A.; Chuan, P.C. Experimental evaluation of laser cut quality of glass fibre reinforced plastic composite. *Opt. Lasers Eng.* **2013**, *51*, 1125–1132. [CrossRef]
30. Wonsuk, C.; Hoon-Young, K.; Jin-Woo, J.; Won-Seok, C.; Sun-Hak, C. Vibration-assisted femtosecond Laser drilling with controllable taper angles for AMOLED fine metal mask fabrication. *Materials* **2017**, *10*, 212.
31. Caprino, G.; Tagliaferri, V. Maximum cutting speed in laser cutting of fiber reinforced plastics. *Int. J. Mach. Tools Manuf.* **1988**, *28*, 389–398. [CrossRef]
32. Klotzbach, A.; Hauser, M.; Beyer, E. Laser cutting of carbon fiber reinforced polymers using highly brilliant laser beam sources. *Phys. Procedia* **2011**, *12*, 572–577. [CrossRef]
33. Leone, C.; Genna, S.; Tagliaferri, V. Fibre laser cutting of CFRP thin sheets by multi-passes scan technique. *Opt. Lasers Eng.* **2014**, *53*, 43–50. [CrossRef]
34. Matthams, T.J.; Clyne, T.W. Mechanical properties of long-fibre thermoplastic composites with laser drilled microperforations: Effect of perforations in consolidated material. *Compos. Sci. Technol.* **1999**, *59*, 1169–1180. [CrossRef]
35. Sola, D.; Grima, L. Laser machining and in vitro assessment of wollastonite-tricalcium phosphate eutectic glasses and glass-ceramics. *Materials* **2018**, *11*, 125. [CrossRef] [PubMed]
36. Pan, C.T.; Hocheng, H. The anisotropic heat-affected zone in the laser grooving of fiber-reinforced composite material. *J. Mater. Process. Technol.* **1996**, *62*, 54–60. [CrossRef]
37. Di Ilio, A.; Tagliaferri, V. Thermal damage in laser cutting of $(0/90)^2$s aramid/epoxy laminates. *Composites* **1989**, *20*, 115–119. [CrossRef]
38. Voisey, K.T.; Fouquet, S.; Roy, D.; Clyne, T.W. Fibre swelling during laser drilling of carbon fibre composites. *Opt. Lasers Eng.* **2006**, *44*, 1185–1189. [CrossRef]

© 2018 by the authors. Licensee MDPI, Basel, Switzerland. This article is an open access article distributed under the terms and conditions of the Creative Commons Attribution (CC BY) license (http://creativecommons.org/licenses/by/4.0/).

 materials

Article

Simulation of Laser Heating of Aluminum and Model Validation via Two-Color Pyrometer and Shape Assessment

Fabrizia Caiazzo * and Vittorio Alfieri

Department of Industrial Engineering, University of Salerno, 84084 Fisciano, Italy; valfieri@unisa.it
* Correspondence: f.caiazzo@unisa.it; Tel.: +39-089-964-323

Received: 4 July 2018; Accepted: 20 August 2018; Published: 22 August 2018

Abstract: The modeling of laser-based processes is increasingly addressed in a competitive environment for two main reasons: Preventing a trial-and-error approach to set the optimum processing conditions and non-destructive real-time control. In this frame, a thermal model for laser heating in the form of non-penetrative bead-on-plate welds of aluminum alloy 2024 is proposed in this paper. A super-Gaussian profile is considered for the transverse optical intensity and a number of laws for temperature-dependent material properties have been included aiming to improve the reliability of the model. The output of the simulation in terms of both thermal evolution of the parent metal and geometry of the fusion zone is validated in comparison with the actual response: namely, a two-color pyrometer is used to infer the thermal history on the exposed surface around the scanning path, whereas the shape and size of the fusion zone are assessed in the transverse cross-section. With an average error of 3% and 4%, the model is capable of predicting the peak temperature and the depth of the fusion zone upon laser heating, respectively. The model is intended to offer a comprehensive description of phenomena in laser heating in preparation for a further model for repairing via additive manufacturing.

Keywords: modeling; laser processing; simulation; pyrometer; aluminum

1. Introduction

Simulation tools are crucial in a competitive environment to prevent a trial-and-error approach to set the optimum processing conditions at a pre-design stage [1]. Moreover, proper modeling of an industrial process is the key to introduce closed-loop real-time monitoring where signals are managed for the purpose of control [2], to correct possible deviations of the main factors with respect to the intended, simulated response. Therefore, two issues must be addressed: the building of a reliable model structure and the arranging of effective equipment for real-time monitoring.

Regarding the former, the need for developing simulation tools to predict the transient temperature fields in laser-based processes has been widely presented in the literature [3]. Indeed, irrespective of the application and the technology, it has been shown that temperature directly affects the mechanical properties of the final component.

Some effort has been made in this field and a wide range of applications are reported in the literature, including but not limited to hardening [4], laser ablation [5], laser cutting [6], laser drilling [7], laser welding [8,9], and additive manufacturing of metal powder [10,11]. Irrespective of the application, the prediction of the temperature field is crucial for many purposes, including but not limited to non-destructive real-time evaluation of the process [2], minimization of residual stresses, and heat accumulation during additive manufacturing [11,12]. In general, advanced complex models are required to consider beam attenuation in the laser-induced plasma plume when higher irradiance is delivered (e.g., in cutting, drilling, and welding). Specific additional references to common assumptions for modeling will be given in the following relevant sections of the paper.

Once a proper simulation tool has been developed to relate the input parameters on the laser thermal cycles, signals must be extracted from the process and continuously compared to the intended output. Therefore, for the purpose of exploiting the transient temperature field in real-time monitoring and control, a cost-effective, fast, and reliable solution should aim to reduce the error in temperature measurements. A number of methods and detection sensors have been proposed in the literature: Thermocouples, photodiodes, and infrared cameras are the main methods that have been tested and compared [13]. Unfortunately, they are generally unsuitable for laser-based processes in an industrial environment, since fast measurements at precise locations approaching the laser path are required. Indeed, sharp temperature gradients as a consequence of fast heating and cooling rates are involved in laser processing; moreover, the acquisition may be significantly affected by laser radiation and plume dynamics, depending on the metal to be processed and the operating window [2].

Instead, fiber-optic pyrometers are a valuable method, being contactless and faster, with a response time in the order of milliseconds [14]. Additional advantages are offered by two-color pyrometers [15] dealing with the ratio of optical powers at two spectral bands to bypass the dependence of emissivity on the temperature. A wide theoretical background about detecting with two-color pyrometers is available in the literature [16].

A model for laser thermal heating in the form of non-penetrative bead-on-plate welds was built in COMSOL Multiphysics in this study, aiming to simulate the creation of the melting pool due to laser heating, in preparation for a further model where impinging metal would be fed for repair via additive manufacturing. Although some effort has been made in the literature to model the process of material deposition for the purpose of additive manufacturing [17,18], many simplifying assumptions are usually made in terms of material properties and boundary conditions, given that many complex phenomena are involved; in this frame, this paper specifically aimed to build a comprehensive model via a methodical approach contemplating several items. The validation of the process of simulating mere thermal heating (i.e., bead-on-plate welds) and the creation of a melting pool was required before moving to a more complex model; with this respect, the results are discussed in this paper. Namely, aluminum alloy (AA) 2024-T3 was chosen as a base metal, as it is widely used in the aerospace and automotive industries for high price-sensitive parts requiring maintenance via focused heat sources, such as laser beams. The reliability of the simulation was assessed in comparison with the experimental data, i.e., the thermal history around the scanning path and the geometrical response in the fusion zone upon cross-cutting. Namely, a fiber-optic two-color pyrometer was used to obtain temperature measurements.

2. Materials and Methods

2.1. Thermal Modeling

2.1.1. Heat Equations

Heat generated by a laser beam above a metal surface is dissipated by means of conduction, convection, and radiation. The theoretical approach to modeling is provided in the literature [11]. Namely, the heat transport equation can be given as:

$$\rho c \frac{\partial T}{\partial t} = \nabla \cdot (k \nabla T) + \alpha Q \tag{1}$$

where ρ is the density, c is the heat capacity, T is the temperature, t is the time, k is the thermal conductivity, α is the absorption coefficient, and Q is the laser heat generation. In addition, convection and radiation losses q_c and q_r are given as:

$$q_c = h(T_\infty - T) \tag{2}$$

$$q_r = \varepsilon\sigma\left(T_{room}^4 - T^4\right) \tag{3}$$

where h is the heat convection coefficient, ε is the emissivity, σ is the Stefan-Boltzmann constant, and T_∞ and T_{room} are the gas medium and room temperature, respectively. In the following sections, losses will be provided in form of boundary conditions, depending on the domain of interest; both T_∞ and T_{room} will be assumed as 22 °C.

2.1.2. Heat Source

The first step to address is a proper description of the laser heat generation. The fundamental mode of a Gaussian beam [19] is generally preferred [11,20] and a Gaussian heat source is provided, accordingly. It is worth noting that although a lean description is gained, the assumption of a true Gaussian beam is not suitable in general, unless high-quality laser beams are considered. Other theoretical formulations are hence proposed in the literature, including a double-ellipsoid power density distribution [17] based on the original model suggested by Goldak [21], and a flat-top beam [19]. The latter is considered in this paper; namely, the heat generated by a super-Gaussian profile (i.e., a smoothed flat-top profile) of transverse optical intensity of order n can be given as:

$$Q(r) = Q_0 \exp\left[-2\left(\frac{r}{w_0}\right)^n\right] \tag{4}$$

where Q_0 is the peak intensity, w_0 is the beam radius over the incident surface, and r is the radial distance from the propagation axis. A conventional Gaussian profile results from a super-Gaussian one of order two; the higher the order, the steeper the edges of the profile. A super-Gaussian intensity profile of order 20 was implemented in the paper (Figure 1), based on actual data acquisition via beam profiler. Under this assumption and for P denoting the operating power, the peak intensity in Equation (4) approaches:

$$Q_0 = \frac{P}{\pi w_0^2} \tag{5}$$

Moreover, since the model is aimed to simulate an application of repairing via metal addition where a defocused beam must be used [22], the laser beam was defocused to a processing diameter of 3 mm.

With x_0 and y_0 being the coordinates of the starting point of the beam path, s the traveling speed of the laser beam along the x-direction, and t the time; a moving heat source was implemented in a Cartesian coordinate system, hence Equation (4) yielding:

$$Q(x,y) = \frac{P}{\pi w_0^2}\exp\left\{-\frac{2\left[(x-x_0)^{20}+(y-y_0)^{20}\right]}{w_0^{20}}\right\} = \frac{P}{\pi w_0^2}\exp\left\{-\frac{2\left[(s\,t-x_0)^{20}+(y-y_0)^{20}\right]}{w_0^{20}}\right\} \tag{6}$$

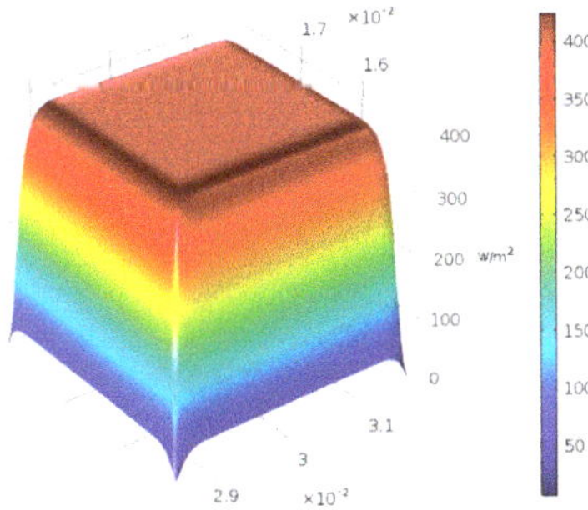

Figure 1. Distribution of transverse optical intensity: Super-Gaussian profile of order 20.

2.1.3. Material Properties

Including temperature-dependent material properties in the model is the key to a reliable prediction of the temperature field. For this purpose, based on the available literature on the characterization of pure aluminum and its alloys, a number of laws have been included in the background for the properties involved in conduction, convection, and radiation equations. AA 2024 of typical composition [23] for wrought products is used; solidus and liquidus temperature of 775 and 911 K, respectively, are given.

At first, a functional form for density (kg·m^{-3}) is borrowed [24], depending on the aggregation status:

$$\begin{cases} \rho_{solid}(T) = 2813 + 0.03 \times T - 7.4 \times 10^{-4} \times T^2 + 10^{-6} \times T^3 - 5.7 \times 10^{-10} \times T^4 & T \leq 775\ \mathrm{K} \\ \rho_{liquid}(T) = 2725 - 0.32 \times T \quad T \geq 911\ \mathrm{K} \end{cases} \quad (7)$$

Within the solidification range between solidus and liquidus temperature, a general rule of mixtures (i.e., a two-phase model) is implemented:

$$\rho = \theta_{solid}\rho_{solid} + (1 - \theta_{solid})\rho_{liquid} \qquad 775 < T < 911\ \mathrm{K} \quad (8)$$

with θ_{solid} denoting the solid volume fraction. Regarding the heat capacity (J·kg^{-1}·K^{-1}), a similar approach is taken [25,26]:

$$c_{solid}(T) = 199 + 3.9 \times T - 7.4 \times 10^{-3} \times T^2 + 5.2 \times 10^{-6} \times T^3 \qquad T \leq 775\ \mathrm{K} \quad (9)$$

Constant extrapolation has been set to extend this law beyond the solidus temperature. An evolution of thermal conductivity (W·m^{-1}·K^{-1}) as a function of temperature in a solid state is available in the literature [26]. Based on this, a functional form has been extracted and implemented:

$$k_{solid}(T) = 137 + 2.9 \times 10^{-4} \times T + 1.3 \times 10^{-6} \times T^2 \qquad T \leq 775\ \mathrm{K} \quad (10)$$

Linear extrapolation has been set to extend this law beyond the solidus temperature.

For an opaque material, the absorption coefficient α is complementary to the reflection coefficient (1–α). Since reflection is one of the main factors affecting the coupling efficiency when processing metals [27], the trend of reflectivity vs. temperature is required: although constant average absorption has been proposed by some authors in the literature [11,18], the reflectivity of aluminum in solid state may decrease from 95 to as low as 60% [28]. Given this, a functional form for reflectivity, for a given operating laser wavelength of a doped YAG (Yttrium Aluminum Garnet) active gain, has been inferred (Figure 2), based on two main assumptions: a reduction of reflectivity and, hence, an increase of absorption, in turn, is reported for increasing temperature [28]; a sharp drop at a measure of 5% is noticed at phase transition [29] for pure aluminum.

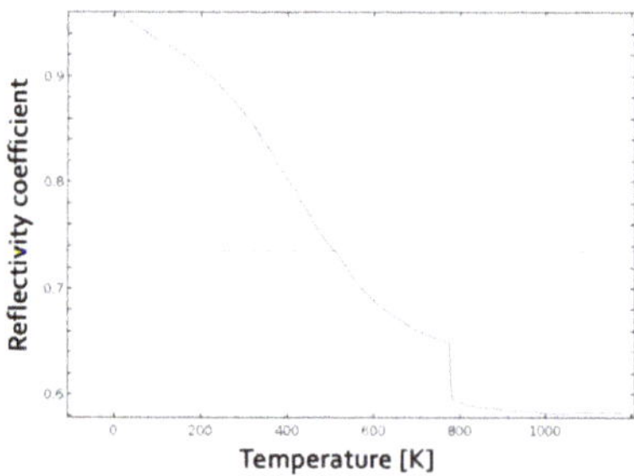

Figure 2. Aluminum reflectivity coefficient as a function of temperature.

In the assumption of natural convection, a constant heat convection coefficient $h = 10\ \mathrm{W \cdot m^{-2} \cdot K^{-1}}$ was fed to the model [17,20]. In the operating range of laser heating, even the dependence of emissivity on the temperature is mild: a constant value $\varepsilon = 0.15$ [30] was set. Eventually, generation of plasma, and, hence, beam attenuation, can be neglected since vaporization is prevented in this application [28].

2.1.4. Virtual Specimen and Meshing

Laser heating was modeled over a virtual specimen, 80 mm long, 60 mm wide, 10 mm thick, this being the size of the plate in the experimental procedure for validation. For the purpose of improving the consistency of the model, the specimen was divided into two domains of interest (Figure 3): a central, 10 mm wide, 0.4 mm deep slot for the laser path (domain 1, D_1) and the remaining (domain 2, D_2). Indeed, it is worth noting that the main issue in the modeling of a laser-based process is a reliable implementation of a processing diameter in the order of tenths of millimeters [19]. As a consequence of this, an ultrafine mesh must be set along the laser path, whereas a coarser mesh is allowed for the purpose of reducing the simulation time: a triangle mesh of variable size was applied, accordingly (Figure 4), the edges being 0.15 mm in size within D_1 then ranging up to 10 mm across D_2 (Figure 5).

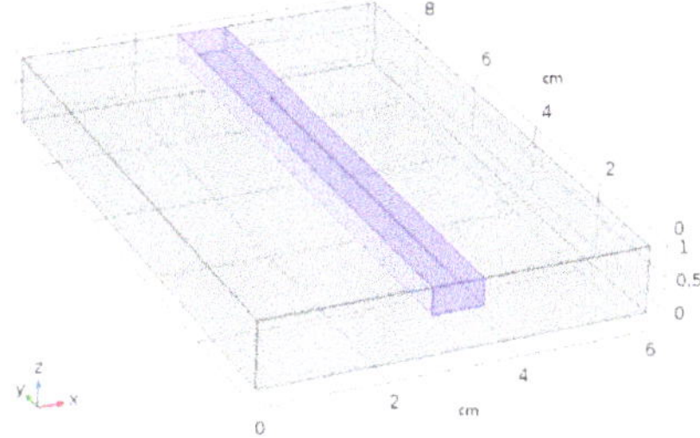

Figure 3. Domains of interest: central slot (blue domain, D_1), and remaining (grey domain, D_2).

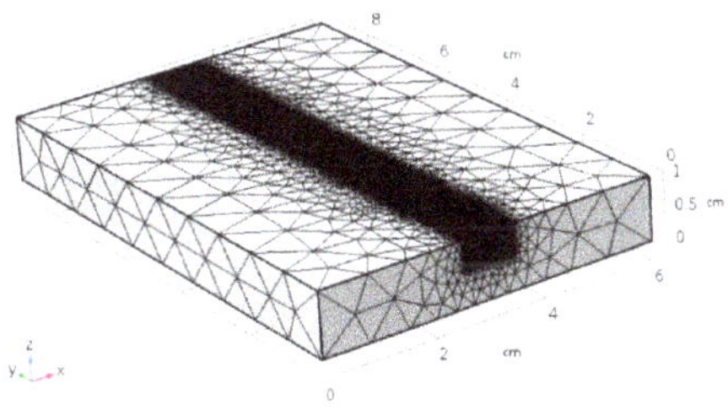

Figure 4. Triangle mesh of variable size on the virtual specimen.

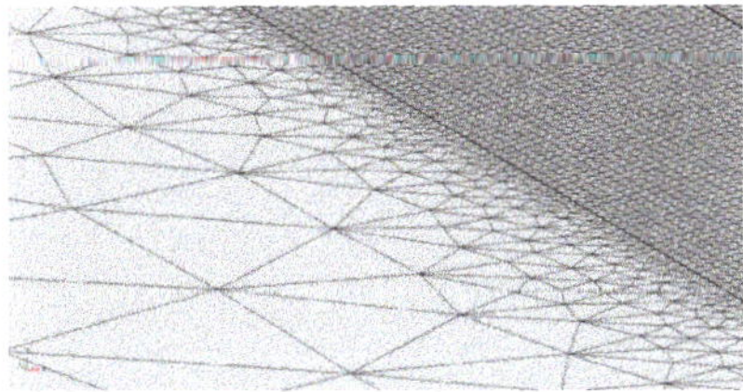

Figure 5. Mesh detail at the interface with the central slot.

Boundary and initial conditions are given for each domain. Regarding D_1, laser heat generation Q is provided along the processing path; convection and radiation losses are experienced at the upper surface. Regarding D_2, convection and radiation losses are experienced at each surface; a condition of thermal continuity is given with respect to D_1 for each shared surface. The initial temperature of the domains is assumed as room temperature.

2.2. Experimental Procedure

Laser heating in the form of bead-on-plate tests was performed. The operating window to achieve effective penetration in the cross-section was borrowed from previous works on the same alloy [22], even aiming for non-penetrative bead-on-plate welds to prevent porosity [31]; traveling speed and power were considered, and the results of six testing conditions have been found (Table 1). For the purpose of an easier comparison of the responses, the length of the scanning path was conveniently set in order to result in 10 s heating, irrespective of the traveling speed.

A thin-disc laser source was used (Table 2). Defocusing of the laser beam was set in order to get a processing diameter of 3 mm on the top surface. Moreover, a tilting angle of 4° was given to the laser head, in agreement with common practice for highly reflective metals, to prevent back-reflections from entering the optics train [27]; although reflectivity depends on the angle of incidence and the plane of polarization of the laser beam, the effect can be neglected at this angle size [28]. The scheme of the processing set-up was composed of a laser head, a clamping device, and a pyrometer (Figure 6).

Table 1. Processing conditions for laser heating.

Speed (mm/s)	Power (W)	Scanning length (mm)
4	2000	40
4	2500	40
4	3000	40
6	2000	60
6	2500	60
6	3000	60

Table 2. Main technical features of the laser source.

Parameter	Value
Maximum output power (kW)	4.0
Operating nominal wavelength (nm)	1030
Beam Parameter Product (mm × mrad)	8.0
M^2 quality factor	24.3
Core diameter of the delivering fiber (μm)	300
Spot size of the laser beam on the surface (mm)	3.0

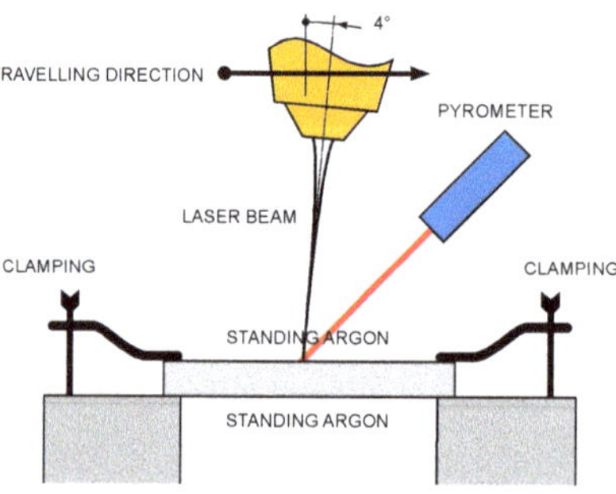

Figure 6. Scheme of the processing set-up; components are not to scale.

Moreover, to prevent oxidation of the base metal, resulting in defects and additional heating as a consequence of energy release, argon for inert shielding was supplied to the working area; a steady shielding atmosphere resulted, giving grounds to the assumptions of natural convection in modeling. The plate was clamped at the edges, so that convection in standing argon was experienced and one may assume conduction is negligible to the purpose of simulating the thermal history at the laser path.

Temperature model validations were performed by means of a fiber-optic two-color pyrometer which was calibrated in a 290–610 °C (i.e., 563–883 K) temperature span to the specific purpose of monitoring AA 2024; as a consequence of this, underflow and overflow may result below and above the lower and higher span limits, respectively. For each given testing condition, the pyrometer was focused halfway in the processing path, 2 mm from the scanning line (Figure 7). The site of interest to focus the pyrometer is suggested by the need to acquire a response within the calibration range of the device; as a consequence, direct acquisition of the thermal history along the laser path is not feasible. A 0.1 ms time step was set for temperature acquisition, thus resulting in 10^5 sample points overall, given a 10 s period of total heating. A post-acquisition smoothing algorithm with a 50-point moving-average was implemented in order to filter noise from the output.

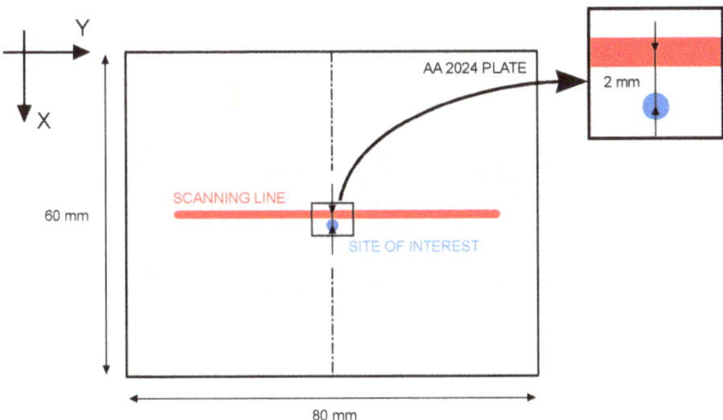

Figure 7. Location and detail of the site of interest (x = 32 mm, y = 40 mm, z = 10 mm) for thermal monitoring.

Further data for model validation were obtained in the transverse cross-section (i.e., parallel to the xz plane). To this purpose, the specimens resulting from laser heating were cross-cut, mechanically ground, and polished to a mirror finish, and chemically etched with a solution consisting of 10% hydrofluoric acid, 15% nitric acid, and water at room temperature [23]. The actual size of the fusion zone was eventually measured via optical microscopy (Figure 8) and compared to the output of the simulation.

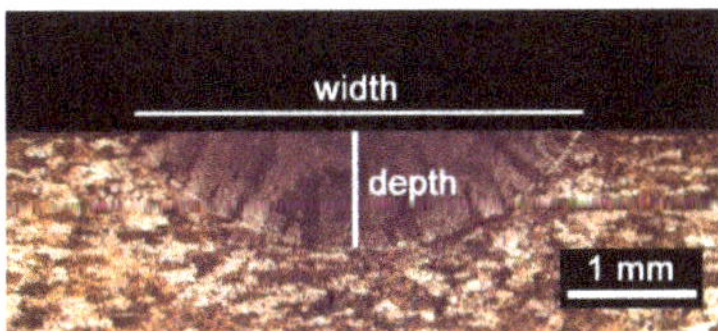

Figure 8. Width and depth of the fusion zone, macrograph resulting at speed of 6 mm/s, power of 2000 W.

3. Results and Discussion

3.1. Local Thermal Cycle

The thermal history at the site of interest during laser heating is a function of the processing parameters (Figure 9); depending on these, fusion may be experienced at the site of interest. The time-scale is started at laser switch-on: since the process duration is dependent on the traveling speed, the time to get the peak temperature at the site of interest (i.e., to cover half the distance) is 5 s.

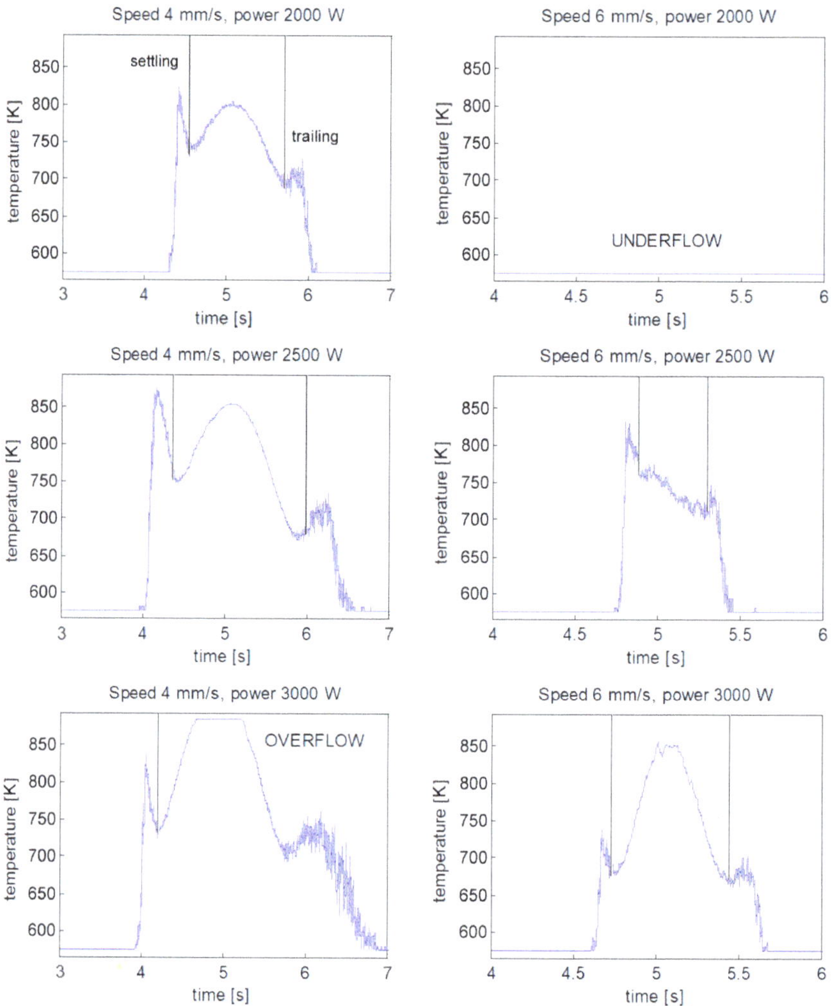

Figure 9. Thermal history of the site of interest as pyrometer output.

A recurring shape was found for the temperature profile: namely, a settling period, resulting in a leading thermal spike, was required by the device when entering the window of calibration (i.e., the operating range of acquisition); a trailing noise was found at the end of the acquisition, due to the air and argon overheating over the site of interest, instead. The thermal evolution shifted below or above the calibration limits under extreme conditions of processing, thus resulting in underflow or overflow, respectively.

For each processing condition, the thermal evolution of the site of interest was simulated (Figure 10) and compared with the acquisition. The peak temperature (Tp) acquired by the pyrometer focus was extracted and compared with the predicted peak temperature in order to quantitatively validate the thermal model (Table 3); the percentage difference between acquisition and simulation was given. An agreement in a measure of 2.7%, absolute, on average, was found in terms of peak temperature; as regarding the conditions of underflow and overflow, the simulated thermal evolution was actually outside of the calibration window of the device.

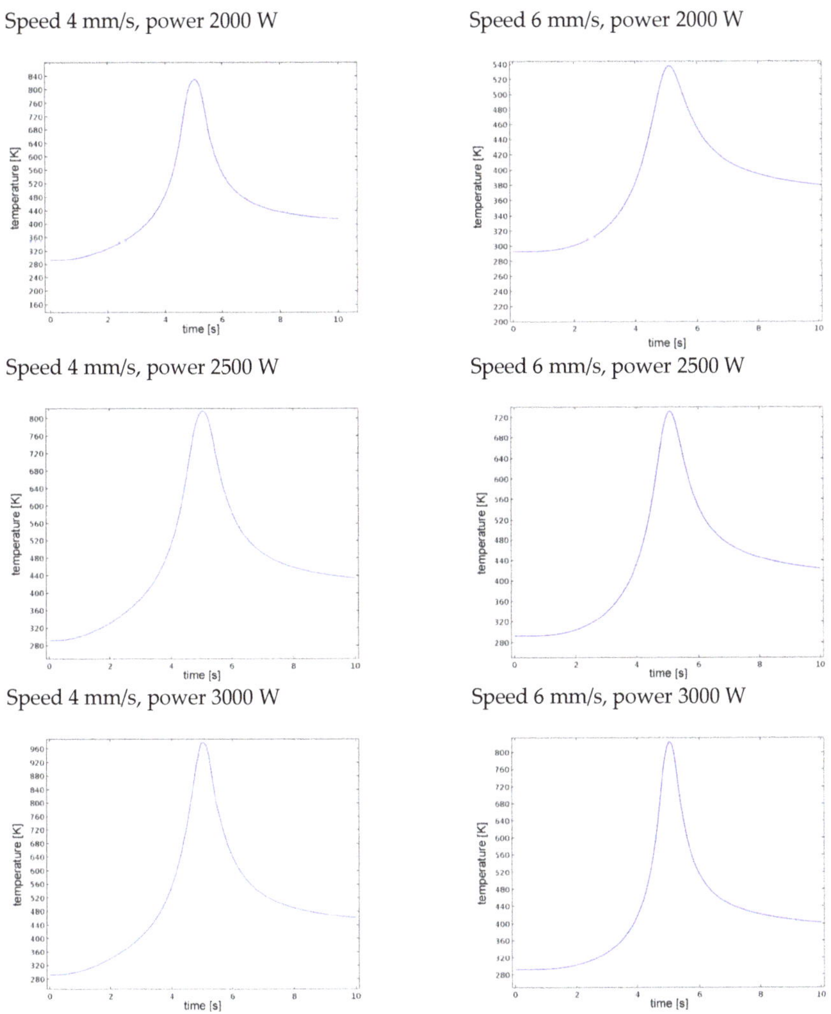

Figure 10. Simulated thermal history of the site of interest for each processing condition.

Table 3. Peak temperatures, actual vs. predicted.

Speed (mm/s)	Power (W)	T_p (K)		
		Actual	Simulated	Difference
4	2000	803	815	1.5%
4	2500	848	828	−2.4%
4	3000	*overflow*	990	*n/a*
6	2000	*underflow*	543	*n/a*
6	2500	753	728	−3.3%
6	3000	851	821	−3.5%

3.2. Geometry of the Fusion Zone

Since the overall size of the fusion zone depends on the thermal history, further information to validate the model was gathered upon inspections in a transverse cross-section with respect to the traveling direction of the laser beam (Figure 8). An indirect measurement of the simulated depth of the fusion zone had to be conducted: namely, a transverse plane at half-length was considered with respect to the traveling direction, then thermal contour lines are drawn (Figure 11). As expected, any increase in the experienced peak temperature yielded a proportional increase in the extent of the fusion zone.

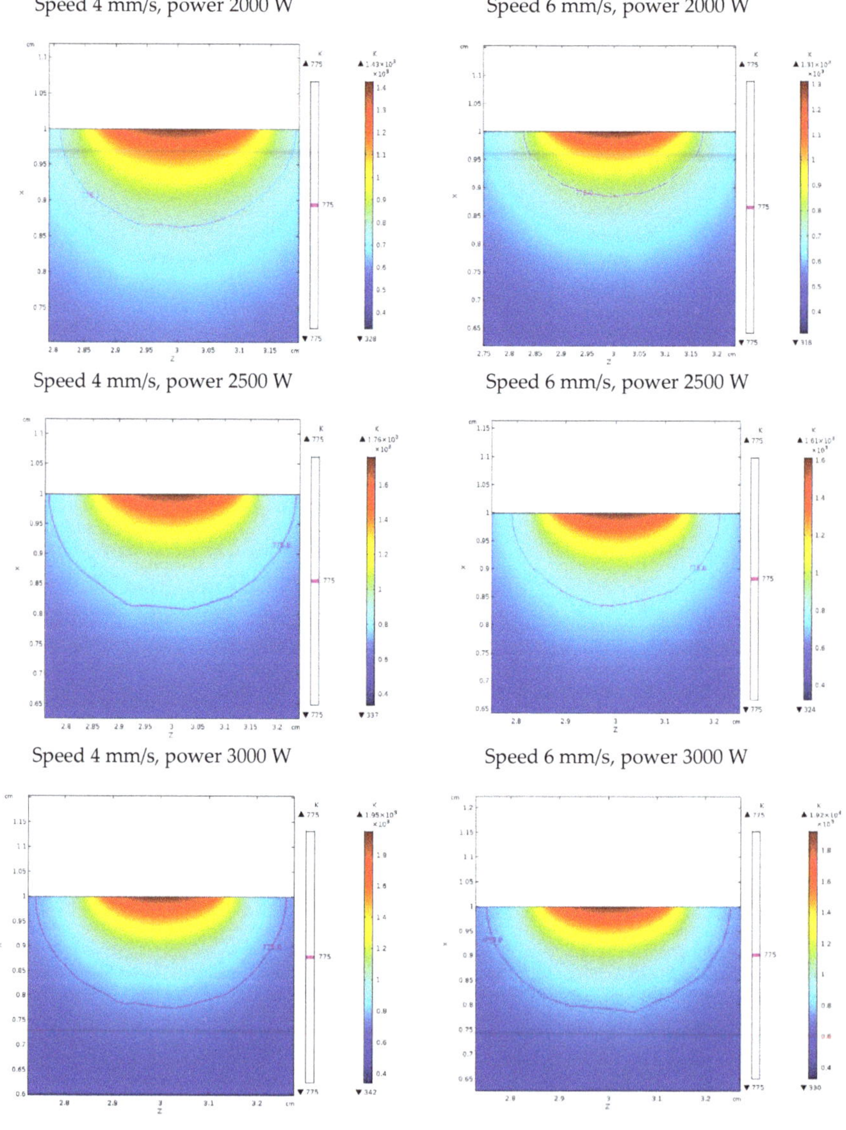

Figure 11. Contour lines to define the extent of the fusion zone for each processing condition.

Based on the solidification range of the parent alloy, depth and width of the fusion zone were inferred. Indeed, since 775 K is the lower limit of the solidification range, fusion was experienced by any point above this temperature limit. For each given processing condition, the depth and width were compared to the corresponding actual geometry (Table 4). An agreement of 3.7% and 16.3%, absolute, on average, was found for depth and width, respectively. In order to improve the reliability of the model in predicting the width of the fusion zone, further investigation must be made and the dependence of the reflectivity on the starting roughness or oxide amount at the exposed surface should be considered.

Table 4. Depth and width of the fusion zone, actual vs. predicted.

Speed (mm/s)	Power (W)	Depth of Fusion (mm)			Width of Fusion (mm)		
		Actual	Simulated	Difference	Actual	Simulated	Difference
4	2000	1.55	1.42	−8.3%	5.17	4.22	−18.4%
4	2500	1.71	1.72	2.8%	5.45	4.53	−16.9%
4	3000	2.11	2.19	1.1%	6.81	5.55	−18.5%
6	2000	1.06	1.09	3.3%	4.06	3.57	−12.1%
6	2500	1.55	1.60	4.2%	5.13	4.38	−14.6%
6	3000	1.95	2.00	2.6%	6.21	5.10	−17.9%

4. Conclusions

A model to simulate laser heating was built and validated. The main elements were discussed and presented, aiming to offer a comprehensive description of the involved variables and phenomena. A super-Gaussian beam profile was implemented as an external thermal source; losses for radiation and convection were considered, whereas losses for plasma attenuation were neglected.

Under these assumptions, with an average error below 3%, the model is capable of predicting the peak temperature upon laser heating in a processing window ranging from 2 to 3 kW power and 4 to 6 mm/s speed, which is suitable to produce a melting pool where metal must be fed in additive manufacturing. With an average error below 4%, the model has been capable of predicting the depth of the fusion zone. Larger errors, of up to 16%, were found for the bead width instead; these will be addressed in future adjustments of the simulation; indeed, the dependence of the reflectivity on the starting roughness or oxide amount at the exposed surface should be considered. Furthermore, the model must be conveniently upgraded with powder or wire feeding to simulate layer-by-layer fabrication.

Interestingly, an industrial environment where a pyrometer output is used in real-time monitoring to match the intended thermal history, and, hence, the intended geometry, can be conceived. Nevertheless, proper actions both at software and hardware stages, must be taken to filter noise from the pyrometer output, depending on the system set-up, the metal to be processed, and the laser wavelength.

Author Contributions: Conceptualizationand Methodology, F.C.; Software, V.A.; Validation, Formal Analysis, Investigation, Data Curation, Writing-Original Draft Preparation, Writing-Review & Editing, F.C. and V.A.; Project Administration, F.C.

Funding: This research received no external funding.

Acknowledgments: The Authors gratefully acknowledge Eng. Antonio Criscitiello for his valuable support in programming the simulation tool.

Conflicts of Interest: The authors declare no conflict of interest.

References

1. Thomas, P. *Simulation of Industrial Processes for Control Engineers*; Butterworth-Heinemann: Oxford, UK, 1999.
2. Muvvala, G.; Karmakar, D.P.; Nath, A.K. Online monitoring of thermo-cycles and its correlation with microstructure in laser cladding of nickel based super alloy. *Opt. Laser Eng.* **2017**, *88*, 139–152. [CrossRef]

3. Mackwood, A.P.; Crafer, R.C. Thermal modelling of laser welding and related processes: A literature review. *Opt. Laser Technol.* **2005**, *37*, 99–115. [CrossRef]
4. Lusquinos, F.; Conde, J.; Bonss, S.; Riveiro, A.; Guintero, F.; Comesana, F.; Pou, T. Theoretical and experimental analysis of high power diode laser (HPDL) hardening of AISI 1045 steel. *Appl. Surf. Sci.* **2007**, *254*, 948–954. [CrossRef]
5. Ohkubo, T.; Sato, Y.; Matsunaga, E.; Tsukamoto, M. Three-dimensional numerical simulation during laser processing of CFRP. *Appl. Surf. Sci.* **2017**, *417*, 104–107. [CrossRef]
6. Yilbas, B. Laser cutting quality assessment and numerical methods for modeling of cutting. In *The Laser Cutting Process—Analysis and Applications*; Elsevier: New York, NY, USA, 2017; pp. 149–203.
7. Abidou, D.; Yusoff, N.; Nazri, N.; Awang, M.; Hassan, M.; Sarhan, A.A. Numerical simulation of metal removal in laser drilling using radial point interpolation method. *Eng. Anal. Bound. Elem.* **2017**, *77*, 89–96. [CrossRef]
8. Chen, G.; Zhang, M.; Zhao, Z.; Zhang, Y.; Li, S.S. Measurements of laser-induced plasma temperature field in deep penetration laser welding. *Opt. Laser Technol.* **2013**, *45*, 551–557. [CrossRef]
9. Pocorni, J.; Han, S.; Cheon, J.; Na, S.; Kaplan, A.; Bang, H. Numerical simulation of laser ablation driven melt waves. *J. Manuf. Process.* **2017**, *30*, 303–312. [CrossRef]
10. Leitz, K.; Singer, P.; Plankensteiner, A.; Tabernig, B.; Kestler, H.; Sigl, L. Multi-physical simulation of selective laser melting. *Met. Powder Rep.* **2017**, *72*, 5. [CrossRef]
11. Kundakcioglu, E.; Lazoglu, I.; Rawal, S. Transient thermal modeling of laser-based additive manufacturing for 3D freeform structures. *Int. J. Adv. Manuf. Tech.* **2016**, *85*, 493–501. [CrossRef]
12. Vastola, G.; Zhang, G.; Pei, Q.; Zhang, Y. Controlling of residual stress in additive manufacturing of Ti6Al4V by finite element modeling. *Addit. Manuf.* **2016**, *12*, 231–239. [CrossRef]
13. Bi, G.; Gasser, A.; Wissenbach, K.; Drenker, A.; Poprawe, R. Identification and qualification of temperature signal for monitoring and control in laser cladding. *Opt. Lasers Eng.* **2006**, *12*, 1348–1359. [CrossRef]
14. Doubenskaia, M.; Bertrand, P.; Smurov, I. Pyrometry in laser surface treatment. *Surf. Coat. Tech.* **2006**, *201*, 1955–1961. [CrossRef]
15. Yu, J.; Lin, X.; Wang, J.; Chen, J.; Huang, W. Mechanics and energy analysis on molten pool spreading during laser solid forming. *Appl. Surf. Sci.* **2010**, *256*, 4612–4620. [CrossRef]
16. Tapetado, A.; Diaz-Alvarez, J.; Miguelez, M.; Vazquez, C. Two-color pyrometer for process temeprature measurement during machining. *J. Lightwave Technol.* **2016**, *34*, 4. [CrossRef]
17. Heigel, J.; Michaleris, P.; Reutzel, E. Thermo-mechanical model development and validation of directed energy deposition additive manufacturing of Ti-6Al-4V. *Addit. Manuf.* **2015**, *5*, 9–19. [CrossRef]
18. Fan, Z.; Frank, L. Numerical modeling of the additive manufacturing (AM) process of titanium alloy. In *Titanium Alloys—Towards Achieving Enhanced Properties for Diversified Applications*; Nurul Amin, A.K.M., Ed.; InTech: Hong Kong, China, 2012; pp. 3–28.
19. Paschotta, R. *Encyclopedia of Laser Physics and Technology*; Wiley-VCH: Berlin, Germany, 2008.
20. Suresh Kumar, K.; Sparks, T.; Liou, F. Parameter determination and experimental validation of a wire feed Additive Manufacturing model. In Proceedings of the 29th Annual International Solid Freeform Fabrication Symposium—An Additive Manufacturing Conference, Austin, TX, USA, 13–15 August 2015.
21. Goldak, J.; Chakravarti, A.; Bibby, M. A new finite element model for welding heat sources. *Metall. Trans. B* **1984**, *15*, 299–305. [CrossRef]
22. Caiazzo, F.; Alfieri, V.; Argenio, P.; Sergi, V. Additive manufacturing by means of laser-aided directed metal deposition of 2024 aluminium powder: Investigation and optimization. *Adv. Mech. Eng.* **2017**, *9*, 1–12. [CrossRef]
23. Davis, J.R. *Aluminum and Aluminum Alloys*; ASM International: Geauga County, OH, USA, 1993.
24. Kurochkin, A.; Popel, P.; Yagodin, D.; Borisenko, A.; Okhapkin, A. Density of copper–aluminum alloys at temperatures up to 1400 °C determined by the gamma ray technique. *High Temp.* **2013**, *51*, 197–205. [CrossRef]
25. Sarmast, A.; Serajzadeh, S.; Kokabi, A. A study on thermal responses, microstructural issues, and natural aging in gas tungsten arc welding of AA2024-T4. *J. Eng. Manuf.* **2013**, *228*, 413–421. [CrossRef]
26. Hrbek, J. Induction heating of thin nonmagnetic sheets in transverse time-variable magnetic field. *Acta Tech.* **2015**, *60*, 15–29.

27. Alfieri, V.; Caiazzo, F.; Sergi, V. Autogenous laser welding of AA2024 aluminium alloy: Process issues and bead features. *Procedia CIRP* **2015**, *33*, 406–411. [CrossRef]
28. Steen, W.; Mazumder, J. *Laser Material Processing*; Springer: Berlin, Germany, 2010.
29. Marla, D.; Bhandarkar, U.V.; Joshi, S.S; Marla, D.; Bhandarkar, U.V.; Joshi, S.S. Models for predicting temperature dependence of material properties of aluminum. *J. Appl. Phys.* **2014**, *10*, 1–13. [CrossRef]
30. Wen, C.; Mudawar, I. Experimental investigation of emissivity of aluminum alloys and temperature determination using multispectral radiation thermometry (MRT) algorithms. *J. Mater. Eng. Perform.* **2002**, *11*, 551–562. [CrossRef]
31. Alfieri, V.; Cardaropoli, F.; Caiazzo, F.; Sergi, V. Investigation on porosity content in 2024 aluminum alloy welding by Yb:YAG disk laser. *Adv. Mater. Res.* **2012**, *383*, 6265–6269. [CrossRef]

© 2018 by the authors. Licensee MDPI, Basel, Switzerland. This article is an open access article distributed under the terms and conditions of the Creative Commons Attribution (CC BY) license (http://creativecommons.org/licenses/by/4.0/).

Concept Paper

Accelerating High-Throughput Screening for Structural Materials with Production Management Methods

Alexander Bader *, Finn Meiners and Kirsten Tracht

Bremen Institute of Mechanical Engineering (BIME), University of Bremen, Badgasteiner Str. 1, 28359 Bremen, Germany; meiners@bime.de (F.M.); tracht@bime.de (K.T.)

* Correspondence: bader@bime.de; Tel.: +49-421-2186-4839

Received: 5 July 2018; Accepted: 30 July 2018; Published: 1 August 2018

Abstract: High-throughput screenings are widely accepted for pharmaceutical developments for new substances and the development of new drugs with required characteristics by evolutionary studies. Current research projects transfer this principle of high-throughput testing to the development of metallic materials. In addition to new generating and testing methods, these types of high-throughput systems need a logistical control and handling method to reduce throughput time to get test results faster. Instead of the direct material flow found in classical high-throughput screenings, these systems have a very complex structure of material flow. The result is a highly dynamic system that includes short-term changes such as rerun stations, partial tests, and temporarily paced sequences between working systems. This paper presents a framework that divides the actions for system acceleration into three main sections. First, methods for special applications in high-throughput systems are designed or adapted to speed up the generation, treatment, and testing processes. Second, methods are needed to process trial plans and to control test orders, which can efficiently reduce waiting times. The third part of the framework describes procedures for handling samples. This reduces non-productive times and reduces order processing in individual lots.

Keywords: micro-manufacturing; manufacturing systems; production planning

---

1. Introduction

Steels and aluminum alloys can be used as structural materials, however, they may not have satisfactory material properties for a given use. For these areas of application, adapted materials must be developed so that components can fulfil their intended structural function. The mechanical properties result from interactions between alloy composition and microstructure, which must be coordinated in many investigation steps. In conventional research and development, the product is adapted to the material properties [1]. Therefore, it is necessary to define alloy composition and the setup of microstructure to meet required performance profiles. In this context, Springer and Raabe have developed a combinatorial process with which varying alloy compositions and heat treatments are systematically investigated by means of tensile tests [2]. Research into these new compositions of alloy and microstructure with conventional, standardized methods of materials testing requires a great amount of time and material effort. For speeding up the testing of materials and identification, Ellendt and Mädler introduced a new approach for the development of structural materials that is based on the identification of materials in high-throughput. This high-throughput system can perform examinations in a few hours or days that would take weeks or months using conventional methods. In order to increase the productivity of high-throughput screening and further shorten the development time, the targeted use of screening strategies is necessary [3]. Within the evolutionary high-throughput system 'Farbige Zustände', metallic materials are developed [4]. This system enables the identification of new alloys, which meet specific performance profiles, defined by mechanical material properties, such as the hardness, the elongation at fracture, the young's module, and other factors. Through statistical and heuristical processes, an experimental plan is calculated, which shows all steps for the production and setup of the alloy's microstructure as well as for the examination to determine the characteristics of the alloy (Figure 1). To achieve these properties, in addition to the alloy composition, the targeted thermal and mechanical treatment of the samples is defined [5].

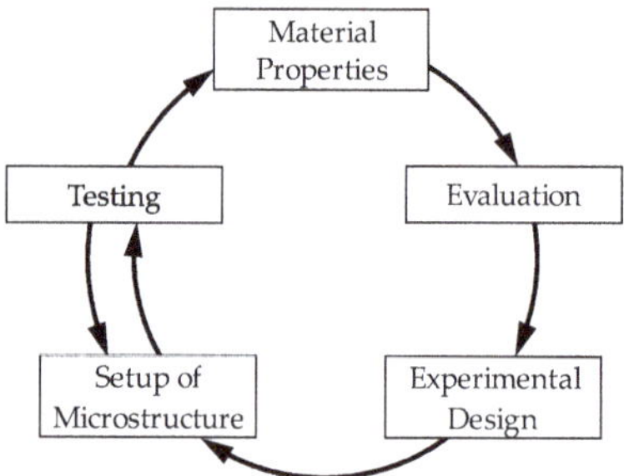

Figure 1. Schematic procedure of the method 'Farbige Zustände' [4].

The high-throughput system 'Farbige Zustände' greatly differs from conventional screenings regarding its structure. Conventional screenings of pharmaceutical products or catalyst materials usually have a linear material flow because these samples are not able to go through a processing station more than once. High-throughput screening consists of interlinked individual stations, each of which carries out investigations on samples. These tests are fully automated and optimized for high throughput. On this basis, high-throughput systems became established in the late 1990s [6]. However, no intelligent experiments are carried out or combinatorial strategies are applied [7]. In this system, adapted methods reduce examinations, which usually last weeks or months, to one or two hours with the same application rate [8]. This is possible because components, temperature, pressure, and other parameters in the high-throughput experiments can be changed. Especially important for successful implementation of high-throughput systems are fast data processing and automated systems, which enable fast sample processing. This makes it possible to process up to 100,000 samples per day for so-called ultra-high-throughput analyses [9]. Research and development is being industrialized to further increase output and reduce development costs in the development of active ingredients. Industrialization is characte rized in particular by the use of high-throughput systems. By means of automation, parallelization, and minimization of processes, these systems enable a faster acquisition of as much data as possible from experiments with the lowest possible material consumption [10]. Within high-throughput screenings for mechanical material testing, an individual composition of test plans for each target of analysis is required. Therefore, the high-throughput screening for structural materials has a high resemblance with job shop production. In such production systems, machines and working systems are arranged according to their functions and not according to the production order of one single product. Orders reach machines in a previously planned sequence for the completion of a component or assembly [11]. Due to the flexible material flow and the complexity of the system, which is also caused by the handling of micro samples, the material flow and the processing of the samples must be controlled. For this purpose, approaches from production technology are examined that lead to a reduction in throughput time in production systems. The focus is on the evaluation of individual processes as well as the entire system. The aim is to significantly increase throughput in order to be able to compare the performance of the high-throughput system with conventional high-throughput screening and to reduce nonproductive times by handling of parts.

2. Enhancement of Throughput Time in Production Systems

In recent years, the production of customized products and the associated flexibility of order processing have come into focus for production systems [12]. In this context, it is often referred to as one-offs [13]. Flexibility and reaction to short-term changes can be achieved primarily by reducing order throughput times [14]. According to Nyhuis and Wiendahl, there is a conflict of objectives with the throughput time and performance. Both show a clear change with the system's inventory. Increasing inventory leads to both increased performance and longer throughput times, so a system-specific optimum operating point must be determined [15]. The main components of the throughput time are queuing times, transport times, and setup times. In this context, the proportion of processing time is significantly lower [16]. Nevertheless, the optimization of production systems often focuses on reducing the processing time. In general, processing times can be planned and offer a reliable savings potential. Transport times, but above all queuing times, are characterized by a large stochastic proportion of time, which makes it difficult or even impossible to plan these time-shares. However, it was shown by Nyhuis and Wiendahl that reduction of the stock has an especially strong positive effect on the resting time [15].

Production systems are factories or production lines in which a value-adding activity is carried out. At the end of these activities, usually finished components or assemblies and products are produced. These systems are supported by methods and strategies to ensure the optimal organization of the system. The methods describe an integrated production organization that consists of effective and efficient collaborative subsystems, work systems, concepts, methods, and tools. The focus is on customer benefit and should be achieved while avoiding waste [17]. For the successful organization of production systems, their characteristics and interaction with the customer are crucial. There are different types of orders for production systems, such as make-to-order or make-to-stock, and they are supplemented by mixed forms (see Figure 2). This shifts the customer decoupling point depending on the order type. According to the order type, different types of production can be selected, from shop floor production to paced continuous flow production. The order type determines the basic structure of production and the arrangement of machines and working systems. Therefore, they have an influence on the material flow and process type as well as on the entire order processing. In this context, orders involving a large number of a serial products must be organized in a different way in production than the manufacturing one-offs. This concerns both the arrangement and connection of machines, and the scheduling of production orders.

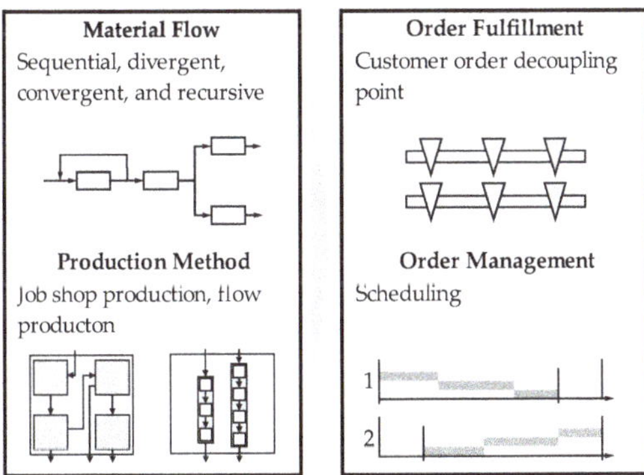

Figure 2. Tasks of production management.

Johnson has designed a framework for reducing throughput time in production systems and described various methods that lead to a reduction in throughput time. Figure 3 shows Johnson's classification of component changes that will reduce the Manufacturing Throughput Time per Part

(MTTP). These changes are reduction of setup time, processing time, transportation time, and queuing time [18]. To reduce setup times, entire production lines are sometimes converted to enable faster set-ups. The use of standardized interfaces or a standardization of work piece carriers are conceivable options. In the automotive industry, entire product classes are combined on a modular platform, which also avoids the set-up of machines or the build-up of several parallel production lines. A typical procedure to fast equipment setups is the Single Minute Exchange of Die (SMED) method [19]. This reduces batch sizes and increases productivity [20,21].

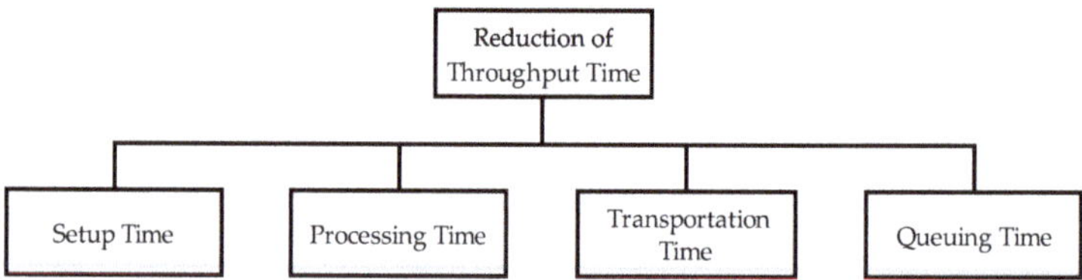

Figure 3. Reduction of manufacturing throughput time per part [18].

Previous research describes various approaches to reduce process times in production. For example, other fast technologies can be used for production or assembly times can be reduced by integrating functions. A learning process for the workers can also lead to an improvement in process time. In addition to reducing machining time through other machine parameters, the number of operations can also be reduced by revising components and processes [18]. In 1994 Suresh and Meredith described in this context an increase in productivity that reduces throughput times by forming product families that are processed together in production [22]. The transportation time can be reduced by accelerating transport systems or reducing distance [18]. Likewise, a systematic determination of lot sizes can influence the transport time per part.

The queuing time makes up a particularly large proportion of the throughput time. Wiendahl has already found that queuing time represents the largest share of lead time at 75%, but processing time is less than 10% [16]. In production planning and control, the focus is therefore on reducing these waiting and idle times before processing. In this context, the dilemma of production was already formulated in the 1960s. Gutenberg describes this as a contradiction between maximizing utilization and minimizing throughput times [23]. Thus, it is not possible to achieve short throughput times and high performance together in one system. An explanation model for this connection is the funnel model. The stock of a working system can be represented as a funnel whose lower opening represents the current performance [11]. This means that an inventory in the funnel is required as a buffer so that capacity utilization can be ensured across performance fluctuations. Too much inventory fills the funnel and increases throughput times. However, if the inventory is too low, the maximum performance cannot be reached [24]. This correlation becomes clear in Nyhuis' and Wiendahl's characteristic curve theory. A direct connection between throughput time, performance, and inventory was established and mathematically documented for the first time. It became clear that the throughput time increases with growing inventory. At the same time, performance reaches its maximum value with increasing inventory [15].

The funnel model forms the framework for many methods of loading planning and production control in job shop production. Logistical control methods achieve short throughput times and high schedule reliability, resulting in higher flexibility and short delivery times [25]. The result is a customer-oriented system that needs to be capable of implementing one-offs. Production control is essential for a flexible and fast production, which is characterized by a high proportion of flowing material [26]. The aim of these methods is to set the inventory as the control parameter and use it to achieve logistical target criteria such as throughput time, utilization, and schedule reliability. The inventory is the total work that is in the system. Depending on the production systems and the company objectives, logistical target criteria are weighted, so that the method used for loading planning varies for each production system. For example, with the very simple Kanban principle, a shorter throughput time can be achieved than that of a control system using Constant Work-In-Process (CONWIP) [27]. Workload Control (WLC), on the other hand, has a positive influence on the schedule reliability and performance of a production system [28]. In the procedure by Bechte, called Load Oriented Order Release (LOOR), a stock list is maintained that lists both the direct stock of a work system and the indirect stock. The indirect stock includes work orders that reach the relevant work system in the future and are included with a discount factor. An order is only released in this method if the direct and indirect stock does not exceed a previously defined limit value [29]. Even today, many order release procedures are still based on the concept of controlling inventories.

Another approach to reduce waiting and queuing times is lean production. This method was developed in the 1980s at Toyota in Japan and has become known worldwide as the Toyota Production System (TPS) [30]. The aim is to reduce non-value-adding activities and minimize waste. The focus is exclusively on value-adding work [31]. Lean production also focuses on reducing inventories and waste as these are seen as a great potential for hiding inefficient processes. In this way, resources can be better used. A further development of lean production was further expanded to support the volatile and competitive business environment [32,33]. Continuous Improvement (CI), the introduction of pull principles and a zero-defect strategy in production by simplifying processes, are of particular importance in lean production. Employee orientation and goal-oriented leadership through visual management are also tools of lean management [34]. Thus, a Lean Production System (LPS) is created that provides for a company-specific orientation of the production system using the eight basic principles [35].

3. Acceleration of High-Throughput Systems

A classification between the characteristics of production systems and high-throughput systems shows comparable theories in organization, order management, material flow, order type, manufacturing type, and process organization. Similar to production organization, orders in high-throughput systems are controlled by a process chain, and processed at various workstations. Orders initiate activities in these systems. One kind of activity by order release is the definition of a fixed sequence. In addition, due dates and the number of samples are also specified while releasing the order [36]. Due to a similar behavior between production systems and high-throughput systems, many approaches from the production organization can be transferred in their principles to high-throughput systems (see Figure 4). In addition to the development of methods for process improvement, enhancements in transportation between stations and control as well as in handling contribute to a reduction in throughput time.

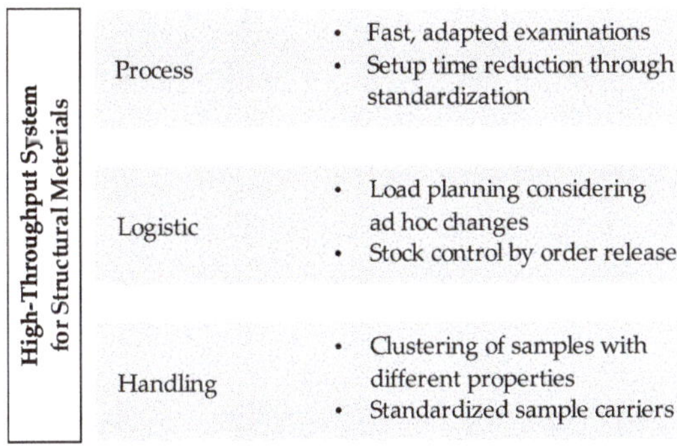

Figure 4. Framework for high-throughput system acceleration.

3.1. Development of Testing Processes

The 'Farbige Zustände' method requires the development of processes for short-term characterization that are suitable to describe structural materials in terms of their suitability for a given performance profile. In their approach to develop this high-throughput method, Ellendt and Mädler describe that previous, partially standardized methods for sample generation and material testing can be applied to a limited extent only, since the material and time requirements exceed the possibilities of the high-throughput system. Therefore, methods for the generation of metallic samples as well as for the treatment and testing of samples are developed or adapted for the high-throughput system [1]. This process is capable of producing several thousand individual samples per day. These must then be subjected to rapid processes to change and characterize the properties of the sample. These methods together form the basis for multi-stage screening. Micro samples are produced in a droplet generator for microscopic spheres, in which the sample size can be specifically adjusted by a controlled solidification process [37]. Thus, the investigation of new alloy compositions is based on the use of micro samples with a diameter of less than one millimeter. This combination saves both time in alloy evaluation and the amount of material [38]. The particular challenge is to determine the micro-properties, although the macro-properties differ significantly from the intended areas of application. The use of micro-samples results in special size effects that affect both the microstructure [39,40] and the macrostructure [40]. These are taken into account by a predictor-based optimization approach in the determination of micro process parameters, so that a high quality of information can be expected [5]. In this context, the use of microspheres offers the advantage that they exhibit homogeneous microstructure without macro segregations immediately after generation [4]. This structure is specifically adjusted in a subsequent heat treatment, which takes the scalability into account. Laser-deep-alloyed samples are also available. This process offers the generation of different alloy compositions and graded compositions in short intervals [41].

In addition to the alloy composition, the microstructure of materials is decisive for the mechanical properties of structural materials. These can be specifically modified by thermal and mechanical treatments. An example of this is the multistage contact deep rolling that produces an increase in surface integrity. For example, the rolling process can cause partial strain hardening of material and thus specifically adapt the mechanical properties of the sample [42]. These modified properties are investigated in various short-term characterization processes that allow conclusions to be made on the material properties of the samples [38]. In addition to methods of nanoindentation, X-ray diffraction and hardness measurement based on laser induced shock waves, a method for dilatometry, are part of the characterization process [4,43,44]. High-throughput dilatometry can be used to examine micro samples that have a high cooling rate due to their small volume and large surface area. Cramer et al. have developed a special dilatometer tailored to the needs of the high-throughput system, with which the thermal expansion of samples can be investigated [45]. These methods, which are developed with the aim of short process times, standardized sample transfer, and a specific testing target for high-throughput systems, form the basis for the development of a high-throughput system for materials testing.

3.2. Efficient Logistical Control for High-Throughput Systems

Logistics control already plays a central role in production today and essentially involves the coordination of industrial production. Acceleration of high-throughput systems cannot be achieved solely through technological improvements. Due to the complexity of the system, the enhancement of the logistical targets has to be expanded by organizational methods. Procedures for logistical control for high-throughput testing in materials testing are developed on the model of production control. The findings of the production are applied to high-throughput systems and adapted to their requirements. This way, a high throughput and a short throughput time of test orders will be achieved. For the use in high-throughput systems, the order generation, order release, sequencing, and capacity control are of great importance, these assume the central tasks for order organization and have a high influence on the logistical targets. The aim is to achieve an optimum of short throughput times, high throughput, and high utilization when planning the schedule of high-throughput tests.

The objective for the use of planning and control methods in high-throughput systems for materials testing is to improve reactivity, flexibility, and throughput. This sometimes results in boundary conditions that are untypical for other high-throughput systems. Schneider et al. describe that stations in conventional material testing are decoupled and not connected by an automatic logistics unit. Instead, samples are often transported manually between the test stations [36]. This results in an undirected material flow, which is also present in a variable process sequence. An individual test plan is created for each sample, which only contains tests that are assigned to a specific test objective. These can also be influenced by the sample characteristics and the test results already available. Furthermore, it may be necessary for material tests that stations have to work within temporarily paced sequences. This results in a variable pace in these sections, where orders have to be processed in certain time slots [46]. It is also necessary to provide decision patterns for re-tests and partial tests in the logistical control in which a test is carried out a second time or carried out partially [47]. Thus, special requirements for logistical controls for high-throughput systems mainly comprise the material flow complexity, the number of variants, and the production principle (see Table 1). In addition, Nyhuis and Wiendahl described criteria to be observed for the implementation of a production control system [15]. These criteria must also be taken into account when selecting a suitable control system due to the similar structure and organization of high-throughput and production systems. In particular, a distinction between make-to-order and make-to-stock or job shop and flow production is fundamental. In addition, the qualification and motivation of employees must also be taken into account as so-called soft factors in the development of management [15].

Table 1. Logistical control requirements [47].

Characteristics	Job Shop Production	Screening for Material Testing
Material flow	Undirected	Undirected
Process chain	Depending on order	Depending on order and aim of analysis
Variety	High	High
Re-routing	Rare case, should be avoided	Common, standard process
Ad hoc changes	Processing according to order content	Schedule changes depending on measured values

The control method for high-throughput systems is capable of minimizing the effects of ad hoc changes, sample re-routing, temporary paced sequences, and quantity changes in order throughput. Thus, the throughput and flexibility of the system are to be maintained. In the case of ad hoc changes, the test plan of orders is changed spontaneously so that these samples follow a different process chain through the system. This can occur, for example, due to an overload of the originally planned work systems or by changing the test plan on the basis of measured values. Figure 5 shows process chains of different orders in an exemplary system. The blue, green, purple, and yellow jobs have a different order through the system so that they represent ad hoc changes, rerouting of samples, and temporary concatenation. The purple order is re routed on the basis of a repeat check. The order is forwarded to working system (WS) 6 after being processed at WS 3. Then, the step at WS 3 and the test at WS 6 are repeated. The blue order represents the throughput of a paced order. It is assumed that the machining operations on working systems 2, 4, and 6 must take place in a paced sequence. Therefore, these working systems are treated as virtual flow production and the order is controlled accordingly. As an example of ad hoc changes, the orange order controls the working systems 5 and 7 at short notice, while the green order explains the effects of changes in unit quantities.

In addition to the tasks of short-term control, logistical planning and control must also include medium-term planning for a machine schedule. Schedule planning tasks include scheduling orders and machines. This results in detailed scheduling taking the current capacity into account. Based on the order and system data, forward scheduling is carried out for each order, without considering the order's situation. The earliest possible start and end date is determined. Conventional throughput time scheduling has been supplemented by additional block times to take dynamic paced sequences into account. As soon as a station is processed in a paced sequence, all other stations of this sequence must be reserved for a seamless connection. In addition, the orders are sorted according to urgent and normal orders to ensure fast processing of higher-priority orders in overloaded work systems. Normal orders are sorted according to the First-in-System-First-out (FiSFo) rule to generate even throughput times. In subsequent capacity requirements planning, capacity buffers must be planned in order to be able to react to unforeseeable changes in the demand program. The capacity requirements planning can be used to identify overloads [48].

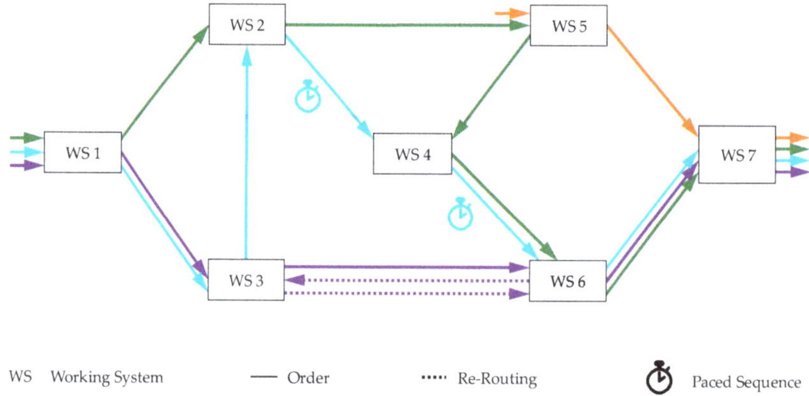

Figure 5. Structure of high-throughput systems for material testing [46].

Based on this planning status, the orders are released for processing and checking in the short-term logistical control system. To do so, a high-throughput-specific sequence of orders is first created. A typical procedure for this is the use of the planned due date from the schedule. The closer the planned due date is to the current time, the further forward the order is taken into account. The structured orders can then be checked for order release. A special challenge for order release has turned out to be the consideration of repeat and partial tests, as well as the efficient control of temporarily timed sequences. Onken et al. presented an approach, in which a reduction in throughput time and the proportion of blockages in the system was achieved through specific order release. The use of a modified Decentralized Work in Process-oriented Manufacturing Control (DEWIP) also significantly increased the output of individual work systems [47]. A further development of the logistic control regarding the consideration of temporarily paced sequences also shows a further reduction of the throughput time compared to order release using the FiSFo. The investigations have shown that even with the use of an order release regulation, the throughput time is reduced by more than 50 percent [47]. A machine-specific analysis of capacity utilization has shown that it fluctuates widely and is in some cases below ten percent. For other work systems, good capacity utilization of over 90 percent is achieved [46]. Schedule planning has not been taken into account, which can be expected to improve the utilization significantly.

3.3. Multimodal Handling

The beginning of technologies for enabling high-throughput can be marked with the invention of the micro titer plate in the 1950s, a carrier for a multitude of liquid samples that fast became a standard in the field. In the following decades, the efficiency was increased by increasing the automation, with automated plate readers and multichannel pipettes [49]. The size of the sample was more and more reduced by introducing 96-, 384-, and 1536-well plates, which reduce the cost, time, and amount of material. While a 96-well plate carried 100–200 μL, the sample size on a 384-well plate was reduced to typically around 50 μL and to 5 μL on a 1536-well plate. Even 3456-well plates have been reported, but these small samples have not been established in the field [50].

The example of HTS in chemistry and biology show the requirements and catalyzers for high-throughput systems, which are easy handling on standardized carriers, automation, parallelization, and miniaturization. These basic concepts of high-throughput systems have also been applied in other fields, as for example materials research. Examples include the development of thermoelectric materials [51] and the development of abrasion resistive surfaces [52]. In these cases, the objects under test are functional materials, which can be tested by their surface characteristics. This allows for easy sample creation by using surface technologies like physical vapor deposition (PVD). The samples are created directly on a substrate, which acts as a sample carrier. This means that many examinations can be carried out on such large-area samples.

For the development of structural materials, most tests cannot be performed on thin films on a carrier. Tests like nanoindentation or hardness measurement based on laser-induced shock waves require a sample as bulk material, due to changes made to the geometry and microstructure of the processes. To provide these, the generated microspheres are used as samples. Similar to the process in chemistry, these samples can be created in high masses with gradually changing characteristics. The most important difference to previous high-throughput systems is the existence of single samples, which have to be prepared individually for the different testing stations within the high-throughput system. Due to the size of the spheres and the required low handling times, an individual handling of the samples raises new challenges for the automation. In order to achieve a system with high throughput, methods from production engineering have to be applied. A particular challenge is that the adhesion forces in the micro area can exceed the weight force. Adhesion forces are Van der Waals forces, capillary forces, and electrostatic forces [53]. Two different concepts are known for the handling of micro components. Similar to macro and meso components, micro components can be handled individually. They can be handled by grasping and moving them with the help of micro grippers, or without grasping them with contactless methods, for example with electrostatic or electromagnetic forces [54]. Fantoni et al. lists magnetic, suction, and Van-der-Waals grippers in addition to conventional frictions grippers [55], which use the size effects of micro components in a targeted manner [56]. This makes it difficult to release and place the components, for which active and passive systems can be used. The challenge is precise and controlled depositing, which can, however, be compensated with additional technology such as measuring systems or targeted introduction of force into the component [55,57].

If micro parts are made from semi-finished products, the application of linked parts can be reasonable. During production, micro-components are left in the semi-finished product that is used as conveying and positioning equipment. It enables an easier handling of the parts in flow productions. As long as the parts are linked together, they can be handled as macro material. Depending on the connection between the individual parts, the linkage can be described as line-compound, ladder-compound, or comb-compound [58]. In the case of a totally linear processing, the linked parts could run through the whole system without any costly reordering. In the case of a multi dimensional factorial experiment, the effort for separating and reconnecting the linked parts would grow fast. In this case, the system cannot benefit from its strengths.

In order to optimize the throughput of the system, it has to be possible to create optimized batches for different processes. Therefore, a core requirement is the possibility of relocating individual samples and grouping them together to benefit from a parallelized handling in batches. Because of the multitude of different generating, treatment, and testing processes with different requirements, a multimodal handling system is used, which can transfer between different preparation methods and relocate samples for optimized batch processing. Examples for different modes are bulk handling, separation in trays, and fixated samples on a carrier. For the fixated handling on carriers, there are different types, depending on the requirements of the actual process. These may be requirements to the resistance to vertical and lateral forces, temperatures, chemicals, or other requirements as for example electrical conductivity. A constant identification of the components is ensured in order for the sphere's position to be linked to the obtained measurement results.

In order to allow relocation, any fixation method must be releasable. A promising method for the fixation of microspheres, which allows fast fixation and release with a small temperature impact, is the embedding in low melting alloys. These are alloys, which have melting points as low as room temperature [59]. This ensures damage-free dissolution of the samples, without changing their mechanical properties and microstructure in a high-speed handling, in order to prevent bottlenecks caused by embedding and re-bedding. Thus, complex process chains with different sample preparation requirements are achieved for each treatment or testing process.

4. Discussion

The acceleration of high-throughput systems for testing structural materials will be achieved through a holistic approach. Aspects of production, setup of microstructure, and test processes as well as aspects of logistics and handling must be considered. Actions must be taken in each of the three areas to accelerate these areas individually. This is the only way to introduce high-throughput screening for testing structural materials. The validation of fast methods for material characterization is particularly important for implantation, since conventional material testing methods cannot be performed on microspheres. However, due to their rapid generation in the droplet generator and their homogeneous microstructure, these are well suited as samples within high-throughput screening.

The acceleration of the entire high-throughput screening process by adapting logistical concepts from production management is promising. Simulations have shown success in reducing throughput times and increasing capacity utilization by using individual components of logistical control. However, further research needs to be addressed in this context, as the improvement achieved does not yet allow for highly productive operation. Further methods and targets need to be developed that reflect the order requirements of individual work systems more clearly. Overall, a fully developed high-throughput system for testing structural materials can thus be used in all areas of design and development. Especially in combination with the development of highly stressed components or the need for special component properties, material development with the 'Farbige Zustände' method is useful.

The completed high-throughput system will give new insights into the principle relations between microstructure and mechanical properties of the material. On the one hand, the predictor function itself will point out connections between measured characteristics and microstructure. These connections may help the scientists to better understand the basic principles of the material's behavior. On the other hand, the high-throughput system 'Farbige Zustände' as a tool can not only be used to find purpose specific materials, but also to get a better overview of the possibilities of material and structure combinations. The fast process sheds light on the white spots in the homogenous parameter space of material composition and treatment. The high-throughput system will allow researchers to find interesting compositions and to investigate them faster than ever before. For the industry, the fully developed high-throughput system will enable the use of specifically fitted materials for their requirements. This results, for example, in lighter or stringer parts, which will help to save energy or make products more durable.

5. Summary and Outlook

High-throughput systems for testing structural materials offer an approach to rapid characterization of material properties. The use of micro-samples allows for the investigation of a large number of alloy compositions with a small material and time input. In addition, various mechanical and thermal treatment processes are available to change the microstructure and thus the material properties. A fast generation of defined metallic samples in combination with fast tests enables the combination to a high-throughput screening. In addition, initial studies show that a logistical optimization can reduce throughput times by more than 50 percent. In contrast to conventional high-throughput screenings, systems for the examination of structural materials show a dynamic, non-directional material flow. Overall, the complexity of the system is increasing, so that actions must be taken to organize the system. This is a well-known problem in the field of production planning and organization. In industrial production systems, the focus is usually on economical operation and high added value. In this context, many processes have been developed that enable a productive and flexible production system. Transferring these approaches to high-throughput systems enables an increase in performance and a reduction in throughput time.

The framework for the acceleration of high-throughput systems makes partial use of these methods. A reduction of the throughput time in high-throughput systems can be achieved through the areas of process, logistics, and handling acceleration. The improvement of the individual processes in terms of throughput time is achieved by developing methods that are adapted to the aim of the analysis and the used samples. In this context, standardized sample carriers are used, which have uniform interfaces and are used for various types of examinations. This also provides an advantage when handling the samples to be tested. A standardized interface enables fast onward transport without any re-embedding processes in the meantime. In addition, these sample carriers offer the possibility of combining samples of different properties in one batch and reducing handling effort. As a third component, the framework for the acceleration of high-throughput systems includes measures for the efficient organization of the system. The focus is on load planning, taking ad hoc changes like re-tests, partial testing, and temporarily paced sequences into account. Inventory control through the use of a logistical control system is also part of order processing.

The development of the high-throughput system has mainly been carried out on individual components. Especially for the logistic approaches, an evaluation of the overall system is important. In this context, future work will create an organizational connection between the individual processes in order to induce interactions between work systems and logistical control. It is to be expected that order combinations will be generated by the test planning and logistic control, which will achieve synergy for better system utilization. In this way, the system's productivity will be increased overall.

Author Contributions: Conceptualization, K.T., A.B., F.M.; Methodology, K.T., A.B.; Investigation, A.B.; Resources, K.T.; Writing—Original Draft Preparation, A.B., F.M.; Writing—Review and Editing, A.B., F.M., K.T.; Visualization, A.B.; Supervision, K.T.; Project Administration, K.T.; Funding Acquisition, K.T.

Funding: This research was funded by Deutsche Forschungsgemeinschaft (DFG) (grant number 276397488).

Acknowledgments: Financial support of subproject P03 'Planning and Control of Defining and Testing Processes of Microstructures' and subproject P04 'High Throughput Handling of Micro Samples' of the Collaborative Research Centre SFB1232 'Farbige Zustände' by the German Research Foundation (DFG) is gratefully acknowledged.

Conflicts of Interest: The authors declare no conflict of interest. The funders had no role in the design of the study; in the collection, analyses, or interpretation of data; in the writing of the manuscript, and in the decision to publish the results.

References

1. Charles, J.A.; Crane, F.A.A.; Furness, J.A.G. *Selection and Use of Engineering Materials*, 3rd ed.; Butterworth-Heinemann: Oxford, UK, 1997.
2. Springer, H.; Raabe, D. Rapid alloy prototyping: Compositional and thermo-mechanical high throughput bulk combinatorial design of structural materials based on the example of 30Mn–1.2C–xAl triplex steels. *Acta Mater.* **2012**, *60*, 4950–4959. [CrossRef]
3. Duff, D.G.; Ohrenberg, A.; Voelkening, S.; Boll, M. A Screening Workflow for Synthesis and Testing of 10,000 Heterogeneous Catalysts per Day–Lessons Learned. *Macromol. Rapid Commun.* **2004**, *25*, 169–177. [CrossRef]
4. Ellendt, N.; Mädler, L. High-Throughput Exploration of Evolutionary Structural Materials. *HTM* **2018**, *73*, 3–12. [CrossRef]
5. Mädler, L. Is High-throughput screening for structural materials/metals possible? In Proceedings of the 4th International Conference on Nanomanufacturing (nanoMan2014), Bremen, Germany, 8–10 July 2014.

6. Hertzberg, R.P.; Pope, A.J. High-throughput screening: New technology for the 21st century. *Curr. Opin. Chem. Biol.* **2000**, *4*, 445–451. [CrossRef]
7. Maier, W.F.; Stöwe, K.; Sieg, S. Combinatorial and high-throughput materials science. *Angew. Chem. (Int. Ed. Engl.)* **2007**, *46*, 6016–6067. [CrossRef] [PubMed]
8. Rehm, H.; Letzel, T. *Der Experimentator: Proteinbiochemie/Proteomics*, 7th ed.; Springer Spektrum: Berlin/Heidelberg, Germany, 2016.
9. Bensch, M.; Schulze Wierling, P.; von Lieres, E.; Hubbuch, J. High Throughput Screening of Chromatographic Phases for Rapid Process Development. *Chem. Eng. Technol.* **2005**, *28*, 1274–1284. [CrossRef]
10. Sewing, A.; Winchester, T.; Carnell, P.; Hampton, D.; Keighley, W. Helping science to succeed: Improving processes in R&D. *Drug Discov. Today* **2008**, *13*, 227–233. [PubMed]
11. Wiendahl, H.-P. *Load-Oriented Manufacturing Control*; Springer: Berlin/Heidelberg, Germany, 1995.
12. Bochmann, L.; Gehrke, L.; Böckenkamp, A.; Weichert, F.; Albersmann, R.; Prasse, C.; Mertens, C.; Motta, M.; Wegener, K. Towards Decentralized Production: A Novel Method to Identify Flexibility Potentials in Production Sequences Based on Flexibility Graphs. *Int. J. Autom. Technol.* **2015**, *9*, 270–282. [CrossRef]
13. Bessant, J.; Haywood, B. Flexibility in manufacturing systems. *Omega* **1986**, *14*, 465–473. [CrossRef]
14. Glock, C.H. Lead time reduction strategies in a single-vendor–single-buyer integrated inventory model with lot size-dependent lead times and stochastic demand. *Int. J. Prod. Econ.* **2012**, *136*, 37–44. [CrossRef]
15. Nyhuis, P.; Wiendahl, H.-P. Logistic Production Operating Curves–Basic Model of the Theory of Logistic Operating Curves. *CIRP Ann.* **2006**, *55*, 441–444. [CrossRef]
16. Wiendahl, H.-P. *Fertigungsregelung: Logistische Beherrschung von Fertigungsabläufen auf Basis des Trichtermodells;* Carl Hanser Verlag: München, Germany, 1997.
17. Schuh, G. (Ed.) *Produktionsplanung und-Steuerung: Grundlagen, Gestaltung und Konzepte*, 3rd ed.; Springer: Berlin, Germany, 2006.
18. Johnson, D.J. A framework for reducing manufacturing throughput time. *J. Manuf. Syst.* **2003**, *22*, 283–298. [CrossRef]
19. Dave, Y.; Sohani, N. Single Minute Exchange of Dies: Literature Review. *Int. J. Lean Think.* **2012**, *3*, 27–37.
20. Grewal, C.S.; Enns, S.T.; Rogers, P. Performance Effects of Setup Time Reduction with and without Decision Variable Re-Optimization: A Simulation-Optimization Study. In Proceedings of the 2009 Winter Simulation Conference, Austin, TX, USA, 13–16 December 2009; pp. 2135–2144.
21. Nyhuis, P.; Mayer, J. Modelling the influence of setup optimized sequencing on lateness and productivity behaviour of workstations. *CIRP Ann.* **2017**, *66*, 421–424. [CrossRef]
22. Suresh, N.C.; Meredith, J.R. Coping with the Loss of Pooling Synergy in Cellular Manufacturing Systems. *Manag. Sci.* **1994**, *40*, 466–483. [CrossRef]
23. Gutenberg, E. *Grundlagen der Betriebswirtschaftslehre: Die Produktion*, 11th ed.; Springer: Berlin/Heidelberg, Germany, 1965.
24. Lutz, S.; Lödding, H.; Wiendahl, H.-P. Logistics-oriented inventory analysis. *Int. J. Prod. Econ.* **2003**, *85*, 217–231. [CrossRef]
25. Lödding, H. *Handbook of Manufacturing Control: Fundamentals, Description, Configuration*; Springer: Heidelberg, Germany, 2013.
26. Hartmann, W.; Fischer, A.; Nyhuis, P. The Impact of Priority Rules on Logistic Objectives: Modeling with the Logistic Operating Curves. In Proceedings of the World Congress on Engineering and Computer Science, San Francisco, CA, USA, 20–22 October 2009.
27. Pettersen, J.-A.; Segerstedt, A. Restricted work-in-process: A study of differences between Kanban and CONWIP. *Int. J. Prod. Econ.* **2009**, *118*, 199–207. [CrossRef]
28. Land, M.J.; Gaalman, G.J.C. The performance of workload control concepts in job shops: Improving the release method. *Int. J. Prod. Econ.* **1998**, *56–57*, 347–364. [CrossRef]
29. Bechte, W. Theory and practice of load-oriented manufacturing control. *Int. J. Prod. Res.* **1988**, *26*, 375–395. [CrossRef]
30. Krafcik, J.F. Triumph of the Lean Production System. *Sloan Manag. Rev.* **1988**, *30*, 41–52.
31. Karam, A.-A.; Liviu, M.; Cristina, V.; Radu, H. The contribution of lean manufacturing tools to changeover time decrease in the pharmaceutical industry. A SMED project. *Procedia Manuf.* **2018**, *22*, 886–892. [CrossRef]

32. Allen, M. Thinking Lean in Compound Management Laboratories. In *Management of Chemical and Biological Samples for Screening Applications*; Wigglesworth, M., Wood, T., Eds.; Wiley-VCH Verlag GmbH & Co. KGaA: Weinheim, Germany, 2012; pp. 83–105.
33. Sundar, R.; Balaji, A.N.; Kumar, R.S. A Review on Lean Manufacturing Implementation Techniques. *Procedia Eng.* **2014**, *97*, 1875–1885. [CrossRef]
34. Warnecke, H.J.; Hüser, M. Lean production. *Int. J. Prod. Econ.* **1995**, *41*, 37–43. [CrossRef]
35. Dombrowski, U.; Crespo, I.; Zahn, T. Adaptive Configuration of a Lean Production System in Small and Medium-sized Enterprises. *Prod. Eng. Res. Dev.* **2010**, *4*, 341–348. [CrossRef]
36. Schneider, D.; Funke, L.; Tracht, K. Logistische Steuerung von Hochdurchsatzprüfungen: Steuerung von Mikroproben in einem System mit mehreren Prüfstationen. *Wt-Online* **2015**, *105*, 818–823.
37. Xu, Y.; Ellendt, N.; Li, X.-G.; Uhlenwinkel, V.; Fritsching, U. Characterization of cooling rate and microstructure of CuSn melt droplet in drop on demand process. *Trans. Nonferr. Met. Soc. China* **2017**, *27*, 1636–1644. [CrossRef]
38. Drechsler, R.; Eggersgluß, S.; Ellendt, N.; Huhn, S.; Madler, L. Exploring superior structural materials using multi-objective optimization and formal techniques. In Proceedings of the IEEE 2016 Sixth International Symposium on Embedded Computing and System Design, ISED 2016, Bihar, India, 15–17 December 2016; pp. 13–17.
39. Kals, T.A.; Eckstein, R. Miniaturization in sheet metal working. *J. Mater. Process. Technol.* **2000**, *103*, 95–101. [CrossRef]
40. Vollertsen, F. Categories of size effects. *Prod. Eng. Res. Dev.* **2008**, *2*, 377–383. [CrossRef]
41. Vetter, K.; Freiße, H.; Vollertsen, F. *High-Throughput Material Development Using Selective Laser Melting and High Power Laser*; 7. WGP-Jahreskongress: Aachen, Germany, 2017; pp. 511–518.
42. Kämmler, J.; Wielki, N.; Meyer, D. Surface integrity after internal load oriented multistage contact deep rolling. *Procedia CIRP* **2018**, *71*, 490–495. [CrossRef]
43. Huhn, S.; Sonnenberg, H.; Eggersglus, S.; Clausen, B.; Drechsler, R. Revealing properties of structural materials by combining regression-based algorithms and nano indentation measurements. In Proceedings of the 2017 IEEE Symposium Series on Computational Intelligence (SSCI), Honolulu, HI, USA, 27 November–1 December 2017; pp. 1–6.
44. Czotscher, T.; Veenaas, S.; Vollertsen, F. Possibilities to Characterise Laser Induced Shock Waves. *J. Technol. Plast.* **2017**, *42*, 1–7.
45. Cramer, L.; Saddei, P.; Surm, H.; Steinbacher, M.; Zoch, H.W. Investigation of metallic adapter sleeves for high throughput dilatometry on spherical micro samples. *Materialwissenschaft und Werkstofftechnik* **2018**, *49*, 101–112. [CrossRef]
46. Bader, A.; Onken, A.-K.; Tracht, K. Order Release for Temporary Paced Sequences in Flexible High Throughput Systems. *Procedia CIRP* **2018**, *72*, 689–694. [CrossRef]
47. Onken, A.-K.; Bader, A.; Tracht, K. Logistical Control of Flexible Processes in High-throughput Systems by Order Release and Sequence Planning. *Procedia CIRP* **2016**, *52*, 245–250. [CrossRef]
48. Bader, A.; Lorenz, J.; Tracht, K. Logistische Steuerung für Hochdurchsatzsysteme: Struktur einer Belegungsplanung für strukturell dynamische Hochdurchsatzsysteme. *Wt-Online* **2017**, *107*, 847–851.
49. Noah, J. New developments and emerging trends in high-throughput screening methods for lead compound identification. *IJHTS* **2010**, *1*, 141–149. [CrossRef]
50. Mayr, L.M.; Bojanic, D. Novel trends in high-throughput screening. *Curr. Opin. Pharmacol.* **2009**, *9*, 580–588. [CrossRef] [PubMed]
51. Wong-Ng, W.; Yan, Y.; Otani, M.; Martin, J.; Talley, K.R.; Barron, S.; Carroll, D.L.; Hewitt, C.; Joress, H.; Thomas, E.L.; et al. High Throughput Screening Tools for Thermoelectric Materials. *J. Electr. Mater.* **2015**, *44*, 1688–1696. [CrossRef]
52. Chisholm, B.; Potyrailo, R.; Shaffer, R.; Cawse, J.; Brennan, M.; Molaison, C. Combinatorial chemistry methods for coating development. *Prog. Org. Coat.* **2003**, *47*, 112–119. [CrossRef]
53. Sánchez, A.J. Handling for Micro-manufacturing. In *Micromanufacturing Engineering and Technology*; Elsevier: New York, NY, USA, 2015; pp. 637–659.
54. Weikert, F.; Tracht, K. Handling of Micro Parts. In *Micro Metal Forming*; Vollertsen, F., Ed.; Springer: Berlin/Heidelberg, Germany, 2013; pp. 331–342.

55. Fantoni, G.; Santochi, M.; Dini, G.; Tracht, K.; Scholz-Reiter, B.; Fleischer, J.; Kristoffer Lien, T.; Seliger, G.; Reinhart, G.; Franke, J.; et al. Grasping devices and methods in automated production processes. *CIRP Ann.* **2014**, *63*, 679–701. [CrossRef]
56. Fantoni, G.; Hansen, H.N.; Santochi, M. A new capillary gripper for mini and micro parts. *CIRP Ann.* **2013**, *62*, 17–20. [CrossRef]
57. Brecher, C.; Weinzierl, M. New approaches for an automated production in ultra-precision machining. *Int. J. Adv. Manuf. Technol.* **2010**, *47*, 47–52. [CrossRef]
58. Tracht, K.; Schenck, C.; Weikert, F.; Kuhfuß, B. Förderung mikrokaltumgeformter Teile im Verbund: Überwindung von Größeneffekten durch Verknüpfen der Bauteile. *wt Werkstattstech. Online* **2010**, *100*, 864–868.
59. Meiners, F.; Tracht, K. Clamping of Microspheres with Low Melting Metals. In Proceedings of the 2nd World Congress on Micro and Nano Manufacturing (WCNMN) 2018, Portorož, Slovenia, 18–20 September 2018; in press.

© 2018 by the authors. Licensee MDPI, Basel, Switzerland. This article is an open access article distributed under the terms and conditions of the Creative Commons Attribution (CC BY) license (http://creativecommons.org/licenses/by/4.0/).

 *materials*

Article

Design and Simulation of Production and Maintenance Management Applying the Viable System Model: The Case of an OEM Plant

Sergio Gallego García [1] and Manuel García García [2,*]

[1] Department of Construction and Fabrication Engineering, National Distance Education University (UNED), 42005 Soria, Spain; sgallego118@alumno.uned.es

[2] Department of Construction and Fabrication Engineering, National Distance Education University (UNED), 28015 Madrid, Spain

* Correspondence: mggarcia@ind.uned.es; Tel.: +34-682-880-591

Received: 2 July 2018; Accepted: 31 July 2018; Published: 3 August 2018

Abstract: Designing, changing and adapting organizations to secure viability is challenging for manufacturing companies. Researchers often fail to holistically design or transform production systems. Reasons are often the conflict of interests between production and maintenance, the temporal divergence of their activities and their organizational structure. Thus, the aim of this study is to propose a holistic approach of how production and maintenance can be designed, changed or managed. Hereby, the Viable System Model was applied. This structure can be applied to any kind of structured organization and for its management with goals to be achieved in modern society; however, focus of the research is the coordination of production and maintenance management. The goal of the developed model is to be able to react to some potential production environments by taking coordinated decisions correctly and in the right moment based on the needed information. To ensure this, standardized communication channels were defined. In conclusion, this proposed approach enables production systems to have internal mechanisms to secure viability depending on all potential environment scenarios.

Keywords: cybernetics; system dynamics; production management; maintenance management; Viable System Model; automotive industry

---

1. Introduction

The global logistics flows have increased dramatically in recent years due to a globalized economy that introduces inherent challenges to the establishment of international business [1] (p. 10). This evolution is combined with the demands of customers who want to be served with shorter delivery times [2] (p. 1) as well as with the increasing product variants in manufacturing and assembly processes that expose planning and control logistics to new challenges [3] (p. 797). The reasons are mainly the variability of final customers favored by competition activities. Other factors that generate uncertainty effects in the planning within the supply chain are, for example, minimum quantities in production as well as deviations of delivery lead times that make more difficult the production planning reducing the planning quality. The conventional answer to this challenge is to increase safety stocks to ensure the expected service level [4] (p. 179).

The success of the actors involved in a cooperative supply chain depends in an essential way on the extent to which they are able to deal with the dynamic market requirements [1] (p. 101)—together with efficient supply processes in costs, a reliable planning within the supply chain as well as robust production and logistics systems play a fundamental role for the long-term success of the actors within the supply chain. Robust production systems avoid negative effects on production processes by identifying potential disruptions and enabling an early adaptation [5] (p. 575). The result of a survey within a study of the German Association of Logistics ratifies this statement detailing that the reliability of logistics and production systems is the factor with greatest relevance considering logistics costs, reaction capabilities, flexibility and use of resources [6]. The logistics goal of robustness defined as the capacity of a system to deal with breakdowns, deviations and changes of the system environment without necessary changes in the structure or in production capacities [7] (p. 34) is gaining more and more importance as strategic feature to secure competitiveness.

Information in real time and its processing for production planning and control can give the correct answer how to face increasing dynamic requirements. However, the current planning and control logics do not provide the necessary support for this, which causes unexpected deviations in planning which cannot be compensated in the short term [7] (p. 160). The turbulences in production and in the supply chain create uncertainties that generate the need for coordination of planning processes, which has not been considered in a methodical way for the current concepts of production planning and control [8] (pp. 36–37). In this way, the insufficient collection and use of planning information lead to delays and deficits in the transmission of information that must be compensated through additional costs [9] (pp. 3–5). Moreover, due to the increasing automation and implementation of process control systems, more and more decentralized maintenance units are located next to production areas. Based on these facts, the common task for both production and maintenance must be developed to optimize the availability and condition of production plants to guarantee the satisfaction of the final customer. From a production point of view, maintenance has evolved from the "auxiliary need" to the holistic maintenance management, in which production and maintenance share a common goal [10] (p. 9).

According to a survey carried out by the VDMA (Mechanical Engineering Industry Association) organization of 240 European companies of all sizes and of different industries, in 2006, 78.5% of managers claimed that the importance of maintenance has increased significantly in recent years and 67.1% that it will continue to increase in the future [11] (p. 38). Considering this, it is prioritized between maintenance and production tasks. The negative long-term effects of insufficiently implemented maintenance measures are often accepted due to short-term production needs. Without denying the character of service, the role of maintenance is different if it is more widely understood as holistic maintenance management. Maintenance management is an essential component of production management but should not be used only to meet the production objectives (production of goods and services), but also the objectives of generating value, human objectives and environmental aspects of the company [10] (pp. 9–10).

Many approaches have been considered to make production and maintenance planning as efficient and flexible as possible. All pursue and many theoretically achieve the partial or total optimization of global production systems by making them efficient. However, when applying these theoretical internal logistics management models to practice, information systems should be used as a vehicle to communicate the necessary information to be managed within the organization. The reasons why the majority of applications of such concepts in information systems, as in ERP (Enterprise Resource Planning) systems, have failed are: the delay or lack of information, the misuse of production planning tools and maintenance, lack of coordination or conflict of interest between the production and maintenance departments, the non-consideration of the environment and the requirements of the final customer in a dynamic process of continuous improvement, leading all of them to take strategic, tactical or operational decisions at an optimal time or in an optimal way.

Research focus is the design and simulation of a coordination model for production and maintenance management using as a basis the current state of the art and pursuing high adaptability to environment changes. To achieve this, the Viable System Model (VSM) is used as a methodological structured approach and system dynamics is used to describe interrelationships between parameters and control loops in production and maintenance. The goal of the research is to generate an approach of how to design and develop production systems to be successful in all potential future environment scenarios considering developments in other areas of the company and in external fields of a company such as customer and supplier needs and conditions, politics, economy, technology, environment and energy regulations and society. In this context, simulation serves as tool to prove the approach in some potential future scenarios for production and maintenance coordination.

The initial hypothesis is that a production system built using the structure of the VSM will be able to react faster to environment changes and therefore to improve short-, medium- and long-term goals of every producing company.

2. Materials and Methods

This chapter contains a literature review of the viable system model, system dynamics as well as a description of the simulation technique and software employed.

2.1. The Viable System Model (VSM)

The Viable System Model (VSM), a cybernetic management model, was developed by Stafford Beer throughout his life [12] (p. 57). Beer deduced the VSM by taking the central nervous system of the human being and cybernetics as a basis to deal with complex systems. The VSM is built on three main principles: viability, recursivity and autonomy [13] (p. 434). The cybernetic model of every viable system consists always in a structure with five necessary and sufficient subsystems that are in relation in any organism or organization with the ability to conserve its identity with independency of its environment [12] (pp. 21–22).

To validate the research methodology, research and practical applications using the VSM were searched. Many authors have used the VSM as basis to describe and develop models how to deal with complex challenges in social and industry fields. Some of the topics worked and that give an indication of the scientific value of the approach are: organizational models for companies [14] (pp. 74–76), lean methods in terms of attenuating and amplifying variety [15], production system focused on "make-to-order" manufacturing [16], optimization of patient care in university hospitals [17], order booking process in mass production companies [18], production master program during launch processes [19], production planning in real time in the consumer goods industry [20] and integrated planning of distribution networks [21].

As described in the literature, the VSM is an unmatched conceptual and methodological tool for the modeling and design of organizations and its areas with the goal of being viable [22] (p. 16). Thus, the aim of the research is to propose a self-regulating approach of how designing and coordinating production and maintenance within manufacturing companies. For this reason, the Viable System Model is applied for this purpose.

2.2. System Dynamics (SD)

System dynamics is a computer-guided approach for studying, managing and solving complex feedback problems with focus on policy analysis and design [23] (p. 342). The origin of system dynamics [24] is the field developed by Forrester called "Industrial Dynamics" [25]. It proposes a methodology for the simulation of dynamic models by studying the characteristics of the information feedback of industrial systems.

SD has been applied to a great set of systems from corporate strategy to the dynamics of diabetes as well as for the cold war arms race between the USA and USSR. System dynamics can be applied to any dynamic system, with any time and spatial scale [24] (pp. 41–42). In a firm context, SD addresses three important issues: it helps to determine which policies should be used to control the behavior of the firm over time and, when the circumstances change, how these policies should be designed in order to have a robust response against change and how the organization can design its information feedback structure to assure the correct implementation of effective policies [26] (p. 3). The use of System Dynamics Modeling in Supply Chain Management has re-emerged in recent years after a long-stagnated period [23] (p. 342). The application of systems dynamics for the coordination of production management and maintenance makes sense since the cause–effect relationships show the interrelationships between the elements of the system and help to evaluate the influences of the different decisions in the global system. Therefore, it is selected for the research purpose.

2.3. Simulation Software

According to "VDI-Richtlinie", simulation is "the reproduction of a system with its dynamic processes in an experimental model capable to gain knowledge that can be transferred to reality" [27] (p. 48). Simulation models are mainly used to support decision-making because they show the dynamic behavior of a system [28] (p. 28). Simulation is the only practical way to test models because our mental models are dynamically deficient, omitting feedback, time delays, accumulations and nonlinearities [24] (p. 37). For all these reasons, simulation is used to reproduce the conceptual model and to validate initial hypotheses. In the market, there are different software packages that enable system dynamics modeling such as: AnyLogic, DYNAMO, iTHINK, POWERSIM, STELLA and VENSIM [29] (p. 43). From all of them, VENSIM simulation software was selected for the research work. VENSIM is a registered trademark of Ventana Systems Inc., Harvard, MA, USA) serves as platform to build stock and flow model diagrams as well as causal loop diagrams. VENSIM also provides very powerful tools for analysis and validation of results and model structure as well as to determine most convenient policy options.

3. Literature Review on Production and Maintenance Management

This chapter provides the basic terminology needed to successfully perform this work.

3.1. Production Management

Production is the foundation of human activity. Natural resources are transformed into useful products through production processes to meet the needs of society [30] (p. 319). Production management contains the tasks of design, planning, monitoring and control of the productive system and business resources such as people, machines, material and information [31] (pp. 249–273). The multi-dilemma of production planning originates discussions repeatedly in the context of divergent objectives. This conflict of goals is shown in Figure 1 [32] (p. 36).

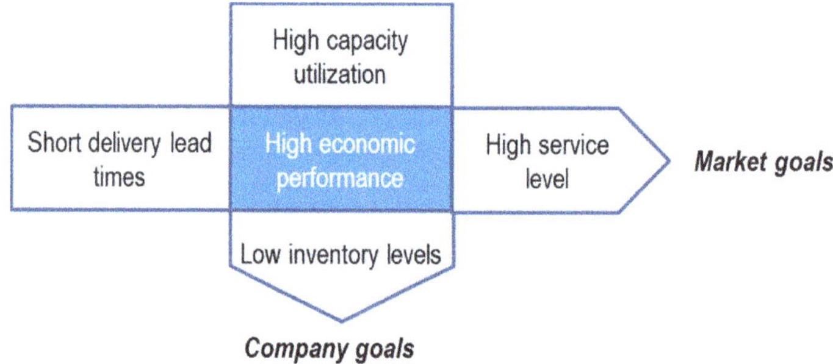

Figure 1. Multi-dilemma of production planning [32] (p. 36).

To analyze the tasks of production management, the Aachener Production Planning and Control (PPC) model, which is a reference model for its analysis, evaluation and design, is used [33] (p. 29).

All tasks are distinguished vertically in Figure 2 according to their strategic, tactical or operational nature [33] (pp. 30–32). In the past, the focus was on operational and tactical problems; however, to successfully manage logistics in the future, an active level of strategic planning is also required [34] (p. 1).

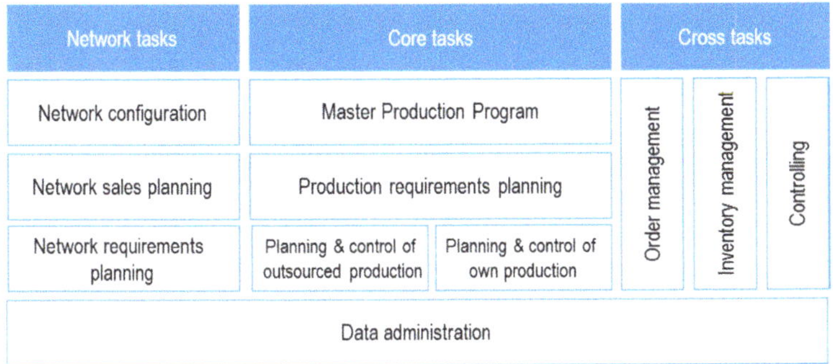

Figure 2. Production management tasks acc. to the Aachener PPC [33] (p. 30).

3.2. Maintenance Management

Industrial maintenance is defined according to DIN (German Institute for Standardization) 31051 as the "combination of all technical, administrative and management measures during the life cycle of an observation unit in order to maintain the functional status or restoring it so that it can fulfill the required function" [35] (p. 26). The greater degree of complexity of technologies, systems and processes has considerably increased the demands on operating time. The increase of maintenance importance is a logical consequence since maintenance processes affect the quality, delivery time and costs and therefore company's performance [15] (pp. 16–17). An optimal strategy must be found between the costs of preventive maintenance and the costs of machine failures [35] (p. 25). The basic maintenance strategies are corrective, periodic, preventive oriented to the condition of the object and predictive [35] (p. 104). The three "classic" maintenance strategies are still justified as seen in Figure 3.

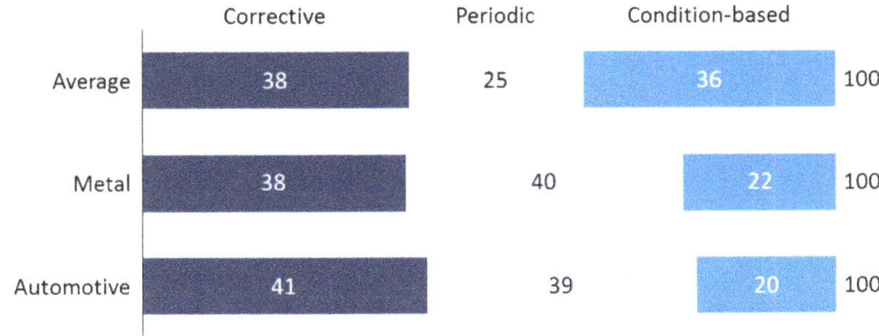

Figure 3. Survey result of 240 European companies on maintenance strategies [11] (p. 41).

The "ideal maintenance organization" does not exist [35] (p. 65) but depends on optimally combining the organizational forms with their advantages and disadvantages for the respective company [11] (p. 23).

According to experts, German companies spend around 140,000 million euros annually on maintenance of machines and installations. This, together with the fact that between 75% and 80% of the execution time of a maintenance order are activities without added value, defines a great potential for savings in production companies. Many of the time losses are in the interfaces between the different areas. The management of maintenance processes must ensure the change from a functional orientation to an orientation towards processes [35] (p. 60). For this purpose, tasks of maintenance management are: planning, control, analysis and measures execution [36] (p. 173).

3.3. Coordination Concepts of Production and Maintenance Management

Maintenance management of production units affects not only the maintenance personnel but also all the employees of the whole company [35] (p. 104). In the last two decades, there has been a general rethinking of industrial maintenance philosophy, from the maintenance of functions to a philosophy of value creation. However, for many maintenance tasks, there is a lack of adequate methods and instruments as well as information technology solutions [11] (p. 69).

Logistics can be described as an interdisciplinary strategy to optimize production. Availability is, therefore, a key factor for logistics and is also the link between logistics and industrial maintenance [35] (pp. 4–6). Maintenance should be "integrated" and dependent on all the functional areas involved in the added value [36] (pp. 9–10). This maintenance concept refers to the function and not just to the maintenance department within an organization. This is characterized by a distribution of the maintenance function "on several shoulders" [11] (p. 38). For all the above, some of the approaches to integrate or manage maintenance within the productive system are [36] (pp. 4–9):

- Total Productive Maintenance (TPM),
- Lean Maintenance,
- Total Lifecycle Cost Strategy (TLC),
- Reliability Centered Maintenance (RCM),
- Knowledge Based Maintenance.

4. Results

Within the research work, a conceptual model is developed and simulated for the integrated planning of production and maintenance based on the Viable System Model (VSM).

4.1. Conceptual Model Design Applying the VSM

4.1.1. Methodology

This chapter describes the components of an integrated planning model for production and maintenance management. In a first step, the planning tasks are presented according to its planning level (strategic, tactical, and operational). Later, the recursion levels of the VSM are described. Then, the planning tasks are associated with the different recursion levels and the information flows are defined between recursion levels and systems of the VSM.

4.1.2. Production and Maintenance Management Tasks Acc. to Their Planning Horizons

Production systems are considered important in relation to aspects of quality, time and costs [37] (p. 1). As explained before, planning tasks can be classified into strategic, tactical and operational planning depending on the respective planning horizon. Therefore, this classification was performed for the production management tasks in Figure 4. Moreover, with the increase of automation levels and the decrease of personnel in production, the importance of maintenance is increasing. With the same methodology, maintenance management tasks are also classified into the different time horizons in Figure 5:

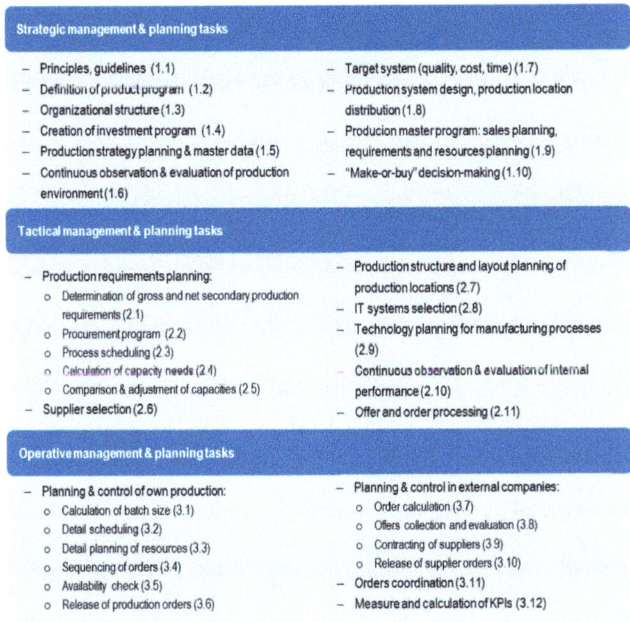

Figure 4. Production management and planning tasks acc. to time horizons (own elaboration).

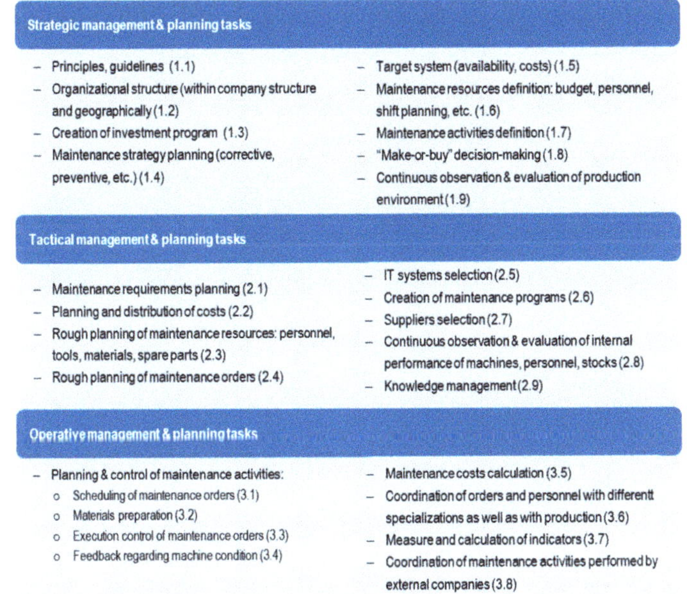

Figure 5. Maintenance management and planning tasks acc. to time horizons (own elaboration).

4.1.3. Recursion Levels Definition

A company is assumed as a viable system that is the first level of recursion in which the five systems neededto ensure viability are found. Therefore, in this research work, four levels of recursion can be differentiated as it is shown in Figure 6:

- The highest level, company (n − 1),
- The recursion level of production (n). In the same recursion level it can be found finance, human resources, IT, research and development, etc.
  - The recursion level of plant or production workshop, for example production management activities in an automotive production or assembly shop (n + 1),
  - The level of machine group, machine or installation with the associated activities for the different production activities such as preparation of the machine, change of tools, operation, control of production, etc. (n + 2),
- The recursion level of maintenance (n). In the same recursion level can be found finance, human resources, production, etc.
  - The recursion level of plant or production workshop—for example, maintenance management activities in an automotive production or assembly shop (n + 1),
  - The recursion level of installation or machine with the associated activities for the different maintenance activities: preventive, planned repairs, corrective, etc. (n + 2).

The activities in the recursion level n + 2 are no longer viable systems in contrast to the higher recursion levels because they do not contain a structure like that of the VSM, since they are the elements of production or maintenance execution.

Within this first level of recursion, company, the different functions of a company can be found, such as production, maintenance, commercial, finance, research and development, information systems, etc. In this research project, production and maintenance tasks will be analyzed in detail, recursion level n, but also considering the function of system 2 at the company level, n − 1, whose function is to coordinate the different functional areas of a company.

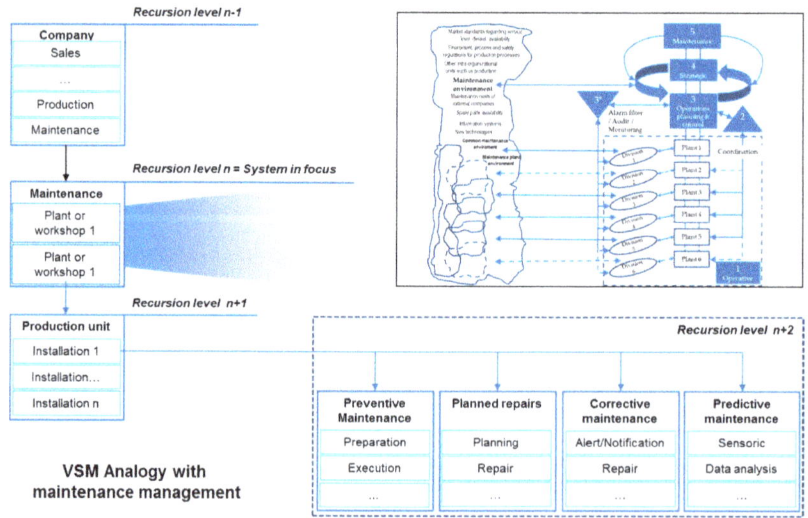

Figure 6. VSM recursion levels for company, production and maintenance (own elaboration).

System 2 at company level plays the role of coordinator between the functional areas of the company trying to solve conflicts between them. Moreover, systems 1 at company level are all functional areas of every company such as production or maintenance. At the recursion level of production and maintenance (n), it is assumed that the different production plants or workshops will be the respective system 1, which also contains a viable system in each of these locations. The VSM of production and maintenance management within a company is described in the next paragraphs by the tasks performed by its five necessary systems:

- System 5 establishes the production and maintenance objectives and communicates them to the other management systems, systems 3 and 4.
- System 4 observes and collects essential information from the external environment of maintenance and production. Production environment is mainly represented by the demands of customers, but also by other factors such as information systems offered by the market for the management, planning and control of production, new manufacturing technologies and, in general, all factors affecting the production system such us market standards, delivery times, production strategies, delays, production costs in external companies for example to help in making decisions about outsourcing or to not manufacture certain parts or about the assembly of certain sets, etc. Maintenance environment could be described by the costs in the market for maintenance activities, average reaction times in the market as well as service level standards, information systems for maintenance, technology development, environmental and work safety regulations and customer delays. With these and other information from the external environment and information from system 5, system 4 creates a vision of what the production and maintenance areas have to be, and which should be the measures to be followed to reach that state. This vision is validated internally with system 3 so that system 4 makes the decision and system 3 makes the changes internally.
- System 3 is responsible for maintaining the internal stability of the model by optimizing the use of internal resources using the information received by system 4 about the clients as well as the information on the different divisions of system 1 through system 2. It would be related to functions such as operative production and maintenance management and control, information management, quality management, operative logistics planning and control, etc. Moreover, system 3* allows a quick response to possible emergencies in the manufacturing process or in the production control and monitoring by acting before information flows through system 2. It is capable of performing actions in real time if something happens outside of normal limits such as making changes in sequencing and production scheduling to avoid stopping production flow.
- System 2 is represented by the functions of coordination between the different production locations in daily activities. This system receives all the information of the different production plants and acts as a filter so that only the necessary information reaches system 3. The difference between both is in the time horizons of action. While system 2 performs functions in daily activities, the tactical system optimizes the performance of the internal system over a longer time horizon.
- System 1: each plant or workshop within the production system is an operational unit that includes the management of the unit and the division that performs the operational activities. An example could be an assembly workshop that contains the planning and production control departments responsible for the components of the workshop with their team leaders together with the operators that finally perform the production tasks.
- Environment: represents all the external factors that influence the production and maintenance management in a company. Figure 7 shows the environment of the entire maintenance area as well as of each plant or workshop.

Figure 7 was developed as a result of applying the VSM for maintenance management. In the same way it was done for production management following the descriptions of the previous paragraphs:

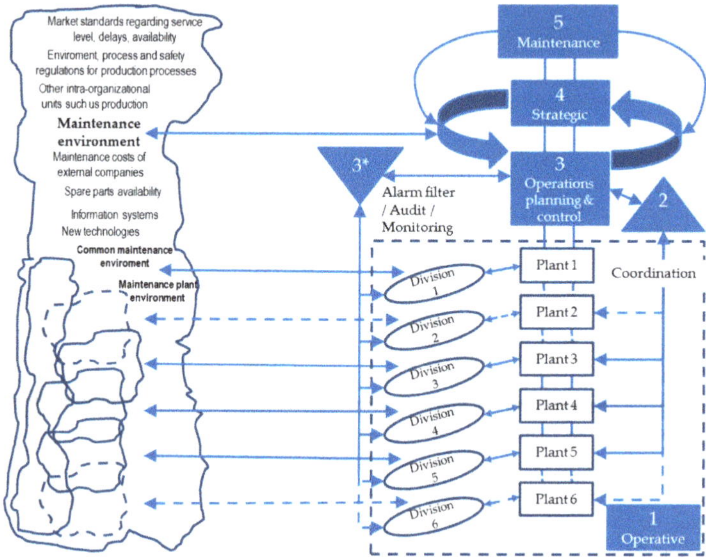

Figure 7. VSM analogy with maintenance management (own elaboration).

4.1.4. Association of Planning Tasks to Recursion Levels

Production and maintenance management tasks were assigned to the VSM systems at recursion levels n and n + 1 defined before. In Table 1, the strategic production management tasks are classified and the tactical maintenance tasks and its classification to the VSM systems are shown in Table 2. It was done for all other production and maintenance tasks in the same way:

Table 1. Strategic production management tasks and its classification to VSM systems (own elaboration).

Key Performance Indicator	Production Recursion Level			Plant Recursion Level		
Systems of the VSM	S5	S4	S3	S5	S4	S3
Principles, guidelines (1.1)	X	-	-	X	-	-
Definition of product program (1.2)	X	-	-	-	-	-
Organizational structure (1.3)	-	X	-	-	-	-
Creation of investment program (1.4)	-	X	-	-	-	-
Production strategy planning & master data (1.5)	-	X	-	-	-	-
Continuous observation & evaluation of production environment (1.6)	-	X	-	-	-	-
Target system (quality, cost, time) (1.7)	-	X	-	-	-	-
Production system design, production location distribution (1.8)	-	X	-	-	-	-
Production master program: sales planning, requirements and resources planning (1.9)	-	X	-	-	-	-
"Make-or-buy" decisions (1.10)	-	X	-	-	-	-

Table 2. Tactical maintenance management tasks and its classification to VSM systems (own elaboration).

Key Performance Indicator	Maintenance Recursion Level			Plant Recursion Level		
Systems of the VSM	S5	S4	S3	S5	S4	S3
Maintenance requirements planning (2.1)	-	-	X	-	-	-
Planning and distribution of costs (2.2)	-	-	-	-	-	X
Rough planning of maintenance resources: personnel, tools, materials, spare parts (2.3)	-	X	-	-	-	-
Rough planning of maintenance orders (2.4)	-	X	-	-	-	-
IT systems selection (2.5)	-	-	-	-	X	-
Creation of maintenance programs (2.6)	-	X	-	-	-	-
Suppliers selection (2.7)	-	-	X	-	-	-
Continuous observation & evaluation of internal performance of machines, personnel, stocks (2.8)	-	-	-	-	-	X
Knowledge management (2.9)	-	-	-	-	X	-

4.1.5. Identification of Information Flows between Recursion Levels

Current technical literature agrees that the connection interfaces between recursion levels is extremely important [38] (p. 59). The goal is to determine basic links that can be transferred to any VSM in any company. The intensity of this connection between the levels varies according to the company [38] (p. 59). An exchange of information within the company and between levels of recursion is necessary to control the corporate environment, which generally has more information than can be processed in the company [14] (p. 287). Between the recursion levels, the following communication flows for both production and maintenance can be found:

- Between the company environment and system 4 at the production/maintenance recursion level,
- Between systems 5 of company and production/maintenance,
- Between systems 4 of company and production/maintenance,
- Between systems 3 of company and production/maintenance,
- Between systems 2 of company and production/maintenance,
- Between the operating units, systems 1, of company and production / maintenance,
- Between the alarm/monitoring filter (System 3*) of the company's recursion level and system 4 of production/maintenance.

Between the two normative systems of company and production/maintenance, there is a flow of information that defines the degree of freedom of decision-making in which production/maintenance recursion level can act. Specifically, it means that the decisions taken by the management of the company are communicated to production and maintenance management defining its guidelines for autonomous decision-making within the respective areas. These guidelines can be financial, on personnel, on affectation to other areas, etc. In the same way, the objective levels such as production in term, production quality and production costs and adaptation capacity are influenced by decisions from the management, defining the priorities and the limits for the coordination among production areas. An example could be: the direction of the company in its strategic plan establishes the target production volume for the following years as well as the required flexibility in percentage on the production as well as the decrease in target costs. Of course, these decisions would influence the decision-making framework for production and maintenance that should adapt their methods and tools to be able to optimize costs, times and quality based on that flexibility also securing the required availability and adapting its maintenance strategy.

As explained during the research work, basic communication flows were defined. In total, 88 information connections were defined for the production recursion level specifying if the communication goes from company's recursion level to production recursion level or between systems in production recursion level. Moreover, for the recursion level of maintenance, a total number of 77 information connections were defined specifying if the communication goes from company's recursion level to maintenance recursion level or between systems in production recursion level. An extract is shown in Table 3 for the communication flows of the production recursion level:

Table 3. Extract of information flows in the production recursion level (own elaboration).

No.	Information on the Production Recursion Level	From ... to ...
1	Information about stopped facilities	From System 1 to 2/3
...	...	...
28	Number of orders and quantity produced	From System 1 to 4/5
29	Number of orders that have met the required deadlines and quantity	From System 1 to 4/5
30	Average Delivery time of the products	From System 1 to 4/5
31	Number of defective Deliveries	From System 1 to 4/5
32	Number of Deliveries with claims	From System 1 to 4/5
33	Total number of changes made to production schedules	From System 1 to 4/5
...	...	...
88	Information about the economic environment of a productive plant	From environment to 4/5

4.2. Applying the Conceptual Model to an OEM Production System

The automotive supply chain is composed of suppliers at three levels (Tier 1–3), OEM (car manufacturers), distributors. For the work, we are going to focus on the production process of an OEM [39] (pp. 23–32):

1. Press shop
2. Bodywork shop
3. Paint shop
4. Engines shop
5. Assembly shops
6. Final assembly shops

Efficient logistics management is becoming a reality to survive in the automotive sector. The fragmentation and segmentation of vehicle models (such as hatchbacks, sedans, vans and pick-ups, cross-over coupes, roadsters, two-seater vehicles, SUVs, etc.) are growing. The complexity of customized models and variants is on the rise, especially as regards the way individual vehicles are equipped. The key trend in automotive production is the standardization of construction modules in common platforms. The modules refer to groups of components and related systems. The diversity of models is an important sales argument and delivery time is the key factor for the automotive market and in the manufacturing process. These requirements involve a change in assembly operations that need to be more flexible and agile [40] (pp. 24–25).

In the automotive industry, there are two approaches depending on when to start the assembly process. In the current market, most OEM producers use both to a greater or lesser extent:

1. Build-to-forecast: the assembly process starts before you have a customer for that car configuration.
2. Build-to-order: the assembly process is started only when there is a customer associated with that variant that is to be assembled.

On the other hand, and regarding the importance of maintenance in the automotive industry, it can be foreseen that, with the increase of automation and mechanization in the automotive industries, the production processes are becoming very sensitive to the machines and the human factor. Consequently, the role of equipment maintenance in the automotive industry as a mean to control and reduce costs while achieving the highest standards of reliability plays a fundamental role [41] (p. 514).

4.3. Model Simulation Using System Dynamics for an OEM Production System

In this step, the conceptual model for a manufacturing process is specified. To do this, the interrelations between indicators are defined by Causal Loop Diagrams (CLD). Later, the concepts are modeled in the VENSIM commercial simulation program. The simulation models present a simplified depiction of a real production and maintenance process to quantify the effect of decisions in a production system. Moreover, it contains the validation of the models and the presentation of the results obtained. Finally, the main conclusions are summarizedas well as a critical reflection of the work including a perspective for future research activities.

4.3.1. Casual Loop Diagrams (CLD)

The possibility of developing simulation models for production and maintenance management starts from the knowledge of the interrelations between its elements. Using casual loop diagrams (CLD), the effects of the changes of certain factors in their dependent parameters are shown. Five CLDs are developed from production management to maintenance planning. Figure 8 shows the CLD for the production system as an example:

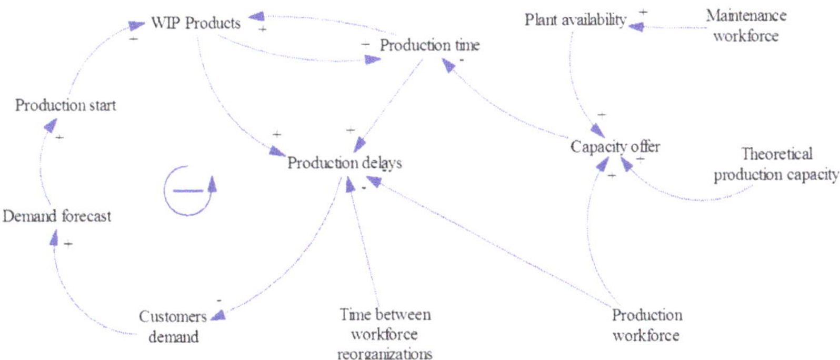

Figure 8. CLD for the production system.

4.3.2. Methodology, Assumptions and Comparison Conditions

The aim of the simulation is to observe the impact of the delay in decision-making in a chain of manufacturing processes from the steel stamping process to the end of assembly in the automotive industry. The hypothesis is that a simulation model applying the VSM will present better results in terms of study parameters compared to the one that does not apply the VSM or that has delays in internal decision-making. The methodology for the simulation design is:

1. Definition of the objective, hypothesis and methodology
2. Number of simulation models
3. Simplification of the complexity of the conceptual model through assumptions
4. Criteria that make possible a comparison between models
5. Definition of quantitative parameters to obtain results and compare models
6. Definition of the production flow
7. Creation of the model based on the CLD (Casual Loop Diagrams)
8. Validation of the behavior of the simulation models
9. Determination of scenarios, simulation and extraction of results
10. Evaluation of the results and derivation of conclusions

First, assumptions are defined to simplify the model with focus on the simulation goal:

- Production times for car models are not variable
- Times for material transport and employee movements are not variable
- Distribution of finished products as given

- Procurement of raw material as given
- Demand does not change if customer service is better or worse
- Each order has a production unit
- Bill of materials are not considered

Some elements are equal in all models to make possible a comparison between the models under the same conditions:

- Same demand, same patterns of demand, demand replication
- Same production capacities per production step
- Same breakdowns per production step
- Same reaction times depending on the location
- Same task execution times
- Same workforce capacity
- The same maintenance capacity is considered for the same class of worker

4.3.3. Key Performance Indicators (KPIs) for the Simulation Model

The objectives can be qualitative or quantitative. The research goal is to study the behavior of the different models in different situations of demand and configuration of maintenance and production areas. The results are quantified to evaluate the response according to the following key performance indicators:

1. Production on time (% per week)
2. Total production (# thous. cars)
3. Total downtime (days for all workshops)
4. Availability of final assembly workshop (%)
5. Capacity utilization of final assembly workshop (%)
6. Cumulated stocks (# Mill. cars)
7. MAD (Mean Absolute Deviation) (# thous. cars)
8. Cumulated demand (# thous. cars)

4.3.4. Definition of the Production Flow in the Simulation Model

The production process consists of a process from steel stamping to the final revision shop. The plants in the process are shown in Figure 9:

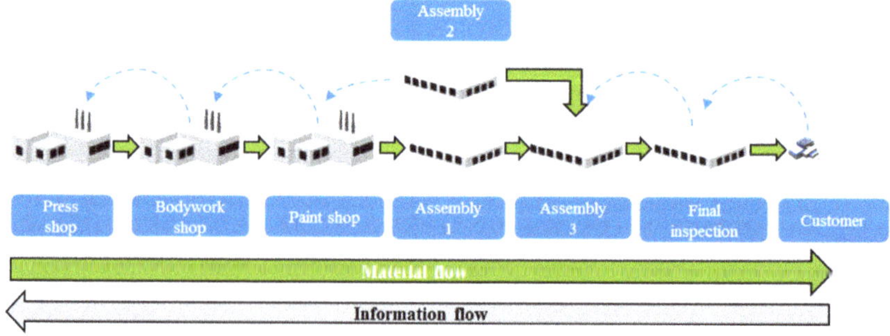

Figure 9. Simulation production flow (own elaboration).

- Press shop
- Bodywork shop
- Paint shop
- Pre-assembly shop—assembly 1
- Mechanical assembly shop—assembly 2
- Final assembly shop—assembly 3
- Final inspection shop

All of them are in different shops and have production and maintenance units associated to each one of them.

4.3.5. Design of the Intra-Organizational Simulation Model for Production and Maintenance Management

As shown in Figure 10 the model is designed according to the following criteria:

1. Time restrictions: first, the modeler must define a time horizon and units of time. It is easy to fulfill that step by asking to what extent the simulation should be considered. In the case of the study, it has been decided to simulate four working years to evaluate influences in the medium and long term.
2. Production strategy: the model offers the possibility to manufacture according to the push or pull principle. A mixed strategy can also be chosen. It assumes the existence of a single product or model in the production flow.
3. Maintenance strategy: the simulation allows defining a preventive maintenance ratio so that it reduces the corrective according to a factor that is assumed.
4. Capacity and production methods: in the model there are two methods: CONWIP (CONstant Work in Progress) in which the quantity of products in production process is always the same and BOA (Workload-dependent order release) that initiates production of an order based on the current and expected workload for the plants.
5. Organizational structure and tasks distribution between production and maintenance: the simulation allows personnel changes in the areas of maintenance and production as well as between them. In addition, assumptions about the influence of the changes are made, for example, if there are more maintenance personnel, it will take less time to perform preventive maintenance tasks.
6. Demand characteristics and forecast: the model makes different forecasts using two different methods: moving average and linear regression.

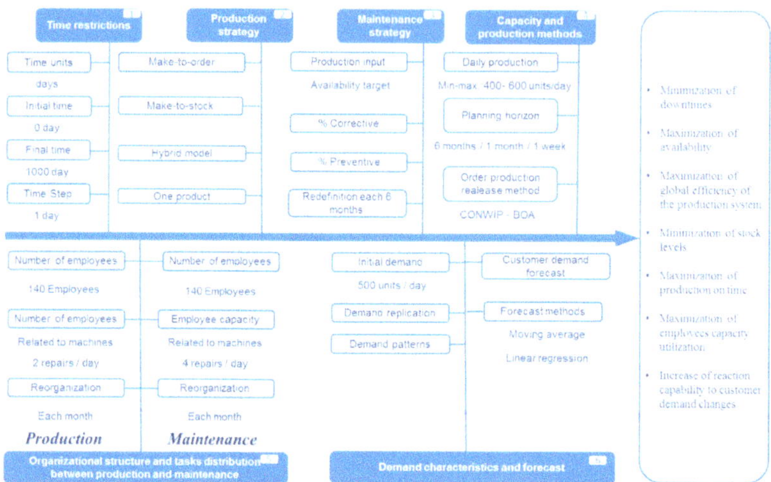

Figure 10. Ishikawa Diagram of Hypothesis for the simulation model (own elaboration).

4.3.6. Simulation Model Validation

The validation of the simulation models can be done using different methods. In this process, some simulation variables will be used to observe their behavior and to evaluate if the models will be validated. Sterman defined 12 possible methods to validate system dynamics models [42] (p. 6). One of them, the test of extreme values, is used to validate the simulation model that shows that the response of the model is plausible when taking extreme values for different input parameters. For all models, the same input and output variables are chosen to analyze and validate the models. These input variables are the total number of employees, the initial stock in work in progress (WIP) and the production strategy, make-to-order vs. make-to-stock.

- For a lower number of employees, production on time (%) must be lower and the total stock should be higher because the production facility provides more products than employees can process. Moreover, utilization of shop capacities should be lower and therefore also the production output.

As shown in Figure 11, it can be observed how the model behaves as expected. With 20 employees of maintenance and production per shop, the results for production on time (%) are 25% higher than with 15 employees and 50% higher than with 10 employees. In addition, total production and capacity utilization are higher for 20 employees than for 15 or 10 employees. Finally, cumulated stock over time is higher for 10 employees than for 15 and 20 employees because, with a lower number of production and maintenance employees, production volumes decreased.

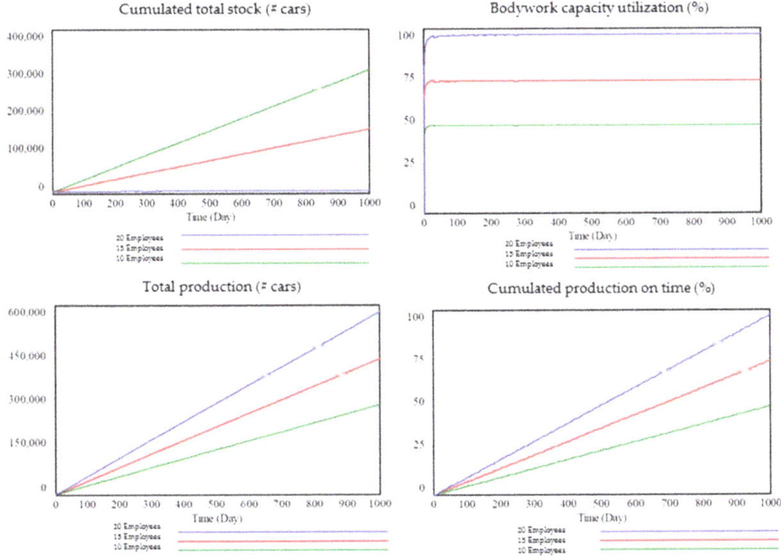

Figure 11. Validation with extreme values: number of employees.

- For higher initial stocks in WIP (work in progress) at initial time, production on time (%) must be higher and total stock should be higher in the first weeks because the production facility has more products within production process at the beginning. Moreover, utilization of shop capacities should be higher and therefore also the production output in the first weeks. Afterwards, due to the CONWIP method, the production WIP converges to a CONWIP goal that is equal to one week of demand and therefore all other indicators also converge.

In Figure 12, it is shown the model results for 500, 300 and 100 WIP products as initial WIP stocks in all intermediary stocks. The results show how initially the total stock, capacity utilization, production on time and total production at the beginning are higher for 500 WIP products at initial WIP; however, after 20–40 days, all of these performance indicators are equal for all set-ups of the initial WIP stock because the model initiates production of more products if the initial stock is low due to the CONWIP method set-up.

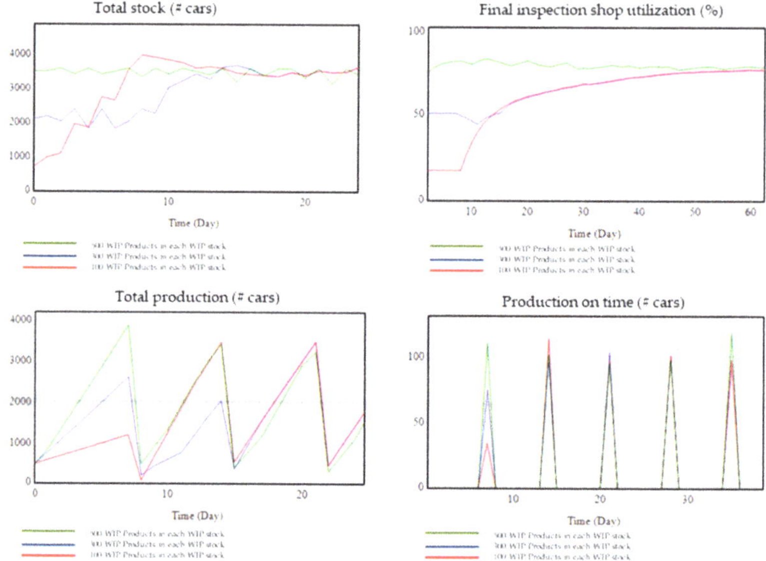

Figure 12. Validation with extreme values: WIP Initial stock.

- For a make-to-order production, it is expected to have less stock during the production process, less capacity utilization and production output as well as less production on time than with a make-to-stock production.

As shown in Figure 13, it can be seen how the model results are as expected. In a make-to-order production stock, the stock after painting is lower than in make-to-stock production. Capacity utilization is 20% lower for make-to-order than for make-to-stock. Total production is higher for make-to-stock than for make-to-order because, in this configuration, only the customer orders are produced and if these orders are lower than the capacity, the extra-capacity is not utilized. Moreover, in a make-to-stock production, more products are served on time.

The model is validated because logical expected values are obtained for three different input parameters influencing multiple key performance indicators of the simulation.

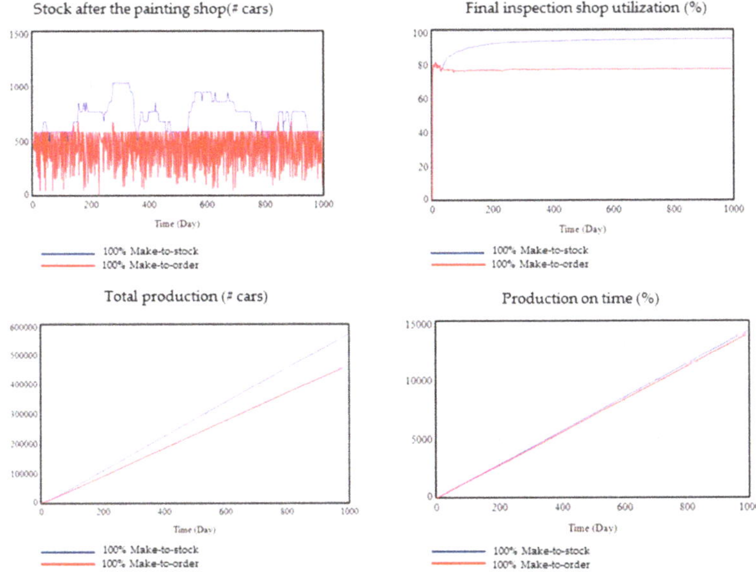

Figure 13. Validation with extreme values: Make-to-order vs. Make-to-stock.

4.3.7. Scenario Definitions

Four case studies are proposed in which two different model configurations are simulated, trying to reflect the importance of decision-making. The VSM model takes earlier these decisions and therefore can benefit earlier from the new production system configuration:

1. Case study 1—reorganization of production and maintenance employees: the production area transfers workers to the maintenance area when there is a low customer demand. With this scenario, how the utilization of production workforce for maintenance activities could increase global production performance, if the decision is made in an early moment when demand starts to be lower than production capacities, is studied.
2. Case study 2—reorganization of production and maintenance employees: the maintenance area transfers workers to production when there is a high customer demand. With this scenario, how the utilization of maintenance employees in production activities could increase global production performance, if the decision is made in an early moment when demand starts to be almost at the maximum level of production capacities, is analyzed.
3. Case study 3—comparison between the CONWIP and BOA methods: for the same customer demand, the order initiation method varies. CONWIP (CONstant Work in Progress) initiates production of new orders maintaining the same WIP cars while BOA initiates production of new orders depending on current and future equipment workload. BOA tries to optimize utilization and CONWIP in the model tries to maintain a week of production demand always as WIP.
4. Case study 4—comparison between preventive and corrective maintenance: for the same demand, a low level of preventive maintenance activities and a high level of maintenance activities is simulated. With this scenario, the effect of the preventive level of maintenance activities can be simulated. In the models, a parametrization is done considering that preventive activities require a certain employee capacity but reduced the unexpected downtimes of the production shops.

4.3.8. Simulation Results

For the four scenarios, the following results are obtained:

1. Case study 1—reorganization of production and maintenance employees with a low demand:
2. Case study 2—reorganization of production & maintenance employees with a high demand:
3. Case study 3—comparison between the CONWIP and BOA methods:
4. Case study 4—comparison between preventive and corrective maintenance:

5. Interpretation of the Simulation Results

1. Case study 1: the production area transfers workers to the maintenance area when there is a low demand for production. This decision is only taken in the VSM simulation model and moves three employees of a total of 20 employees from each production area to its respective maintenance area. The logic reason for this change is, since production employees cannot be fully utilized, they can help the maintenance units to perform maintenance activities in order to improve the production system performance. As it can be seen in Table 4, downtimes are more than 30% lower and production on time is almost 1% higher.
2. Case study 2: the maintenance area transfers workers to production when there is a high production demand. In this case, the Non-VSM model presents in one case 20 maintenance and production workers per plant or workshop while the VSM model changes to 23 workers for production and 17 for maintenance. Therefore, maintenance activities have less capacity while production capacity is increased. As it can be seen in Table 5, the VSM model presents downtimes almost 30% lower, total production is higher by 2000 cars, production on time is 0.2% higher and stocks are 26% lower.
3. Case study 3: it can be observed in Table 6 how the BOA method has better results in terms of quantity produced by almost 5%, in terms of production on time by 0.8% and in terms of capacity utilization by 4.3% since it starts the production of orders according to the workload. However, the BOA method has more cumulated stocks, 17% more than the CONWIP method, because orders are more time in production flow while securing capacity utilization. Therefore, in a VSM model of a manufacturing company, it can be decided whether to use BOA or CONWIP depending on company goals.
4. Case study 4: in the simulation model, there is a parameter that indicates the level of intensity of preventive maintenance. As it can be seen in Table 7, when making the decision of increasing preventive maintenance, availability increased by 3.7% and unexpected downtimes decreased by 39%.

Table 4. Case study 1—Results.

KPI-No.	Key Performance Indicator	Without Reorg. of Employees (non-VSM)	With Reorg. of Employees (VSM)
1	Production on time (% per week)	98.4	99.2
2	Total production (# thous. cars)	458	458
3	Total downtime (days for all workshops)	139	97
4	Availability of final inspection shop (%)	94.1	94.1
5	Capacity utilization of final inspection shop (%)	76.3	76.3
6	Cumulated stocks (# Mill. cars)	3.3	3.3
7	MAD (# thous. cars)	23	23
8	Cumulated demand (# thous. cars)	458	458

Table 5. Case study 2—Results.

KPI-No.	Key Performance Indicator	Without Reorg. of Employees (non-VSM)	With Reorg. of Employees (VSM)
1	Production on time (% per week)	96.9	97.1
2	Total production (# thous. cars)	579	581
3	Total downtime (days for all workshops)	143	100
4	Availability of final inspection shop (%)	94.2	94.2
5	Capacity utilization of final inspection shop (%)	96.5	96.8
6	Cumulated stocks (# Mill. cars)	5.7	4.2
7	MAD (# thous. cars)	12.3	12.3
8	Cumulated demand (# thous. cars)	582	582

Table 6. Case study 3—Results.

KPI-No.	Key Performance Indicator	CONWIP	BOA
1	Production on time (% per week)	96.3	97.1
2	Total production (# thous. cars)	542	568
3	Total downtime (days for all workshops)	143	100
4	Availability of final inspection shop (%)	94.2	94.2
5	Capacity utilization of final inspection shop (%)	90.4	94.7
6	Cumulated stocks (# Mill. cars)	4.6	5.4
7	MAD (# thous. cars)	12.3	12.3
8	Cumulated demand (# thous. cars)	577	577

Table 7. Case study 4—Results.

KPI-No.	Key Performance Indicator	10% Preventive	100% Preventive
1	Production on time (% per week)	97.1	97.1
2	Total production (# thous. cars)	579	579
3	Total downtime (days for all workshops)	192	117
4	Availability of final inspection shop (%)	91.3	95.0
5	Capacity utilization of final inspection shop (%)	96.5	96.5
6	Cumulated stocks (# Mill. cars)	5.1	4.9
7	MAD (# thous. cars)	12.3	12.3
8	Cumulated demand (# thous. cars)	579	579

Based on the results presented, it can be said that the VSM simulation model improves relevant KPIs when making the decision to move employees and can help production systems when deciding which production methods should be used as well as deciding the optimal percentage of preventive maintenance to be performed.

6. Conclusions

After completion of the research work, the importance of the design and management of a productive system that integrates the planning tasks of production and maintenance of a company has been proved. In addition, the following points can be successfully concluded:

- Identification of the objectives, key performance indicators and functions of production and maintenance management as well as basic methods and tools
- Clear evidence of the need for new approaches in the coordination of production and maintenance management after reviewing the state of the art
- Confirmation of the need for new communication and coordination systems as well as integral management of internal logistics as one of the biggest potentials to deal with current and future challenges in the manufacturing sector
- The application of the structure of the Viable System Model (VSM) for the design of the conceptual model for production and maintenance management provided a structure that has allowed the definition of information flows, planning levels as well as mechanisms of autonomy and escalation between levels within a company and within the areas of production and maintenance.
- System Dynamics has provided the notation and functions necessary for the design of simulation models using VENSIM as commercial software.

- The comparison of the different models for the four cases presented has shown the relevance of the organizational structure decisions and methods for production and maintenance management. In addition, it was proved how production and maintenance performance can be optimized by making decisions at the right moment. Key performance indicators such as total production, service level, downtime, etc. improved significantly when moving employees, changing production method or the maintenance strategy.

According to the results of the comparison, it has been demonstrated that correct and early decision-making through the application of the VSM gives a better response of all the key performance indicators. A similar response of the models without application of the VSM can occur when there are no external or internal changes that affect the productive system, that is, if the production system would be in a static situation. As environments and companies are always dynamic, the usefulness of the VSM as a principle for the definition of responsibilities, interfaces, escalation methods and decision-making are confirmed.

In conclusion, this proposed approach can increase the efficiency of internal logistics models such as the productive system in its interaction with industrial maintenance. Finally, it is important to point out new ways of research or new ways to continue improving the project carried out:

- Improve the conceptual model with all the different methods and tools by level of recursion and task as well as its inclusion in the simulation model
- Extend the study for all internal logistics processes to generate a tool to help management, since, as we have seen, the markets require flexibility and right now the companies do not have mechanisms to make the right decisions in most of the cases. Many of them centralize maintenance units and a few years later they decentralize it again, which suggests that there is no strategic or tactical direction.
- Apply the simulation model based on the VSM for a real productive system

Author Contributions: Conceptualization, M.G.G. and S.G.G.; Methodology, M.G.G. and S.G.G.; Software and programming, S.G.G.; Validation, S.G.G.; Data Analysis, S.G.G.; Writing (Review and Editing), S.G.G.

Funding: This research received no external funding.

Conflicts of Interest: The authors declare no conflict of interest.

References

1. Frazelle, E. *Supply Chain Strategy: The Logistics of Supply Chain Management*; McGraw Hill: New York, NY, USA, 2002.
2. Tu, Y.; Dean, P. *One-of-a-Kind Production*; Springer Science & Business Media: Berlin/Heidelberg, Germany, 2011.
3. Auerbach, T.; Bauhoff, F.; Beckers, M.; Behnen, D.; Brecher, C.; Brosze, T.; Esser, M. Selbstoptimierende Produktionssysteme. In *Integrative Produktionstechnik für Hochlohnländer*; Springer: Berlin/Heidelberg, Germany, 2011; pp. 747–1057.
4. Jodlbauer, H. *Produktionsoptimierung*; Springer Science & Business: Berlin/Heidelberg, Germany, 2008.
5. Sobaszek, Ł.; Gola, A.; Kozłowski, E. Application of survival function in robust scheduling of production jobs. In Proceedings of the 2017 Federated Conference on IEEE Computer Science and Information Systems (FedCSIS), Prague, Czech Republic, 4–7 September 2017; pp. 575–578.
6. Frank, S.; Hans-Christian, P. *Studie Trends und Strategien in der Logistik 2008: Die Kernaussagen*; Bundesverband Logistik BVL: Bremen, Germany, 2008.
7. Schikora, A. *Anforderungen an Die Unternehmensführung im Turbulenten Umfeld*; Igel Verlag RWS: Hamburg, Germany, 2014.
8. Fischäder, H. *Störungsmanagement in Netzwerkförmigen Produktionssystemen*; Deutscher Universitäts-Verlag: Wiesbaden, Germany, 2007.
9. Fleisch, E.; Christ, O.; Dierkes, M. Die betriebswirtschaftliche Vision des Internets der Dinge. In *Das Internet der Dinge*; Springer: Berlin/Heidelberg, Germany, 2005; pp. 3–37.

10. Der Arbeitsdirektionen Stahl, A.E.M. *Integration von Produktion und Instandhaltung;* Arbeitspapier, Hans-Boeckler-Stiftung: Düsseldorf, Germany, 2003.
11. Kuhn, A.; Schuh, G.; Stahl, B. *Nachhaltige Instandhaltung: Trends, Potenziale und Handlungsfelder Nachhaltiger Instandhaltung;* Ergebnisbericht der vom BMBF geförderten Untersuchung Nachhaltige Instandhaltung; VDMA-Verlag: Hanover, Germany, 2006.
12. Espejo, R.; Harnden, R. (Eds.) *The Viable System Model: Interpretations and Applications of Stafford Beer's VSM;* Wiley: Hoboken, NJ, USA, 1989.
13. Schuh, G.; Stich, V.; Brosze, T.; Fuchs, S.; Pulz, C.; Quick, J.; Bauhoff, F. High resolution supply chain management: Optimized processes based on self-optimizing control loops and real time data. *Prod. Eng.* **2011**, *5*, 433–442. [CrossRef]
14. Herold, C. *Ein Vorgehenskonzept zur Unternehmensstrukturierung Heuristische Anwengung des Modells Lebensfahiger Systeme;* Difo Druck: Bamberg, Germany, 1991.
15. Herrmann, C.; Bergmann, L.; Halubek, P.; Thiede, S. Lean production system design from the perspective of the viable system model. In *Manufacturing Systems and Technologies for the New Frontier;* Springer: London, UK, 2008; pp. 309–314.
16. Brosze, T. Kybernetisches Management Wandlungsfähiger Produktionssysteme. Ph.D. Thesis, RWTH Aachen, Aachen, Germany, 2011.
17. Erbsen, A. *Krankheit im Zentrum: Gestaltung von Krankheitsorientierten Spitalstrukturen aus Kybernetisch-Konstruktivistischer Sicht;* Springer: Berlin/Heidelberg, Germany, 2012.
18. Kompa, S. *Auftragseinlastung in überlastsituationen in der Kundenindividuellen Serienfertigung;* Apprimus-Verlag: Aachen, Germany, 2014.
19. Schürmeyer, M. *Kybernetische Produktionsprogrammplanung;* Apprimus Verlag: Aachen, Germany, 2014; p. 19.
20. Hering, N.S. *Echtzeitfähig-Koordinierte Produktionsplanung in Supply-Chains der Verbrauchsgüterindustrie;* Apprimus-Verlag: Aachen, Germany, 2014.
21. Groten, M. *Integrierte Planung von Distributionsnetzwerken auf Basis des Viable-System-Models;* Apprimus Verlag: Aachen, Germany, 2017.
22. Schwaninger, M. *Intelligent Organizations: Powerful Models for Systemic Management;* Springer Science & Business Media: Berlin/Heidelberg, Germany, 2008.
23. Angerhofer, B.J.; Angelides, M.C. System dynamics modelling in supply chain management: Research review. In Proceedings of the 32nd Conference on Winter Simulation, Society for Computer Simulation International, Orlando, FL, USA, 10–13 December 2000; pp. 342–351.
24. Sterman, J.D. *Business Dynamics: Systems Thinking and Modeling for a Complex World;* Irwin/McGraw-Hill: New York, NY, USA, 2000.
25. Forrester, J.W. *Industrial Dynamics. 4. Auflage;* MIT Press: Cambridge, UK, 1969.
26. Coyle, R.G. *System Dynamics Modelling: A Practical Approach;* Chapman & Hall: London, UK, 2008.
27. Brunner, A. *Simulationsbasierte Bewertung von Supply-Chain-Management-Konzepten Apprimus;* Apprimus Verlag: Aachen, Germany, 2011.
28. Reggelin, T. *Schneller Entscheiden. Log.Kompass;* DVV Media Group: Hamburg, Germany, 2012; p. 5.
29. Campuzano, F.; Bru, J.M. *Supply Chain Simulation: A System Dynamics Approach for Improving Performance;* Springer Science & Business Media: Berlin/Heidelberg, Germany, 2011.
30. Zelenović, D.M. Flexibility—A condition for effective production systems. *Int. J. Prod. Res.* **1982**, *20*, 319–337. [CrossRef]
31. Nyhuis, P. *Produktionskennlinien—Grundlagen und Anwendungsmöglichkeiten;* Springer: Berlin/Heidelberg, Germany, 2008.
32. Friedli, T.; Schuh, G. *Wettbewerbsfähigkeit der Produktion an Hochlohnstandorten;* Springer: Berlin/Heidelberg, Germany, 2012.
33. Schuh, G.; Brosze, T.; Brandenburg, U.; Cuber, S.; Schenk, M.; Quick, J.; Hering, N. *Grundlagen der Produktionsplanung und-Steuerung;* Produktionsplanung Und-Steuerung 1; Springer: Berlin/Heidelberg, Germany, 2012; pp. 9–293.
34. Schuh, G.; Stich, V. *Logistikmanagement. Handbuch Produktion und Management 6. 2., Vollst. Neu Bearb. und Erw. Imprint;* VDI-Buch, 6; Springer: Berlin/Heidelberg, Germany, 2013.
35. Matyas, K. *Instandhaltungslogistik: Qualität und Produktivität Steigern*, 5th ed.; Carl Hanser Verlag, Ed.; München, Wien: München, Germany, 2013.

36. Pawellek, I.G. Planung der Instandhaltung. In *tegrierte Instandhaltung und Ersatzteillogistik*; Springer: Berlin/Heidelberg, Germany, 2013; pp. 285–324.
37. Dombrowski, U.; Mielke, T. *Ganzheitliche Produktionssysteme: Aktueller Stand und Zukünftige Entwicklungen*; Springer: Berlin/Heidelberg, Germany, 2011.
38. Ríos, J.M.P. *Design and Diagnosis for Sustainable Organizations: The Viable System Method*; Springer Science & Business Media: Berlin/Heidelberg, Germany, 2012.
39. Bozalongo Santander, L. Implantación de un Plan de Producción fijo Para el Taller de Prensas de Volkswagen Navarra. 2013. Available online: https://hdl.handle.net/2454/7443 (accessed on 1 July 2018).
40. Lešková, A. Logistics Concept of Supply Chain in Automotive Production. 2012. Available online: http://web2.vslg.cz/fotogalerie/acta_logistica/2012/3-cislo/4_leskova.pdf (accessed on 1 July 2018).
41. Pophaley, M.; Vyas, R.K. Plant maintenance management practices in automobile industries: A retrospective and literature review. *J. Ind. Eng. Manag.* **2010**, *3*, 512–541. [CrossRef]
42. Wensing, T. *Periodic Review Inventory Systems. Performance Analysis and Optimization of Inventory Systems within Supply Chains*; Springer: Berlin/Heidelberg, Germany, 2011.

© 2018 by the authors. Licensee MDPI, Basel, Switzerland. This article is an open access article distributed under the terms and conditions of the Creative Commons Attribution (CC BY) license (http://creativecommons.org/licenses/by/4.0/).

Article

Feature-Based Framework for Inspection Process Planning

Fernando Romero Subirón, Pedro Rosado Castellano, Gracia M. Bruscas Bellido * and Sergio Benavent Nácher

Department of Industrial Systems Engineering and Design, Jaume I University, 12071 Castellón de la Plana, Spain; fromero@uji.es (F.R.S.); rosado@uji.es (P.R.C.); benavens@uji.es (S.B.N.)

* Correspondence: bruscas@uji.es; Tel.: +34-964728189

Received: 30 June 2018; Accepted: 19 August 2018; Published: 22 August 2018

Abstract: Feature-based approaches have been profusely used in the last decades to incorporate domain-specific knowledge in the design and development of technical systems that, according to the new Concurrent Engineering approaches, involves not only the definition of the product, but also of the required manufacturing/inspection/assembly process and the corresponding production system. Although the ability of feature-based modeling to ease and integrate knowledge intensive processes has always been recognised, in practise the different feature-based modeling proposals are strongly dependent on the domain and on the development stage of the solution (conceptual, detailed, etc.). On the other hand, inspection process planning, including the design and selection of the technical system to realize the dimensional and geometrical verification of the manufactured artefacts, has been traditionally considered separately from the rest of the manufacturing process planning, and even also from the product functional specification tasks. In this work, a feature-based framework for inspection process planning, based on a similar approach to the one applied in GD&T (Geometrical Dimensioning & Tolerancing) specification, analysis and validation of product artefacts, is presented. With this work, the proposed framework and feature concept ability to model interaction components belonging to both the product and the inspection system (inspection solution) is proved. Moreover, to facilitate the Collaborative and Integrated Development of Product-Process-Resource, the Inspection Feature has been conceived as a specialization of a generic Feature previously proposed by the authors.

Keywords: feature-based modeling; inspection planning; dimensional and geometrical specification; process specification; collaborative and integrated product-process development

1. Introduction

To face nowadays intense global competition, companies require manufacturing systems to be more flexible, adaptable, reactive and interoperable. This circumstance, together with the development of new information and communication technologies, such as Service-Oriented Computing (SOC)/Service-Oriented Architecture (SOA) or Cloud Computing and Web Services, has given rise to the emergence of several manufacturing technical and operational paradigms such as Digital Manufacturing, Reconfigurable Manufacturing, Service Manufacturing or Cloud Manufacturing, among others [1–5].

To be able to cope with the above-mentioned characteristics and to encourage collaboration in complex current manufacturing systems, a more reactive, adaptable and distributed Process Planning with an enhanced connection to production scheduling and product design is required [6–8]. The need that Process Planning has these characteristics and that is configured as a central element in an integrated product-process-resource system, was already stated many years ago, from one of the first proposals by the authors of [7] up to the proposal of a framework for the Collaborative and Integrated Development of Product, Process and Resource (CIDP^{2}R) process [8,9]. In order to reach high levels of integration and adaptability, the proposal by the authors of [8,9] is based on a service-oriented architecture and locates Process Planning as a central activity interacting with Design and Production Planning and Control activities. In addition, this last proposal fosters an approach based on (central part of Figure 1): (a) a unified activity model valid for any process planning activity; and (b) a unified product-process-resource information model. This information model is based on a feature concept able to consistently support the development of solutions to meet functional requirements in the different involved domains (product design, manufacturing process planning, inspection process planning, etc.) at any abstraction/specialization level.

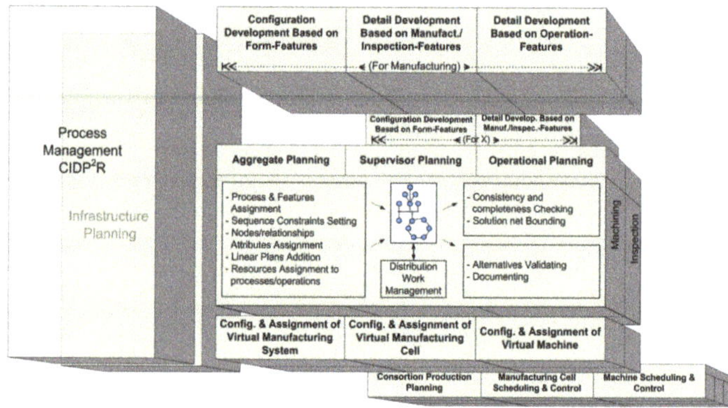

Figure 1. Functional architecture of the Collaborative and Integrated Development of Product, Process and Resource (CIDP^{2}R) process [8].

These feature-based approaches are also the basis of recent works in the newest and current Cloud-based Design and Manufacturing contexts. The need of a feature-based approach together with service-oriented architectures for data exchange in Cloud-Based Design and Manufacture contexts is stated in [10]. Similarly, cloud and feature-based Functional Blocks (FB) technologies to develop a Cloud Distributed Process Planning system that works as a central service aimed at increasing responsiveness and adaptability in current collaborative environments is adopted in [3]. However, all these proposals make use of a very specific feature concept and highlight the need of a generic feature concept able to support frameworks such as the previously mentioned CIDP^{2}R one.

The unified activity model developed in the CIDP^{2}R (Figure 1) considers the integration of the activities and their relationships in two dimensions [11]. One of the dimensions refers to the development process maturity and distinguishes three levels: aggregate, supervisory and operational. The second dimension refers to the perspective and takes into account the product, the needed process plans and the required resources. In [11], the supervisory level and process planning activities are thoroughly described, and particularly the manufacturing and inspection processes integrationin order to encourage the use of new in line inspection (in and post process) capabilities, especially on-machine measurement, to obtain even real-time performance information and improve system reactivity. This has been in increasing need in recent years, due to the appearance of hybrid machines that combine processing technologies (e.g., subtractive and additive manufacturing) with new measurement technologies.

The requirement of a generic feature concept to support the CIDP^{2}R process, led to the proposal of the Unified Application Feature (UAF) framework, which includes the definition of a generic Application Feature, and which will be briefly reviewed in the next section [12]. In addition, a Specification Feature, such as a specialization of the Application Feature, was proposed in [13].

This Specification Feature considers geometry with defects to support all the activities of the CIDP^{2}R process where the consistent representation and treatment of dimensional and geometrical variations is essential: product specification, process (manufacturing and inspection) specification, and resource assignment. Additionally, in the same work, a system-oriented and tolerance-driven artefact model was also defined, where the workpiece is understood as a part of an assembly (assembly model), valid for all the product life cycle phases (final product assembly, machining process assembly, inspection process assembly), which is required to achieve unification.

According to the above, and in addition to the definition of a specific feature for the inspection domain, this work aims to prove that the UAF framework, based on the proposed Application Feature, has the sufficient generality to provide the required flexibility in order to define feature specializations. These Application Feature specializations are not only in the product domain, but also in the process planning and resource assignment and configuration domains.

The rest of this paper is organised as follows. In Section 2, a generic specification feature model developed by the authors in previous work is briefly summarized. Section 3 presents the proposal of an Inspection Feature, as a subtype of Specification Feature, which enables Supervisory Inspection Plan specification and validation based on the inspection assembly, resulting from the assembly of the subject part for inspection and the measurement resource (including fixture, probe, equipment, etc.). Finally, Section 4 concludes with a summary of the main contributions and indicates some future work.

2. Background and Methodology

Traditionally, the dimensional and geometrical specification exercises are carried on the assembly corresponding to the final product and their objective is to establish and validate product functionality through the geometric specification of all its individual parts. However, along the different product lifecycle stages, each of these parts participates in other assemblies required for its realization (manufacturing and inspection). These process assemblies (manufacturing and inspection) that are established for process plan specification, analysis and validation (inspection blueprints), in addition to the part, include the resources on which the part is fixtured in the different process set-ups. Therefore, a feature-based framework for specification is necessary to enable, in a dual and consistent manner, a uniform product and process plan specification considering, analyzing and validating two different types of assemblies: product and process assemblies (manufacturing and inspection).

Before presenting the Feature-based Framework for Geometric Specification, in the first part of this section, a general review of feature concept and feature-based modeling frameworks is carried out. One of the generic featured-based frameworks used for geometric specification and aimed at fulfilling the requirements of a consistent product and process plan specification, is summarized in the second part of this section. This framework has been presented in prior published authors' work [12,13]. The third part of this section presents the parts of the framework for geometric specification that includes a geometry model, a specification feature model and an assembly model. Finally, the section ends with a summary of the methodology used to develop and validate the proposal of an Inspection Feature.

2.1. Literature Review on Features Definitions and Modeling Frameworks

The feature is a concept that was incorporated in the design and development of technical systems by the end of the last century, especially in the Computer Aided Design and software product line engineering fields. In the first one, a feature represents the engineering meaning of the geometry of a part or assembly [14]. In the second one, feature modeling is a common approach to manage variability supporting the establishment of a product line configuration that meets multiple, and often contradictory, requirements [15]. However, in recent decades this approach from the software domain was progressively applied to the management of technical systems, and particularly to mechatronics systems [16]. This fact has been fostered by the customised mass production paradigm, since feature modeling offers a transparency for capturing and visualizing optional and alternative conceptual design solutions that the traditional requirement specification process does not provide [17].

If the feature concept and definition in specialised literature is analyzed, it can be noted that the feature usually depends on the context of the application domain and that, additionally, the concepts used remain still ambiguous and very often contradictory, even when the domain context is perfectly established. The consistency of the meaning given to the feature in different engineering areas was analyzed in [18], concluding that the feature concept has been employed with very different representation purposes such as an abstract concept, a set of properties, the material it is constituted of, a component structure, etc. Although the authors of [18] reveal this reality, the reason for it is not explained. From this paper author's perspective, the reason for this reality is that during the design exercise (intent) the engineer needs the support of different entities, although they should be unified to ensure consistency.

In addition, feature generic definitions and frameworks aimed at unifying the concept and the development of feature-based models and to support the interoperability among the applications can be found in all domains, from the most specific to the most general ones. A very general definition describes the feature like *anything about the thing being designed that's from interest* [19]. Based on this definition [19] establishes three types of feature: *Functional, Behavioural and Structural Feature*. Other authors define more specifically the feature as: (1) *An information object (feature type), always related to an artefact, that specifies engineering intent* [18]; (2) *A property that is relevant to some stakeholders and is used to discriminate between concept instances.* In the case of technical systems, these properties can be structural (e.g., shape, size), behavioural (e.g., an operation mode) and functional (e.g., cruise control of a car) [20]; or (3) *As abstractions or groupings of requirements describing structural, behavioural or functional properties of a system that are relevant and understandable for different stakeholders* [21].

Considering all the previous definitions, it can be concluded that the feature must always be understood as something that facilitates the specification and therefore, in order to do so, it must be perceived as an informational object that belongs to the design solutions space. In this design solutions space, two different sub-spaces can be distinguished: the design (functional) rationale one and the design (structural) components one. Both sub-spaces can be established at different abstraction levels. The feature is able to describe solutions either in one of them or in both (mapping the functional solution with the structural solution).

Moreover, instanced feature objects must always be described in a simplified way (label), according to [17] by a single word or a short line of text. This last circumstance and the fact that the feature is always related with the function (Functional Feature), with the technical product (Structural Feature), or with both, is the reason why the feature is frequently mistaken with a function and/or with a product. A car, a car impeller or a car body are instances of (functional) features at different conceptual levels present in the feature (tree) model representing the design intent (rationale) analysis. Likewise, an engine, a piston or a rod are also examples belonging to more specific analysis levels and that are linked with their embodiment, facilitating the relation with the components and their (structural) features, which represent the virtual product structure.

According to [22], the engineering community lacks a common way to represent features, which is suitable to support data sharing and interoperability between systems and communities. Two feature frameworks, both using Unified Modeling Language (UML) class diagrams, are highlighted by [22]: the Generic Feature proposed by [23] and the Unified Application Feature proposed by [12]. The Unified Application Feature (UAF) is the basis for this work and, therefore, it will be briefly described in the next section.

2.2. *Unified Application Feature Framework*

The UAF framework is based on the essential need that the feature encapsulates the design intent, as proposed by de PPO (Product-Process-Organization)model [24,25], and is enhanced with ideas from feature modeling techniques in software domain, such as FODA (Feature-Oriented Domain Analysis), and particularly, with the approach proposed by [15] who considers different types of feature. The UAF framework has been represented using UML class diagrams. The choice of UML as representing language is due to two main reasons. Firstly, UML is normally used in many engineering domains to describe a data model. Secondly, UML conceptual descriptions translate well to OWL, which is a language commonly employed to develop ontologies.

This framework defines a generic feature, the Application Feature [12,13], as a container of different feature categories, likewise other authors had already done [15,16]. The Application Feature is able to support any design solution, including the mapped functional and structural design solutions and independently of the abstraction/specialization level.

The proposed Application Feature, which is described in Figure 2 using a UML class diagram, is defined as an aggregation of other features (Object Feature). An Object Feature, which represents a valuable aspect for the stakeholders, is an information entity of one of the following four categories: (a) *Functional Feature*, which represents the way in which the artefact may interact with other systems. Therefore, it represents a functional solution for a functional requirement (resulting from, for example, product or process specification); (b) *Interface Feature*, which represents the artefact elements that play a port role in the interaction with other systems; (c) *Structural Feature*, which represents the artefact configuration, that is, the artefact components and their structural relationships (e.g., part of, composed of, formed by, etc.); and (d) *Parameter Feature*, which contains all other non-functional properties with required quantification by values or quality assignment (e.g., colour, weight, volume, etc.).

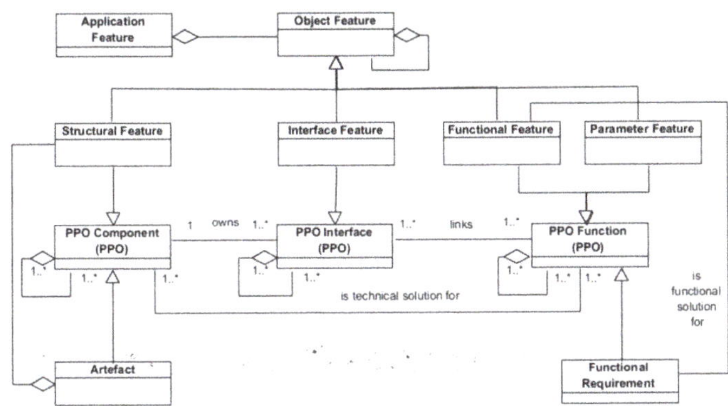

Figure 2. Updated basic structure of the Application Feature Model and its relation with the Product-Process-Organization (PPO) model.

Figure 2 also illustrates that the four Object Features are specializations of the three fundamental classes (*Component*, *Interface* and *Function*) of the PPO model. In this way, the fundamental classes inherit the properties of the PPO, easing collaborative work [24,25]. Additionally, the labels of the association relations established between the PPO classes determine the relationships typical of an engineering process driven by the design intent.

In contrast to [23], the UAF model also supports the feature-based description of the design solutions and, as can be seen in Figure 2, includes the relationship with the components hosting the features (*Artefact*) and the relationship with the functions (*Functional Requirement*) fulfilled by the features. In this sense, the proposed definition of the Application Feature is also in line with the concepts behind of the so-called Configurable Components used in product platform design that are understood as autonomous knowledge-carrying configurable generic subsystems [26]. Another shared characteristic of the Application Feature with the Configurable Components is that they allow component links through their interface and interaction elements. Thus, the Application Feature is an information object that enables product and manufacturing system (technical artefact) design based on the definition of the corresponding platforms. In addition, similarly to the Configurable Components, the Application Feature makes no distinction between the product artefact and the manufacturing system artefact, considering both in the same way (technical artefact). This allows for the establishment of interrelationships between product and manufacturing system Application Features by means of an Interaction Feature relating Interface Features of the Product and the Manufacturing System.

2.3. Feature-Based Framework for Geometric Specification

As has been mentioned, the geometric specification exercises are carried out on assemblies (product or process). These assemblies, which represent technical solutions for product or process functional requirements, are made of parts interacting through their geometric interfaces. The representation of these geometric interfaces for the specification exercise requires of appropriate geometry models in order to consider and limit the geometric variability. These geometry models are described in the first part of this section. The proposed geometry models are used in the Specification Feature Model presented in the second part of this section. Finally, the Assembly Model, which enables the establishment of a chain for each functional requirement, is presented.

2.3.1. Geometry Model for Specification

During the product specification task, the designer works with imaginary geometries with defects of the parts of the product assembly. Based on these imaginary geometries and considering the geometrical conditions of the final product function, the designer carries out several simulation exercises on the product final assembly, with the aim of specifying permissible geometric deviations (tolerances) for each individual part of the assembly. Likewise, during the inspection process specification task, the planner works with imaginary geometries of the components of the measurement assemblies, devised solutions to measure the subject part of the inspection, and carries out different simulation exercises in order to specify the permissible uncertainties for the planned measurement assembly. In this way the appropriateness of the measurement assembly solutions (reference surfaces, fixture, probes, etc.) established to determine the GD&T (Geometrical Dimensioning & Tolerancing) characteristics specified for the subject part of the inspection is validated. The types of supported geometric defects need to be compatible with the selected simulation tool used and with the type of deviation that the measurement instrument or equipment is capable of extracting. Alike simulation exercises are also present in the specification of any manufacturing process, such as the machining process, with the aim of specifying the manufacturing systems (dimensional and geometrical) capabilities.

From all the above-mentioned, it can be gathered that process (manufacturing or inspection) specification exercises, similar to what the GPS (Geometrical Product Specification) standard establishes for product specification and its verification, are also based on the distinction between the real world, where several and different realizations of the part exist, and the imaginary models (surface models), used to represent those realizations. The GPS standard establishes a similar distinction for product specification and its verification [27] defining three types of surface models (nominal, skin and extracted).

Figure 3 shows the digital models with defects considered in this work that can be used in the different simulation tools and that are linked to the conceptual skin model. Two types of these surface models are considered: ideal models, which are defined by a parametrized equation, and non-ideal models, which can be defined by a set of surface patches (continuous) or by a set of points, segments or tessellation elements (discrete). In practice, it is unfeasible to obtain the non-ideal continuous model, since it would require a large amount of complex information. Therefore, discrete models, which are obtained by sampling on the real part, are used in the specification exercise. This discrete model is the one considered by GPS, and hence always assuming a measurement method based on discrete digitalization. The non-ideal models can be simplified to different ideal models. If the simplification process neglects the form and orientation defects of the surfaces, models with dimensional (linear) defects and models with angular (position) defects are obtained respectively. The simplification process can lead to skeleton models with defects, when the geometries participating in the functional condition of the assembly are geometric elements derived from surface elements. If the simplification process is applied to ideal models, substitution and/or reduction operators are involved. Extraction operators are the ones involved when simplifying from a non-ideal continuous model to a discrete model.

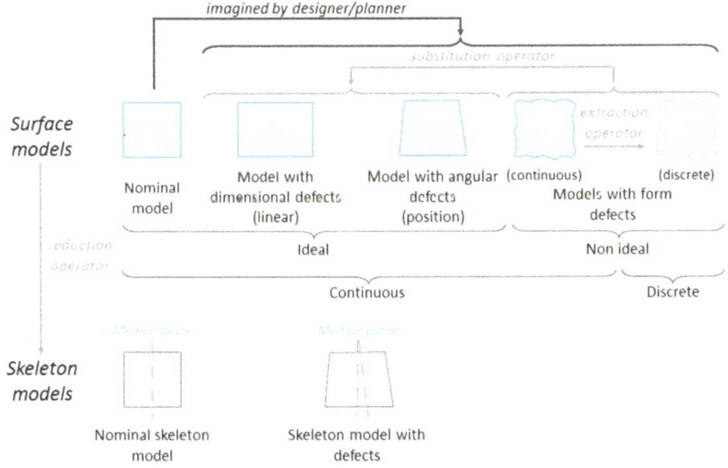

Figure 3. Geometry models for product/process specification.

Although in recent literature discrete models to represent the geometry with defects have been proposed [28], the majority of the analysis methodologies and tools use ideal continuous geometries and geometric tolerancing models based on variational geometries that do not include form defects (Degree of Freedom–D.o.f., Small Displacement Torsor–SDT, etc.). Therefore, if the analysis is carried out using one of these techniques, the representation of the geometry with defects (skin model) is either an ideal surface, including location, orientation and size defects, or a skeleton model. During specification, transformation between different geometric models might be required. These transformations are ruled by an operator consisting of a set of GPS basic operations such as partition, extraction, filtration, association, reduction, etc.

2.3.2. Specification Feature Model

The three specification exercises (product, manufacturing plan and inspection plan) involve the management of geometrical variability, although different names are used in each specification field: tolerance in product design, natural process tolerance (capability) in manufacturing and uncertainty in inspection. For that reason, the three specification exercises should be based on a unified feature model where the geometrical interface is represented using the same geometry model as described in the previous section. Based on this assumption, authors of this work proposed a unified Specification Feature (Figure 4) that will be briefly summarized in this section [13]. This feature aggregates the three types of object feature considered in any Application Feature: *Geometry Feature* (geometric interface), *Specification Structure* (structural elements) and *Condition* (functional geometrical condition for which the structural elements are a solution).

As Figure 4 shows, that the geometric interface contains, the nominal geometry (*Nominal Feature*) and additionally the representation of the deviations for this geometry (*Geometry with defects Feature*) in any of the tolerance models (GPS, TTRS, etc.) necessary to support the corresponding specification exercise. The *Geometry with defects Feature* aggregates three features: (1) *Extracted Feature*, which represents the geometry in the form in which it is extracted from part surfaces with defects; (2) *Substitute Feature*, which represents an ideal and continuous geometry related to the geometry with defects; and (3) *Reference Feature*, commonly known as Datum Feature, which represents an ideal geometry that positions extracted and substitution geometries. An *Extracted Feature* can be of two types: *Discrete Geometry* or *Envelope*. When it is a discrete geometry, it is made up of a set of points, segments or tessellation elements. Otherwise, if the extracted geometry is of type envelope, it is made up of a set of (two) trimmed ideal and continuous lines or surfaces enveloping, internally and/or externally, the real geometry with defects. This second type of extracted geometry is not considered by the GPS standard, where only extracted models able to support the way in which coordinate measurement machines take measures are considered.

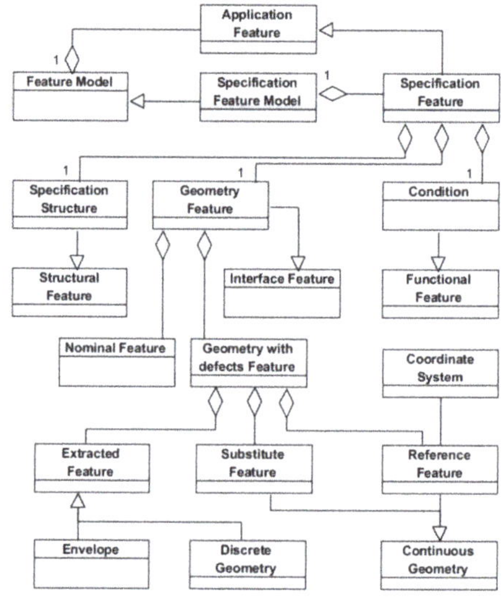

Figure 4. Specification Feature Model (updated from [13]).

2.3.3. Assembly Model for Specification

The geometry with defects (*Geometry with defects Feature*) of the Specification Feature, seen in Figure 5, is the central element of the Assembly Model for Specification. This is a key model in order to establish conditions on kinematic loops associated with a mechanical assembly (product or process simulation exercises). These loops are determined according to the different assembly configurations established by the set of joints between the geometric interfaces of the different components. Therefore, for the specification exercise a model including both the assembly architecture and the chains and functional conditions is required. The links in these loops establish the relationships between the geometric interfaces that may belong to the same or to different parts. These interfaces will be represented by the corresponding geometry with defects included in the Specification Feature previously described.

The model establishes the relationships between this geometry with defects and other concepts involved in the simulation exercises, such as the specification architecture and loops (*Specification Assembly Architecture* and *Chain*). In particular, an assembly is characterised by an architecture defined as an aggregation of all contact conditions (*Contact Condition*) between the geometry with defects of all features, either in the product or in the process assembly. The types of contacts considered in the model are *Floating*, *Fixed* and *Sliding Contact* [29]. The model also considers the non-contact conditions (*Non-Contact Condition*) that establish either a condition within the same part or a separation condition between two different parts. In addition, a *Chain* aggregates all the associations between the features including the information about the geometry with defects required to close the functional loop. From all links included in the *Chain*, just one of them is associated with the condition (*Condition*) to be fulfilled (either product or process condition), and the rest of the links will be associated with other conditions (contact or non-contact).

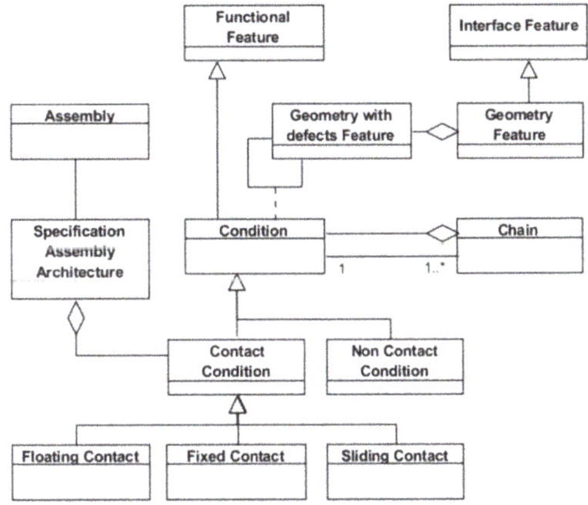

Figure 5. Assembly Model for Specification (updated from [13]).

2.4. Methodolgy

The methodology used to develop and validate at a conceptual level the proposal of the Inspection Feature can be summarized as follows:

1. Development of a functional model for inspection process planning in an integrated product and process (machining an inspection) development context, especially fostering in-line inspection. In this way, part quality inspection plans can feed product quality assurance and the resulting activation of management strategies. These strategies allow for smooth defect propagation throughout the process stages and to the final customer. The functional model, developed using IDEF0, enables to identify the main information requirements and shows at the aggregate level the relationships between the tasks involved in inspection process planning, machining process planning and product design. Furthermore, in order to ease the integration of all these planning tasks, a dual activity model for both process planning tasks is established. This activity model is supported by a common part representation based on a single feature concept, the Specification Feature.
2. Study of the following topics:
    - Tolerance information models used in CAT (Computer-Aided Tolerancing) applications, both for the interpretation models (such as vector equation model, variational surface model, kinematic model, degree of freedom (DOF) model, etc.) and for the representation models (such as surface graph model, technologically and topologically related surface (TTRS) models, category theory model, GeoSpelling model, ontology-based model, etc.). In particular, the concepts considered by the Geospelling language and the GPS standard are revisited.
    - Measurement processes and systems. More particularly, the ways in which the part can be situated (oriented and/or located) in relation to the geometries of the measurement resource are studied. Additionally, the alignment operations, either physical or by means of calculations (verification operator), that are performed during the verification process, are also analyzed.

- The role of tolerancing in the context of the uncertainty management, in order to ensure that the product meets its functional requirements.

3. Development of a proposal for the specification exercises carried out in inspection process planning, which is dual to the one established for the specification exercises in product design. Accordingly, inspection plan specification (including analysis and validation) is addressed using similar assembly models, geometry models, which incorporate the representation of defects, and tools and techniques for variability management.
4. Analysis of the general UAF framework and Specification Feature Model to determine their suitability to provide a specific solution for inspection planning.
5. Development of an Inspection Feature Model and an Inspection Assembly Model based on the general UAF framework and Specification Feature Model. In particular, the models for the inspection planning domain should be adequate to support the definition, analysis and validation of the set-ups included in the inspection plan and the allocation of the inspection resources.
6. Categorization of Measurement Resources in generic types that include all type of measurement equipment, ranging from basic instruments to coordinate-based machines. The generic types of Measurement Resources have been established based on the degrees of freedom characterising the movement axes of the inspection equipment and the axes including sensors to register measurement data.
7. Development of an Inspection Feature Library. The library classes are based on the study, from measurement viewpoint, of the different geometry types that can be present in mechanical parts. The definition of the different types of features considers the way each type of feature interacts with the resource interface features corresponding to the defined Categories of Measurement Resources. The knowledge about compatibility between the part and resource interfaces is essential for the inspection planner in order to allocate the most appropriate resource. This knowledge is embedded in the form of compatibility constraints and properties of interaction
8. Validation of the proposed models by the application to several case studies. The aim is just to validate that the concepts supporting the Inspection Feature Model are adequate to select and analyze an inspection solution. A developed graph-based methodology that supports the inspection chains representation corresponding to each characteristic to be verified in one set-up is used in order to facilitate analysis and validation exercises of the inspection solution.

In this work, only stages 3, 4 and 5 are covered due to space limitations, although a simple case study is presented so that the reader can see how the Inspection Feature can support the involved tasks in the inspection process plan specification. The rest of the stages are out of the scope of this work and they will be the object of future publications.

It should be noted that author's interest does not aim the development of an object-oriented application for inspection planning. Therefore, although UML notation has been used to show the concepts of the proposed model, the UML classes have not been detailed with their attributes and operations.

3. Results: Feature-Based Framework for Inspection

As already mentioned, product specification involves:

a. The definition of an assembly (product artefact) that can be a technical solution for the required functionality expressed as functional conditions. The technical solution is a set of parts with their particular geometrical shapes that are kinematically related through their geometrical interfaces.
b. Since the part geometrical interfaces will have defects (intrinsic or extrinsic), different characteristics limiting them have to be specified. A specified characteristic is a characteristic with the permissible (maximum and minimum) limits, where a characteristic is a linear or angular distance defined between geometric elements (ideal or non-ideal) [27]. Each specified

characteristic requires the definition of a GPS operator that establishes the procedure to obtain it from the data of the involved geometrical elements.

c. The validation whether the total assembly performance (tolerance) meets the functional condition. This is calculated through a chain that considers the characteristics of the assembly components (individual parts) and the contact conditions.

Similarly, for each specification to be verified, the inspection plan specification involves:

a. The definition of an assembly (inspection artefact), formed by the subject part of the inspection and a set of components (measurement resource, fixture, probe, etc.). This assembly must be a technical solution capable of extracting part geometric information needed for the verification of the specified characteristic.
b. Since the extracted part geometry will have defects, the planner, similarly to the designer, uses ideal geometry models that enable him/her to represent the measured geometry with defects. Working on these imaginary ideal geometries with defects, which belong to both the part and the rest of the assembly components (measurement resource, fixture, probe, etc.), the planner establishes the GPS operator to obtain characteristics to be measured corresponding to the specified characteristics.
c. The validation whether the total uncertainty (method and implementation) of the inspection assembly meets the requirements of the verification of the specified characteristic. This uncertainty is calculated through a chain that considers the characteristics of the assembly components (part and inspection resource) and contact conditions.

As is clear, in product specification and inspection plan specification very similar tasks must be undertaken. Furthermore, both specification exercises work on a common representation of the geometry with defects of the part, either of the conceived or of the real one. GPS operators are applied to the geometry with defects in order to quantify the characteristics and their variability. When the operators used for both exercises are coincident (duality), then uncertainty is minimised.

Hence, the specification of the inspection process plan involves establishing GPS operators on the verification geometries of both the part and the components of the inspection resource and analyzing the contacts between the previous geometries. Therefore, in the first part of this section the specific geometries needed for verification (part and inspection resource) are going to be studied. In the next two parts of this section, an inspection feature and assembly models are proposed. These models are based on the verification geometries. Finally, a case study is presented.

3.1. Geometry Model for Verification

Usually, in product specification the designer considers skin and/or skeleton models with defects for a GD&T analysis process based on simulation [29]. These models are constructed from the nominal model based on ideal geometries with imagined dimensional and angular defects. As shown in Figure 6, these imagined geometries with defects are represented by the Substitute and Extracted Features defined in relation to a reference geometry represented by the Reference Feature, which usually is the same as the Nominal Feature.

However, in the specification of the inspection plan the planner considers skin and/or skeleton models with defects imagined as a result of the extraction process. The type of these imagined extracted geometries depends on the type of extracted geometry that the inspection resource can provide.

The Geometry of the Part

In order to verify a part characteristic, the inspection planner must have an adequate representation of the part real geometry. This representation should enable the planner to establish the verification GPS operator, as a set of several GPS operations (partition, extraction, association, etc.). This GPS operator will include a last evaluation operation to allow for the verification of the characteristic. The representation of the part geometry, gathered during the measurement process, is referred in this work as "verification primitive model". Hence, the GPS operator established by the planner will operate on this verification primitive model transforming it into simpler ones from which the required linear and angular distances to complete the evaluation operation of the characteristic can be computed.

The verification primitive model for inspection plan specification, unlike product specification, is very often a discrete model, obtained by sampling a finite number of points, segments or tessellation elements on real part surfaces. The vast majority of specialised literature, including the GPS standard, assume that inspection plan specification starts with this type of verification primitive model (discrete model).

However, this situation in only present when metrological systems based on coordinate measuring processes equipped with (mechanical and optical) probes are used. When basic metrology, such as a calliper, is employed, the verification primitive model is a much simpler one, since from the available information only an ideal profile model with dimensional defects (due to linear and/or angular variations) can be obtained. As Figure 6 shows, this profile model does not consider part form and orientation defects, since these are neglected by the contact surfaces of the inspection resource assumed geometrically perfect.

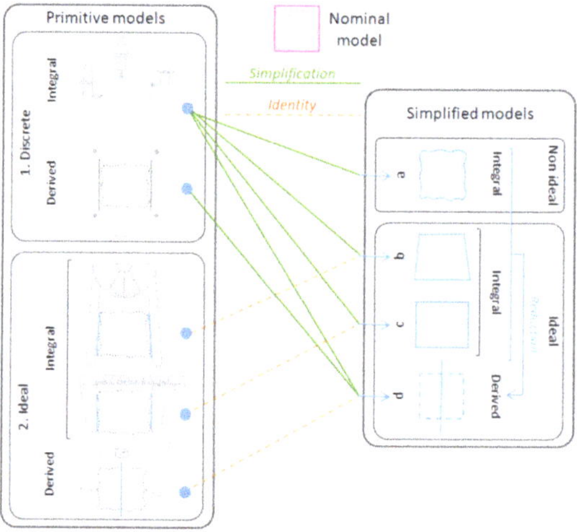

Figure 6. Part geometry models for verification.

Then, the verification primitive model is a representation of a real instance of the part geometry that depends on the extraction method and the inspection resource used, and can be of two main different types (Figure 6, left):

1. Discrete models with defects (integral or derived profiles/surfaces) consisting of sets of points, segments or tessellation elements (with a particular pattern). These models are obtained when measurement is performed by equipment that provides coordinate information, such as CMM (Coordinate Measurement Machine), optical equipment, surface form/texture metrology, etc. The coordinate information is referred to the equipment coordinate system that is realised by the movements of its guideways. To make this equipment very flexible, its guideways can be linear, resulting in rectangular coordinate systems, or a combination of linear and angular movements, resulting in spherical, cylindrical, etc., coordinate systems. There is equipment with two guideways that can be used to obtain two dimensional discrete models and others with three or more guideways that can be used when three dimensional discrete models are required.

2. Ideal models with dimensional defects that keep the nominal form. These models are obtained when measurement is performed either by conventional equipment (calliper, micrometer, goniometer, etc.) or by equipment and set-ups used in comparison measurements. The first ones provide a specific linear or angular distance between two ideal geometries that are embodied by the measurement equipment. The second ones provide two linear or angular distances (maximum and minimum deviations) that enable the construction of two ideal geometries (surfaces or profiles) that are internally or externally enveloping the real part geometry. The construction of these two ideal enveloping geometries is performed by the movement of the measurement equipment guideways (sweeping movement). When surface models (3D) are required, it will be necessary to use two axes for the sweeping movement (two isoparametric lines), resulting in an enveloping surface. However, if plane profile models (2D) are desired, just one axis for the sweeping movement will be required, resulting in an enveloping line. When surface models of a complete partitioned geometry using any of these two types of measurement processes (conventional equipment or set-ups for comparison measurements) are desired, measurements in several planes (parallel, coaxial, etc.) will be required in order to cover the whole partitioned surface. Obviously, the uncertainty of these surface models with dimensional defects will depend on the possibility of coincidence of the reference geometry with these profiles, as it will be explained in the next section.

From these primitive models, the verification GPS operator can establish other simplified models (Figure 6, right) required to assess the part specified characteristics. In particular, the verification operator can establish the following simplified continuous surface/profile geometric models with defects: (a) Non-ideal models, which are generated by reconstruction operations (fitting and interpolation) from the primitive models with the aim of obtaining the points that match with the sampling points established in the specification; (b) Ideal models with angular defects; (c) Ideal models with linear dimensional defects; and (d) Ideal derived models that can be obtained either by a GPS derivation operation from the previous simplified models or directly from a derived primitive model resulting from a measurement process.

Figure 6 also shows that when the primitive models are ideal, they are the same as the corresponding simplified models (b–d). Although it is often unnoticed, very often the primitive geometry itself already contains information (measurements) about the specified characteristics and, therefore, no subsequent transformation of the geometry will be necessary to obtain these characteristics. This is the case of many dimensional characteristics associated with a specific geometrical element that are obtained by direct measurements of dimensions (angle, diameter, width, etc.) or by sweeping processes. When using sweeping processes, the measurement process or equipment does not register deviations of specific points of the geometrical element of the part, and provides only the total deviation produced in the sweeping process of specific profiles or of the complete surface. A classic example of this type is the measurement of the straightness of a plane using a rule and an indicator.

The Geometry of the Inspection Resource

In addition to the real part geometry, in verification, an inspection resource is also involved in the measurement process. This inspection resource has real geometries of high quality that are always assumed to be ideal, neglecting their defects, since they are usually very small. Examples of these geometries are the surface of a plate, the axis of a chuck, etc. Based on this assumption, the real geometries of the inspection resource are represented using ideal models (without defects) for all the reasoning and computing processes required to obtain the measured characteristic. The uncertainty of the measured characteristic is influenced by the quality of the real geometries assumed as perfect.

Generally, in order to obtain a measured characteristic between two geometries (target and datum) the comparison of the real target geometry in relation to the datum frame geometry (specification reference geometry) is required. This comparison involves obtaining linear and angular distances in relation to this specification reference geometry. In turn, this specification reference geometry can also be considered as a target geometry, whose measurement involves comparing it with another datum frame geometry (measurement reference geometry).

Therefore, every measurement reference geometry can be used as a specification reference geometry. The measurement reference geometry, in relation to which linear and angular distances are obtained, is always realised by the inspection resource. This realization, as it will be explained later in this section, can be of different types, such as a flat surface contact, an axis of a revolved surface, etc. On the other hand, the specification reference geometry, which is always required for the measurement of a specified characteristic, is obtained either by a measurement process comparing it with a measurement reference geometry, or by doing it coincident with a measurement reference geometry embodied by the inspection resource using an alignment process.

In general, these three geometries (tolerance geometry, specification reference geometry and measurement reference geometry) are involved in the measurement of a specified characteristic (Figure 7). According to the specified characteristics and the selected measurement process, some of these geometries are the same. For example, when form characteristics are verified, the specification reference and the target reference can be the same. When orientation and location characteristics are verified, the specification reference and the measurement reference can be the same.

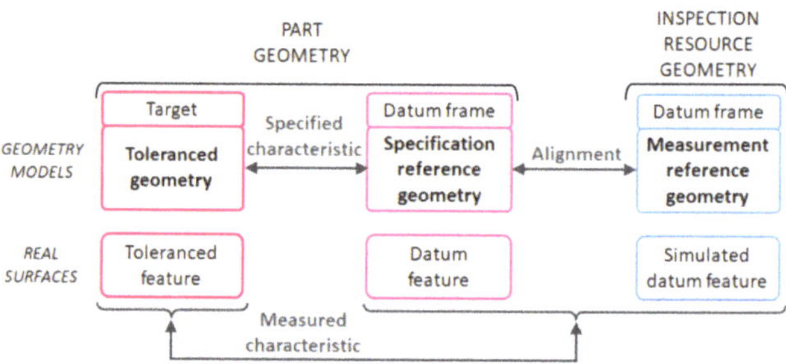

Figure 7. Geometries involved in verification.

The real geometries of the inspection resource considered as ideal models are normally known as embodiments in the metrological domain. The linear or angular distance values obtained by the inspection resource are always referred to these embodiments that are the reference for the measurements. Embodiments to establish the measurement reference can be also other ideal geometries that are of the same type to the previous ones (real geometries of the inspection resource). They are usually an offset of the real ones and are established during the equipment set-up process. For example, when a parallelism specification is inspected by means of a set-up using a surface plate, a height gage and a dial indicator, the reference measurement can be the surface plate itself contacting the specification reference. However, an imaginary plane parallel to the surface plate with a specific offset controlled by the height gage could also be used as the measurement reference.

The embodiment of the "measurement references" by the inspection resource can be of one of the following types:

a. Positioning embodiment, when the reference is realised by physical contact with surfaces of the equipment or set-up (e.g., gusset plates, mandrel, precision jaws, precision fixture, etc.) or is realised as an offset of the previous ones by gauges used during the setting or calibration process of the equipment.
b. Kinematic embodiment, when the reference is realised by the movement of the measurement equipment guideways. Obviously, this reference is located in the inspection resource, since the guideways used to generate it have a specific location in the equipment. The number of

measurement equipment guideways has to provide the minimum number of independent axes required by the type of tolerance geometry.

c. Calculated embodiment, when the reference is obtained by mathematical association operations using the part extracted points, segments or tessellation elements and appropriate criteria such as least square, minimum outer diameter, etc.

Not all types of references embodied by the inspection resource can be used with all types of primitive models of the tolerance geometry. In particular, the reference as calculated embodiment (c) leads to discrete part primitive models of the tolerance geometry, which can be simplified to ideal geometries if appropriate. On the other hand, kinematic embodiments (b) or positioning embodiments (a) lead to an ideal part primitive model of the tolerance geometry. More specifically, the ideal part primitive model obtained using a kinematic embodiment is a set of two ideal geometries enveloping the real geometry. These two ideal enveloping geometries are of the same type and are generated simultaneously with the kinematic embodiment geometry. However, the positioning embodiment leads to an ideal part primitive model that is an ideal geometry establishing a single boundary (external or internal) of the real geometry.

As has been mentioned, the measurement reference is embodied by the inspection resource, whereas the tolerance geometry to be extracted exists on the part. In addition, to obtain the measured characteristic a specification reference also existing on the part is required. This specification reference must be located (usually by coincidence) in relation to the reference embodied by the inspection resource. This is the so-called alignment process that always introduces an additional uncertainty in the inspection process. If a misalignment between the real geometry of the part and the reference geometry appears, a misalignment error is also present.

The aim of the alignment process is basically to make two geometries, one of the parts and one of the inspection resources, coincident (orientation and situation). The measurement reference geometry is realised by real geometries of the resource (high precision surface plates, gusset plates, mandrels, etc.) known as simulated datum. The defects of these real geometries of the resource are neglected compared to the part geometry defects and, therefore, they are considered to be ideal geometries. It must be noticed that the effect of this assumption is included in the resource uncertainty obtained during the calibration process. The lower the quality of the real geometries of the inspection resource, the higher the measurement (implementation) uncertainty. The part geometry must contact with these real (assumed ideal) geometries of the inspection resource. However, since part geometry is not ideal, there is no one single stable solution for the contact. Due to the significant effect of this circumstance on the uncertainty, the use of some requirements to rule the relative location is required, such as the minimum requirement or the minimum rock requirement [30].

Very often, the alignment process is realised locating the part by physical contacts with the inspection resource minimising the deviations between part and inspection resource geometries. In these alignments by physical contact, two cases can be distinguished depending on whether the part contact surface is the same or not as the specification reference. An example of the first case is when a part flat surface directly contacts with the surface plate that orients the part and is used as reference. An example of the second case is the clamping process of a cylindrical part using a roundness measuring instrument where dial indicator values on the cylindrical surface when turning the part around the equipment axis are minimised.

In coordinate-based measurement processes, the alignment process is the calculation of the measurement reference. In this case, the alignment process involves calculating an ideal geometry that is used as specification/measurement reference and probing on its normal direction.

3.2. Inspection Feature Model

In this section, and based on the concepts related to the geometries with defects explained in the previous section, an Inspection Feature Model is proposed and described using UML diagrams.

The Inspection Feature is defined as a subtype of the Application Feature considered in the UAF framework outlined in section.

The *Inspection Feature* (InspF) shown in Figure 8, as subclass of the *Specification Feature* class, is an aggregation of the classes with the information about the structure (*Inspection Structure*), the geometric interface (*Inspection Geometry Feature*) and the functional geometric condition (*Inspection Condition*), which are established as requirements on the characteristics to be measured (*Characteristic Measurement*) in order to obtain the values of the specified characteristics (*Inspection Requirement*) that points to the self-geometries of the InspF. These specified characteristics have been established along the product design stating their GPS specification operators and their variation limits (tolerances).

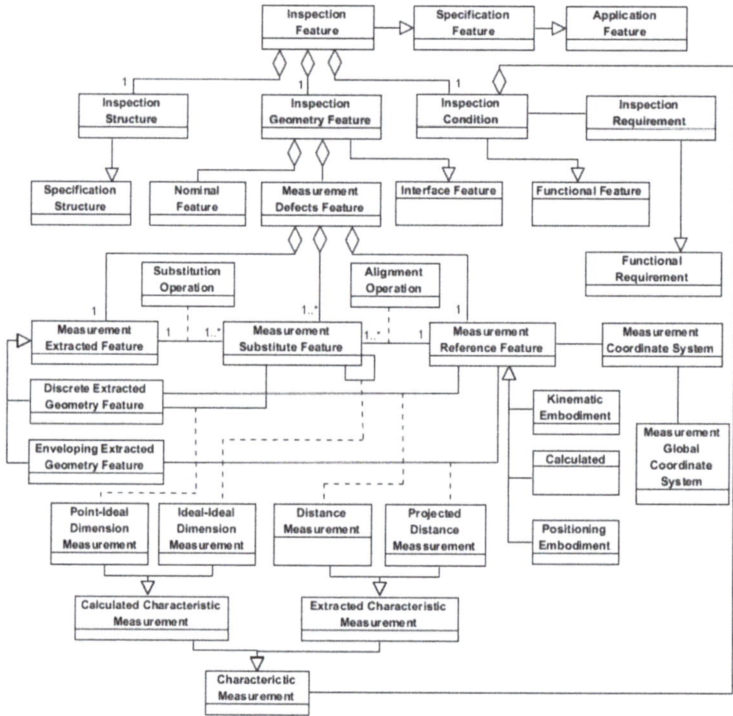

Figure 8. Inspection Feature (InspF) Model.

The inspection process plan specification starts analyzing those specified characteristics in order to define the part geometry using the feature types from the InspF Library (feature recognition) and to establish the *Inspection Condition*. The part recognition using the InspF Library developed as stated in the methodology section, is essential to ensure: (1) that is possible to extract the measurement data for the specified characteristic calculation and (2) that there exists inspection resource type able to execute the data extraction. These inspection resource types facilitate the selection of one or more technical solutions to carry out the InspF measurement.

In the same way that *Inspection Condition* relates the InspF with the product functional structure, the *Inspection Structure* relates the InspF, and more specifically its *Nominal Feature*, with the component structure of the planned inspection assembly, in which part participates. For this, the *Inspection Structure* contains the topological structure of the InspF and positions it in the part framework.

The *Inspection Geometry Feature* aggregates two feature: (1) The *Nominal Feature*, which represents the nominal geometries of the feature that are defined as ideal geometries; and (2) The *Measurement*

Defects Feature, which is used to represent the real geometries participating in the measurement process as ideal geometries that model form and location (orientation and situation) defects. The *Measurement Defects Feature* aggregates three features: (1) The primitive geometries extracted in the measurement process (*Measurement Extracted Feature*); (2) The reference geometries (datum frame) used to obtain the previous ones (*Measurement Reference Feature*); and (3) The required geometries resulting from simplification processes applied on the primitive geometries (*Measurement Substitute Feature*).

The extracted geometry with defects (*Measurement Extracted Feature*), which is a representation of the real geometry obtained as described in Section 3.1, can correspond to discrete primitive models (*Discrete Extracted Geometry Feature*) or to ideal primitive models (*Enveloping Extracted Geometry Feature*) as an envelope model, consisting of one or two ideal geometries limiting the real one.

As it has been previously explained in Section 3.1, the *Measurement Reference Feature*, which is the reference for the measured values, can be a *Positioning*, a *Kinematic* or a *Calculated Embodiment*. The *Measurement Reference Feature* can be any of the invariance classes geometries [27,29,31,32].

As Figure 8 shows, the *Inspection Condition* aggregates the characteristics to be measured (*Characteristic Measurement*). The *Characteristic Measurement* is an associative class that, in general, characterises the relation between *Measurement Defects Features*. This characterization is expressed according to Geospelling language as a set of sequenced GPS operations to establish and obtain the value of a characteristic (linear or angular distance) between any of the three components of the *Measurement Defects Feature* (Extracted, Substitute and/or Reference). The *Characteristic Measurements* can be of two main types: (a) *Extracted Characteristic Measurement*, which are characteristics between a *Measurement Extracted Feature* and a *Measurement Reference Feature* directly obtained by the inspection resource as linear or angular distances; and (b) *Calculated Characteristic Measurement*, which are characteristics between a *Measurement Substitute Feature* and either another *Measurement Substitute Feature* or a *Measurement Extracted Feature* obtained as linear or angular distances after applying mathematical/geometrical operations to values given by the inspection resource. The second type (b) of characteristic measurements are the most common ones when using inspection resources that provide a big amount of part geometrical data, such as the widely used coordinate-based measurement equipment. The latest standard developments in this field mainly focus on this type of measurement equipment.

Two types of *Extracted Characteristic Measurements* can be distinguished:

a. The *Distance Measurement* is the relation between an extracted discrete geometry (*Discrete Extracted Geometry Feature*) and a reference geometry (*Measurement Reference Feature*). For example, the measurement of a distance between a point and a plane.
b. The *Projected Distance Measurement* is the relation between an extracted enveloping geometry (*Enveloping Extracted Geometry Feature*) and a reference geometry (*Measurement Reference Feature*). For example, the measurement of an angular distance using a goniometer where both instrument probes contact part surfaces resulting in two ideal geometries (straight lines) whose included angle is the characteristic measurement. Both ideal geometries are enveloping extracted geometries (only one limit in this case) from the part and result in two substitute geometries (*Measurement Substitute Feature*) through an operator (*Measurement Substitution Feature*) that in this case is as simple as the identity. It must be noticed that one of them is used as the measurement reference (*Measurement Reference Feature*) being the alignment (*Alignment*) in this case the identity. Obviously, the type of reference in this case is established by the part-instrument contact (*Positioning Embodiment*). In the case of enveloping geometries with two limits, these will have the same form and location than the reference surface used for the measurement and are obtained by a sweeping process on that reference surface. This sweeping process is performed using the measurement equipment guideways.

Similarly, two types of *Specified Characteristic Measurements* can be distinguished:

a. The *Point-Ideal Dimension Measurement* establishes the relation between an extracted discrete geometry (*Discrete Extracted Geometry Feature*) and a substitute geometry (*Measurement Substitute Feature*) as a sequence of GPS operations that results in the quantification of the characteristic to be verified. For example, when for the verification of a parallelism between two planes a surface plate and a height gauge are used. In this case, one of the planes contacts the surface plate (*Positioning Embodiment*) establishing the measurement reference (*Measurement Reference Feature*) and the ideal substitute geometry (*Measurement Substitute Feature*) by an alignment (*Alignment*) that is the identity. The other plane is sampled with the height gauge obtaining a discrete geometry (*Discrete Extracted Geometry Feature*) as a set of points. The GPS operator to verify the specified characteristic is the result of the difference between the maximum and minimum height (measured from the reference plane) of the set of sampled points.
b. The *Ideal-Ideal Dimension Measurement* establishes the relation between two ideal substitute geometries (*Measurement Substitute Feature*) as a sequence of GPS operations between those ideal geometries resulting in the quantification of the characteristic to be verified. For example, using the previous example of a parallelism specification between two planes, but now using a CMM. In this case, both planes are sampled as a set of points in relation to the same reference measurement (*Measurement Reference Feature*). From the extracted geometry of both planes (*Discrete Extracted Geometry Feature*) the corresponding ideal substitute geometry (*Measurement Substitute Feature*) is obtained by an appropriate substitution operator (*Measurement Substitution Feature*). Between the two ideal substitute geometries, a GPS operator containing basically construction and evaluation operations is used to quantify the specified characteristic.

As has just been described, a key entity of the *Inspection Feature* is the *Measurement Defects Feature* that represents the real geometry of the part with defects through a combination of three geometries: the *Measurement Extracted Feature* and *Measurement Substituted Feature*, representing the defects on the part, and the *Measurement Reference Feature*, required in every measurement process for verification in order to orient and/or locate the first two. In addition, the model includes several associative classes to characterise, through GPS operators, the relationships between these three geometries, either for simplification and alignment purposes (*Substitution Operation* and *Alignment Operation*) or for the evaluation of the characteristic to be verified (all subtypes of *Characteristic Measurement*). The latter is related to the *Inspection Condition*, which is also included in the *Inspection Feature*.

3.3. Inspection Assembly

As mentioned in Section 2, the inspection planner task for the verification of a specific characteristic consists of defining an assembly, made up of the part and the inspection resource. This assembly must be able to extract the part geometric information required for the evaluation of the characteristic by a GPS operator. In addition, the planner must also validate that total uncertainty of the selected assembly is adequate for the limit established for the inspection condition. The extraction of the part geometric information, as it has been explained in the former section, involves the selection of reference surfaces in relation to which deviations, as linear or angular distances, are measured. On the other hand, the use of dual verification and specification operators will reduce the uncertainty.

The complete inspection process plan specification will include all the assemblies required to measure the InspF involved in the verification of all specified characteristics of the part. Obviously, in order to optimise the inspection process, the number of assemblies used should be minimised. Each assembly will require a set-up including the orientation and location of the part in the inspection resource, which has been previously referred as the alignment process. This alignment process can be more or less time consuming depending on the type of inspection resource and alignment and will have an influence on the uncertainty.

As previously mentioned, the established inspection assemblies are made up of two components (the part and the inspection resource) and two interactions exist between them. Each interaction includes all associative classes that describe the relations between part and resource features. The two interaction types are: (a) the location interaction, which holds, orients and/or positions the part in relation to the equipment reference system; and (b) the measurement interaction, which generates the stimulus, by contact or without contact between the probe and the part, for the registration of the sensors signals. Although the inspection resource or equipment itself is a mechanical assembly made up of several components and their interfaces, it will be considered as a whole (black box), characterized by a global uncertainty accompanying all the values of measurements carried out using that resource.

Figure 9 shows the model for the inspection assembly that enables the planner to analyze and specify the inspection process by reasoning on the assembly chains or loops (*Inspection Chain*). An *Inspection Chain* aggregates *Inspection Contacts*, which represent all the fixed location interactions between the part and the inspection resource defining the assembly architecture, and *Inspection Conditions*, that aggregate one or more *Characteristic Measurement*. Each *Inspection Chain* is useful to analyze one of the *Inspection Condition* that corresponds to an *Inspection Requirement*. Usually an *Inspection Condition* is related to measurement operations that results in measurement data. This type of *Inspection Condition* is a *Measurement Condition*. However, when the part is inspected using gages, only the conformance is checked, but no measurement data is available. This type of *Inspection Condition* is a *Gage Condition*.

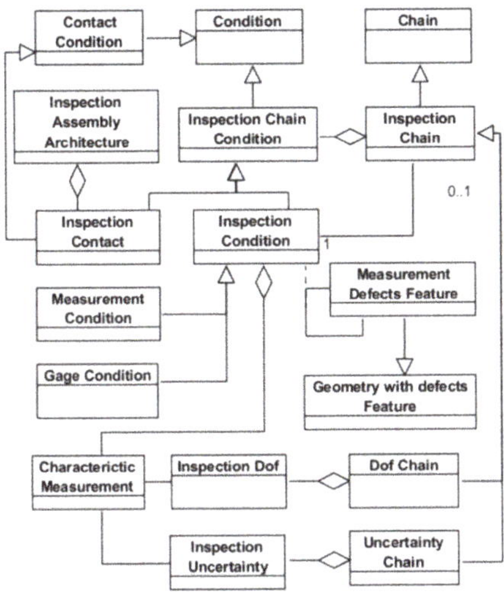

Figure 9. Inspection Assembly Chain Model.

These chains allow for the planner the establishment and validation of the final solution through the analysis of the required D.o.f. (*Dof Chain*) and the uncertainties (*Uncertainty Chain*) introduced by the different involved elements. The uncertainty chain includes the uncertainties (*Inspection Uncertainty*) of all the relationships between geometrical measurement defects features of the part and of the inspection resource as a whole that have to be stacked up to fulfil the inspection condition. The D.o.f. chain includes the information about the required active and inactive D.o.f. (*Inspection Dof*) for part location, sweeping and measurement.

3.4. Case-Study

In this subsection, a simple case study is described with the aim of showing how the proposed Inspection Feature Model supports the reasoning carried out in some of the tasks typical of the inspection process planning. The example considers a very simple part (see central part of Figure 10) with just one key characteristic. This characteristic has been established using a standard position tolerance specification that restricts the deviation of the hole axis in relation to a datum defined by plane A and plane B.

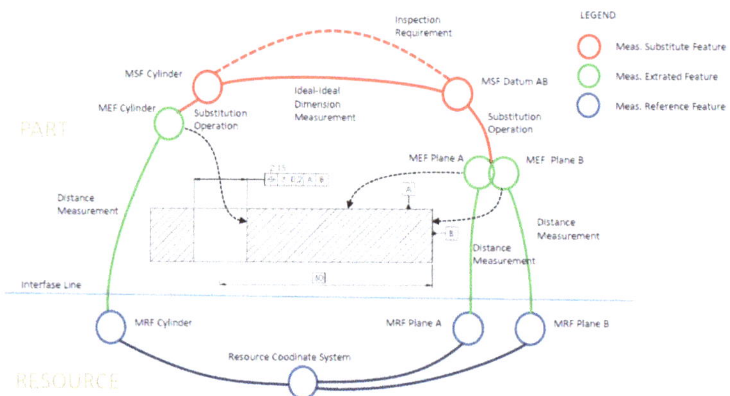

Figure 10. Graph including the InspF for the case-study.

The specification of the process plan begins with the recognition of the toleranced geometry (cylinder and planes A and B) based on the InspF types established in the Library. In this case, the planner identifies the hole surface as one Cylinder InspF type and the two plane surfaces as two Plane InspF type. Additionally, taking the tolerance of the specified characteristic (0.2 mm) as basis, the planner establishes the Inspection Requirement with the statement: "To measure the deviation of the hole position with a maximum uncertainty of 0.03 mm". This uncertainty value complies with the 1/6 relation usually established between the specified tolerance and the uncertainty of the measurement process.

Once the functional requirement (Inspection Requirement) has been established, the planner must find a solution to measure the characteristic. Previously, however, he/she will have to define the Measurement Substitute Features (MSF) that are capable of obtaining and evaluating the measurement of position characteristic by the application of the required construction, calculation and evaluation GPS operations. In this case, the MSF defined are two, one corresponding to the Cylinder InspF and another one aggregating the two Planes InspF of the compound datum AB. When these MSF have been determined and taking into consideration the requirements compelled by the InspF types they belong to, different inspection solutions can be examined.

For this, several alternative *Measurement Reference Features* (MRF) for each MSF can be considered. Next, for each of these alternatives, a series of requirements must be established on the *Characteristic Measurements* necessary to fulfil the Inspection Requirement. These *Characteristic Measurements*, which constitute the *Inspection Condition*, are established in terms of uncertainties and D.o.f.

In particular, for this case study, the MDF that could be linked to the MSF corresponding to the hole and to the datum AB could be any of the types considered in the model (*Kinematic Embodiment*, *Calculated*, *Positioning Embodiment*). However, some of these MDF would be difficult to realise and should be disregarded. Furthermore, if as usual the MDF are kinematic embodiment or calculated, the *Measurement Extracted Features* (MEF) required to obtain the MSF should also be defined.

Usually, since several alternative MRF will have been defined for each InspF, the planner should study whether an MRF corresponds to more than one InspF, because the existence of MRF common to several InspF helps to minimize the number of required inspection assemblies (set-ups). In the case at hand, given its simplicity, it is clear that the two MSF can be obtained using a single inspection assembly and the following alternative solutions could be considered:

1. Measurement of the cylinder using a resource of type "measurement on axis" (e.g., center bench). Planes A and B would be used to locate the part on the resource by means of a location gage 3–2.
2. Measurement of the cylinder and of plane B using a resource of type CMM. Plane A would be used to locate the part on the resource.
3. Measurement of the cylinder and of planes A and B using a resource of type CMM. No specific location of the part is required in this case.

Although the specification of any of the three alternatives could be object of study, only the third is going to be analyzed. The analysis will be supported by the construction of the graph shown in Figure 10. Following the previously described procedure, MSF are first placed in the graph and later the MRF and the coordinate system of the resource are also placed. In this case, as the selection of solution has already been made, only one MRF is represented for each MSF, all of them of *Calculated* type. Thus, three MEF are also incorporated in the graph. These MEF correspond to the cylinder and the two planes. Proceeding with the graph construction, the thus far represented entities are linked by lines that symbolise the relationships established among the entities. In this case, two types of relationships can be established, i.e., *Distance Measurement* and *Substitution Operation*. The whole set of links is a graphical representation of the *Inspection Chain* that supports the identification of the involved uncertainties and D.o.f. chains.

As the graph shows, there are some entities that belong to the part (placed above the interface line) and others that belong to the resource (placed underneath the interface line). It can also be noted that there are some links that cross the interface line. These links are instances of the *Extracted Characteristic Measurement* and represent the measurement interaction between the part and the resource. In this case, there are no links representing the location interaction, since part location is not involved in the measurement.

Finally, although the objective of the present research work is not the development of an object-oriented application for process planning, as already mentioned in the methodology subsection, Figure 11 shows some instances of the entities and relationships defined in the case study for the Cylinder InspF using a UML object diagram. The aim is to help the reader in the comprehension of the case by detailing some of the attributes of the classes of the InspF Model. As can be seen in Figure 11, among the object attributes those required to build and analyze the D.o.f. and uncertainty chains can be found.

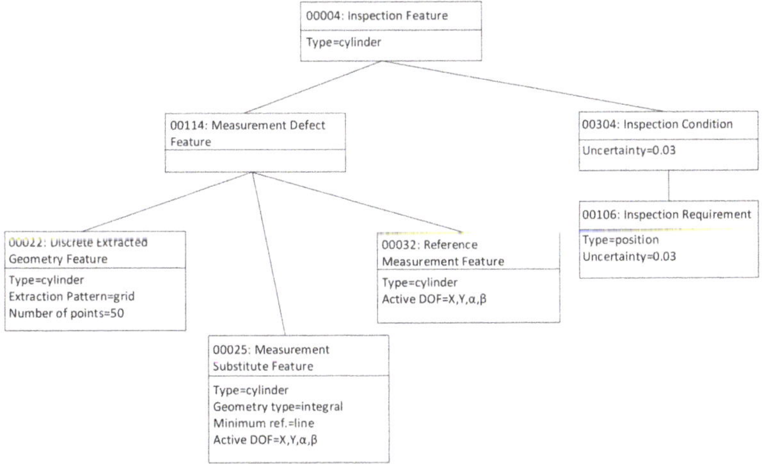

Figure 11. Some object instances and relations of the case study.

4. Conclusions and Future Work

In this work, a feature-based framework for inspection has been proposed. This framework is a specialization of a more general feature-based framework that supports the specification, analysis and validation of any technical solution (artefact). In this general framework, the Application Feature plays a key role since it is an informational object that carries the mapped functional and the structural solutions.

The development of the proposed feature-based framework for inspection has enabled to prove that the general feature-based framework is adequate not only for the specification, analysis and validation of GD&T characteristics on components of product artefacts (assemblies), but also for process artefacts (assemblies), more particularly for inspection assemblies. These inspection assemblies participate in the execution of the operations included in a set-up of the inspection plan. An inspection assembly (set-up) is made of two components: the subject part of inspection and all measurement devices (chucks, rules, plates, gages, probes, guideways, etc.) that together constitute the measurement resource.

As part of the feature-based framework for inspection, the Inspection Feature (InspF) is an essential element because it contains the necessary information to check the compatibility between the part and resource features allowing, as exposed in the included case-study, the specification and validation of inspection assemblies.

The results of this research show the possibilities of the proposed Inspection Feature for the development of knowledge-based applications in the field of inspection planning. The proposed model supports the design/selection of inspection solutions in collaborative production contexts, described in the introduction. However, from a conceptual point of view, additional work to validate the proposed approach is still needed. To that end, it is proposed, on one hand, to study in depth the inspection interaction from the resource perspective, and, on the other hand, to test the model consistency by stating an ontological model implemented in OWL (Ontology Web Language) and SRWL (Semantic Web Rule Language). In addition, the ontological approach will allow the incorporation of knowledge required to support process planning tasks, enabling the automated reasoning, the capture of new knowledge through the addition of new rules, etc.

Author Contributions: Conceptualization, F.R.S., P.R.C. and G.M.B.B.; Formal Analysis, F.R.S., P.R.C. and G.M.B.B.; Literature review, F.R.S. and S.B.N.; Supervision, F.R.S.; Validation, F.R.S., P.R.C. and G.M.B.B.; Writing-Original Draft Preparation, P.R.C. and G.M.B.B.; Writing-Review & Editing, G.M.B.B. and S.B.N.

Funding: This research received no external funding.

Conflicts of Interest: The authors declare no conflict of interest.

References

1. Lu, Y.; Morris, K.C.; Frechette, S. Current Standards Landscape for Smart Manufacturing Systems; NISTIR 8107. 2016. Available online: http://nvlpubs.nist.gov/nistpubs/ir/2016/NIST.IR.8107.pdf (accessed on 1 October 2017).
2. Zhang, Y.; Zhang, G.; Liu, Y.; Hu, D. Research on services encapsulation and virtualization access model of machine for cloud manufacturing. *J. Intell. Manuf.* **2015**, *28*, 1109–1123. [CrossRef]
3. Wang, X.V.; Givehchi, M.; Wang, L. Manufacturing system on the cloud: A case study on cloud-based process planning. *Procedia CIRP* **2017**, *63*, 39–45. [CrossRef]
4. Sang, Z.; Xu, X. The framework of a cloud-based CNC system. *Procedia CIRP* **2017**, *63*, 82–88. [CrossRef]
5. Mourtzis, D.; Vlachou, E.; Xanthopoulos, N.; Givehchi, M.; Wang, L. Cloud-based adaptive process planning considering availability and capabilities of machine tools. *J. Manuf. Syst.* **2016**, *39*, 1–8. [CrossRef]
6. Givehchi, M.; Haghighi, A.; Wang, L. Cloud-DPP for distributed process planning of mill-turn machining operations. *Robot. Comput.-Integr. Manuf.* **2017**, *47*, 76–84. [CrossRef]
7. Sormaz, D.N.; Arumugam, J.; Rajaraman, S. Integrative process plan model and representation for intelligent distributed manufacturing planning. *Int. J. Prod. Res.* **2017**, *42*, 3397–3417. [CrossRef]

8. Rosado Castellano, P.; Romero Subirón, F.; Vila Pastor, C. A model for collaborative process planning in a engineering and product network. In Proceedings of the 13th International Research/Expert Conference "Trends in the Development of Machinery and Associated Technology", Hammamet, Tunisia, 16–21 October 2009.
9. Romero Subirón, F.; Estruch Ivars, A.; Rosado Castellano, P. A framework for the development of an IT platform for collaborative and integrated development of product, process and resources. In Proceedings of the 13th International Research/Expert Conference "Trends in the Development of Machinery and Associated Technology", Hammamet, Tunisia, 16–21 October 2009.
10. Wu, Y.; He, F.; Zhang, D.; Li, X. Service-Oriented Feature-Based Data Exchange for Cloud-Based Design and Manufacturing. *IEEE Trans. Serv. Comput.* **2015**, *11*, 341–353. [CrossRef]
11. González Contreras, F.; Romero Subirón, F.; Bruscas Bellido, G.M.; Gutiérrez Rubert, S.C. Modelado de actividades para el desarrollo integrado de planes de mecanizado e inspección en entornos distribuidos y colaborativos. In Proceedings of the 3rd Manufacturing Engineering Society International Conference, Alcoy, Spain, 17–19 June 2009.
12. Romero, F.; Rosado, P.; Bruscas, G.M. Application Feature Model for Geometrical Specification of Assemblies. *Procedia Eng.* **2015**, *132*, 1128–1135. [CrossRef]
13. Romero, F.; Rosado, P.; Bruscas, G.M. Application Feature Models for Product and Process Analysis Based on Assembly Models. *J. Manuf. Technol. Res.* **2016**, *8*, 135–151.
14. Shah, J.J.; Mäntylä, M. *Parametric and Feature-Based CAD/CAM*; John Wiley & Sons: Toronto, ON, Canada, 1995; ISBN 0-471-00214-3.
15. Riebisch, M. *Towards a More Precise Definition of Feature Modelling Variability for Object-Oriented Product Lines*; Riebisch, M., James, D.S., Coplien, O., Eds.; BookOnDemand Publ. Co.: Norderstadt, Germany, 2003; pp. 64–76.
16. Zha, X.F.; Sriram, R.D. Feature-based component model for design of embedded systems. In *Intelligent Systems in Design and Manufacturing V*; International Society for Optics and Photonics: Philadelphia, PA, USA, 2004. [CrossRef]
17. Thörn, C.; Sandkuhl, K. Feature Modeling: Managing Variability in Complex Systems. In *Complex Systems in Knowledge-Based Environments: Theory, Models and Applications. Studies in Computational Intelligence*; Tolk, A., Jain, L.C., Eds.; Springer: Berlin/Heidelberg, Germany, 2009; Volume 168, pp. 129–162. ISBN 978-3-540-88074-5.
18. Sanfilippo, E.M.; Borgo, S. What are features? An ontology-based review of the literature. *Comput.-Aided Des.* **2016**, *80*, 9–18. [CrossRef]
19. Brown, D.C. Functional, Behavioral and Structural Features. In Proceedings of the ASME 2003 International Design Engineering Technical Conferences and Computers and Information in Engineering Conference, 15th International Conference on Design Theory and Methodology, Chicago, IL, USA, 2–6 September 2003. [CrossRef]
20. Czarnecki, K.; Eisenecker, U.W. *Generative Programming: Methods, Tools and Applications*; Addison-Wesley: Reading, MA, USA, 2000.
21. Kühn, A.; Bremer, C.; Dumitrescu, R.; Gausemeier, J. Feature models supporting trade-off decisions in early mechatronic systems design. In Proceedings of the NordDesign 2014 10th Biannual NordDesign Conference—Creating Together, Espoo, Finland, 27–29 August 2014.
22. Sanfilippo, E.M. Ontological Foundations for Feature-Based Product Modelling. Ph.D. Thesis, Department of Information Engineering and Computer Science, Università Degli Studi Di Trento, Trento, Italy, 2016.
23. Tang, S.H.; Chen, G.; Ma, Y.S. Fundamental Concepts of Generic Features. In *Semantic Modelling and Interoperability in Product and Process Engineering*; Ma, Y., Ed.; Springer Series in Advanced Manufacturing; Springer: London, UK, 2013; ISBN 978-1-4471-5072-5.
24. Ballu, A.; Dufaure, J.; Teissandier, D. An adaptive tolerance model for collaborative design. In *Digital Enterprise Technology. Perspectives and Future Challenges*; Cunha, P.F., Maropoulos, P.G., Eds.; Springer: Boston, MA, USA, 2007; ISBN 978-0-387-49864-5.
25. Noël, F.; Roucoules, L. The PPO design model with respect to digital enterprise technologies among product life cycle. *Int. J. Comput. Integr. Manuf.* **2008**, *21*, 139–145. [CrossRef]
26. Edholm, P.; Johannesson, H.; Söderberg, R. Geometry interactions in configurable platform models. In Proceedings of the DESIGN 2010 the 11th International Design Conference, Dubrovnik, Croatia, 21–24 May 2010.

27. *ISO17450-1: Geometrical Product Specifications (GPS)—General Concepts—Part 1: Model for Geometrical Specification and Verification*; International Organization for Standardization (ISO), Technical Committee: Geneva, Switzerland, 2011.
28. Schleich, B.; Anwerb, N.; Mathieu, L.; Wartzack, S. Skin Model Shapes: A new paradigm shift for geometric variations modelling in mechanical engineering. *Comput.-Aided Des.* **2014**, *50*, 1–15. [CrossRef]
29. Costadoat, R. Contribution à la Recherche de Spécifications Pour la Gestion des Variations au Plus tôt Dans le Cycle de Conception. Ph.D. Thesis, Université de Paris Sud 11, École Normale Supérieure de Cachan, Paris, France, 2010.
30. Henzold, G. *Geometrical Dimensioning and Tolerancing for Design, Manufacturing and Inspection*, 2nd ed.; Elsevier: Oxford, UK, 2006; ISBN 978-0-7506-6738-8.
31. Kandikjan, T.; Shah, J.J.; Davidson, J.K. A mechanism for validating dimensioning and tolerancing schemes in CAD systems. *Comput.-Aided Des.* **2001**, *33*, 721–737. [CrossRef]
32. Shen, Z.; Shah, J.J.; Davidson, J.K. Analysis neutral data structure for GD&T. *J. Intell. Manuf.* **2008**, *19*, 455–472. [CrossRef]

© 2018 by the authors. Licensee MDPI, Basel, Switzerland. This article is an open access article distributed under the terms and conditions of the Creative Commons Attribution (CC BY) license (http://creativecommons.org/licenses/by/4.0/).

Article

Joint Optimization of Process Flow and Scheduling in Service-Oriented Manufacturing Systems

Joe Vargas and Roque Calvo *

ETS Ingeniería y Diseño Industrial, Universidad Politécnica de Madrid, Ronda de Valencia, 3, 28012 Madrid, Spain; joe.vargas.villa@alumnos.upm.es

* Correspondence: roque.calvo@upm.es; Tel.: +34-910-677-7683

Received: 25 June 2018; Accepted: 24 August 2018; Published: 29 August 2018

Abstract: Customer-oriented management of manufacturing systems is crucial in service-oriented production and product service systems. This paper develops the selection of dispatching rules in combination with alternative process flow designs and demand mix, for a maintenance, repair and overhaul center (MRO) of turbo shaft engines, both for complete engines and engine modules. After an initial systematic screening of priority dispatching rules, the design of experiments and discrete-event simulation allows a quantitative analysis of the better rules for the alternative process flows with internal and service metrics. Next, the design of experiments with analysis of variance and the Taguchi approach enables a search for the optimal combination of process flow and dispatching rules. The consideration of extra costs for overdue work orders into the costing breakdown provides a quantitative evaluation of the optimum range of load for the facility. This facilitates the discussion of the significant trade-offs of cost, service, and flexibility in the production system and the operational management alternatives for decision-making.

Keywords: priority dispatching rules; simulation optimization; job shop scheduling; flexible manufacturing systems; service-oriented manufacturing systems; maintenance; aircrat engine repair and overhaul (MRO)

---

1. Introduction

In a global competitive market, service-oriented manufacturing systems follow a path where gaining and retaining the customer becomes fundamental for sustainability. From product service systems (PSS) [1] to classical flexible manufacturing systems [2], passing through service-oriented flexible manufacturing systems to a different extent [3], all of these taxonomies refer to system configurations where classical or new manufacturing competitive factors trade off system performance for service through operational decision-making. Maintenance, repair and overhaul (MRO) highly qualified manufacturing activities are offered with strong competition in the current global economy. MRO systems include the industrial activities of assembly and disassembly, components repair, inspection, as well as replacement or fault correction of an aircraft or its components, in order to preserve the airworthiness conditions and to guarantee aircraft safety operations [4]. MRO companies follow the technical requirements that aeronautical authorities establish [5]. The MRO sector includes the original manufacturer of the equipment (OEM), airlines or aircraft operators, and MRO independent companies [6]. In this market, OEMs are the biggest players through their after-sale services, with 56% share achieved by 2014 and still growing [7].

Fleet maintenance cost of aircraft operators reached USD 61,100 MM in 2014 [8], and it represented from 10% to 15% of the total operational cost of the aircraft operations [9]. A trend of 3.8% annual growth is expected, reaching about USD 90,000 MM by 2024 [8]. Engine MRO represents about 40% of the total. The MRO activities expansion evolves with heavy competence [10], so MRO companies must focus on excelling their operational effectiveness by reducing costs, stocks and job shop times [9], while holding high quality and reliability, at the level of the aeronautical regulations and standards. In a competitive market, operational flexibility and resources utilization is important in accordance with the business goals in MRO [11]. At the MRO facility level, these requirements are highly qualified experienced personnel together with a proper integration of scheduling and provisioning in the supply chain, both upstream (suppliers) and downstream (customer) [5,12]. In this path, different operational techniques and strategies formerly used only in conventional manufacturing organizations are becoming important in MRO production activities. For instance, conventional repair shop layouts have been transformed into cellular systems in order to improve MRO performance [13].

In this context, both the operations practice and its research can benefit from discrete-event simulation of manufacturing system models. The real complexity of the system can hardly be represented through analytical models and it is only partially tackled by the simulation model [14], but it provides an off-the-line test bench to foresee potential trends of performance and to help in decision making or to validate analytical models [15]. Manufacturing system simulation bears the additional advantage of low cost experimentation close to the configuration of real systems [16], and it can help to improve decision making on new manufacturing systems design. Discrete-event simulation techniques have a broad scope [17]. Four different fields of work can be considered: simulation model development, model use, field application, or discrete-event simulation with other simulation techniques. In the manufacturing research area [18], initially three main different areas of work are the design of the manufacturing system, manufacturing operations study, and software development. In addition, the following can be included [19], the programming of maintenance activities, job shop task scheduling, and the research on simulation of meta-models and optimization.

The increasing importance of manufacturing service activities requires a proper combination of production internal operations with customer service goals. Operations quality improvement in a broad sense enables the increase of sales and the reduction of production costs [20]. Beyond the conformism of an adequate performance, current paradigms of improvement look for waste elimination (lean manufacturing). The Taguchi approach to quality combines the classic meaning of conformity with the aim of optimizing system performance [21] by considering that the gap with the optimum is a waste. Moving from the conceptual approach to the real operations in a MRO facility, customer satisfaction can be influenced by repair quality in a highly regulated activity, but also by the lead-time of maintenance and due date compliance. The concurrent analysis of programming, scheduling, system capacity management, and their metrics follow-up is a complex task [22]. Operations management becomes an integrated effort of customer care and of internal process and resources management [23,24]. In consequence, dispatching rules are important decision-making criteria in the daily operation management, with impact on the overall system results and its logistic operating curves [25].

Dispatching rules define job shop scheduling by priority rules based on performance metrics. There are many different designed and studied dispatching rules. A first classification [25] includes four groups: rules based on the processing time, rules based on the due date, combined rules from the former groups and a group including all the rest. Other classifications have been proposed [26], based on the input data to the rules and the priority rule itself, sorting more than 300 existing rules into nine groups. It can be inferred that choosing the proper rule is neither immediate nor easy. It would depend on the operational context, the own production system, and the more valuable performance metric. A multi criteria approach of processing time, work order tardiness, and utilization of resources [27–30] seems to be suitable for production systems focused on service.

Sequencing can be considered a decentralized process in the decision problem. Specific research in scheduling service-oriented job shops [31] not only considers the mean time to completion of work orders, but also the overdue orders tardiness accumulation. This last study is based on discrete-event simulation on a schematic facility. Even without detailed description of the processes, it shows that when using finite resources at different degrees of utilization, the ordinary need of tracing the serial number of specific parts adds extra complexity to the repair process. This increases the effort of coordination in final assembly, adding extra time and inventories involved in operations. The priority rule "shortest processing time first" was identified as the best option towards reducing mean flow time. Focused on the service metric of overdue work orders, scheduling by "earliest due date first" was more effective. Also, rules based on the complexity of the bill of materials (BOM) to speed assembly, reparation, and disassembly activities seem to be only effective for simple product structures, with difficulties arising of the serial part number matching. A significant simulation based on field data inputs from a capital goods industry [32] can be considered that partially approaches the scheduling constraints of MRO, in particular for spare parts production. It concludes with a better behavior of the "most remaining operation first" rule for spare parts at product level, but the "shortest operation first" rule is better at component level. In the particular field of aircraft engine MRO, this last result suggests that the system could require for schedule optimization different priority rules in the cases of complete engines overhaul maintenance or only modules maintenance.

MRO is a multi-project scheduling problem due to the diversity of tasks that cannot be fully determined a priori at the work order entrance. Studies that revisited the overall problem of multi-project scheduling [33] show the convenience of different priority rules for different local or global objectives.

A more specific recent study of aircraft engine maintenance [34] includes the relevance of the unsteady flow of work orders content in MRO and the main barrier in reaching due dates with origin in parts procurement and part repair. This study is limited to only one process flow for priority rules analysis, based on simulation and design of experiments. The main result suggests the application of the "slack" rule (higher priority in work orders with the lowest difference between the remaining time to completion date and the remaining processing time), in a decentralized application. Real field tests of the rule gave positive results, maintaining system performance while facing a significant load increase.

A recent research contribution seeks an integrated approach combining the study of dispatching rules with parts pooling sourcing for service [35]. In addition, recent research prospects look for an overall optimization of service-oriented production and business management [36]. This study uses a simple example of theoretical sequencing to illustrate the integration of cost penalties of tardiness in the operation decisions. Work planning is shown in two levels: project and scheduling. Better results are reached in project performance giving priority to jobs with small resource workload. The "minimum worst case slack" rule gives the better results at scheduling level. Former research trends in the integration of planning and logistic factors converge with the recent outlooks of a main aircraft manufacturer [37]. It outlines the increasing importance of inventory pooling and the increasing use of maintenance planning tools by the aircraft owner, together with a need of MROs improvement in speed turnaround time, reducing the time in the job shop.

Improving and extending former studies, this paper presents a combined optimization of alternative process flows with a selection of significant priority rules. Different from the former studies on priority rules that are based on fictitious generic processes or a single process as a bench case, this paper approaches the design of the MRO system to real options configuration, facing the problem of mix work orders (overhaul and modules maintenance) in aircraft engine MRO. In addition, the study includes the analysis of overall costing results of service penalties, a real situation only conceptually treated in former studies.

This paper is organized as follows: in Section 2, the process flow of the MRO is analyzed. Section 3 presents the MRO system models development for discrete-event simulation. Next, Section 4 includes the selection of a set of dispatching rules by testing them through simulation, and a scoring process. System behavior and its implications in operational curves are analyzed. Section 5 covers the analysis looking for the best combination of process flow and dispatching rules in relationship with system metrics, with the Taguchi techniques, and cost analysis including charges for overdue work orders, and presents a synthesis of discussion and analysis of the former results with concluding remarks.

2. Materials and Methods

2.1. MRO Production System: Layout and Process Flow

Turbo shafts are aircraft engines ordinarily used as helicopter power plants. They are characterized by their compact size, low weight, high power, and high reliability [38]. Current aircraft engines, in particular the turbojet engines and their variants, are based on modular design and manufacturing. The modular conception has also implications for operation, fault diagnosis, and maintenance [39]. The typical configuration of a turbo shaft engine includes a cold section module, made up of the compressor; the hot section module, including the combustion chamber, and the high-pressure turbine, which drives the compressor, and finally the power section module that delivers the power to the main shaft.

The more complex and deeper maintenance program is the overhaul. It includes the overall engine inspection and repair. The modular build of the engine allows conduction of the maintenance of every module independently, so personnel, tools and shop layouts are focused on each module. There are eight main chained stages of maintenance: (1) engine shop income; and a (2) pre-analysis that allows a proper planning of the tasks in relationship with the customer's requirements and engine condition records; (3) initial inspection, with general cleaning that allows a proper visual inspection looking for faults and component damage (corrosion, wear, etc.); (4) engine disassembly, to separate modules from each other; (5) module repair, based on its condition, including cleaning, disassembly sub-sets and detail inspections with proper recovery of condition, followed by re-assembly; (6) engine assembly from modules; (7) bench test; and (8) engine conditioning and storing before the delivery to the customer. These steps can be more clearly identified in the Figure 1 process chart.

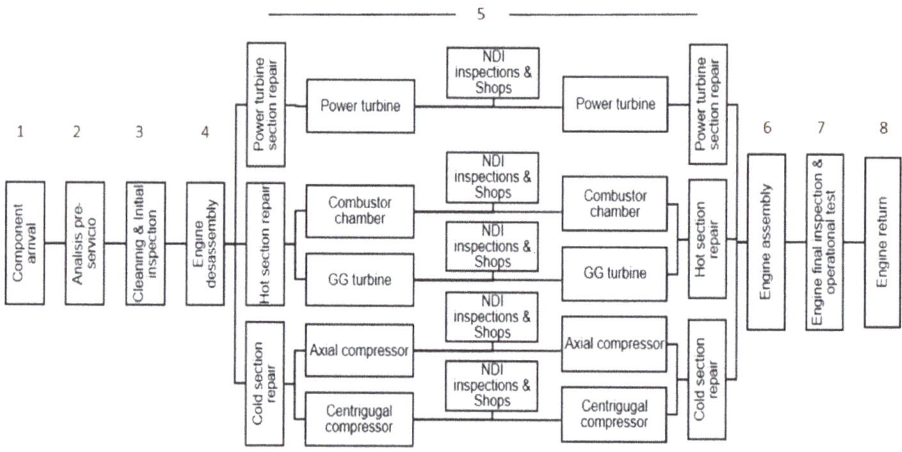

Figure 1. Cell-oriented overhaul process flow.

Across the different overhaul steps, quality assurance includes an airworthiness guarantee of engine components. The physical layout of the plant is an important productive factor of flexibility, response capability and finally competitiveness of the MRO organization. The shop includes reception and shipping areas, assembly/disassembly zones, special inspections (non-destructive inspection and testing) area, and in-house repairs of different techniques.

The alternatives of production organization can be oriented to the process or to the product [40]. These orientations influence production plant layouts, so production lines follow the sequence of the processes, and production cells typically support families of products (engine modules in this case).

Production with layout process-oriented is in relationship with a high volume and low variety of the product. The division and specialization of labor and a proper line of balancing provides high labor utilization. In addition, a proper design of capacity for a stable demand is required to reach high efficiency. In the extreme case, a production with forecasted demand without uncertainty could be optimized in dedicated lines. In the case of modular aircraft engines of a particular model or family, the maintenance and production can be planned in a sequence where every module is refurbished in every work center of the line. Because the necessary repair operations at each stage are not the same for every engine, it appears the unavoidable imbalance in queues at each work center of the production line, with the result of longer lead times.

In the case of production layout and product-oriented planning, human and material resources are organized by thinking of the variety of products and customer needs. Instead of the same rigid sequence of operations of the production line, production systems organized by product can allocate a great variety of different sequences (routing across the layout) depending on the product configuration. Nevertheless, a general big drawback could be a lower utilization of resources. This type of production system configuration can be applied in engine overhaul to every engine module, so its maintenance is planned and executed independently.

Another current main type of production system is the manufacturing cell. It is oriented to product, but taking into account the process commonalities of the different products of a family. This means that the families share processes, more than the product functionality. The organization around production cells requires the identification of those commonalities across the product portfolio, placing high skilled workers and transforming the layout design around the cells. Efforts to reduce waste in all forms guide their configuration. This organization provides good results in throughput, productivity, and quality [41]. Inspired in the principles of waste reduction of lean manufacturing, MRO facilities have initiated the adoption of this approach, but less generalized than other ordinary manufacturing systems. That is because MRO is frequently focused on customer satisfaction over the product or process constraints [13].

In the case of aircraft engines, its modular architecture gives commonality in the operations of maintenance across engine models. That is, for instance, the maintenance tasks for the hot section of different engines share similar processes. A process of engine overhaul under such premises is depicted in Figure 1. It contains cells by the two different types of compressors, the combustion chamber and the two turbines. This configuration in cells requires extra working shop area and investment, but it obtains some independence and specialization in the processes, from which a shortening of production time with better service is expected. This will be the basis of the production layout considered hereafter for study.

The layout determines the routing of the engine or its components across the plant. For every engine, its modules or part of them (just components in general), can be repaired by routing them to every step (disassembly, inspection non-destructive testing, repair, and assembly) and after task completion they wait in a meeting area until the three modules have been finished at that step, in order to pass to the next one. This flow process guarantees the engine (work order) evolves through the facility together as a whole entity, with easier flow control of work orders. In the case of work orders of modules, they do not wait, so they pass through the system without visiting the waiting area. For complete engines or modules, we call this process flow synchronous, model S.

Alternatively, the components can be routed to the proper repair shop independently and once the tasks have been finished, they can wait in a meeting area downstream to the rest of the components, just prior to final assembly. We call this asynchronous flow, model A.

2.2. MRO Model and the Discrete-Event Simulation of the System

The base model of Figure 1 under two workflows, synchronous S and asynchronous A, is considered for discrete-event simulation, using ARENA software, Rockwell Automation. In the simulation model, the components (modules or part of them, also named modules for simplicity) are entities of four types: motor AA (full engine including its three modules), module TP (power turbine), module SC (hot section), and module SF (cold section). The entity attribute declares the properties to route or make decisions about every module. The following attributes are associated with each entity (work order): STEP (routing step), NSTEP (locate at meeting area), PRODUCT (product identification at each step entrance), TPROCESS (time of process at every process step), TINSPECTION (inspection time at non-destructive testing NDT), PRIORITY (priority assignation at the queue).

2.2.1. Maintenance Sequence

The sequences defined in the model are used to establish automatically the order of process execution in inspection and repair and to establish the processing times and priorities, Tables 1 and 2. The assignation is done by accounting for every entity and its status, at every step of the maintenance process.

2.2.2. Production Process, Process Time, and Resources

One of the advantages of discrete-event simulation is an easy modelling of process duration variability and the events occurring by probabilities distributions [42].

In order to choose the better probability distribution to fit the variability of the process times or events occurrence, the proper behavior of the modelled feature is essential. Some functions are used to approach real behavior: triangular distribution for the process time, Table 2, at each workstation or exponential distribution to simulate the arrival of entities and uniform distribution to assign the type of component.

Table 1. Sequences of inspection and repair for each entity.

Command	Function	Sequence	
Sec Mod TP	Inspection route for module TP	dye penetrant inspect—waiting area [(1)]—repair zone	
Sec Mod SC	Inspection route for module SC	eddy current inspect—waiting area [(1)]—repair zone	
Sec Mod SF	Inspection route for module SF	magnetic particles inspect—waiting area [(1)]—repair zone	
Command	Function	Sequence	Assignation
MEC1	Special repair route assignation	machining—welding—machining—heat treatment—bonding—waiting area [(1)]—assembly area	Mod. TP: 36% Mod. SC: 15% Mod. SF: 46%
MEC2	Special repair route assignation	machining—welding—machining—heat treatment—bonding—waiting area [(1)]—assembly area	Mod. TP: 30% Mod. SC: 0% Mod. SF: 15%
SOL1	Special repair route assignation	machining—welding—machining—heat treatment—bonding—waiting area [(1)]—assembly area	Mod. TP: 14% Mod SC: 25% Mod. SF: 16%
SOL2	Special repair route assignation	welding—machining—bonding—waiting area [(1)]—assembly area	Mod. TP: 9% Mod. SC: 20% Mod. SF: 12%
REC1	Special repair route assignation	bonding—waiting area [(1)]—assembly area	Mod. TP: 11% Mod. SC: 40% Mod. SF: 11%
Engine assy	Assembly engine route assignation	waiting area—assembly area	

[(1)] Applicable only to simulation in model A.

Table 2. Process times.

PROCESS TIME [man hour]	Min	Mean	Max	PROCESS TIME [man hour]	Min	Mean	Max
Initial inspection	10.6	12	14.3	Welding repair (SC)	6.1	7	8.3
Disassembly/Inspection (TP) module	22	26	29.8	Welding repair (SF)	19.3	23	26.1
Disassembly/Inspection (SC) module	20	24	27	Heat treatment (TP)	8.4	10	11.4
Disassembly/Inspection (SF) module	40.7	48	55.1	Heat treatment (SC)	4.9	6	6.6
Engine disassembly	14.9	18	20.1	Heat treatment (SF)	13.8	16	18.6
Eddy current inspection (SC)	3.6	4	4.8	Bonding repair (TP)	10.1	12	13.7
Magnetic particles inspection (SF)	30.1	35	40.7	Bonding repair (SC)	9.8	12	13.3
Dye penetrant inspection (TP)	25.6	30	34.6	Bonding repair (SF)	33	39	44.7
Dye penetrant inspection (SC)	4.2	5	5.6	Assembly TP	51.8	61	70
Dye penetrant inspection (SF)	35.6	42	48.1	Assembly SC	37.1	44	50.2
General repair (TP) module	85.6	101	115.9	Assembly SF	136.4	160	184.5
General repair (SC) module	24.5	29	33.2	Balancing (TP)	13.1	15	17.7
General repair (SF) module	283.1	333	383.1	Balancing (SC)	9.5	11	12.9
Machining repair (TP)	13.1	15	17.8	Balancing (SF)	8.9	11	12.1
Machining repair (SC)	3.7	4	5	Engine assembly	22.3	26	30.2
Machining repair (SF)	55.3	65	74.8	Inspection and final test	10.7	13	14.5
Welding repair (TP)	5.1	6	6.8	Delivery to customer	2.6	3	3.5
Engine assembly total [man hour]				1265.17			
TP module total [man hour]				292.68			
SC module total [man hour]				160.64			
SF module total [man hour]				787.42			

Processes and resources are both companions in the modelling process. Every process is accomplished by its machinery and/or human resource assignations. The simulation framework computes their use associated with the task so the utilization can be calculated. The duration of the tasks together with the limited number of resources assigned to them, both bound the capability of the workstations to tackle every work arrival, so the queues couple workflow and workstation capabilities. Different types of specialized human resources are assigned to the different processes including technicians for general tasks, specialists in different processes, and propulsion engineers like supervisors and/or decision-making [43].

2.2.3. Simulation Runs

In addition to the setup of the simulation model described above, a proper simulation requires the minimization of bias in the results by the effect of initial conditions. The initial behavior of a system awaiting and empty is not the stable state that is sought through simulations studies. The method adopted to determine a stabilization time [44] consists of two steps. First, the number of run replicates necessary to reach the half-width of desired maximum variability, by (1). Where η is the number of necessary replicates, η_0 the number of initial replicates (set to 10), h_0 the initial half-width reached, h the desired half-width. Next, graphically the point is established where the transient period finishes.

$$\eta = \eta_0 \cdot {h_0}^2/h^2 \tag{1}$$

The calculations show that at least 22 replicate runs for model S and 15 for model A are necessary to come up with the desired half-range in the worst case. The next step requires fixing the overall conditions of the simulation. The particular setup conditions include: 22 warm up replicates for model S and 15 for A, 400 days for stabilization plus 1825 days of simulation run, an average of eight units per month with random demand under exponential distribution, seasonality with high demand in winter and low demand in summer following a cosine curve. Labor days of 16 h are used with continuous availability of resources, and neglecting at first approach the transportation times inside the facility. The full model programmed in Arena is run to determine the overall behavior, as in Figure 2 [43], in order to estimate the stabilization period of the different parameters across the baseline time.

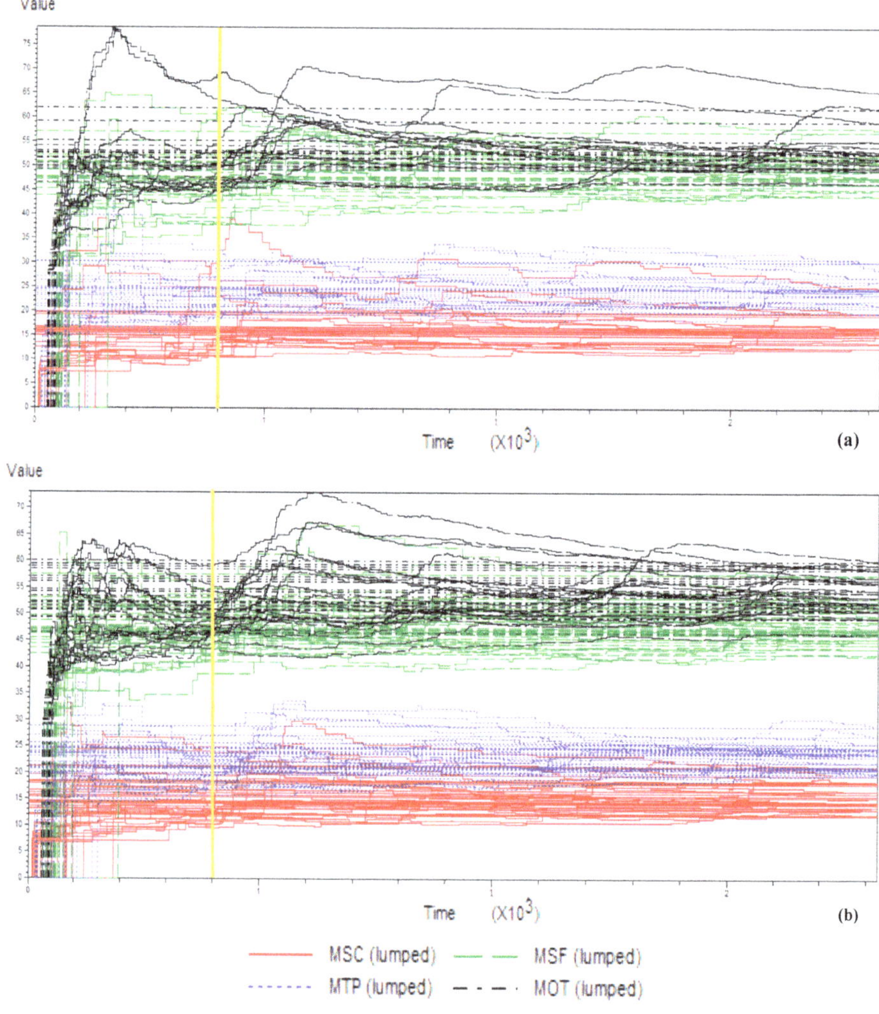

Figure 2. Simulation warm-up period estimation: (**a**) For model S, (**b**) for model A.

3. Results

3.1. Integrated Analysis of Process Flow and Dispatching Rule

Operations management in an MRO facility holds two main global objectives: customer service and internal performance. Customer service is defined as the proper quality level, including the technical proficiency that complies with regulations and achieves the agreed short lead times at a competitive cost. The internal objective requires operations with a proper resource assignation, so that the full capacity of the facility should be used effectively under variable demand. The implications of the operations under uncertainty involves trade-offs of service ratios and manufacturing flexibility [45], including the external position (market share) and internal performance (maximum capacity compatible with the intended service ratios and proper resources utilization at partial load).

Dispatching rules are tools for operations management and decision-making. In the customer service field, this includes reaching the due dates to avoid extra costs for delay, but taking into account the complex relationship of multiple work orders in processes of different customers, with different task times, number of operations or engine delivery due date.

In the MRO operational context, choosing the more convenient priority rule is not at all evident. In a first screening process, eight noteworthy dispatching rules from the literature are studied, Table 3. They are selected based on the objective, but also on the practical applicability. The simulation model of the MRO facility is run under these different rules and for different demand profiles.

Table 3. Selected priority rules.

Rule Name	Formulation
Critical Ratio and Shortest Processing Time (CR+SPT) [27]	$Z = \text{processing time} \times \max\left\{\frac{\text{due date}-\text{current date}}{\text{remaining processing time}}, 1\right\}$
Critical Ratio (CR)	$Z = \frac{\text{due date}-\text{current date}}{\text{remaining processing time}}$
Slack Time Remaining (STR)	$Z = \text{due date} - \text{current date} - \text{remaining processing time}$
Slack Time Remaining/Operation (STR/OP)	$Z = \frac{\text{due date}-\text{current date}-\text{remaining processing time}}{\text{remaining operations}}$
Earliest Due Date (EDD)	$Z = \text{due date} - \text{current date}$
First In, First Out (FIFO)	Ordering according to entrance time to the system
Largest Processing Time (LPT)	Ordering according to processing time
Shortest Processing Time (SPT)	Ordering according to processing time

The first five rules have the main objective of reducing the delay time and the number of delayed deliveries, which is to say, coping with customer service. They include operation parameters like due date, processing times, as well as the total process time or remaining operations at each step. The rules largest processing time (LPT) and shortest processing time (SPT) take global consideration of the work order content. The rule LPT prioritizes work orders of engines under overhaul, instead of engine modules, taking care of the higher costs to the customer due to the engine out-of-service cost (including leasing or insurance charges). Meanwhile SPT gives precedence to work orders of engine modules before the overhaul of engines. Engine modules require less repair time and the delivery times are shorter. Finally, the FIFO (first in, first out) rule is an initial reference baseline for comparison purposes.

Considering the setup of the simulation model of Section 3, the results are included in Table 4. The average values allow a comparison of synchronous (S) versus asynchronous (A) models. It can evaluate the general limitations of the facility model in a pair wise comparison across priority rules. The model A behaves slightly better in the metrics associated with work orders and time. Model A performs better than model S for work orders accomplishment, about 8% in wait time, reduces 11% due work orders and 3.6% in work in process (WIP). In relationship with time metrics, Model A improves 4.8% value added and a 4.5% in total lead-time, while utilization is similar for both A and S flow processes.

This result brings a first consideration that in a broad approach across priority rules and with the portfolio of engine modules and overhauled engines, the asynchronous model A performs better than S. It improves service ratios in work orders completion and time metrics. The A flow process is more appropriate in terms of service, speeding up work orders better across the system, and in an exchange with the reduction of inventories (work in process), with similar dedicated resources. That is, the service is improved by reducing intermediate stocks with the same resources over the period. The overall capacity remains practically the same (see the number of complete work orders and utilization).

The relative performance of every priority rule is different for every different metric. The selection of the most appropriate priority rule is a multi-criteria decision-making problem. In order to compare the relative performance between them, the simple well-known FIFO rule can be a reference level, frequently used in queues studies and stock management. Note that the rules LPT and SPT do not reach a simulation solution under the limiting condition of a maximum number of work orders in process, so they are left aside initially from any further investigation. Counting the number of performance criteria that are over those obtained by the FIFO rule allows a simple initial ranking of priority rules that outperforms FIFO. The score weights equally the work orders (WO) throughput, average cycle time, and the performance across customer WO. As a result, the rules CR+SPT, EDD, and SPT (this one only for the synchronous flow S) obtain total positive scores, outstanding over the FIFO reference. In addition, the CR rule performs better for overhauled engines than FIFO. Therefore, the rules CR+SPT, EDD, and CR (for full engine overhaul) are selected for further study by a design of experiments.

Table 4. Overall performance under selected priority rules: work orders and time.

Priority Rule	# Completed Works		# Due Works		% Late Works		Work in Progress (WIP) (units)	
	S	A	S	A	S	A	S	A
CR	413	425	152	153	36.8	36.0	10.7	10.8
CR+SPT	413	414	152	136	36.8	32.9	10.6	10.2
STR/OP	412	410	163	143	39.6	34.9	11.1	10.4
STR	421	415	169	150	40.1	36.1	11.2	11.0
EDD	415	417	153	128	36.9	30.7	10.9	10.5
LPT	421	NA	186	NA	36.9	NA	12.7	NA
SPT	415	NA	151	NA	36.9	NA	10.5	NA
FIFO	412	414	159	148	36.9	35.7	10.7	10.4
average	415.3	415.8	160.6	143.0	37.6	34.4	11.0	10.6

NA: Due to Arena software restrictions, it was not possible to take data

Priority Rule	Value Added (days)		Wait Time (days)		Lead Time (days)		Value Added Time (%)		Utilization (%)	
	S	A	S	A	S	A	[%]	A	S	A
CR	48.3	48.3	71.5	66.0	48.2	46.5	40.3	42.2	67.4	69.1
CR+SPT	48.2	48.3	67.6	62.6	46.8	44.9	41.6	43.5	67.9	68.1
STR/OP	48.2	48.2	74.1	67.2	49.3	46.6	39.4	41.8	67.8	67.4
STR	48.3	48.3	71.9	71.5	48.9	48.5	40.2	40.3	68.4	68.2
EDD	48.2	48.2	72.8	68.6	48.1	46.5	39.8	41.3	67.5	68.1
LPT	48.1	NA	87.6	NA	54.8	NA	35.5	NA	68.6	NA
SPT	48.3	NA	66.7	NA	46.3	NA	42.0	NA	68.1	NA
FIFO	48.3	48.3	69.5	65.9	47.6	46.6	41.0	42.3	67.6	67.5
average	48.2	48.2	72.7	66.9	48.8	46.6	40.0	41.9	67.9	68.1

NA: solution not reached under setup constraints due to inventory overflow
S = synchronous model; A = asynchronous model

3.2. Logistic Operating Curves

Even when in the case of MRO its service metrics play the main role in decision making, conventional internal manufacturing trade-offs can be surveyed by the logistic operational curves [25]. They are put into relationship output with the product level inventory (work orders in process). Note that the spare parts inventory and its availability is another face of the whole MRO operational scenario, not tackled directly in this study. For given resources, a priori operation with high output and low level of inventory contributes to internal efficiency and service.

We can consider the slope value of the operating curves as a sign of performance per unit of inventory (WIP) for the completed work orders (QC). Conversely, the quantity of late work orders (QR) and the lead-time in the system (TTS) are sought to be reduced, so the lower the slope of the curve, the better performance with respect to inventories. These operating curves, Figure 3, show the system performance for a demand mix with 70% engines for overhaul.

The results show that the synchronous flow S performs slightly better under the rule CR+SPT for the maximum total output (QC). In terms of lead time (TTS) and late work orders (QR), the rule CR+SPT works also better over 10 units of WIP, and all three rules have a similar behavior under that level.

Under the asynchronous flow A, the system reaches the higher output (QC) managed by the CR rule, but at the cost of worse (TTS) and (QR) metrics, while the best is EDD.

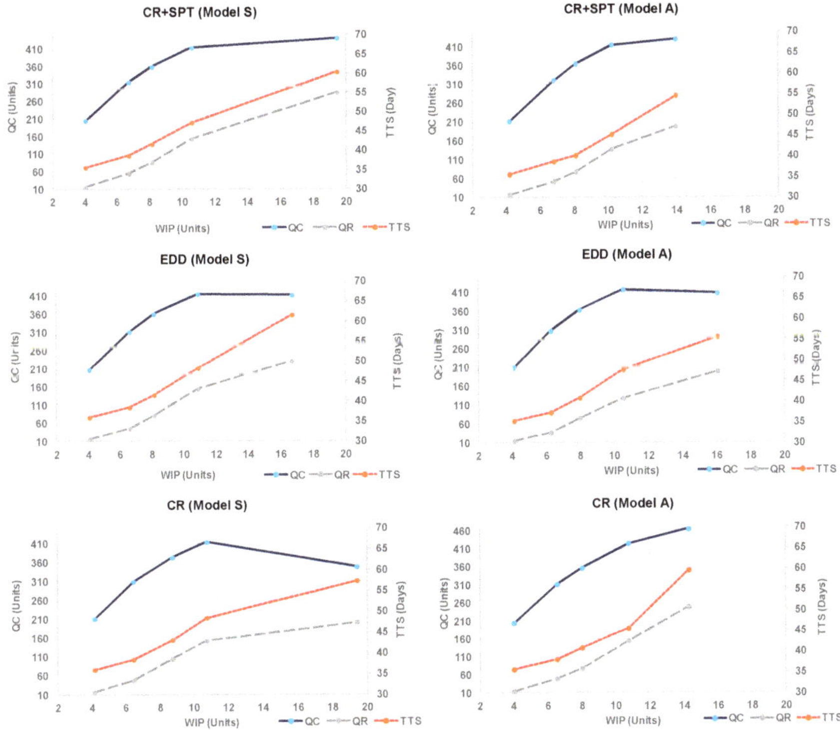

Figure 3. Logistic operating curves under selected priority rules and process flow models.

Comparing the results between models, the output the asynchronous model A reaches higher output ruled by CR, followed by CR+SPT. In general, the operating curves show higher slopes under 10 units of inventory than over the 10 units. That level of work-in-process can be considered an operative reference level convenient for the size of the system under study. Under that level, the growth of inventory serves to increase output, but after that level only marginal increases of output are obtained, and always at the cost of increasing lead-time in the system (TTS) and late work orders (QR).

4. Analysis and Discussion

4.1. Operations Variability and Evaluation through the Taguchi Approach

Even when the simulation model includes variability in the internal conditions of operations of processing times, the variability of controlled and not controlled factors can be studied in a more systematic manner through the design of experiments (DOE) and the Taguchi approach. The goal of this analysis is to find the more robust internal control factors opposed to the external uncontrolled factors that cause variability. Operational metrics are the numbers of late work orders, average total time of work orders in the system, inventories (work-in-process), and the relative cost evaluated through the utilization of resources. In particular, for the experiments, the work order mix (percentage of overhauled engines versus engines modules) and the priority rule are controllable factors, Table 5. Meanwhile, the demand level, the learning or progress curve and the percentage of rework in the shops are variability factors included as noise factors bounded in each simulation.

Table 5. Design of experiments (DOE) factors, levels and ranges.

Factor	Type	Range	Number of Levels (#)
Dispatching rule (A)	Controllable	CR-EDD-CR+SPT	CR (1)-EDD (2)-CR+SPT (3)
Type of component (B)	Controllable	Engines: 0–100%	60 (1)–70 (2)–80 (3)
Monthly demand (C)	Noise	6–10 units	6 (1)–8 (2)
Learning curve (D)	Noise	95%–85%	95 (1)–85 (2)
Rework (E)	Noise	5%–15%	5 (1)–15 (2)

The response variables correspond to the operational metrics that are sought to be robust, Table 6. This represents the optimization of its value (maximum or minimum objective) versus its variability (noise); that is, the Taguchi approach of maximizing the signal-to-noise ratio. This DOE considers the analysis of four different response variables: Total time in the system (TT), late or delayed works (DW), relative cost from the utilization of resources (BRC), and in process inventories (work-in-process, WIP).

Table 6. DOE response variables and set-up.

Response Variable	Magnitude	Objective	Maximizing
Total Time (TT)	days	Smaller-the-better	$S/N[db] = 10 \cdot Log\left[\frac{1}{\frac{1}{n}\sum_{i=1}^{n} TT_i^2}\right] = 10 \cdot Log\left[\frac{1}{TT^2+\sigma_{n-1}^2}\right]$
Due Works (DW)	work orders	Smaller-the-better	$S/N[db] = 10 \cdot Log\left[\frac{1}{\frac{1}{n}\sum_{i=1}^{n} DW_i^2}\right] = 10 \cdot Log\left[\frac{1}{DW^2+\sigma_{n-1}^2}\right]$
% Busy resource cost (BRC)	$BRC\% = \frac{\text{Busyresourcecost}}{\text{Total resourcecost}}$	Larger-the-better	$S/N[db] = 10 \cdot Log\left[\frac{1}{\frac{1}{n}\sum_{i=1}^{n} BRC_i^2}\right] = -10 \cdot Log\left[\frac{1}{n}\sum_{i=1}^{n}\frac{1}{BRC_i^2}\right]$
Work-in-process (WIP)	work orders in process	Smaller-the-better	$S/N[db] = 10 \cdot Log\left[\frac{1}{\frac{1}{n}\sum_{i=1}^{n} WIP_i^2}\right] = 10 \cdot Log\left[\frac{1}{WIP^2+\sigma_{n-1}^2}\right]$

L_9 (3^2) Inner Array			L_4 (2^3) Outer Array			
Treatments	Controllable factors		Treatments	Noise factors		
	A	B		C	D	E
1	1	1	1	1	1	1
2	1	2	2	1	2	2
3	1	3	3	2	1	2
4	2	1	4	2	2	1
5	2	2	**DEGREES OF FREEDOM (DOF)**			
6	2	3				
7	3	1	factor	Quantity	Level	Number of DOF
8	3	2	Controllable	2	3	2 × (3 − 1) = 4
9	3	3	Noise	3	2	2 × (2 − 1) = 2

The criteria of the smaller-the-better minimizes the mean and the standard deviation of the response when maximizing the signal-to-noise ratio [46]. The case of larger-the-better criteria, maximizes the mean and minimizes the standard deviation. Finally, the criteria nominal-the-best minimizes the standard deviation or variation. The experimental setup and array are also detailed in Table 6. It includes internal and external orthogonal arrays for the controlled and noise factors, and the number of levels that allow discrimination of the influences.

The results of the experiments through simulation are summarized in Table 7. The column signal-to-noise ratio shows the results of the best level of the controllable factors (LEV) together with the contribution to the variability (VAR) of the mix and the priority rule. The contribution result is included only when the variability contribution of the priority rule is significant. In the column signal-to-noise the optimum levels show the results through the criteria of smaller-is-better or larger-is-better, as per Table 6. As labelled, in the column signal-to-noise ratio nominal the best, the optimization criterion minimizes only the response variability. Finally, the column mean response gives the results for the sensitivity of the response parameters.

Table 7. DOE and ANOVA analysis of results.

Process Flow Model	Resp.	Signal-to-Noise Ratio				Mean Response				Signal-to-Noise Ratio Nominal the Best			
		Mix		Rule		Mix		Rule		Mix		Rule	
		LEV	VAR	LEV	VAR	LEV	VAR	LEV	VAR	LEV	VAR	LEV	VAR
S	TT	60	90			60	90			60	80	CR+SPT	10
	DW	60	60	EDD	30	60	50	EDD	40	60	75	EDD	20
	BRC	80	99			80	99			60	85	CR+SPT	10
	WIP	60	90			60	90			60	80	CR+SPT	10
A	TT	60	90			60	90			60	70		
	DW	60	70	CR+SPT	25	60	65	EDD	30	60	80	CR+SPT	15
	BRC	80	99			80	99			60	75		
	WIP	60	99			60	90			60	70		

Process flow model: S-synchronous; A-asynchronous; Mix: % of full engines in the work orders mix; LEV: experiment setup level; VAR: % contribution to variance; Resp.: response variable; TT: total time in the system; DW: due work orders; BRC: busy resource cost ratio; WIP: work-in-process.

The synchronous (S) process flow model exhibits a better general behavior for the mix of 60, because is the best level for TT, DW, and WIP. Note that the criteria of larger-the-better for the relative utilization in the synchronous model require a facility working with 80% work orders for overhaul engines; an exigent situation from the point of view of customer's service management. Simultaneously the criteria nominal-the-better presents the best response (minimum variability) for the 60% mix in combination with the CR+SPT priority rule for three of the criteria, but DW is always better for EDD. In overall terms, the synchronous flow process (S) presents the best response for a middle mix (60% overhaul engines and 40% engine modules) in service metrics (TT, DW), and also WIP. In addition, this mix is favorable for internal facility management because it corresponds to the better level for internal inventories (WIP) and minimizes the variation of relative cost of utilization BCR of internal resources (nominal-the-best) that is a convenient departing point for ulterior internal improvement actions. Focused exclusively on customer service response and in particular on the potential customer extra costs for late work orders, the rule EDD seems to be more convenient in the synchronous flow process together with the mix level 60. The rule has a remarkable influence (up to 40%) in the variability of the mean response of due or delayed work-orders (DW).

The asynchronous flow process A behaves in a similar qualitative overall way, presenting better customer-focused response in service by working at level of 60. The mix level 60 is better for the response metrics TT, DW, and WIP. In addition, onlyt focusing on the costs of facility operations, the higher the level of overhaul engines resulting, the better the performance. Although, the dominating rule for DW optimization is in this case CR+SPT instead of EDD.

The results of the DOE are compatible and complementary with those obtained in the logistic operating curves. In these curves, the mix under consideration was at the intermediate level of the experiments, 70% engines for overhaul, in between 60% and 80%. The operational curves show the mean results of simulations, but with a fixed setup. Conversely, the experiments evaluate mainly the variability behavior with alternative setups, more interesting for the design of the system than for its operations. Both approaches complement each other for system design and operation.

The results of the DOE show little influence of the priority rule when the percentage of full engines for overhaul increases in the mix. In accordance with the analysis of variance, the optimum results at the level of mix 80% (utilization ratio, BCR) are practically determined by the mix itself and with very little influence from the priority rule.

Both models' results, S and A, show reasonable trade-off trends: the optimization of cost in the system can be associated with a higher utilization of available resources through full service (overhaul engines versus modules repair, in an MRO facility). Nevertheless, service to the customer is associated with the service ratios DW and TT. Therefore, flexibility for a better response requires the non-full utilization of internal resources, coping with a mix that contains not only engines, but also modules. That is, for a good customer response, flexibility maintains a trade-off with the internal resources utilization. The energetic behavior of internal efficiency—the greater the amount of full service work orders, the better the utilization that results—seems to be out of the path of service metrics (DW and TT) and even of the general principles of current facilities management for inventories minimization (WIP).

These results are consistent with previous studies, and they expand former research. Focused only on service, the dominance of the rule EDD (earliest due date) first for the synchronous model S is also in the findings of Guide et al., 2000, even when their research is based on a generic non-detailed process. In the same sense, a recent contribution based on just one process flow [34] gives the better result in terms of service using the slack rule. This is the case for the asynchronous model A in the present study. Note that the combined rule CR+SPT is a hybrid rule of two factors that included the slack rule rated by the remaining processing time (see Table 3). Since both former studies used different process designs, the difference between them has not been previously discussed nor their differences interpreted, so the importance of the coupling of process flow with a priority rule remained hidden, conversely to the findings of the present study.

4.2. Cost Sensitivity Analysis

The general results show a better behavior of the system in terms of service when a significant participation of modules in the work orders exists. Conversely, the optimization in terms of cost for proper resources utilization, BCR, asks for more work orders of engine overhaul. In general, scheduling a service-oriented manufacturing system must pursue customer satisfaction [47]. Due to the high value of the availability of aircraft engines, the costs of the operation of an MRO cannot disregard the contractual charges due to delays or overdue work orders. Airlines exploit aircraft engines through flight operations, so each day the aircraft is on the ground due to maintenance impacts with revenues losses [48]. The MRO activity shares these impacts, so charges are agreed for overdue work orders. The MRO manufacturing service facility must consider them in the operative analysis for risk management and decision-making.

A cost of 20% contribution of direct overheads in the total cost breakdown is considered. Its quantification is taken directly from the simulation results. Other direct costs are the materials and subcontracted repairs that can be up to 70% and 10% of the direct cost, respectively [6]. Indirect costs associated with the MRO facility include power, consumables, general service maintenance or real state expenses. Those costs can be up to 40% of the total cost of MRO operations [2].

The charges of delayed work orders should be a function of the value of the hardware and the amount of the delay. Rates of 20,000 EUR per day for full engines and 15,000 EUR per day of delivery delay in the case of modules are considered [49]. In fact, those penalty charges are only a small portion of the real operational cost impact to the customer. The repercussion on customer loyalty and business reputation might go beyond those figures for the MRO facility.

The cost study is setup with a range of demand from 4 to 10 work orders per month. Profiting from the best results from the DOE, the model S is run under the rule EDD with a 60% of overhaul. In the case of model A, the dispatching rule is CR+SPT, and with a 60% of engine overhaul.

The results of the simulations are represented in Figure 4. In the x-axis the total work orders or units processed are presented. The y-axis shows the different components of unitary cost. The total cost is composed of direct and indirect components plus the variable cost with origin in the overdue work orders charges. This extra cost impacts at high load level and establishes an optimal minimum cost in the range between 300 to 350 work orders in a simulation horizon of 5 years, or an equivalent average annual demand from 60 to 70 work orders for the system under analysis. In the comparison the behavior of both process flow models, the asynchronous flow A responds slightly better to high load, in terms of the expected extra charges for delays. The better behavior of the process flow A together with the compound slack rule (CR+SPT) is consistent with the better results found in recent research [36] where the penalty in MRO operations is explored based on a simple generic process.

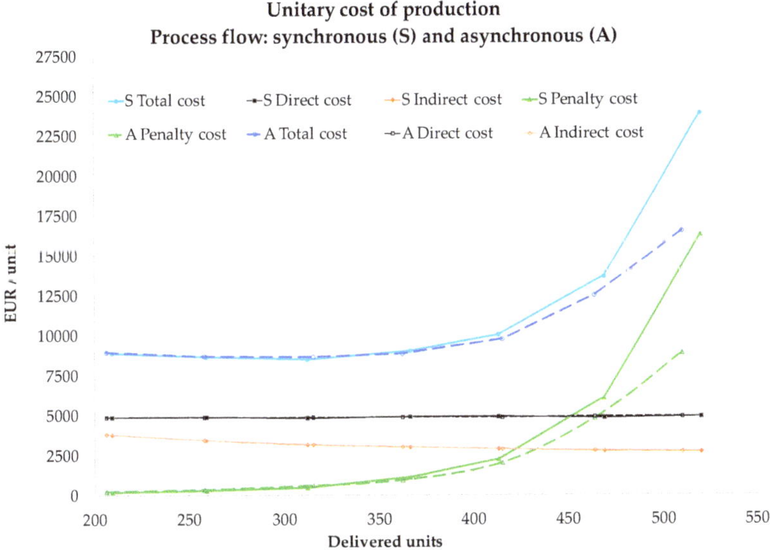

Figure 4. Unitary cost analysis breakdown in a 5 year simulation period.

The situation of overload might not be an ordinary state and it can be considered to establish facility design boundaries or expansion flexibility [45]. Focusing on the ordinary range of operations, the total unitary cost is similar for both process flows. The convenient range of lower cost per delivered unit is from 60 to 70 annual work orders. In this range, the extra costs per overdue work orders could be around 8% to 12% of the total cost, but operating at around 80 work orders per year can bear about 25% of extra cost. In addition, note that working at middle load with 40 work orders per year, with charges only representing 3% of total cost, is a range where the underutilization of resources dominates the unitary cost, due to the low load of the system.

5. Conclusions

System simulation and the design of experiments allow the quantitative evaluation of the significant effects in operations and the convenient dispatching rules for each process flow. System operating curves can be estimated, so the throughput and inventory trade-offs can be assessed. The influence of work in process inventories in service ratio and in system flexibility are important reasons for paying attention to the adequate level of inventories in the overall performance in service-oriented systems. In addition, simulation allows the evaluation of the convenient range of work load into the system.

The work order mix of the demand (engines and modules) is demonstrated to be an important influence factor to obtain good operational results. The optimization of the resource utilization leads to full capacity use through work orders of engine overhaul, but this situation does not represent the better scenario for service metrics. Being conscious that demand mix cannot be fully controlled, customer service managers might conduct customer portfolio for the more convenient mix combination of engine overhaul and modules through demand management. For the detailed model of the MRO facility under study, the dispatching rules analysis results show a better performance for Earliest Due Date (EDD) rule and Critical Ratio and Shortest Processing Time (CR+SPT) rule in a system with a mix of engine overhaul (60% in the system under study). When each flow is managed with the proper dispatching rule, there is no conclusive advantage in performance based on synchronous and asynchronous flow. Synchronous flows appears to be better run through the EDD rule, while the asynchronous flow better by the the CR+SPT rule. Considering the current maintenance trend to improve life cycle management and pooling of components, the asynchronous flow with the (CR+SPT) rule can be an advantage, approaching better the pooling operation of components.

In service oriented manufacturing systems, supplier's metrics include quality levels and service ratios, and overdue work orders can generate charges to the supplier. Including the charges of overdue work orders can become significant for MRO centers operational assessment. The quantitative results show that with an increase of overdue work orders extra costs would balance the benefits of increasing load by the reduction of internal cost of the better resource utilization. Other additional effects of overdue orders are also possible, such as the position in the market or customer loyalty. The methodology has allowed the assessment of the convenient range of load that minimizes the unitary cost. In the system under study, when surpassing that range the extra costs of charges increase rapidly. Below that range, operations can benefit from operational flexibility with a stable unitary cost over a wide interval of load. Considering the two process flow type alternatives, slightly better behavior has been shown with high load by the asynchronous flow process in the operating curves and the unitary cost. When considering mainly the service performance and its variability, the synchronous model has shown general more robust behavior in the simulation experiments. Beyond the absolute quantitative differences, it is important that the selection of the proper priority dispatching rule associated with the flow process, be in accordance with the relative management importance given to each metric, in a complex multi criteria decision-making process.

The combined methodology of discrete-event simulation and design of experiments (DOE) through the Taguchi approach allows the detailed consideration of the main influential factors in system operations in relationship to a particular process flow. The operational decision-making improves under the complex relationships between system design and operations management (priority or dispatching rules), looking for the optimal priority rules for a particular manufacturing system. With the same decision-making metrics, the optimal dispatching rule appears to be associated with a particular process flow. In the particular case of the engine MRO center under analysis, many processes and tasks are established by the airworthiness standards, so the system process flow options can be designed by taking into account an integrated analysis of the internal and service ratios. Discrete-event simulation combined with the design of experiments is a convenient technique to evaluate scalable models of manufacturing systems focused on service, including their design alternatives. In general, the trade-off between productive factors and service ratios recommends to look for a process flow design and dispatching rules with favorable results in cost, but in tight relationship with the operational competitive advantages of flexibility and a proper service.

Former results suggest that future research works can benefit from discrete-event simulation, in addition to other method use (constraint programming, meta-heuristics, agents, etc.). They should combine the analysis of process flow and priority rules with the flexibility under partial resource utilization, in order to cope with load fluctuations while reaching the desired service. In direct connection with the results, the current trend in life cycle management of capital goods (e.g., aircraft engines) encourages pooling systems of equipment or components. Therefore, the study of external coupling of MRO scheduling with hardware management might be probably better studied through asynchronous process flow.

Author Contributions: Conceptualization, J.V. and R.C.; Methodology, R.C.; Software, J.V.; Validation, J.V. and R.C.; Formal Analysis, J.V. and R.C.; Writing—Original Draft Preparation, J.V.; Writing—Review & Editing, R.C.

Funding: This research received no external funding.

Conflicts of Interest: The authors declare no conflict of interest.

References

1. Mont, O.K. Clarifying the concept of product–service system. *J Clean. Prod.* **2002**, *10*, 237–245. [CrossRef]
2. Netland, T.H.; Frick, J. Trends in manufacturing strategies: A longitudinal investigation of the International Manufacturing Strategy Survey. In *International Manufacturing Strategy in a Time of Great Flux*; Brennan, L., Vecchi, A., Eds.; International Publishing: Cham, Switzerland, 2017; pp. 1–16, ISBN 978-3-319-25350-3.
3. Calvo, R.; Domingo, R.; Sebastián, M.A. Systemic criterion of sustainability in agile manufacturing. *Int. J. Prod. Res.* **2008**, *46*, 3345–3358. [CrossRef]
4. EASA. Easy Access Rules for Continuing Airworthiness (Regulation (EU) No 1321/2014). European Union, 2017. Available online: https://www.easa.europa.eu/document-library/general-publications/easy-access-rules-continuing-airworthiness-regulation-eu-no (accessed on 28 August 2018).
5. Aubin, B. *Aircraft Maintenance—The Art and Science of Keeping Aircraft Safe*; Society of Automotive Engineers International: Warrendale, PA, USA, 2004; ISBN 9780768012422.
6. Department for Business, Innovation and Skills. *UK Aerospace Maintenance, Repair, Overhaul (MRO) & Logistics*; Crown: London, UK, 2016. Available online: https://assets.publishing.service.gov.uk/government/uploads/system/uploads/attachment_data/file/502588/bis-16-132-uk-mrol-analysis.pdf (accessed on 20 June 2018).
7. Aeronautical Repair Station Association. Global Fleet and MRO Market Economic Assessment. Available online: http://arsa.org/wp-content/uploads/2017/03/ARSA-OW-MROMarketAssessment-20170225_ES.pdf (accessed on 20 June 2018).
8. IATA. Airline Maintenance Cost Executive Commentary. Available online: https://www.iata.org/whatwedo/workgroups/Documents/MCTF/MCTF-FY2016-Report-Public.pdf (accessed on 20 June 2018).
9. Ackert, S. Engine Maintenance Concepts for Financiers. Available online: http://www.aircraftmonitor.com/uploads/1/5/9/9/15993320/aircraft_mx_handbook_for_financiers_v1.pdf (accessed on 27 August 2018).
10. Rodrigues, D.; Lavorato, P. Maintenance, Repair and Overhaul (MRO) Fundamentals and Strategies: An Aeronautical Industry Overview. *Int. J. Comput. Appl.* **2016**, *135*, 21–29. [CrossRef]
11. Ayeni, P.; Baines, T.; Lightfoot, H.; Ball, P. State-of-the-art of 'Lean' in the aviation maintenance, repair, and overhaul industry. *Proc. Inst. Mech. Eng. Part B: J. Eng. Manuf.* **2011**, *225*, 2108–2123. [CrossRef]
12. Uhlmann, E.; Bilz, M.; Baumgarten, J. MRO-Challenge and Chance for Sustainable Enterprises. *Procedia CIRP* **2013**, *11*, 239–244. [CrossRef]
13. Reopel, M. Smarter MRO-5 Strategies For Increasing Speed, Improving Reliability, and Reducing Costs—All at the Same Time. Available online: https://www2.deloitte.com/content/dam/Deloitte/us/Documents/consumer-business/us-avitran-thl-smartermro-072612.pdf (accessed on 20 June 2018).
14. McLaughlin, P.; Durazo-Cardenas, I. Cellular manufacturing applications in MRO operations. *Procedia CIRP* **2013**, *11*, 254–259. [CrossRef]
15. Law, A.M.; Kelton, W.D. *Simulation Modelling and Analysis*; McGraw-Hill: Singapore, 1991; ISBN 978-0073401324.
16. Banks, J.; Carson, J.S., II; Nelson, B.L.; Nicol, D.M. *Discrete-Event System Simulation*; International Edition Pearson Education Limited: Harlow, UK, 2014; ISBN 978-1-292-02437-0.
17. Rossetti, M.D. *Simulation Modelling and Arena*; John Wiley & Sons: New Jersey, NJ, USA, 2016; ISBN 978-0131446793.
18. Robinson, S. Discrete-event simulation: From the pioneers to the present, what next? *J. Oper. Res. Soc.* **2005**, *56*, 619–629. [CrossRef]
19. Smith, J.S. Survey on the use of simulation for manufacturing system design and operation. *J. Manuf. Syst.* **2003**, *22*, 157–171. [CrossRef]
20. Negahban, A.; Smith, J.S. Simulation for manufacturing system design and operation: Literature review and analysis. *J. Manuf. Syst.* **2014**, *33*, 241–261. [CrossRef]
21. Heizer, J.; Render, B. *Dirección de la Producción y de Operaciones. Decisiones Estratégicas*; Pearson: Madrid, Spain, 2015; ISBN 978-8490352878.
22. Taguchi, G.; Chowdhury, S.; Wu, Y. *Taguchi's Quality Engineering Handbook*; John Wiley & Sons: Hoboken, NJ, USA, 2005; ISBN 978-0471413349.
23. Graves, S.C. A review of production scheduling. *Oper. Res.* **1981**, *29*, 646–675. [CrossRef]

24. Slack, N.; Chambers, S.; Johnston, R. *Operations Management*; Pearson: Harlow, UK, 2010; ISBN 978-0-273-73046-0.
25. Nyhuis, P.; Schmidt, M. Logistic Operating Curves in Theory and Practice. In *Advances in Computer Science and Engineering*; Schmidt, M., Ed.; Intech Open Publisher: London, UK, 2011; pp. 371–390.
26. Holthaus, O.; Rajendran, C. Efficient dispatching rules for scheduling in a job shop. *Int. J. Prod. Econ.* **1997**, *48*, 87–105. [CrossRef]
27. Kemppainen, K. *Priority Scheduling Revisited: Dominant Rules, Open Protocols, and Integrated Order Management*; Helsinky School of Economics: Helsinky, Finland, 2005; Available online: https://aaltodoc.aalto.fi/bitstream/handle/123456789/11236/a264.pdf?sequence=1&isAllowed=y (accessed on 20 June 2018).
28. Barman, S. Simple priority rule combinations: An approach to improve both time and tardiness. *Int. J. Prod. Res.* **1997**, *35*, 2857–2870. [CrossRef]
29. Anderson, E.J.; Nyirenda, J.C. Two new rules to minimize tardiness in a job shop. *Int. J. Prod. Res.* **1990**, *28*, 2277–2292. [CrossRef]
30. Haupt, R. A survey of priority rule-based scheduling. *OR Spektrum* **1989**, *11*, 3–16. [CrossRef]
31. Guide, V.D.R., Jr.; Srivastava, R.; Kraus, M.E. Priority scheduling policies for repair shops. *Int. J. Prod. Res.* **2000**, *38*, 929–950. [CrossRef]
32. Hicks, C.; Pongcharoen, P. Dispatching rules for production scheduling in the capital goods industry. *Int. J. Prod. Econ.* **2006**, *104*, 154–163. [CrossRef]
33. Browning, T.R.; Yassine, A.A. Resource-constrained multi-project scheduling: Priority rule performance revisited. *Int. J. Prod. Econ.* **2010**, *126*, 212–228. [CrossRef]
34. Reményi, C.; Staudacher, S. Systematic simulation based approach for the identification and implementation of a scheduling rule in the aircraft engine maintenance. *Int. J. Prod. Econ.* **2014**, *147*, 94–107. [CrossRef]
35. Denkena, B.; Dittrich, M.A.; Georgiadis, A. Combining in-house pooling and sequencing for product regeneration by means of event-driven simulation. *Procedia CIRP* **2017**, *62*, 153–158. [CrossRef]
36. Hoffmann, L.S.; Kuprat, T.; Kellenbrink, C.; Schmidt, M.; Nyhuis, P. Priority rule-based planning approaches for regeneration processes. *Procedia CIRP* **2017**, *59*, 89–94. [CrossRef]
37. The Boeing Company. Services Market. Outlook. 2017. Available online: http://www.boeing.com/resources/boeingdotcom/commercial/market/services-market-outlook/assets/downloads/2017-commercial-smo-final.pdf (accessed on 10 February 2018).
38. Soares, C. *Gas. Turbines: A Handbook of Air, Land, and Sea*; Elsevier: Burlington, MA, USA, 2008; ISBN 978-0750679695.
39. Rolls-Royce. *The Jet Engine*; Wiley: Derby, UK, 1996; ISBN 978-1119065999.
40. Chase, R.B.; Aquilano, N.J.; Jacobs, F.R. *Production and Operations Management*; Irwin/McGraw-Hill: Boston, MA, USA, 1998; ISBN 0-256-22556-7.
41. Black, J.T. *The Design of the Factory with a Future*; McGraw Hill: New York, NY, USA, 1998; ISBN 978-0070055506.
42. Kelton, W.D.; Sadowski, R.P.; Sadowski, D.A. *Simulation with Arena*; McGraw Hill: New York, NY, USA, 1998; ISBN 978-0073401317.
43. Vargas, J. Análisis y Simulación de la Configuración y Operaciones de un Centro de Mantenimiel nto de Motores Turboeje de Aviación. MSc Production Engineering Dissertation, Universidad Politécnica de Madrid, Madrid, Spain, 2017.
44. Starks, D. Steady-State and Warmup. Arena Blog, 23 January 2017. Available online: https://www.arenasimulation.com/support/entry/steady-state-and-warmup (accessed on 10 October 2017).
45. Calvo, R.; Domingo, R.; Sebastián, M.A. Operational flexibility quantification in a make-to-order assembly system. *Int. J. Flex Manuf. Sys.* **2007**, *19*, 247–263. [CrossRef]
46. Jugulum, R.; Samuel, P. *Design for Lean Six Sigma: A Holistic Approach to Design and Innovation*; John Wiley & Sons: New Jersey, NJ, USA 2010; ISBN 978-0470007518.
47. Kim, J.M.; Zhou, Y.D.; Lee, D.H. Priority scheduling to minimize the total tardiness for remanufacturing systems with flow-shop-type reprocessing lines. *Int. J. Adv. Manuf. Technol.* **2017**, *91*, 3697–3708. [CrossRef]

48. Saltoglu, R.; Humaira, N.; Inalhan, G. Scheduled maintenance and downtime cost in aircraft maintenance management. *Int. J. Mech. Aerosp. Ind. Mechatron. Manuf. Eng.* **2016**, *10*, 580–585. [CrossRef]
49. Ferguson, M.; Nelson, S. *Aviation Safety: A Balanced Industry Approach*; Cengage Learning: Delmar, DE, USA, 2014; ISBN 978-8131523018.

© 2018 by the authors. Licensee MDPI, Basel, Switzerland. This article is an open access article distributed under the terms and conditions of the Creative Commons Attribution (CC BY) license (http://creativecommons.org/licenses/by/4.0/).

 *materials*

Article

In-Process Measurement for the Process Control of the Real-Time Manufacturing of Tapered Roller Bearings

Francisco Javier Brosed [1,*], A. Victor Zaera [2], Emilio Padilla [2], Fernando Cebrián [2] and Juan José Aguilar [1]

[1] Design and Manufacturing Engineering Department, Universidad de Zaragoza, 3 María de Luna Street, Torres Quevedo Bld, 50018 Zaragoza, Spain; jaguilar@unizar.es

[2] FERSA, 18 Bari Street-PLAZA, 50197 Zaragoza, Spain; victor.zaera@fersa.com (A.V.Z.); emilio.padilla@fersa.com (E.P.); fernando.cebrian@fersa.com (F.C.)

* Correspondence: fjbrosed@unizar.es; Tel.: +34-876-555-456

Received: 19 June 2018; Accepted: 2 August 2018; Published: 7 August 2018

Abstract: Tapered roller bearings can accommodate high radial loads as well as high axial loads. The manufacturing process consists of machining processes for ring and component assembly. In this contribution, the parameters of influence on the measurement procedure were studied. These parameters of influence were classified as environmental, process, and machine parameters. The main objective of this work was to optimize the process using real-time measurements, which required the study of the influence of several parameters on the measurement uncertainty and how to correct their effects.

Keywords: in-process measurement; geometric accuracy; grinding process; tapered roller bearings

1. Introduction

Bearing manufacturing is a high-precision technology where the material composition, hardness, and micrometric dimensions need to be ensured to meet the product requirements [1,2].

The quality of the product is an important strategic factor for the competitiveness of the European manufacturing industry in the global market [3]. In this context, process control, automation, and optimization are key to having the best quality at a competitive cost [4,5]. Process control ensures quality and reduces scrap and rework, but requires dimensional measurements (the effect on the temperature of the grinding process needs to be considered). Automation is the key to achieving competitive cost by using machines with enough accuracy and cycle time, and automatic adjustment of the manufacturing process in real time. Current data collection and inspection technologies allow data to be collected online along the process chain and can significantly increase quality control and improvements in current dynamic and modifiable environments [6,7]. The real challenge facing companies is the problem of synthesizing highly heterogeneous data to gain in-depth understanding of the correlations between the variables throughout the stages of a multi-stage system. This is aimed at achieving the generation of zero defects at the single process level, and the propagation of zero defects at the system level through the proactive control of the process [8].

Tapered roller bearings can accommodate high radial loads as well as high axial loads. They have four main components: inner ring, outer ring, rollers, and cage. In general and in the case under investigation, these components are metallic, although the cage can be plastic depending on the application. The manufacturing process consists of the machining processes of rings and component assembly. The expected quality standard needs to measure each individual part using an appropriate measurement instruments. These instruments have an uncertainty in their measurements. Ambient temperature, the temperature of the system, and the temperature of the part under inspection also influences the precision of the components and of the mounted bearing [9]. Different methods can be used to measure the size of the parts: touch probes [10,11], air pressure [12,13], and laser systems [14,15]. Most of the developments that can be found in the literature regarding bearing inspection are methods for fault diagnosis and service life estimation. Many experiments and studies have been performed to explore the nature of bearing defects with the help of several monitoring techniques such as vibration, acoustic emissions, oil-debris, ultrasound, electrostatic, shock-pulse measurements, etc. [1,16,17], although some of them estimate the size of the manufacturing defects of tapered roller bearings with vibration measurement [18].

As stated before, process optimization and online monitoring and control are key factors for improving efficiency and quality in machining [19]. With the development of intelligent machining, the optimization of cutting process configuration during actual production has become more accessible, and optimizing the volume of removed metal by adjusting the grinding time for each part can decrease the process time and improve tool life [20,21]. Finally, the grinding machine control system can receive the measurement results of the machined part as feedback information for process verification.

The work described in this article focused on the "in process" verification [22,23] of a tapered roller bearing. The magnitude under inspection was the outside diameter of the outer ring (D in Figure 1). Its design tolerance was ± 0.025 mm. The measurement result was needed to feedback the real-time control of the grinding process used to manufacture the outside diameter of the outer ring. The authors describe the model and analysis of a measurement system and the effects of its main error sources, namely the temperature and the misalignment of the devices or of the work piece and the master piece. The influence of the error sources was studied and an estimation of the uncertainty of the system was provided using simulations programmed using the Monte Carlo method [24,25], and finally the process improvement achieved when the measurement results were fed back into the manufacturing process is shown. A general approach for modelling the uncertainty associated with coordinate measuring systems (CMSs) is given in [26]. Several authors [24,25] have shown a comparison between the estimation of measurement uncertainty using the law of propagation of uncertainty [27] and that using the propagation of distribution using the Monte Carlo method [28]. In this case, the method presented in [28] was the one that better fit our application, as the variables affecting the results of the measurement presented different distributions, some of them with asymmetrical effects. A comparison of the results obtained with the simulation using the Monte Carlo method and the experimental results allowed for the identification of the main error sources and quantified their influence.

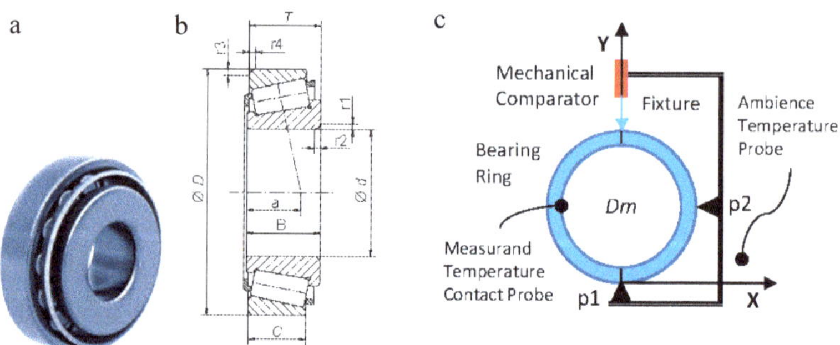

Figure 1. *Cont.*

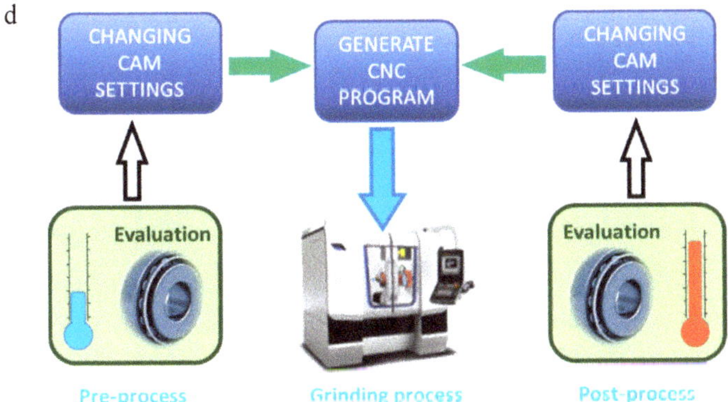

Figure 1. (**a**) Tapered roller bearing; (**b**) Functional dimensions; (**c**) Measurement device; (**d**) Synoptic of operation.

2. Materials and Methods

The measurement of the diameter of the outer ring ("D" in Figure 1) was carried out using a mechanical comparator, a registration tool to position the ring, and two temperature probes, one per contact to measure the temperature of the piece and another to measure the ambient temperature. A scheme of the system taking the measurement appears in Figure 1 and the characteristics of its components are indicated in Table 1.

Table 1. Components of the measurement system.

Equipment	Range	Resolution	Expanded Uncertainty (k = 2)
Mechanical comparator (probe)	−1.5/+1.5 mm	0.001 mm	0.0013 mm
Contact probe thermometer	0 a 250 °C	0.1 °C	0.39 °C
Thermocouple probe thermometer	−50 a 100 °C	0.1 °C	0.40 °C

The tooling was adjusted using a standard part (or master). The adjustment of the tooling consisted of fixing the position in the X direction of the support point p2 (Figure 1). For this purpose, a master piece was used, the previously-calibrated diameter of which corresponded to the nominal diameter to be measured with the system (in this case D_0 = 112.712 mm). The point p2 shifted in the X direction until the maximum indicated by the probe was obtained. In this position, p2 was set. This operation was carried out each time the reference was changed.

Once the tooling had been adjusted, the zero of the comparator with the diameter of the masterpiece (D_0) was set as the reference. This operation was performed every 30 min (approximately every 50 parts checked). Both the machine and the masterpiece were stabilized at room temperature as the master piece was kept in the workshop between measurements.

After these two steps, the system was ready to measure the diameter of the outer rings (D_m) after being rectified so that its temperature ($T_{m,m}$) was higher than the ambient temperature at the time of performing the measurement ($T_{a,m}$). The verification of the outer ring was performed at a rate of approximately 100 parts/h. The system and the masterpiece were kept in the workshop, thus it was considered that both elements were at room temperature (Equations (1) and (2)).

$$T_{m,0} = T_{s,0} = T_{a,0} \quad (1)$$

$$T_{s,m} = T_{a,m} \quad (2)$$

where $T_{x,t}$ is the temperature of the ambience if $x = a$, of the system if $x = s$, and of the measured part if $x = m$. The temperature changes over time, thus $T_{x,t}$ indicates the temperature at the moment of measuring the m-th part if $t = m$ and at the time of measuring the master piece if $t = 0$.

The following system elements were introduced in the model of the measuring system: D_m, measurand; H, tooling height; and L_m, probe length in the measurement of the m-th piece. The influence factors considered and the indicated system elements are summarized in Table 2 together with the nomenclature used in this work.

Factors due to the measurement process were considered, such as the possible inclination of the part due to incorrect support in the tooling (θ_m). This angle depends on the measurement process and therefore may be different when measuring the master piece and the part to be checked.

Finally, we analyzed the effect of factors influencing the configuration and mechanical behavior of the system. Therefore, the effect on the result of the measurement of a possible displacement of the mechanical comparator (λ_m) and the effect of an angular deviation of the probe from the vertical (ψ_m) was studied. It should be noted that when it comes to system factors, it is reasonable to consider that its values will not be modified between the measurement of the master piece and the measurement of the rest of the parts, thus their effect will tend to cancel out, as will be verified later (Section 3.3).

The following sections analyze the effect of these factors in detail. The measurement model used in the mentioned analysis is explained hereafter:

The master piece is measured at a different temperature than the m-th part. In this way, it can be said that in measuring the master at $T_{m,0}$, a diameter $D_{0,0}$ (Equation (3)) is measured and the mechanical comparator at temperature $T_{s,0}$ provides a reading L_0 (corresponding to a length of the probe $L_{0,0}$) (Equation (4)).

$$D_{0,0} = D_0 \cdot (1 + \alpha_0 (T_{a,0} - 20)) \tag{3}$$

$$L_{0,0} = L_0 \cdot (1 + \alpha_L (T_{a,0} - 20)) \tag{4}$$

In addition, since there is a slow variation in plant temperature, both the tooling-comparator system and the master piece are considered at plant temperature, $T_{s,0} = T_{m,0} = T_{a,0}$ (Equation (1)). At this point, the length of the comparator for which the system reference is defined can be written according to the height of the tooling, $H_{s,0}$ (Equation (5)), depending on the nominal value and its temperature $T_{s,0}$ (Equation (6)).

$$L_{0,0} = H_{s,0} - D_{0,0} \tag{5}$$

$$H_{s,0} = H \cdot (1 + \alpha_s (T_{a,0} - 20)) \tag{6}$$

The measurement of the m-th part is performed through a comparison with the master piece. The temperature of the part will be $T_{m,m}$ and that of the equipment $T_{s,m}$, which is taken as the ambient temperature at the time of inspection (Equation (2)). When measuring the diameter $D_{m,m}$ (Equation (7)), the comparator will provide a reading ΔL_m (Equation (8)) (corresponding to a length of the probe $L_{m,m}$ (Equation (9)).

$$H_{s,m} = H \cdot (1 + \alpha_s (T_{a,m} - 20)) \tag{7}$$

$$\Delta L_{m,m} = \Delta L_m \cdot (1 + \alpha_s (T_{a,m} - 20)); \tag{8}$$

$$\Delta L_{m,m} = L_{0,m} - L_{m,m}; \tag{9}$$

where α_x is the thermal expansion coefficient of the m-th part if $x = m$; of the master piece if $x = 0$; and of the measuring system and the probe if $x = s$. In this case, $\alpha_m = \alpha_0 = \alpha_s = 11 \times 10^{-6}\ ^{\circ}\mathrm{C}^{-1}$.

Table 2. Factors influencing the measurement result and system elements. Nomenclature.

Influencing Factors

	$T_{x,t}$	$t = m$	$t = 0$
T, temperature (°C)	$x = a$	$T_{a,m}$	$T_{a,0}$
	$x = s$	$T_{s,m}$	$T_{s,0}$
	$x = m$	$T_{m,m}$	$T_{m,0}$
θ_t (°)		θ_m	θ_0
ψ_t (°)		ψ_m	ψ_0
λ_t (mm)		λ_m	λ_0

Measurement System Elements (H, L, D)

		$t = m$	$t = 0$
H, tooling height (mm)	$H_{s,t}$	$H_{s,m}$	$H_{s,0}$
	H, tooling height at 20 °C		
	-	H	
L, probe length (mm)	$L_{x,t}$	$t = m$	$t = 0$
	$x = m$	$L_{m,m}$	N.a. [1]
	$x = 0$	$L_{0,m}$	$L_{0,0}$
	L_t, probe length at 20 °C		
	-	L_m	L_0
D, measurand (mm)	$D_{x,t}$	$t = m$	$t = 0$
	$x = m$	$D_{m,m}$	N.a. [1]
	$x = 0$	$D_{0,m}$	$D_{0,0}$
	D_t, measurand at 20 °C		
	-	D_m	D_0

x, parameter corresponding to the measurand, $x = m$; to the system, $x = s$; to the ambience, $x = a$; and to the master piece, particular case of measurand, $x = 0$. t, value of the parameter during the measurement of the m-th part, $t = m$; and of the master piece, particularly in the case of measuring, $t = 0$. [1] N.a.: Not applicable.

Substituting Equation (9) into Equation (7) provides the dimension of the diameter (Equations (10) and (11)).

$$D_{m,m} = H_{s,m} - L_{0,m} - \Delta L_{m,m} \tag{10}$$

$$D_{m,m} = D_m \cdot (1 + \alpha_m (T_{m,m} - 20)); \tag{11}$$

$L_{0,m}$ can be calculated by taking the reference each time the temperature changes (Equation (12)). Substituting Equation (12) into Equation (10), the part diameter from the master piece data and the probe reading is calculated (Equation (13)).

$$L_{0,m} = H_{s,m} - D_{0,m}; \text{ with } D_{0,m} = D_0 \cdot (1 + \alpha_0 (T_{a,m} - 20)) \tag{12}$$

$$\begin{gathered} D_{m,m} = H_{s,m} - H_{s,m} - D_{0,m} - \Delta L_{m,m} \Rightarrow \\ \Rightarrow D_{m,m} = D_0 \cdot (1 + \alpha_0 (T_{a,m} - 20)) - \Delta L_m \cdot (1 + \alpha_s (T_{a,m} - 20)) \end{gathered} \tag{13}$$

In the case of not measuring the reference when the temperature changes, it is possible to estimate $L_{0,m}$ (Equation (14)) and $D_{0,m}$ (Equation (15)), however, the result of Equations (13) and (15) will only coincide if the thermal expansion coefficients of the probe and the measurand coincide. In Figure 2, the error of Equation (15) calculated as Equations (15)–(13) is plotted when α_s differs from α_m.

$$L_{0,m} = L_0 \cdot (1 + \alpha_L (T_{a,m} - 20)) \tag{14}$$

$$D_{m,m} = H_{s,m} - L_0 \cdot (1 + \alpha_0 (T_{a,m} - 20)) + \Delta L_m \cdot (1 + \alpha_s (T_{a,m} - 20)) \tag{15}$$

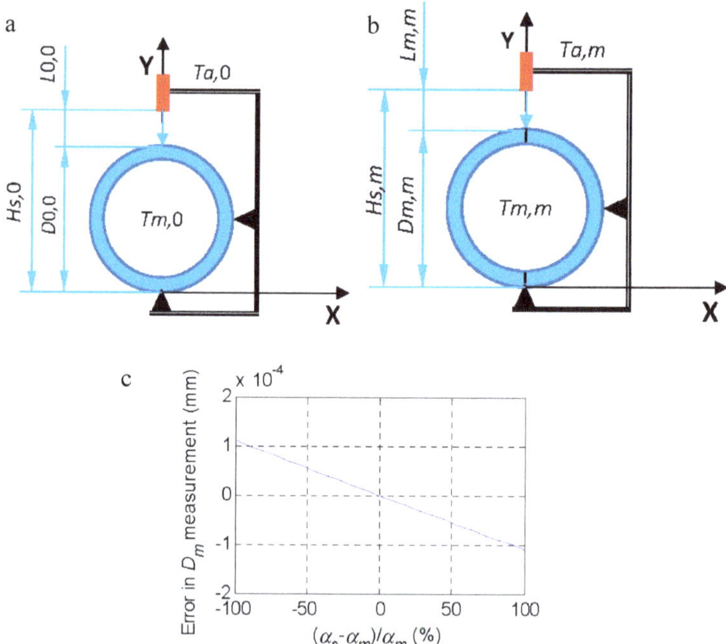

Figure 2. (**a**) Definition of the reference with D_0 at temperature $T_{m,0} = T_{a,0}$. (**b**) comparison with the part under inspection D_m at temperature $T_{m,m}$ where the ambient temperature is $T_{a,m}$. (**c**) Error (15)–(13) as a function of the difference between α_s and α_m for a temperature $T_{m,m}$ = 30 °C and $T_{0,0}$ = 20 °C.

To obtain D_m, the mechanical comparator reading for the m-th part was compared with the reading taken when measuring D_0, thus the result of the measurement (D_m at 20 °C) was obtained from Equation (16).

$$D_m = \frac{H \cdot (1 + \alpha_s(T_{a,m} - 20)) - L_0 \cdot (1 + \alpha_L(T_{a,m} - 20)) + \Delta L_m \cdot (1 + \alpha_L(T_{a,m} - 20))}{1 + \alpha_m(T_{m,m} - 20)} \tag{16}$$

3. Results

3.1. Effect of Temperature on Measurement

The effect of the temperature in the process can be corrected using Equation (16), as the temperature was known. However, if one of the process temperatures (the ambient temperature or the temperature of the m-th part) was not known, the following situations could occur, as described in Table 3.

Table 3. Cases if one or both of the temperature probes are not available. "1" means that the temperature data are available, "0" means that the temperature data are not available.

Diameter (eq.)	T_a	T_c	ΔT for 0.025 mm Error (°C)
D_m with ambient and contact thermometer (16)	1	1	Not applicable
D_m w/o contact thermometer	1	0	20.8
D_m w/o ambient thermometer	0	1	19.6
D_m w/o any thermometer	0	0	19.6

The effect of the ambient temperature and the temperature of the m-th part was studied by varying them independently and evaluating the error (Equation (11)) that they introduced to the calculation of the diameter in the function of the four cases raised (Table 2). In each case, a variation between 10 and 39 °C was introduced for the ambient temperature at the moment of measuring the masterpiece ($T_{a,0}$), the ambient temperature at the moment of inspecting the m-th part ($T_{a,m}$), and the temperature of the m-th part ($T_{m,m}$) (Figure 3). From these results, it was shown that by applying Equation (16), it was possible to measure at any temperature. However, if one of the two probes was not available, an error was introduced in the measurement. For these cases, the temperature increase necessary to obtain a measurement error equal to the tolerance (±0.025 mm) was quantified, and is presented in Table 2. As can be seen, the influence of the ambient temperature at the moment of taking the reference with the master piece had less weight than temperatures at the time of measuring the m-th part.

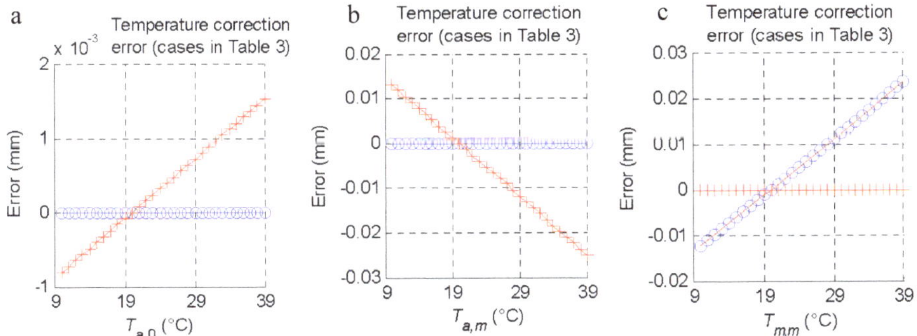

Figure 3. *Cont.*

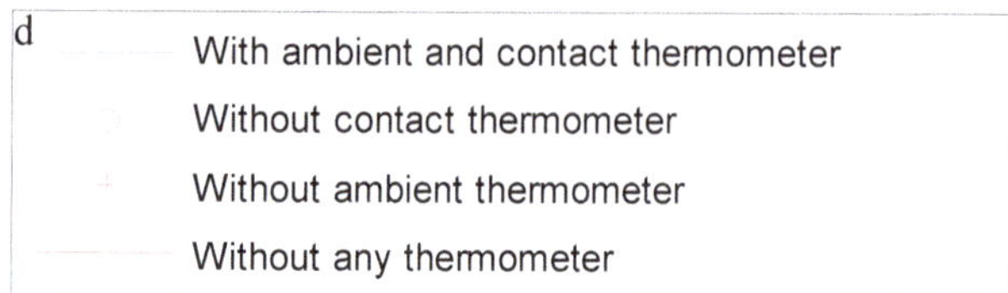

Figure 3. Temperature compensation depending on the measurement process: correction error of the cases described in Table 3 (the first case of Table 3 is the reference); (**a**) Effect of the ambient temperature variation at the moment of measuring the masterpiece ($T_{a,0}$). (**b**) Effect of the ambient temperature variation at the moment of inspecting the m-th part ($T_{a,m}$). (**c**) Effect of the temperature of the m-th part ($T_{m,m}$) variation. (**d**) Legend of the results from using each case of Table 3.

3.2. Effect of Process and Machine Factors on the Measurement

In this section, we analyzed the effect of the incorrect positioning of the part on the tooling (θ_m, process factor) and the effect of a possible deformation of the tooling-comparator system (λ_m displacement and rotation ψ_m, machine factors) (Figures 4 and 5).

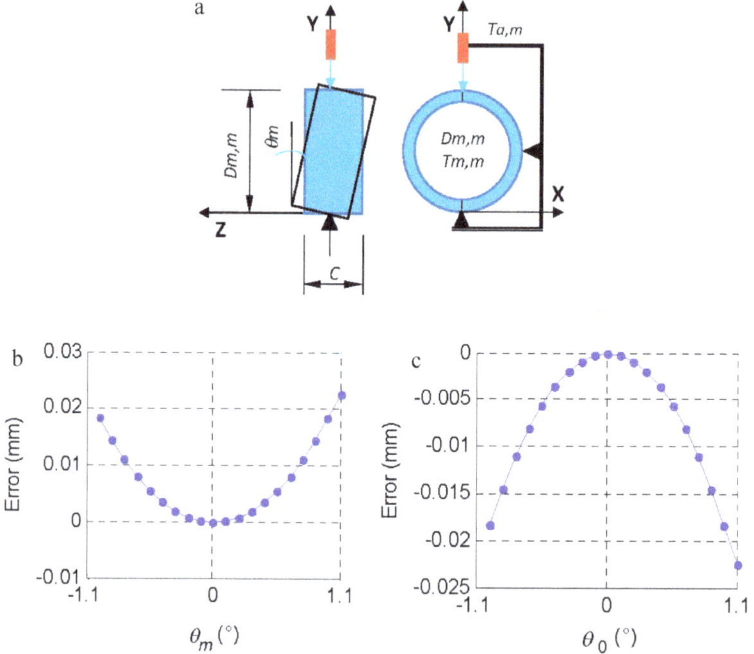

Figure 4. (**a**) θ_m, inclination of the piece when it is placed in the measuring tool. (**b**) Effect of the inclination of the m-th part (θ_m) on the measurement error (comparator reading—$D_{m,m}$). (**c**) Effect of the inclination of the masterpiece (θ_0) on the measurement error (comparator reading—$D_{0,0}$).

The effect of improper support on the tooling was evaluated, thus a sloping part (θ_m) was measured instead of measuring a part that was perfectly horizontal (Figure 4). This can occur both when inspecting the m-th part (θ_m) and when measuring the master piece (θ_0) and may be due, for example, to the presence of a chip in the machining process. The geometry of the piece (dimension C = 23.812 mm) limits the maximum inclination (11°) from which the measurement would not be possible. In the measurement process, a slope close to the indicated maximum would be visually detected, so the slope range studied was lower (± 1.1°, Figure 4b,c).

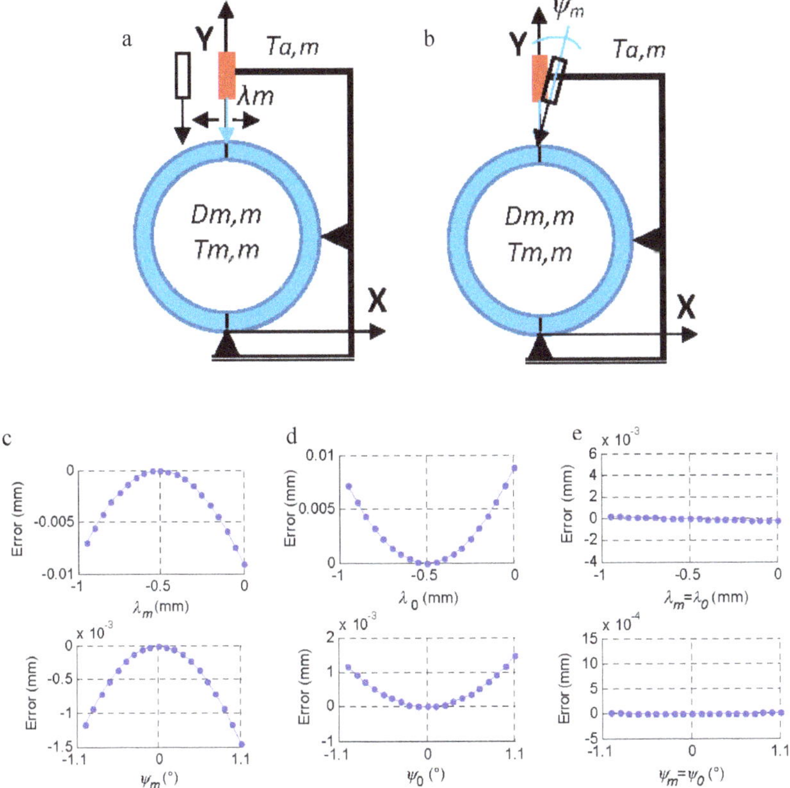

Figure 5. Effect of the system deformation on the measurement error. (**a**) Moving the probe λ_m. (**b**) Turning the probe ψ_m. (**c**) Effect of λ_m and ψ_m on the measurement of $D_{m,m}$. (**d**) Effect of λ_0 and ψ_0 on the measurement of $D_{m,m}$. (**e**) Effect if $\lambda_m = \lambda_0$ and $\psi_m = \psi_0$ in the measurement of $D_{m,m}$.

If the part is measured with a slope of ± 1.1°, the measurement error (comparator reading—$D_{m,m}$) is 0.018 mm.

If the machine suffers some type of deformation, the probe may suffer shifts and rotations that cause incorrect measurement (Figure 5a,b). A displacement (λ_m) on the X axis prevents correct probing at the diametrical point of the part, and a rotation (ψ_m) of the probe on the Z axis introduces a cosine error in the probe reading. These effects were introduced independently in the measurement of the master piece (λ_0 and ψ_0) and the measurement of the m-th part (λ_m and ψ_m), but, as is reasonable to assume, if the system is deformed in the measurement of the master piece, it will also be so in the part, and the effect is the same as in Equations (17) and (18), Figure 5.

$$\lambda_m = \lambda_0, \tag{17}$$

$$\psi_m = \psi_0, \tag{18}$$

The effect of θ_m is inverse to that of λ_m and ψ_m since θ_m tends to increase the reading when its value increases, while λ_m and ψ_m tend to increase the length of the probe when taking the reading and, therefore, reduce the diameter of the ring when they increase their value, as seen in Figure 5. The effect of λ_m and ψ_m tends to be canceled when their values are the same at the time of measurement of the master piece and at the time of measurement of the rest of the parts (Equations (17) and (18), Figure 5e).

3.3. Estimation of the Uncertainty and Contribution of Each Parameter

Simulations were programmed using the Monte Carlo method [24,25] to determine the contribution of each parameter to the final uncertainty. The probability density distributions (PDD) of each factor were defined based on its calibration data or the characteristics of the process. The PDDs associated with the measurement equipment, such as temperature probes or the mechanical comparator, were defined based on their measurement uncertainty. The PDDs associated with the process or tooling factors were defined from a uniform distribution where the limit values were defined from the analysis performed in Section 3 and according to the process and characteristics of the tooling (Table 4).

A simulation of the effect of each parameter was carried out by introducing the PDDs indicated in Table 4. The results of the simulation of the effect of each parameter are shown in Figure 6, where the distribution of the error in the measurement resulting from the variation of the parameter below it are presented.

Simulation using the Monte Carlo method allows for the estimation of the value of the measurement uncertainty according to the "Guide to the expression of uncertainty in measurement" (GUM) and its supplement 1 [28] from the PDD of the influence factors. From the simulation shown in Figure 7, the uncertainty values shown in Table 4 were obtained as a function of the parameter that introduced the variation. At the end of Table 4, the uncertainty obtained when combining the variation of all the factors appeared. The distribution of the error by combining all the factors is shown in Figure 8 together with the evolution of the results of the uncertainty estimated as a function of the number of iterations used. The result stabilized from 10^5 iterations. Other authors [24,29,30] have observed a similar number of iterations to obtain a stabilized result with simulation using the Monte Carlo method.

Table 4. Distribution and range of variation assigned to each parameter. Estimated uncertainty in each case (for each parameter and at the end of the table for all parameters at a time).

Simulation Input Parameters			Simulation (10^6 Iterations) by Varying a Single Input Parameter		
Parameter	**Equipment**	**Distribution**	**Measurement Uncertainty Um (k = 2)**	**Up. Limit**	**Low. Limit**
T_c (°C)	Contact Therm.	Normal ($\mu = 30$; $\sigma = 0.39/2$)	0.0014	112.7356	112.7384
T_a (°C)	Ambient Therm.	Normal ($\mu = 20$; $\sigma = 0.40/2$)	0.0007	112.7363	112.7377
Lm, L_0, D_0 (mm)	Mech.probe	Normal ($\mu = 0$; $\sigma = 0.0013/2$)	0.0018	112.7352	112.7389
θ_m (°)	Parameters of the measurement process and of the system whose variation is estimated with a uniform distribution	Uniform (U.L. = −0.003; L.L. = 0.003)	0.0003	112.7365	112.7370
θ_0 (°)			0.0003	112.7370	112.7375
ψ_m (°)		Uniform (U.L. = −0.005; L.L. = 0.005)	0.0000	112.7370	112.7371
ψ_0 (°)			0.0000	112.7369	112.7370
λ_m (mm)		Uniform (U.L. = −0.2; L.L. = 0.2)	0.0002	112.7370	112.7373
λ_0 (mm)			0.0002	112.7366	112.7370
Simulation (10^6 iterations) varying all parameters			0.0036	112.7334	112.7406

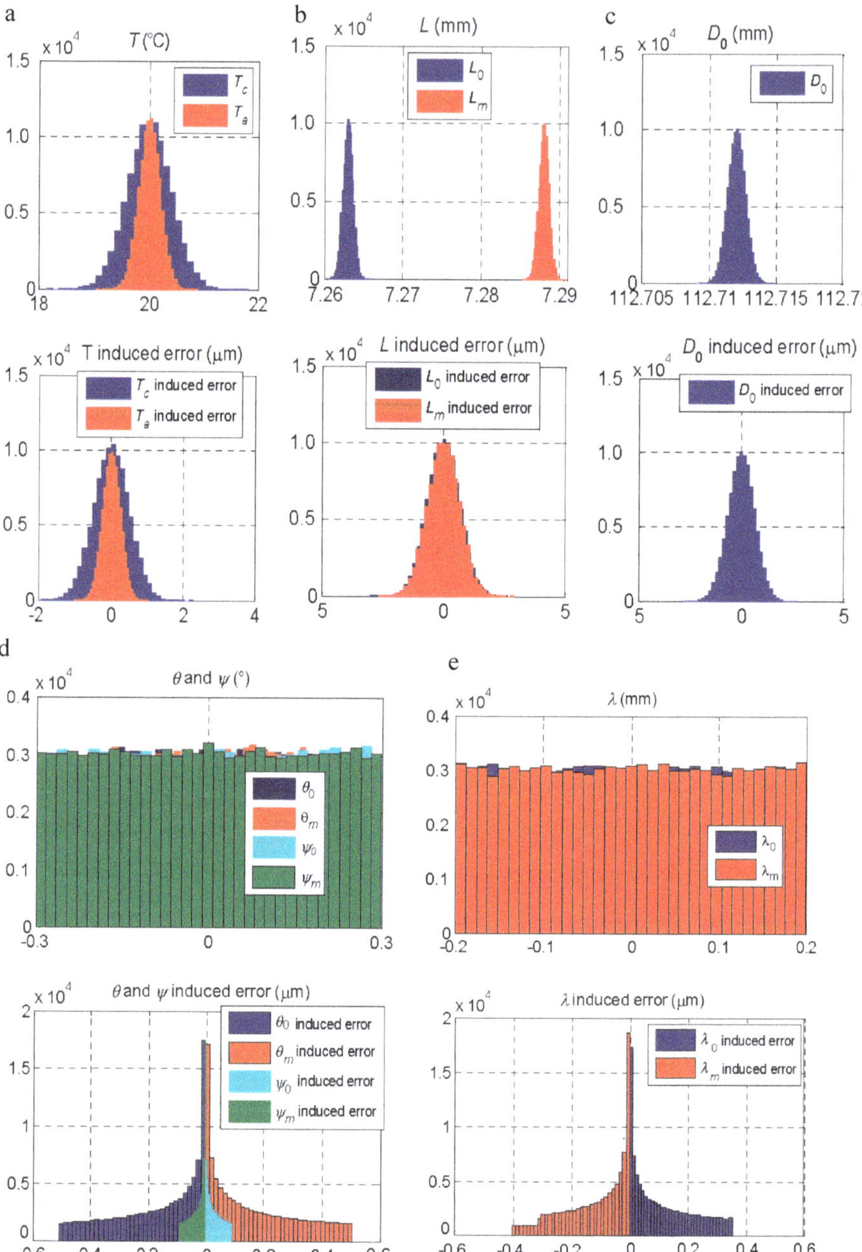

Figure 6. Influence of the variability of the factors on the measurement result. Above, variation introduced in the parameter. Below, the error resulting from the measurement when entering the variation of each parameter. (**a**) Ambience (T_a) and measurand (T_m) temperature. (**b**) Probe length when measuring the master piece (L_0) and the measurand (L_m). (**c**) Master piece diameter (D_0). (**d**) System angles, θ and ψ, when measuring the master piece (θ_0 and ψ_0) and the measurand (θ_m and ψ_m). (**e**) System length, λ, when measuring the master piece (λ_0) and the measurand (λ_m).

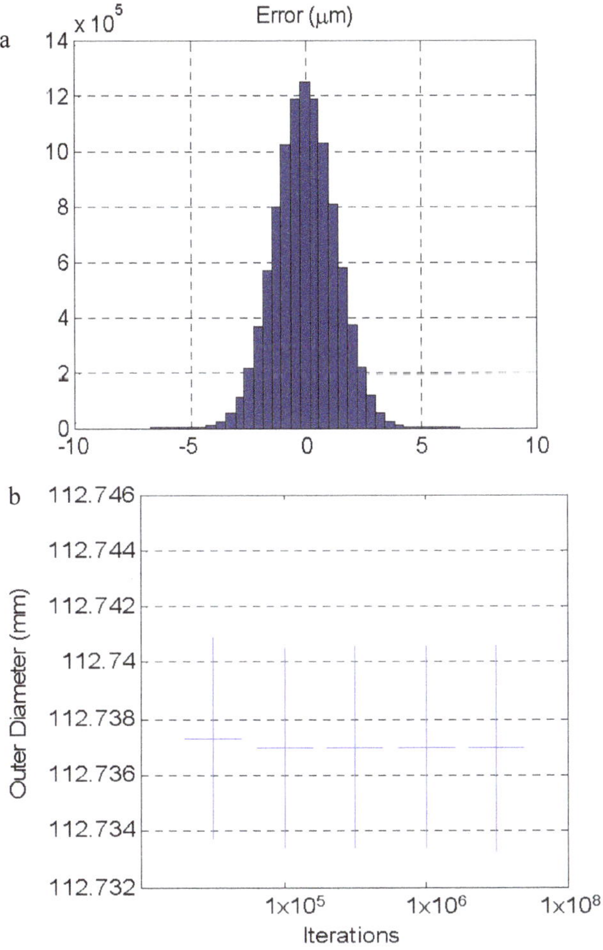

Figure 7. (**a**) Distribution of the error by combining all analyzed parameters with their corresponding distributions. (**b**) Convergence of the result of the simulation.

4. Discussion

The uncertainty obtained through simulation using the Monte Carlo method can be compared with the variability of the experimental results. The contribution of each error source can be extracted from the results presented. In this way, the uncertainty in the measurement of the measurand temperature and that in the measurement of the probe length with the comparator dial are the error sources with the main effect in the final uncertainty. The results obtained from the simulation also allowed us to estimate the number of iterations needed to obtain a stable result of the uncertainty. In this case, if the number of iterations is greater than 10^5, then the value of the uncertainty is stable.

In general, the greater the variability of a factor, the greater its contribution will be to the final uncertainty. For this reason, the variability assigned to each variable affecting the result was taken from the calibration certificate of the instrument, this was the case for the thermometers and the comparator dial, or from an analysis of the geometrical characteristics of the part under inspection and the system. This was the case for the variables of the process and the measurement system, where a study of the possible inadequate use of the system or defect occurrence in the measurand was made.

The dispersion of the experimental results (standard deviation: 0.0019 mm) was similar to that obtained with the simulation by the Monte Carlo method (standard deviation: 0.0018 mm) although the distribution of the values in the experimental results was not as close to a normal distribution as it was to the distribution obtained with the simulation (Figure 8a). From these results, the effect of the temperature on the measurement process and the need to control or monitor it to avoid an increase of the measurement uncertainty was clear, and its influence and the influence of the other variables was quantified.

When applied to the machining process of tapered roller bearings, the process measurement methodology modeled in this work for the external diameter of the outer ring of the bearing and the feedback of the measurement results showed an improvement in process capability (PPK) and a 90% reduction of rework/scrap. This methodology was tested with other dimensions of the bearing. Figure 8b represents the results obtained after applying this methodology to the internal diameter of the inner ring, *d*, and to the thickness, *T* (in addition to *D*, the diameter, the measurement of which was the object of this work).

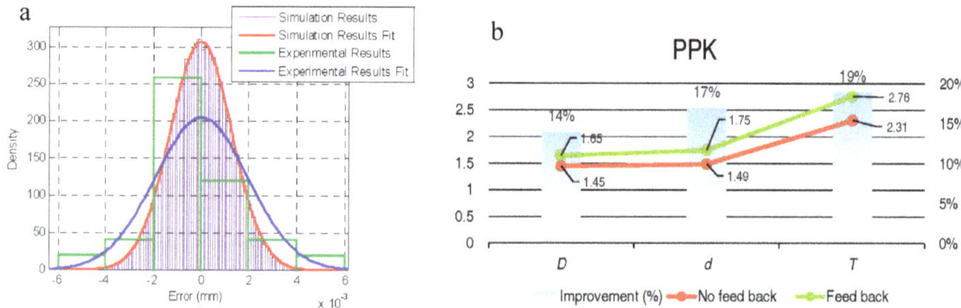

Figure 8. (**a**) Comparison between the results obtained using the Monte Carlo simulation and those obtained experimentally. (**b**) PPK results after applying the process measurement methodology modeled in this work for the external diameter of the outer ring of the bearing (*D*); and results obtained after applying this methodology to the internal diameter of the inner ring (*d*) and to the thickness (*T*). "*D*", "*d*" and "*T*" according to Figure 1b.

5. Conclusions

From the results, it can be concluded that the values obtained experimentally and those obtained by simulation presented a similar dispersion. The Monte Carlo simulation, instead of being a large process, provided reliable results and was a useful method to identify the error sources and quantify their influence by taking into account the correlation between each source (this point could be especially complicated by applying the law of propagation of uncertainty that is also explained in the GUM [27]). Thus, when calculating the uncertainty of a system that has a complex mathematical model, it is preferable to apply Monte Carlo simulation if we can detect and quantify the significant sources that affect the system.

More generally, it can be added that the implemented methodology allows for complex and real-time control of the precision manufacturing process by means of contact sensors and temperature probes, in this case, for dimensions of the tapered roller bearing like "*D*" and "*d*". These measurements are able to control the grinding process in real time and improve the final part quality, scrap and rework, and reduce costs by optimizing the cycle time.

Author Contributions: Conceptualization, A.V.Z. and F.C.; Methodology, F.C. and J.J.A.; Software and Validation, F.J.B. and E.P.; Formal Analysis, F.J.B. and J.J.A.; Investigation, F.J.B. and A.V.Z.; Resources, A.V.Z. and E.P.; Writing-Original Draft Preparation, F.J.B., F.C., and J.J.A.; Writing-Review & Editing, F.J.B.

Funding: This research was funded by the Universidad de Zaragoza and Fundación Ibercaja grant number JIUZ-2016-TEC-05.

Conflicts of Interest: The authors declare no conflict of interest. The funders had no role in the design of the study; in the collection, analyses, or interpretation of data; in the writing of the manuscript, and in the decision to publish the results.

Nomenclature

$T_{a,m}$ (°C)	Ambient temperature when measuring the m-th piece.
$T_{a,0}$ (°C)	Ambient temperature when measuring the master piece.
$T_{s,m}$ (°C)	System temperature when measuring the m-th piece.
$T_{s,0}$ (°C)	System temperature when measuring the master piece.
$T_{m,m}$ (°C)	Measurand temperature when measuring the m-th piece.
$T_{m,0}$ (°C)	Measurand temperature when measuring the master piece.
T_c (°C)	Temperature from the contact thermometer (used in the Monte Carlo simulation)
T_a (°C)	Temperature from the ambient thermometer (used in the Monte Carlo simulation).
$H_{s,m}$ (mm)	Tooling height of the system (see Figure 2) when measuring the m-th piece.
$H_{s,0}$ (mm)	Tooling height of the system (see Figure 2) when measuring the master piece.
H (mm)	Tooling height of the system at 20 °C.
$L_{m,m}$ (mm)	Probe length when measuring the m-th piece.
$L_{0,m}$ (mm)	Probe length from the measurement of the master piece but when measuring the m-th piece.
$L_{0,0}$ (mm)	Probe length when measuring the master piece.
L_m (mm)	Probe length when measuring the m-th piece (dimension at 20 °C).
L_0 (mm)	Probe length when measuring the master piece (dimension at 20 °C).
$D_{m,m}$ (mm)	Measurand (diameter) of the m-th piece at the moment of measure it.
$D_{0,m}$ (mm)	Measurand (diameter) of the master piece at the moment of measure the m-th piece.
$D_{0,0}$ (mm)	Measurand (diameter) of the master piece at the moment of measure it.
D_m (mm)	Measurand (diameter) of the m-th piece (dimension at 20 °C).
D_0 (mm)	Measurand (diameter) of the master piece (dimension at 20 °C).
ΔL_m (mm)	Figure, at 20 °C, of the difference between $L_{0,m}$ and $L_{m,m}$, see Equations (8) and (9).
θ_m (°)	Angle between a vertical line and the front face of the ring (see Figure 4) when measuring the m-th piece.
θ_0 (°)	Angle between a vertical line and the front face of the ring (see Figure 4) when measuring the master piece.
ψ_m (°)	Angle between a vertical line and the axis of the contact prober of the comparator dial when measuring the m-th piece (see Figure 5).
ψ_0 (°)	Angle between a vertical line and the axis of the contact prober of the comparator dial when measuring the master piece (see Figure 5).
λ_m (°)	Linear translation on X-axis direction (see Figure 5) when measuring the m-th piece.
λ_0 (°)	Linear translation on X-axis direction (see Figure 5) when measuring the master piece.
α_m (°C^{-1})	Thermal expansion coefficient of the m-th piece.
α_0 (°C^{-1})	Thermal expansion coefficient of the master piece.
α_s (°C^{-1})	Thermal expansion coefficient of the system.

References

1. El-Thalji, I.; Jantunen, E. A summary of fault modelling and predictive health monitoring of rolling element bearings. *Mech. Syst. Signal Process.* **2015**, *60–61*, 252–272. [CrossRef]
2. Liu, J.; Shao, Y.; Qin, X. Dynamic simulation for a tapered roller bearing considering a localized surface fault on the rib of the inner race. *Proc. Inst. Mech. Eng. Part K* **2017**, *231*, 670–683. [CrossRef]
3. Wee, D.; Kelly, R.; Cattel, J.; Breunig, M. Industry 4.0: How to Navigate Digitization of the Manufacturing Sector. Available online: https://www.mckinsey.com/~/media/McKinsey/Business%20Functions/Operations/Our%20Insights/Industry%2040%20How%20to%20navigate%20digitization%20of%20the%20manufacturing%20sector/Industry-40-How-to-navigate-digitization-of-the-manufacturing-sector.ashx (accessed on 20 July 2018).

4. D'Addona, D.M.; Bracco, F.; Bettoni, A.; Nishino, N.; Carpanzano, N.; Bruzzone, A.A. Adaptive automation and human factors in manufacturing: An experimental assessment for a cognitive approach. *CIRP Ann.* **2018**, *67*, 455–458. [CrossRef]
5. Finlay, B.; Rackwitz, N.; Conerney, B.; Warren, E.; Erdmann, D.; Stoddard, K.; Weber, A.; Scanlon, T. Requirements for First-Time-Right Response in Advanced Manufacturing. In Proceedings of the 29th Annual SEMI Advanced Semiconductor Manufacturing Conference (ASMC), Saratoga Springs, NY, USA, 30 April–3 May 2018; pp. 186–188.
6. Fraunhofer IPA. *Statista—Das Statistik-Portal, Worin sehen Sie konkrete Bedarfe an Intelligenten Lösungen im Produktionsalltag*; Fraunhofer IPA: Hamburg, Germany, 2016. Available online: https://de.statista.com/statistik/daten/studie/605169/umfrage/bedarf-an-intelligenten-loesungen-im-produktionsalltag-im-deutschen-mittelstand/ (accessed on 20 July 2018).
7. Horváth & Partners. *Statista—Das Statistik-Portal, Welche Industrie 4.0-Technologien Sind Heute Bzw. Werden Zukünftig in Unternehmen von Bedeutung Sein*; Horváth & Partners: Hamburg, Germany, 2015. Available online: https://de.statista.com/statistik/daten/studie/606952/umfrage/bedeutung-von-technologien-der-industrie-40-heute-und-zukuenftig-in-der-dach-region/ (accessed on 20 July 2018).
8. Colledani, M.; Fischer, A.; Iung, B.; Lanza, G.; Schmitt, R.; Tolio, T.; Vancza, J. Design and management of manufacturing systems for production quality. *CIRP Ann. Manuf. Technol.* **2014**, *63*, 773–796. [CrossRef]
9. Jurko, J.; Panda, A.; Valíček, J.; Harničárová, M.; Pandová, I. Study on cone roller bearing surface roughness improvement and the effect of surface roughness on tapered roller bearing service life. *Int. J. Adv. Manuf. Technol.* **2016**, *82*, 1099–1106. [CrossRef]
10. Weckenmann, A.; Estler, T.; Peggs, G.; McMurtry, D. Probing systems in dimensional metrology. *CIRP Ann. Manuf. Technol.* **2004**, *53*, 657–684. [CrossRef]
11. Neugebauer, M.; Lüdicke, F.; Bastam, D.; Bosse, H.; Reimann, H.; Töpperwien, C. A new comparator for measurement of diameter and form of cylinders, spheres and cubes under clean-room conditions. *Meas. Sci. Technol.* **1997**, *8*, 849. [CrossRef]
12. Olczyk, A.; Magiera, R.; Kabalyk, K. A 2-port, space-saving, maintenance-friendly pneumatic probe for velocity measurements. *Metrol. Meas. Syst.* **2018**, *1*, 171–185.
13. Crnojevic, C.; Roy, G.; Bettahar, A.; Florent, P. The Influence of the Regulator Diameter and Injection Nozzle Geometry on the Flow Structure in Pneumatic Dimensional Control Systems. *J. Fluids Eng.* **1997**, *119*, 609–615. [CrossRef]
14. Hoffrogge, C.; Mann, R. Measuring Laser Inferometer Device for Inside and Outside Dimensions. *Messtechnik* **1973**, *81*, 1–6.
15. Kim, J.A.; Kim, J.W.; Kang, C.S.; Eom, T.B. An interferometric Abbe-type comparator for the calibration of internal and external diameter standards. *Meas. Sci. Technol.* **2010**, *21*, 75–109. [CrossRef]
16. Chen, Y.; He, Z.; Yang, S. Research on On-Line Automatic Diagnostic Technology for Scratch Defect of Rolling Element Bearings. *Int. J. Precis. Eng. Manuf.* **2012**, *13*, 357–362. [CrossRef]
17. Kumar, R.; Singh, M. Outer race defect width measurement in taper roller bearing using discrete wavelet transform of vibration signal. *Measurement* **2012**, *46*, 537–545. [CrossRef]
18. Deák, K.; Mankovits, T.; Kocsis, I. Optimal Wavelet Selection for the Size Estimation of Manufacturing Defects of Tapered Roller Bearings with Vibration Measurement using Shannon Entropy Criteria. *Stroj. Vestnik J. Mech. Eng.* **2017**, *63*, 3–14. [CrossRef]
19. Chen, M.; Wang, C.; An, Q.; Ming, W. Tool path strategy and cutting process monitoring in intelligent machining. *Front. Mech. Eng.* **2018**, *13*, 232–242. [CrossRef]
20. Zhou, M.; Zheng, G.; Chen, S. Identification and looping tool path generation for removing residual areas left by pocket roughing. *Int. J. Adv. Manuf. Technol.* **2016**, *87*, 765–778. [CrossRef]
21. Gao, X.; Mou, W.; Peng, Y. An intelligent process planning method based on feature-based history machining data for aircraft structural parts. *Procedia CIRP* **2016**, *56*, 585–589. [CrossRef]
22. Tönshoff, H.K.; Friemuth, T.; Becker, J.C. Process Monitoring in Grinding. *CIRP Ann. Manuf. Technol.* **2002**, *51*, 551–571. [CrossRef]
23. Shiraishi, M. Scope of in-process measurement, monitoring and control techniques in machining processes—Part 3: In-process techniques for cutting processes and machine tools. *Precis. Eng.* **1989**, *11*, 39–47. [CrossRef]

24. Brosed, F.J.; Aguilar, J.J.; Santolaria, J.; Lázaro, R. Geometrical verification based on a laser triangulation system in industrial environment. Effect of the image noise in the measurement results. *Procedia Eng.* **2015**, *132*, 764–771. [CrossRef]
25. Jalid, A.; Hariri, S.; El Gharad, A.; Senelaer, J.P. Comparison of the GUM and Monte Carlo methods on the flatness uncertainty estimation in coordinate measuring machine. *Int. J. Metrol. Qual. Eng.* **2016**, *7*, 302. [CrossRef]
26. Forbes, A.B.; Mengot, A.; Jonas, K. Uncertainty associated with coordinate measurement in comparator mode. In Proceedings of the Laser Metrology and Machine Performance XI, Huddersfield, UK, 17–18 March 2015; pp. 150–159.
27. BIPM; IEC; IFCC; ILAC; ISO; IUPAC; IUPAP; OIML. *Evaluation of Measurement Data—Guide to the Expression of Uncertainty in Measurement*; JCGM 100: Bureau International des Poids et Mesures (BIPM): Sèvres, France, 2008.
28. BIPM; IEC; IFCC; ILAC; ISO; IUPAC; IUPAP; OIML. *Evaluation of Measurement Data—Supplement 1 to the "Guide to the Expression of Uncertainty in Measurement"—Propagation of Distributions Using a Monte Carlo Method*; JCGM 101: Bureau International des Poids et Mesures (BIPM): Sèvres, France, 2008.
29. Eichstädt, S.; Link, A.; Harris, P.M.; Elster, C. Efficient implementation of a Monte Carlo method for uncertainty evaluation in dynamic measurements. *Metrologia* **2012**, *49*, 401–410. [CrossRef]
30. Los, A.; Mayer, J.R.R. Application of the adaptive Monte Carlo method in a five-axis machine tool calibration uncertainty estimation including the thermal behavior. *Precis. Eng.* **2018**, *53*, 17–25. [CrossRef]

© 2018 by the authors. Licensee MDPI, Basel, Switzerland. This article is an open access article distributed under the terms and conditions of the Creative Commons Attribution (CC BY) license (http://creativecommons.org/licenses/by/4.0/).

Article

Analysis of Surface Extraction Methods Based on Gradient Operators for Computed Tomography in Metrology Applications

Sinué Ontiveros [1], Roberto Jiménez [2], José A. Yagüe-Fabra [3,*] and Marta Torralba [2]

1 Department of Industrial Engineering, Autonomous University of Baja California, 21100 Mexicali, Mexico; sinue.ontiveros@uabc.edu.mx
2 Centro Universitario de la Defensa, A.G.M. Carretera Huesca s/n, 50090 Zaragoza, Spain; rjimenez@unizar.es (R.J.); martatg@unizar.es (M.T.)
3 I3A, Universidad de Zaragoza, María de Luna 3, 50018 Zaragoza, Spain
* Correspondence: jyague@unizar.es; Tel.: +34-976-762-561

Received: 6 July 2018; Accepted: 14 August 2018; Published: 17 August 2018

Abstract: Among the multiple factors influencing the accuracy of Computed Tomography measurements, the surface extraction process is a significant contributor. The location of the surface for metrological applications is generally based on the definition of a gray value as a characteristic of similarity to define the regions of interest. A different approach is to perform the detection or location of the surface based on the discontinuity or gradient. In this paper, an adapted 3D Deriche algorithm based on gradient information is presented and compared with a previously developed adapted Canny algorithm for different surface types. Both algorithms have been applied to nine calibrated workpieces with different geometries and materials. Both the systematic error and measurement uncertainty have been determined. The results show a significant reduction of the deviations obtained with the Deriche-based algorithm in the dimensions defined by flat surfaces.

Keywords: computed tomography; surface extraction; Canny algorithm; Deriche algorithm

1. Introduction

Today's industry increasingly demands faster and more precise systems and metrological procedures, with better economics of measurement processes and, in some cases, multifunctionality. In recent years, Computed Tomography (CT) has been acquiring an important role and is emerging as a valid option in metrological applications, being the subject of numerous investigations [1–5]. However, a considerable amount of work is still needed to ensure traceability in the measurements. For today's industry, CT is the only technology available to carry out a wide variety of analyses simultaneously and non-destructively [4]. While it is true that CT has great advantages, it is also true that it has some significant limitations.

Among the multiple factors of influence in CT, surface extraction is a significant contributor. This process is mainly affected by the methodology used to locate the surface and by the algorithm used in the process of forming it. The location of the surface for metrological applications is generally based on the definition of a gray value as a characteristic of similarity to define the regions of interest [6]. This gray value threshold is mainly dependent on the properties of the material, the thickness of the piece, and the radiation intensity that is used in the CT data acquisition; therefore, a standard value cannot be determined as it is done in medical applications. A typical method for identifying the threshold value is the ISO 50% method, which has been shown on numerous occasions to result in a displaced surface [7]. Another method uses the histogram of individual cross-sectional slices with a separate determination of the threshold value for each stack, although its use is only recommended in CT images of parts made of a single material [8]. The surface extraction procedure presented in [9] is a method that requires the calibration of a reference workpiece to determine a certain correction for air-material threshold values on the gray scale of the CT volume. An alternative method is based on the local or dynamic threshold, in which a threshold value can be defined for each of the elements or zones of the tomographic volume [10]. There are other investigations focused on the determination of the surface in multi-material pieces [11,12], but they are also based on the determination of a gray value threshold as a starting point and depend on the user's experience for its correct use.

A second source of influence comes from the algorithm used in the formation of the surface, a process for refining the surface location to obtain sub-voxel determination. There are several algorithms that can be employed to achieve this refinement and they can be grouped into three categories: reconstruction algorithms, in which the surface is reconstructed from the discrete values obtained in the surface location step [13]; interpolation algorithms, where the gradient function is interpolated with a neighborhood of the local maximum [14]; and algorithms based on the calculation of moments, which use information of the numerical properties of the image to calculate the final position of the surface (generally, this information comes from the derivative of the gradient of the image [15]).

An approach different to the use of threshold values is to perform the detection or location of the surface based on the discontinuity or gradient. According to this approach, an edge or surface of an image is a significant local change in the intensity of the image that is usually associated with a gray level discontinuity. This principle allows the derivation of a single operator shape that is optimal at any scale. The optimal detector has a simple approximate implementation in which edges are marked at the maximum of a Gaussian-smoothed gradient function [16]. This type of technique uses local operators based on different discrete approximations of the first and second derivatives of the gray levels of the image. In metrological applications, it has been demonstrated that an algorithm whose development is based on the Canny algorithm provides reduced measurement deviations than an algorithm based on a dynamic threshold and is capable of offering similar or reduced measurement uncertainty [17].

In this investigation, a comparison is made between two discontinuity- or gradient-based surface extraction procedures: a Canny-based algorithm and a Deriche-based algorithm. Despite the observed benefits of the algorithm based on Canny, it requires a high computational cost in the first preliminary edge detection step. In addition, saving the complete results of step one requires a significant use of memory. Therefore, in order to optimize the total computational time, the subsequent "Sub-voxel refinement" is computationally lighter, by taking advantage of the information provided by the preliminary edge detection step. The Deriche algorithm allows a faster preliminary edge detection, which permits the "Sub-voxel refinement" step to be more exhaustive, and consequently, may improve the precision obtained. The preliminary edge detection looks for specific gradient values for each surface point, which are calculated along the surface normals of the workpiece. Performing the "Sub-voxel refinement" along the surface normals can increase the independence of the results with respect to the orientation of the workpiece during scanning.

This study compares the performance of the new approach based on Deriche with the performance of the Canny algorithm for different surface types (e.g., spheres, cylinders, planes, and free forms). To perform this analysis, both algorithms have been applied to nine workpieces with different geometries (including complex geometries) and materials, calibrated standard parts, commercial workpieces, and multimaterial workpieces. A complete measurement must have a statement of uncertainty. Therefore, uncertainty from both methods is determined by comparison to calibrated workpieces. A cone beam CT system has been used for measurements in this study.

The article is organized as follows: First, the algorithms used and the results of the measurements (deviation and measurement uncertainty), together with the characteristics of the workpieces and the measurement systems used, are described in Section 2. The results obtained are shown in Section 3. Finally, the conclusions obtained are explained.

2. Materials and Methods

2.1. Surface Extraction Procedure

In this work, a new surface extraction algorithm based on the Deriche operator is presented. In order to evaluate its features, the new algorithm will be compared with the Canny algorithm [17].

2.1.1. 3D Adapted Deriche Algorithm

The 3D adapted Deriche algorithm is a novel approach to surface extraction and is based on the work of Rachid Deriche [18]. This algorithm is based on the principles of detection quality (all existing edges should be marked and no false detection should occur), accuracy (the marked edges should be as close to the edges in the real image as possible), and unambiguity (no multiple responses to one edge in the real image should occur). The Deriche segmentation algorithm was developed for use in two-dimensional images and, despite its robust implementation, must be adapted to three dimensions for use in CT. Furthermore, the use of CT for measurements requires sub-voxel determination.

Taking this into account, the developed algorithm implements the three following steps:

- Preliminary edge detection.
- Sub-voxel refinement.
- Suppression of no-maximum points.

These three stages are explained below:

1. Preliminary edge detection: In this first step, a gradient operator adapted from the Deriche operator is applied. In this adaptation (Figure 1), typical filtering is omitted to reduce the computational cost and avoid loss of information. Therefore, the recursive operator used is the following:

$$\gamma^{+}_{ijk} = X_{ijk-1} + 2{\cdot}e^{-\alpha}{\cdot}\gamma^{+}_{ijk-1} - e^{-2\cdot\alpha}{\cdot}\gamma^{+}_{ijk-2} \quad (1)$$

$$\gamma^{-}_{ijk} = X_{ijk+1} + 2{\cdot}e^{-\alpha}{\cdot}\gamma^{-}_{ijk+1} - e^{-2\cdot\alpha}{\cdot}\gamma^{-}_{ijk+2} \quad (2)$$

$$\theta_{ijk} = -(1-e^{-\alpha})^{2}{\cdot}(\gamma^{+}_{ijk} + \gamma^{-}_{ijk}) \quad (3)$$

where X_{ijk} is the gray value for the voxel ijk and α is a parameter of the gradient operator, the value of which was determined experimentally to be equal to 3.

Applying this operator in each of the three orthogonal XYZ directions of the 3D volume (Figure 1b), the value of the gradient of each voxel is obtained according to those same directions ($\theta^{X}_{ijk}, \theta^{Y}_{ijk}, \theta^{Z}_{ijk}$). Using Equation (4), the three values are combined to obtain a single representative value of the gradient for each point of the volume (Figure 1c).

$$\theta^{XYZ}_{ijk} = \left|\theta^{X}_{ijk}\right| + \left|\theta^{Y}_{ijk}\right| + \left|\theta^{Z}_{ijk}\right| \quad (4)$$

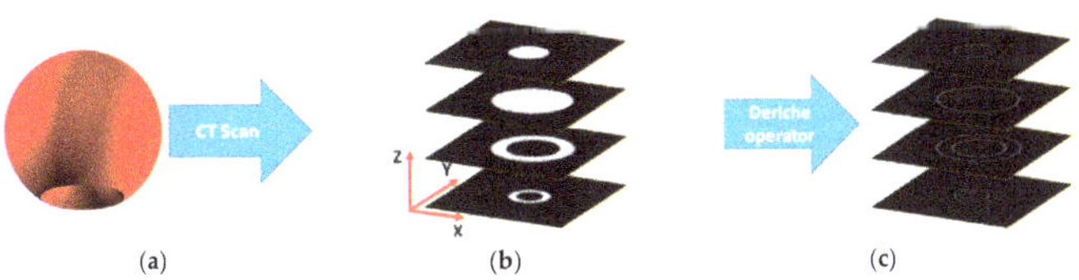

Figure 1. Preliminary edge detection process: (**a**) Scanned part, (**b**) image slices from 3D volume, and (**c**) Deriche edge detection according the three main directions.

2. Sub-voxel refinement: The Deriche and Canny algorithms, as proposed by their authors, define the surface points of the part only according to the coordinates of the voxels that have a maximum

value. Sub-voxel determination of the surface position can be achieved using gradient information of the neighborhood voxels. In this manner, the material transition point can be localized to within 0.01 voxels.

The procedure is as follows: the elements of the matrix θ_{ijk}^{XYZ} with the largest values are taken. As described in the previous step, these values were calculated taking into account only the three main *XYZ* directions. Depending on the orientation of the surface with respect to the reference system, the limitation of preliminary edge detection for the three coordinate axis directions can result in a shift of the obtained contour points and, consequently, a reduced dimensional performance. This problem is solved by applying this sub-voxel refinement along the surface normal.

To calculate the gradient through the surface normal or close to it, the developed Deriche algorithm is applied along 13 directions: The three main directions (+*X*, +*Y*, +*Z*) and the ten diagonals between them are shown in Figure 2b. The value of the gradient in each of those directions will be calculated within a search window of a fixed size and using Equations (1)–(3). The direction with the largest gradient value is considered to be approximately equal to the surface normal (Figure 2c). Once the surface normal is estimated, gradient values are obtained for each voxel in the search window. The center of gravity of the search area gradients is calculated and used to define the *XYZ* surface position to less than a voxel, as shown in Equation (5).

$$X = \frac{\sum_{i=1}^{i=n} (X_i \cdot \theta_x)}{\sum_{i=1}^{i=n} (X_i)} \tag{5}$$

where θ_x is the gradient value of each voxel along the surface normal, and X_i is the respective *XYZ* position of each voxel.

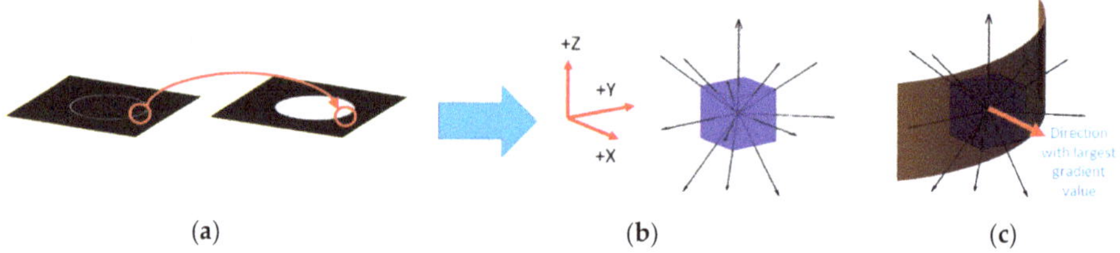

Figure 2. Sub-voxel refinement process: (**a**) Local maximum position detected, (**b**) calculation of the gradient along 13 directions, and (**c**) normal surface obtained from the largest gradient value direction to calculate the center of gravity.

3. Suppression of no-maximum points: In both the Canny and Deriche algorithms, the objective of this step is to identify the voxels contiguous to other voxels with higher values. While in the Canny algorithm, those voxels with lower values were eliminated, in the Deriche algorithm, this information is used to perform sub-voxel refinement. Therefore, the objective in this phase is both to keep the value of the voxels in the matrix θ_{ijk}^{XYZ} (i.e., those with higher values) and to not generate new points too close to other voxels with higher gradient values, which is a risk when those contiguous voxels with lower values are used. Therefore, once a point has been generated based on the previous step, those points which, following the direction of the maximum local slope, are at a distance less than a predetermined limit, will be marked. These marked voxels will be ignored when generating new points regardless of their value (Figure 3). This step is especially important in multi-material parts, where the gradient values may exhibit significant variations depending on the interface surface (e.g., air-part or part-part) (Figure 3a). Since only one gradient limit is used in the previous step, the right value for the part-part interface surfaces (Figure 3b) would generate, in this step, several very close points in the air-part interface surfaces (see Figure 3c). Therefore, it must be considered that, when the

wall thickness or gaps between elements of very few voxel sizes are expected, this limit should be set carefully, due to the risk of eliminating one of the two surfaces.

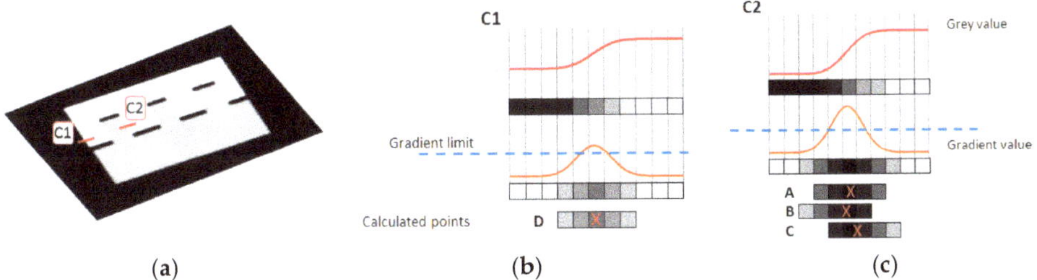

Figure 3. Suppression of no-maximum points in multi-material parts: (**a**) Profile; (**b**) situation where the right gradient limit value generates only one point; (**c**) situation where the same gradient limit generates several close points with high gradient values, justifying the need for non-maximum suppression.

2.1.2. Canny Algorithm (CA)

The Canny-adapted algorithm for the 3D surface extraction process was developed by the authors and is fully explained in [17]. The implementation of this algorithm follows the same three steps as in the Deriche algorithm, but with two main differences:

1. The preliminary edge detection in the Canny algorithm is based on the application of the Canny filter to the images along each of the three Cartesian directions, using a 1×10 convolution mask, oriented along the direction. This process leads to a precise determination of the gradient, but it requires significant computational resources.
2. The second main difference is in the sub-voxel refinement step. In the Deriche algorithm, the point is calculated along a direction perpendicular to the part surface, while in the Canny algorithm, the calculation of the optimal position of the point with sub-voxel definition is carried out by applying Equation (6):

$$X' = \frac{\sum_{i=1}^{i=3} (X_i \bullet G_{X,i})}{\sum_{i=1}^{i=3} G_{X,i}}; Y' = \frac{\sum_{j=1}^{j=3} (Y_j \bullet G_{Y,j})}{\sum_{j=1}^{j=3} G_{Y,j}}; Z' = \frac{\sum_{k=1}^{k=3} (Z_k \bullet G_{Z,k})}{\sum_{k=1}^{k=3} G_{Z,k}} \quad (6)$$

where X_i, Y_j, and Z_k are the coordinates of the voxels inside the search window; and i, j, and k are the voxel indices within the search window. $G_{X,i}$, $G_{Y,j}$, and $G_{Z,k}$, with possible values from 0 to 65,535, i.e., 16 bits, are the gray value transitions obtained in the preliminary surface detection phase for the X, Y, and Z directions, respectively. This refinement is carried out separately and independently along the three directions, providing the three dimensional coordinates of each surface point. In order to obtain each of the coordinates, only the 3D image along that direction, obtained from the previous step, is used.

2.2. Workpieces

In this study, a total of nine calibrated workpieces with different characteristics and materials have been used (Figure 4). Pieces with primitive geometries (spheres), pieces with dimensions dependent on each other (outer and inner cylinder), with parallel planes (grooves), with different materials, and different sizes have been chosen. In addition, a piece with a complex geometry and multimaterial pieces have been used to assess the behavior of surface extraction procedures in these cases. With the pieces used during the investigation, a wide spectrum of geometric and material characteristics are covered, which makes it possible to ensure the generality of the results obtained.

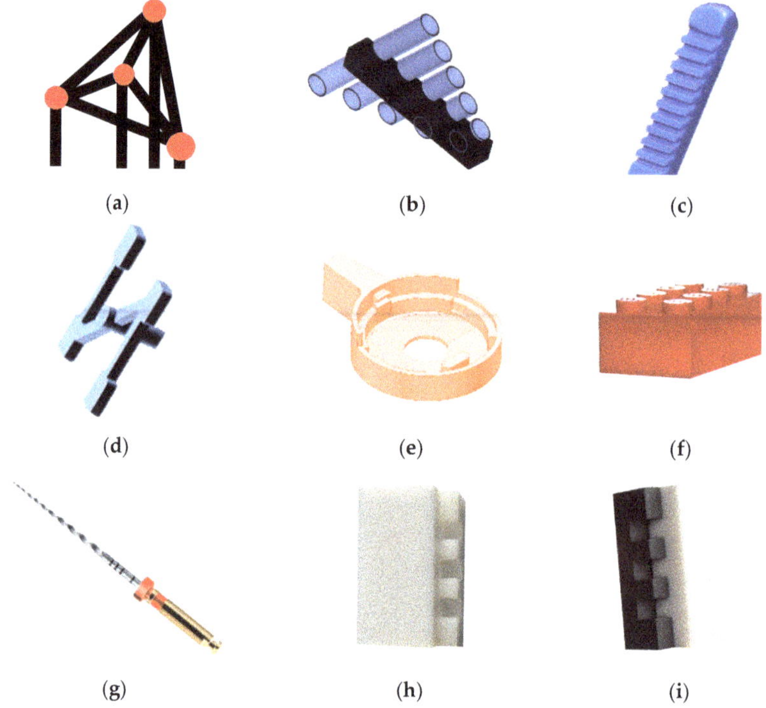

Figure 4. Work pieces: (**a**) CT Tetrahedron, (**b**) Pan Flute, (**c**) Step Gauge, (**d**) Dog bone, (**e**) Toggle, (**f**) Lego, (**g**) Endodontic file, (**h**) Multi-material TEF-POM, and (**i**) Multi-material TEF-PVC.

CT Tetrahedron: This part is called the "CT Tetrahedron" (Figure 4a) and is part of the group of parts used in the "CT Audit" intercomparison [19]. The CT tetrahedron consists of four synthetic ruby spheres held in place by a structure of carbon fiber: the centers of the four spheres coincide with the vertices of a tetrahedron. The dimensions analyzed in this piece are the diameters of each sphere (D) and the distance between them (L). The nominal dimensions are: D1 = 5 mm, D2 = 4 mm, D3 = 4 mm, D5 = 3 mm, and L = 25 mm (Figure 5a). This piece has been manufactured and calibrated by the University of Padova [19].

Pan Flute: The Pan Flute Gauge (Figure 4b), like the CT Tetrahedron, was also used in the intercomparison "CT Audit". This piece consists of five glass tubes supported by a carbon fiber frame. The dimensions analyzed are the inside diameters (ID), outside diameters (OD), and the length (L) of each of the tubes. The five tubes have the same nominal diameters, ID = 1.5 mm and OD = 1.9 mm, while the tube lengths are L1 = 12.5 mm, L2 = 10.0 mm, L3 = 7.5 mm, L4 = 5.0 mm, and L5 = 2.5 mm (Figure 5b). The pan flute has been manufactured and calibrated by the University of Padova [19].

Step Gauge: The Step Gauge (Figure 4c) is a 42 mm long slotted pattern that was specifically designed for the verification of optical systems for micro-fabrication by the Technical University of Denmark—DTU [20]. The step gauge is made of bisacryl, a material used in dental applications. In this piece, the following dimensions have been verified: The distance between each of the slots (R), the distance between each casing (A), and the passage of the slot (P) between each of the slots

(Figure 5c). Each groove and casing has a nominal distance of 2 mm and the pitch between similar features is 4 mm. This part was calibrated at the University of Zaragoza.

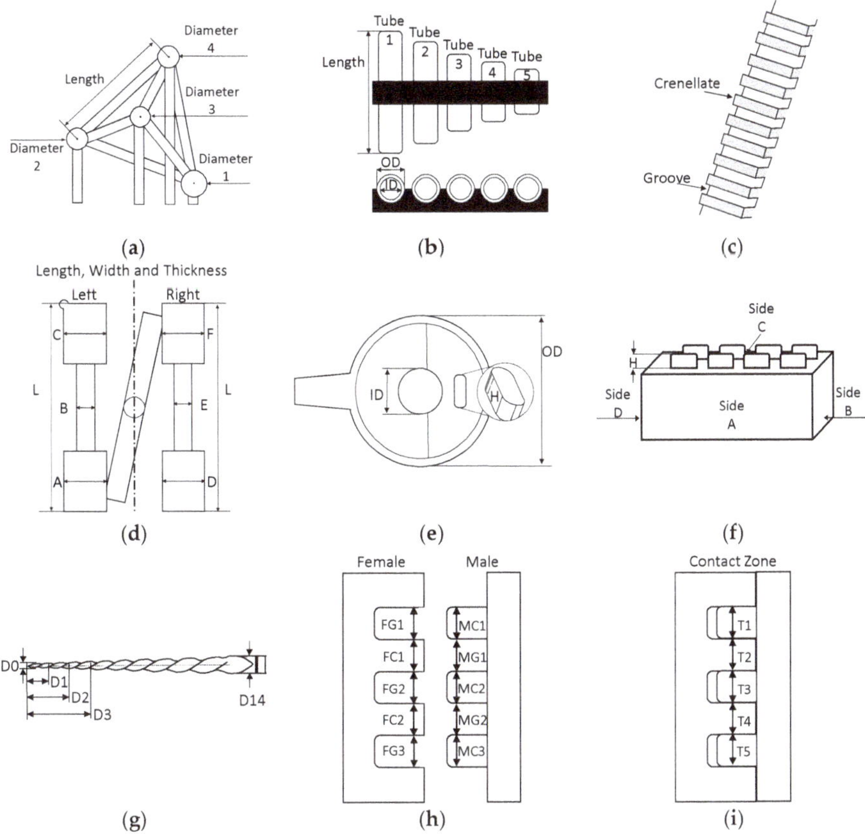

Figure 5. Work piece dimensions: (**a**) CT Tetrahedron, (**b**) Pan Flute, (**c**) Step Gauge, (**d**) Dog bone, (**e**) Toggle, (**f**) Lego, (**g**) Endodontic file, (**h**) male and female in TEF-POM and TEF-PVC, and (**i**) contact zone in TEF-POM and TEF-PVC.

Dog bone: This part (Figure 4d) has been manufactured by micro-injection and is used to perform an analysis of the mechanical behavior of the material. It is made of polyoxymethylene (POM), also called acetal, polyacetal, and polyformaldehyde. It is a technical thermoplastic used in precision components that require a high rigidity, low friction, and excellent dimensional stability. POM is characterized by its high strength, hardness, and rigidity [21]. Two types of dimensions have been verified for this piece: the length of the piece in four different zones on both sides of the piece (L, a, b, and c) and the thickness in six zones of the two lateral bodies (A, B, C, D, E, and F). The nominal dimensions of each of the dimensions are the following: L = 11.80 mm, a and c = 3.00 mm, b = 1.50 mm, d = 1.35 mm, and the thickness from A to F = 1.00 mm (Figure 5d).

Toggle: This part (Figure 4e) has been manufactured in liquid crystal polymer (LCP) by a micro-injection process. Three dimensions of this part have been verified: the outside diameter of the main body (D), the diameter of the central hole (d), and the height of the pillar (A) located inside (Figure 5e). The nominal dimensions are the following: D = 5.40 mm, d = 1.55 mm, and H = 0.38 mm [22].

Lego: This piece is a common block of Lego® (Figure 4f) and has been used in the intercomparison of coordinate measuring machines "Video Audit" led by the University of Padova. The dimensions verified in this work piece are the following: The diameter (D) of each of the buttons, the height (A) of button 1, and the length (L) of each of the sides (LA, LB, LC, and LD). The nominal values of each dimension are the following: D = 5.0 mm, A = 1.7 mm, LA and LB = 31.0 mm, and LB and LD = 16.0 mm (Figure 5f) [23].

Endodontic file: In addition to primitive geometries, we decided to analyze the behavior of the algorithms in a piece with a complex geometry, which is why an endodontic file has been chosen (Figure 4g). It belongs to the ProTaper mechanical instrumentation system developed by the Dentsply Maillefer company. The instruments of the ProTaper system are manufactured with an Ni-Ti alloy [24]. All the pieces of this system have geometrical characteristics of great interest from the point of view of their dimensional characterization: they have a variable conicity that allows the instrument to reduce stress due to torsion, file fatigue and the possibility of fracture, the angle, and a step of the variable helix, with propellers of increasing separation as one advances in the direction of the shank of the instrument. In addition, the ProTapet endodontic files are characterized as having a triangular transversal section that is convex or rounded. The chosen piece is a file for the endodontics of model F2. This piece has a red identification ring that indicates that it is a model characterized by having a greater conicity in the tip, which decreases in the direction of the stem. The dimensions verified in the work piece are the following: (1) Length of the active part (La), which comprises the active cutting surface; and (2) diameter (D), which varies along the length of the entire file and is measured from the center of the tip to the shank of the file, with D0 being the diameter of the tip of the file, and every millimeter moving away from the file being D1, D2, D3, etc. (Figure 5g).

TEF-POM: This is a multi-material part composed of two slotted elements (male and female, Figure 4h). It is made of Polytetrafluoroethylene (PTFE) and POM. The Polytetrafluoroethylene or Teflon is considered to be one of the most versatile among plastic materials due to its many applications, for which other materials cannot be utilized. PTFE is a high temperature resistant material with a low friction coefficient and high resistance to the action of chemical agents and solvents, among other characteristics. PTFE is often considered to be a thermostable polymer and can be used in a continued process at any temperature between −200 °C and +260 °C [25]. POM is a technical thermoplastic used in precision components that require a high rigidity and excellent dimensional stability. POM is characterized by high resistance, hardness, and rigidity up to −40 °C. The most common applications are sectors such as automotive, consumer electronics, heavy loads transport applications, and in low temperatures [26].

TEF-PVC: The second multi-material piece is made of PTFE and Polyvinyl chloride (PVC) (Figure 4i). PVC is a very durable and long lasting thermoplastic construction material that can be used in a variety of applications, can be either rigid or flexible, and can be produced in a wide variety of colors. Due to these characteristics, PVC is used in many industries and is found in many popular and necessary products [27].

Both elements of parts TEF-POM and TEF-PVC have been assembled in a way in which a portion of their interface is not in direct contact. In addition, the assembly tolerance in the contact zone is an adjustment without interference, which does not deform any of the elements. The dimensions that have been evaluated are: The width of the groove (G) and the crellenate (C) and the contact zone between the two pieces (T). G, C, and T have a nominal distance of 5 mm. Therefore, 15 dimensions were assessed in total: Five in the female piece (FG1, FC1, FG2, FC2, and FG3), five in the male piece (MC1, MG1, MC2, MG2, and MC3), and five in the contact zone (T1, T2, T3, T4, and T5) (Figure 5h,i).

2.3. Computed Tomography System

The workpieces were measured on a cone-beam micro-CT eXplore Locus SP machine by General Electric (Boston, MA, USA). The CT machine has an X-ray source that has a power range from 50 to 90 kV, a resolution (minimum voxel size) of 8 μm, and a cylindrical work volume of 44 mm in diameter by 56 mm in height. During the scanning of the workpiece, the temperature was continuously recorded inside the machine, obtaining a temperature range of 20 ± 2 °C for all pieces. Each workpiece was measured 10 times. The selected parameters used for the CT measurements are presented in Table 1.

Table 1. CT scanning parameters.

Parameter	CT Tetrahedron	Pan Flute	Step Gauge	Dog Bone	Toggle	Lego	Endodontic File	TEF-POM	TEF-PVC
Voltage (kV)	90	90	45	80	80	45	90	45	45
Current (μA)	80	80	120	95	95	120	80	120	120
Angular indexing step (degrees)	0.2	0.35	0.4	0.4	0.4	0.4	0.4	0.3	0.3
Voxel size (μm)	45.5	16.5	47	8	8	34	28	28	56

2.4. Calibration Systems

2.4.1. Coordinate Measuring Machines (CMM)

Two CMMs were used to calibrate the workpieces used in this study. The Step Gauge, the Lego, and the TEF-POM and TEF-PVC workpieces [28] were calibrated on a PMC 850S-CNC CMM (Figure 6) by Carl Zeiss (Oberkochen, Germany), which has a measurement volume of 850 × 700 × 600mm and a longitudinal MPE = 2.30 μm + (L/300) μm (L in mm). The probing system used was a Zeiss Vast-XT with an active scanner, allowing working with extensions of up to 500 g in weight and 500 mm in length.

The CT Tetrahedron and the Pan Flute were calibrated at the University of Padova using a tactile CMM (MPE = 1.4 + L/300 μm, with L in mm). The measurements were performed following the procedures indicated during the CT Audit inter-comparison [19].

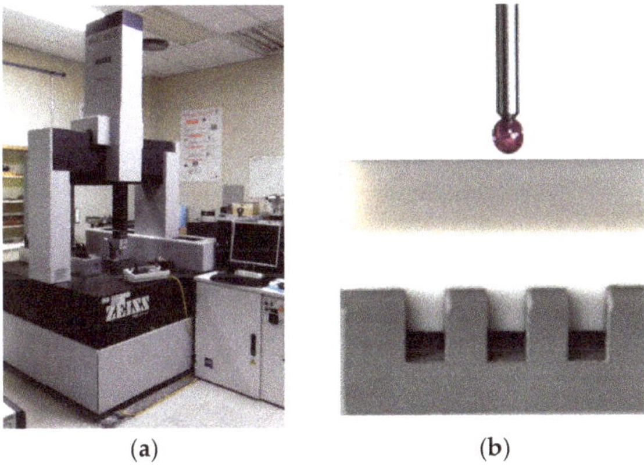

Figure 6. (**a**) CMM Carl Zeiss PMC 850S-CNC used for the calibration of the Step Gauge, the Lego, and the TEF-POM and TEF-PVC workpieces [28]; (**b**) image of the measuring process of the TEF-PVC workpiece with the CMM.

2.4.2. Optical Coordinate Measuring Machines (OCMM)

An optical coordinate measuring machine DeMeet 220 [29] has been used as a reference calibration system for the Dog bone, Toggle, and Endodontic file parts. The DeMeet 220 machine by Schut (Groningen, The Netherlands) has a measuring range of 220 mm × 150 mm × 100 mm, backlight and coaxial light, diascopic illumination with a light ring, and a magnification lens 2×. The DeMeet 220 has tele-centric optics to avoid distortion around the center of the field of view, can achieve magnifications in the range of 40× to 400×, and has a field of view of 3111 μm × 2327 μm (Figure 7).

The OCMM uncertainty for length measurements in the 100–1000 μm range was evaluated, resulting in a maximum permissible error MPE_{OCMM} = 1.7 μm (i.e., suitable for the diameter measurements of the endodontic file). For the measurements of the endodontic file with a length L > 1 mm, the maximum permissible error of the OCMM obtained is: MPE_{OCMM} = 5 μm + (L/150) μm (L in mm).

(**a**) (**b**)

Figure 7. (**a**) OCMM De Meet 220 [29] used for the calibration of the Dog bone, Toggle, and Endodontic file workpieces; (**b**) image of the measuring process of the Dental file workpiece with the OCMM.

3. Results

3.1. CT Comparison Results

CT measurements using each surface extraction procedure are compared to the respective reference values. In Figure 8, the average absolute deviation over 10 repeat measurements of the first seven workpieces (single material parts) is shown. This average deviation takes into account all the dimensions measured for each part. In addition, the maximum and minimum deviations of all those dimensions from their absolute calibrated values are represented by the error bars for each part. Deviations from Deriche-based surface extraction are lower in six of seven pieces when compared to the deviations from Canny-based surface extraction. The largest reduction in deviations is observed for measurements of the Toggle and Dog Bone, where the deviations are reduced by more than 50%. Deviation was 7% higher for Deriche-based surface extraction in the measurement of the Pan Flute. Pan Flute was the only workpiece where the deviation was higher for Deriche than for Canny. Absolute maximum deviations were lower with Deriche-based surface extraction for all the measured parts, observing the largest difference, equal to 7.9 μm, between the two surface extraction methods in the Dog Bone workpiece.

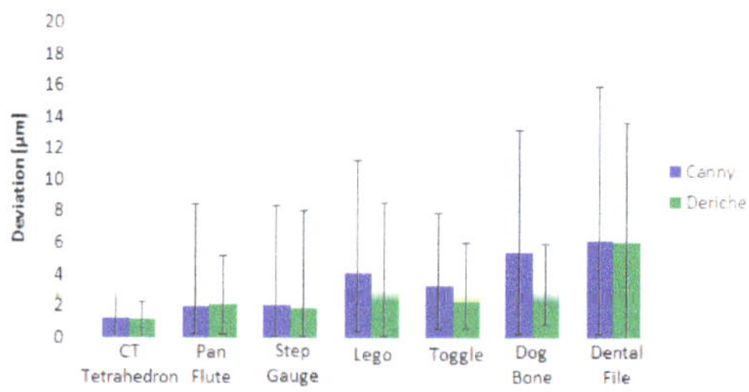

Figure 8. Average deviation for the Canny-based method (blue bars) and Deriche-based method (green bars) and maximum and minimum deviation (error bars) of all the part dimensions from their calibrated absolute values of single material parts.

Figure 9 shows a differentiated analysis by the type of surface. On the left side of the graph, the average deviation of spheres, diameters, or dimensions dependent on these two forms (e.g., distance between spheres) is presented, while on the right side are the dimensions dependent on planes. For dimensions related to spheres and diameters, Deriche-based surface extraction provides lower deviations in four of five workpieces. The most significant reduction is observed in the measurement of the Lego piece. With the Canny-based surface extraction algorithm, the average deviation is 3.6 µm and the maximum deviation is 7.6 µm. Meanwhile, with Deriche-based surface extraction, the average deviation is 2.1 µm and the maximum deviation is 3.6 µm. In the case of the planes-dependent dimensions, deviations are lower with Deriche-based surface extraction. The most significant case is that of the Dog Bone piece, where the average deviation with Canny-based surface extraction is 5.4 µm, while it is only 2.8 µm with Deriche-based surface extraction. Absolute maximum deviations were lower with Deriche-based surface extraction for all the measured parts and all the dimension types.

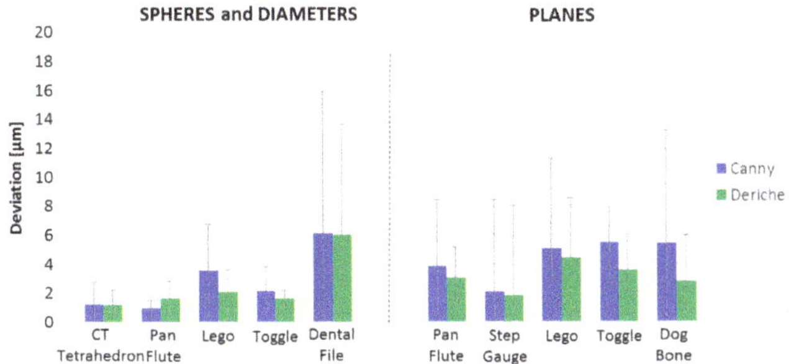

Figure 9. Average deviation for the Canny-based method (blue bars) and Deriche-based method (green bars) and maximum and minimum deviation (error bars) of all the part dimensions from their calibrated absolute values of single material parts grouped by dimension type.

Figure 10 shows the results obtained in the multimaterial workpieces. In the measurement of the TEF-POM, the average deviation with the Canny-based surface extraction is only slightly lower (difference equal to 0.2 µm) than with Deriche-based surface extraction. For the TEF-POM, the maximum deviation with the Canny-based surface extraction is also slightly lower (difference equal to 0.3 µm) than with Deriche-based surface extraction. In the measurement of the TEF-PVC part, the average deviation and the maximum deviation with Deriche-based surface extraction are smaller than the same deviations with Canny-based surface extraction by 3.5 µm and 7.0 µm, respectively.

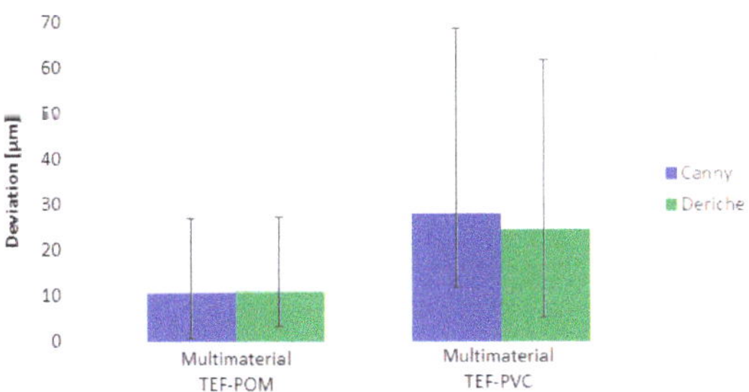

Figure 10. Average deviation for the Canny-based method (blue bars) and Deriche-based method (green bars) and maximum and minimum deviation (error bars) of all the part dimensions from their calibrated absolute values of multimaterial workpieces.

3.2. Uncertainty Estimation for CT Measurements

In order to evaluate the influence of the data processing on uncertainty, the most accepted standard currently available for CT sensors is applied to compare both surface extraction methods, i.e., the Verein Deutscher Ingenieure/Verband Deutscher Elektrotechniker (VDI/VDE) guideline 2630-2.1 [30]. This standard is based on the assessment of the measurement uncertainty by means of a calibrated workpiece. The expression of the uncertainty is given by Equation (7).

$$U95, \text{method}(i), \text{piece}(j), \text{measurand}(k) = 2 \cdot \sqrt{u_{cal}^2 + u_p^2 + u_w^2 + u_b^2}, \quad (7)$$

For expanded uncertainty with a confidence interval of 95.45%, which is expressed as U_{95} in this study, the coverage factor is equal to 2, as shown in Equation (7). The measurement uncertainty value depends on the surface extraction method (i = Canny or Deriche); the analyzed workpiece (j = 1, ..., 7); and the measurands (k), which in this work, are mainly separated as spheres and diameters or planes. The considered error sources in the uncertainty budget are those included in the VDI/VDE 2630-2.1 standard and they are the following:

- The term u_{cal} represents the standard uncertainty of calibration of the workpiece determined by a CMS (coordinate measuring system).
- The term u_p is the standard uncertainty of the measurement procedure (repeatability).
- The term u_w is the standard uncertainty of the material and manufacturing variations of the measured process. It is specifically associated with two uncertainty sources: variations in the mechanical properties of the workpiece (u_{w1}); and variations in the CTEs (coefficient of thermal expansion) of the workpiece (u_{w2}). In this work, the first factor has been previously included in u_p (effects of material composition and shape). A rectangular statistical distribution for 20% of CTE variation has been established for the second term.
- The term u_b is the standard uncertainty associated with the systematic error of the measurement process: the influence of the temperature variation during the CT measurement (u_{b1} when $\Delta T = \pm 2$ °C); and the estimation of the systematic error related to the surface detection technique: Canny and Deriche (u_{b2}).

The average expanded uncertainty, taking into account all the part dimensions, is shown in Figure 11 by workpiece and in Figure 12 by type of measured geometry in each item. The error bars in both Figures represent the maximum and minimum expanded uncertainty of all the part dimensions for each part. As can be seen in both Figures 11 and 12, Deriche-based surface extraction generally provides lower average, maximum, and minimum uncertainties U_{95} for each item.

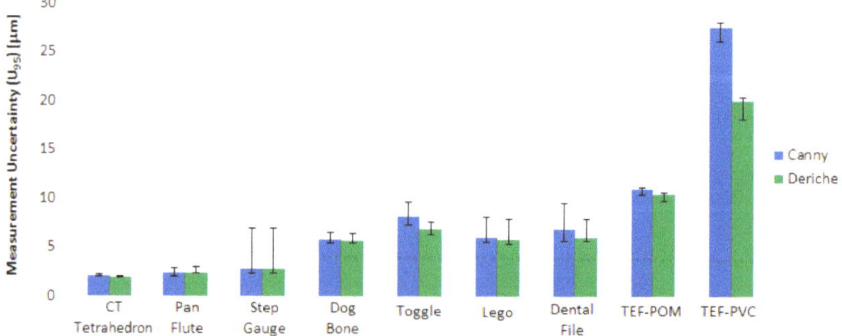

Figure 11. Expanded average measurement uncertainty (U_{95}) for the Canny-based method (blue bars) and Deriche-based method (green bars), and maximum and minimum measurement uncertainties (error bars) of all the part dimensions for each item.

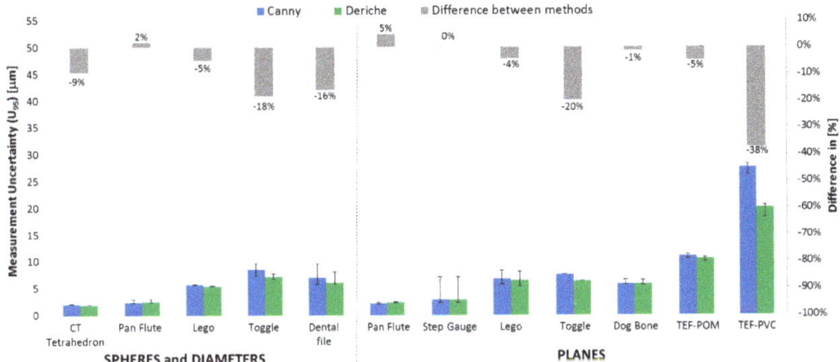

Figure 12. Expanded average measurement uncertainty (U_{95}) for the Canny-based method (blue bars) and Deriche-based method (green bars), differences between them (grey bars), and maximum and minimum measurement uncertainties (error bars) of all the part dimensions for each item by dimension type.

Most of the different uncertainty contributors considered in Equation (7) suppose a similar contribution in the expanded uncertainty for both Canny- and Deriche-based surface extraction. Nevertheless, the systematic error contribution (u_{b2}) associated with the surface extraction method shows the largest difference between U_{95} obtained with Canny- and Deriche-based surface extraction. Additionally, u_{b2} constitutes 30% of U_{95}, so a specific analysis was carried out and shown in Figure 13, where the contribution u_{b2} is represented for both methods for the two surface typologies.

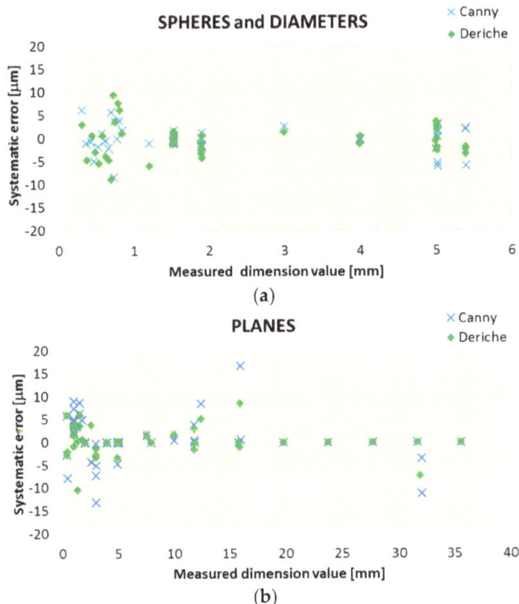

Figure 13. Systematic error source contribution analysis (u_{b2}) by method and by dimension type for (**a**) spheres and diameters and (**b**) planes.

Analyzing the contribution u_{b2} of both methods for the two types of surface, curves and planes, a similar behavior can be observed between both algorithms for the case of spheres and cylinders. However, for the defined dimensions between flat surfaces, the Deriche-based algorithm slightly improves the results obtained for most of the pieces.

4. Discussion

The average deviations with respect to the calibrated values do not present, in a general way, large differences for the two algorithms analyzed, with only a slight improvement in the Deriche-based surface extraction algorithm. However, a clear difference between curved and flat surfaces is concluded from this analysis: the reduction of the deviations obtained with the Deriche-based algorithm is evident, with reductions as large as 50% with the Deriche-based surface extraction for flat surfaces compared to the Canny-based algorithm. Similar results were observed in measurement uncertainty. The results of uncertainty obtained for the Deriche-based algorithm are further reduced in the dimensions defined by flat surfaces.

Both results agree with the main novelty of the Deriche algorithm: the realization of the "Sub-voxel refinement" in a direction perpendicular to the surface. In flat geometries, a non-optimal orientation is more detrimental to deviations, and especially to the uncertainty obtained, in the Canny-based algorithm. In contrast, the Deriche-based algorithm reduces the effect of the orientation in which the piece was measured.

Also noteworthy are the results of the two multi-material pieces, both with defined dimensions between flat geometries. In this case, although in the TEF-PVC workpiece the uncertainties are larger than in any other, so is the reduction that is achieved by applying the algorithm based on Deriche. This multi-material piece is made of two materials with similar attenuation coefficients, Teflon and PVC, which makes it more difficult to distinguish them in the contact area. This fact would justify the greater difficulty in its precise measurement, but it must also be highlighted that the best behavior of the Deriche-based algorithm is displayed in these difficult conditions.

5. Conclusions

In this paper, an adapted 3D Deriche algorithm based on gradient information has been presented and compared with a previously developed adapted Canny algorithm for different surface types. Both algorithms have been applied to the measurement of nine calibrated workpieces with different geometries and materials. Both the systematic error and measurement uncertainty have been determined; a significant reduction of the deviations obtained with the Deriche-based algorithm in the dimensions defined by flat surfaces and a slight improvement in the dimensions defined by spheres and diameters is observed. Therefore, the use of this algorithm could prove pivotal in reducing systematic errors and uncertainty in CT measurements.

The approach used in the Deriche-based algorithm can, in certain cases, be a more efficient alternative for surface extraction. In the Deriche-based surface extraction, the calculations made in the "Sub-voxel refinement" step are independent of the previous step, which does not happen in the Canny algorithm. This may improve the efficiency of the measuring process implemented with a Deriche-based algorithm. For example, in order to use the Deriche-based surface extraction, the information of the "Preliminary edge detection" could be directly obtained from nominal information of the piece supplied (e.g., in CAD format). Provided it is possible to perform an alignment of that CAD model with the position of the piece in the scan, the use of nominal information could further improve the efficiency of Deriche-based surface extraction. Another possibility would be to perform the "Preliminary edge detection" using other types of algorithms more efficient but with limitations in terms of accuracy. This would be the case for the techniques of threshold ISO 50%; although it is demonstrated that they quickly obtain a fast extraction of surfaces, it is also demonstrated that the obtained precision is inferior to the one of the gradient techniques. After the "Preliminary edge detection" step with the threshold ISO 50%, the Deriche-based algorithm "Sub-voxel refinement" and "Suppression of no-maximum points" could be applied to improve the performance.

Author Contributions: S.O., R.J. and J.A.Y.-F. conceived the Deriche-based and the Canny-based algorithms; M.T., R.J. and J.A.Y.-F designed the experiments; S.O. and R.J. performed the experiments; M.T. calculated the measurement uncertainties. All authors wrote the manuscript. All authors contributed to the editing of the manuscript.

Funding: This research was funded by the Research Foundation MINECO (Spain) via projectgrant number DPI2015-69403-C3-1-R.

Acknowledgments: The authors would like to thank Simone Carmignato, from the University of Padova, and Guido Tosello and Leonardo de Chiffre, from the DTU, for the pieces provided to carry out this study.

Conflicts of Interest: The authors declare no conflicts of interest.

Nomenclature

CMM	Coordinate measuring machine
CMS	Coordinate measuring system
CT	Computed Tomography
DTU	Danmarks Tekniske Universitet (Technical University of Denmark)
ID	Inside Diameters
LCP	Liquid Crystal Polymer
OCMM	Optical Coordinate Measuring Machine
OD	Outside Diameters
POM	Polyoxymethylene
PTFE	Polytetrafluoroethylene (Teflon®)
PVC	Polyvinyl chloride
TEF-POM	Multi-material part made of PTFE and POM
TEF-PVC	Multi-material part made of PTFE and PVC
VDI/VDE	Verein Deutscher Ingenieure/Verband Deutscher Elektrotechniker

References

1. Ametova, E.; Ferrucci, M.; Chilingaryan, S.; Dewulf, W. A computationally inexpensive model for estimating dimensional measurement uncertainty due to X-ray computed tomography instrument misalignments. *Meas. Sci. Technol.* **2018**, *29*, 065007. [CrossRef]
2. Carmignato, S.; Pierobon, S.; Rampazzo, P.; Parisatto, M.; Savio, E. CT for Industrial Metrology—Accuracy and Structural Resolution of CT Dimensional Measurements. In Proceedings of the Conference on Industrial Computed Tomography (ICT), Wels, Austria, 19–21 September 2012; pp. 161–172.
3. Carmignato, S.; Aloisi, V.; Medeossi, F.; Zanini, F.; Savio, E. Influence of surface roughness on computed tomography dimensional measurements. *CIRP Ann. Manuf. Technol.* **2017**, *66*, 499–502. [CrossRef]
4. De Chiffre, L.; Carmignato, S.; Kruth, J.P.; Schmitt, R.; Weckenmann, A. Industrial applications of computed tomography. *CIRP Ann. Manuf. Technol.* **2014**, *63*, 655–677. [CrossRef]
5. Kruth, J.P.; Bartscher, M.; Carmignato, S.; Schmitt, R.; De Chiffre, L.; Weckenmann, A. Computed tomography for dimensional metrology. *CIRP Ann. Manuf. Technol.* **2011**, *60*, 821–842. [CrossRef]
6. Carvalho, L.E.; Sobieranski, A.C.; Wangenheim, A. 3D Segmentation Algorithms for Computerized Tomographic Imaging: A Systematic Literature Review. *J. Digit. Imaging* **2018**, *31*, 1–52. [CrossRef] [PubMed]
7. Kiekens, K.; Welkenhuyzen, F.; Tan, Y.; Bleys, P.; Voet, A.; Kruth, J.P.; Dewulf, W. A test object with parallel grooves for calibration and accuracy assessment of industrial computed tomography (CT) metrology. *Meas. Sci. Technol.* **2011**, *22*, 115502. [CrossRef]
8. Obrist, A.F.; Flisch, A.; Hofmann, J. Point cloud reconstruction with sub-pixel accuracy by slice-adaptive thresholding of X-ray computed tomography images. *NDT E Int.* **2004**, *37*, 373–380. [CrossRef]
9. Kowaluk, T.; Wozniak, A. A new threshold selection method for X-ray computed tomography for dimensional metrology. *Precis. Eng.* **2017**, *50*, 449–454. [CrossRef]
10. Shapiro, L.G.; Stockman, G.C. *Computer Vision*, 1st ed.; Prentice Hall: Upper Saddle River, NJ, USA, 2001; ISBN 0130307963.

11. Hermanek, P.; de Oliveira, F.B.; Bartscher, M.; Carmignato, S. Experimental Investigation of new multi-material gap reference standard for testing computed tomography systems. In Proceedings of the 7th Conference on Industrial Computed Tomography, Leuven, Belgium, 7–9 February 2017.
12. De Oliveira, F.B.; Stolfi, A.; Bartscher, M.; De Chiffre, L.; Neuschaefer-Rube, U. Experimental investigation of surface determination process on multi-material components for dimensional computed tomography. *Case Stud. Nondestruct. Test. Eval.* **2016**, *6*, 93–103. [CrossRef]
13. Venkatachalam, V.; Wasswerman, R.M. Comprehensive Investigation of Subpixel Edge Detection Schemes in Metrology. *Mach. Vis. Appl. Ind. Insp.* **2003**, *5011*, 200–212. [CrossRef]
14. Walkenburg, R.J.; Mclvor, A.M.; Power, P.W. An Evaluation of Subpixel Feature Localization Methods for Precision Measurements. *Videom. III Proc. SPIE* **1994**, *2350*, 229–238. [CrossRef]
15. West, G.; Clarke, T.A. A Survey and Examination of Subpixel Measurement Techniques. *Close Range Photogramm. Meets Mach. Vis.* **1990**, *1395*, 456–463. [CrossRef]
16. Canny, J. A Computational Approach to Edge Detection. *IEEE Trans. Pattern Anal. Mach. Intell.* **1986**, *8*, 679–698. [CrossRef] [PubMed]
17. Yagüe-Fabra, J.A.; Ontiveros, S.; Jiménez, R.; Chitchian, S.; Tosello, G.; Carmignato, S. A 3D edge detection technique for surface extraction in computed tomography for dimensional metrology applications. *CIRP Ann. Manuf. Technol.* **2013**, *62*, 531–534. [CrossRef]
18. Deriche, R. Using Canny's Criteria to derive a recursively implemented optimal edge detector. *Int. J. Comput. Vis.* **1987**, *1*, 167–187. [CrossRef]
19. Carmignato, S. Accuracy of industrial computed tomography measurements: Experimental results from an international comparison. *CIRP Ann. Manuf. Technol.* **2012**, *61*, 491–494. [CrossRef]
20. De Chiffre, L.; Carmignato, S.; Cantatore, A.; Jensen, J. Replica calibration artefacts for optical 3d scanning of micro parts. In Proceedings of the 9th Euspen International Conference of the EUSPEN, San Sebastian, Spain, 2–5 June 2009; pp. 352–355.
21. Massey, L.K. *Permeability Properties of Plastics and Elastomers*, 2nd ed.; William Andrew Publishing: Norwich, NY, USA, 2003; ISBN 1-884207-97-9.
22. Gasparin, S.; Tosello, G.; Hansen, H.N.; Islam, A. Quality control and process capability assessment for injection-moulded micro mechanical parts. *Int. J. Adv. Manuf. Technol.* **2013**, *66*, 1295–1303. [CrossRef]
23. Carmignato, S.; Voltan, A.; Savio, E. Metrological performance of optical coordinate measuring machines under industrial conditions. *CIRP Ann. Manuf. Technol.* **2010**, *59*, 497–500. [CrossRef]
24. Ruddle, C.J. The ProTaper technique. *Endod. Top.* **2005**, *10*, 187–190. [CrossRef]
25. Industrias, J.Q. Available online: http://www.jq.com.ar/Imagenes/Productos/Teflon/Virgen/teflonvirgen.htm (accessed on 10 June 2018).
26. Habasit. Available online: http://www.habasit.com/en/pom.htm (accessed on 10 June 2018).
27. British Plastics Federation (BPF). Available online: http://www.bpf.co.uk/plastipedia/polymers/pvc.aspx (accessed on 10 June 2018).
28. Jiménez-Pacheco, R.; Ontiveros, S.; Yagüe-Fabra, J.A. A surface extraction analysis in a multi-material test part for computed tomography in metrology applications. *Procedia Manuf.* **2017**, *13*, 487–494. [CrossRef]
29. Schut. Available online: http://www.schut.com/Products/DeMeet/index_Specifications.htm (accessed on 10 June 2018).
30. The Association of German Engineers. *VDI/VDE 2630 Part 2.1. Determination of the Uncertainty of Measurement and the Test Process Suitability of Coordinate Measurement Systems with CT Sensors*; Verein Deutscher Ingenieure e.V.: Düsseldorf, Germany, 2015.

© 2018 by the authors. Licensee MDPI, Basel, Switzerland. This article is an open access article distributed under the terms and conditions of the Creative Commons Attribution (CC BY) license (http://creativecommons.org/licenses/by/4.0/).

Article

Some Considerations about the Use of Contact and Confocal Microscopy Methods in Surface Texture Measurement

Jesús Caja García [1,*], Alfredo Sanz Lobera [2], Piera Maresca [1], Teresa Fernández Pareja [3] and Chen Wang [1]

1 Department of Mechanical, Chemical and Industrial Design Engineering, ETS Ingeniería y Diseño Industrial, Universidad Politécnica de Madrid, Ronda de Valencia, 3, 28012 Madrid, Spain; piera.maresca@upm.es (P.M.); chen.wang@alumnos.upm.es (C.W.)

2 Department of Aerospace Materials and Production, ETSI Aeronáutica y del Espacio, Universidad Politécnica de Madrid, Plaza del Cardenal Cisneros, 3, 28040 Madrid, Spain; a.slobera@upm.es

3 Department of Engineering Surveying and Cartography, ETSI en Topografía, Geodesia y Cartografía, Universidad Politécnica de Madrid, Campus Sur, Carretera de Valencia, Km 7, 28031 Madrid, Spain; teresa.fpareja@upm.es

* Correspondence: jesus.caja@upm.es; Tel.: +34-910-677-682

Received: 1 July 2018; Accepted: 18 August 2018; Published: 20 August 2018

Abstract: Surface metrology employs various measurement techniques, among which there has been an increase of noteworthy research into non-contact optical and contact stylus methods. However, some deeper considerations about their differentiation and compatibility are still lacking and necessary. This work compares the measurement characteristics of the confocal microscope with the portable stylus profilometer instrumentation, from a metrological point of view (measurement precision and accuracy, and complexity of algorithms for data processing) and an operational view (measuring ranges, measurement speed, environmental and operational requirements, and cost). Mathematical models and algorithms for roughness parameters calculation and their associated uncertainties evaluation are developed and validated. The experimental results demonstrate that the stylus profilometer presents the most reliable measurement with the highest measurement speed and the least complex algorithms, while the image confocal method takes advantage of higher vertical and horizontal resolution when compared with the employed stylus profilometer.

Keywords: surface texture; contact measurement; optical measurement

1. Introduction

The evaluation of a surface texture involves the analysis of a large number of data using complex models [1]. For this purpose, metrology instruments must scan/measure the surface, obtaining a finite digital sample. Leach et al. [2] establish that the surfaces to be evaluated are defined by the measurement method used (measuring principle). The results will be different even when measuring the same surface with different instruments, since different physical properties are being measured. According to this idea, the instruments used in surface texture measurements can be classified into different groups, depending on the measurement technique and the physical property used to obtain the surface coordinates. When surface texture measurements are carried out on a mechanical manufacturing environment, the two main groups of instruments are based on tactile and optical methods.

In this context, tactile methods are mainly based on the use of stylus profilometers (SP), which are currently the most widely used instruments in the mechanical manufacturing industry [3]. In a common stylus profilometer, a probe, which is in contact with the surface, is physically moved over it, so that the vertical movement of the tip allows for characterization of the surface heights. This kind of instrument is preferably used in 2-D measurements based on profiles.

On the other hand, optical methods present a wide number of techniques for performing the measurement without surface contact using light instead of a physical probe to measure the surface [4]. The light reflected on the surface and its subsequent detection allows the evaluation of the surface texture. Conroy et al. [5] pointed out that the most widely used optical methods are interferometry and confocal microscopy (CM). The present work is focused on the second one, due to better adequacy when used in mechanical manufacturing environments.

The physical principle of confocal microscopy is based on eliminating the reflected light coming from the out-of-focus planes. The way to achieve this consists of illuminating a small area of the sample. The light beam from the focal plane is then taken so the beams from the lower and upper planes are removed by using a pin-hole. This confocal probe evaluates each point on the surface to be measured and obtains its height and associated light intensity. A system of lateral scanning allows having a line profile and areal measurement. The metrological characteristics of these devices are similar to those provided by the stylus profilometers and are gradually replacing them in specific metrological approaches [6].

On the other hand, different studies have evaluated and compared different techniques of surface metrology, emphasizing the characteristics of the equipment and the differences between the obtained surface parameters but not going into such detail in other non-metrological aspects related with the measurement.

Conroy et al. [5] measure a specimen consisting of an 80 μm pitch square wave Al-coated etched grating with a nominal step height of 187 nm and use stylus profilometers (SP), confocal microscopy (CM), and interferometric microscopy (IM) in their comparison. They do not provide experimental results of the evaluated surface texture parameters nor employed algorithms and conclude that the use of any technique requires an understanding of the properties of the sample, limitations of the technique used, and the analysis required before carrying out the surface measurement. Vorburger et al. [7] compare four techniques including stylus profiling (SP), phase-shifted microscopy (PSIM), white light interferometric microscopy (WLIM), and confocal microscopy (CM). They find discrepancies between WLIM and the other techniques, obtaining similar results among the other three. Poon et al. [8] compare three techniques—stylus profiler (SP), atomic force microscope (AFM), and non-contact optical profiler (NOP)—and conclude with a recommendation on the use of the analyzed techniques when a glass-ceramic substrate is measured. Nouira et al. [9] focused their work on the development of a high-precision profilometer with both optical (CM) and tactile capabilities and measures a VLSI Step-Height Standard (SHS 880-QC). Obtained results in their work show that the tactile measurements, which include stylus profilometers (SP) and atomic force microscopy (AFM), are more accurate than the optical measurements carried out by confocal microscopy (CM). The comparison of both tactile techniques reveals that the SP and the AFM measurements produce very similar results. Piska and Metelkova [10] analyze the relations between 2-D (profile) and 3-D (areal) surface parameters of the same measured surface, and they observed that both methods, SP and CM, give very comparable results only if the surface has a good reflection value. Nielsony et al. [11] analyze differences between a stylus profilometer (SP) and confocal microscopy (CM) in measuring a cladded surface, concluding that CM values of the roughness parameters are higher than SP values. The same result is obtained by Merola et al. [12] analyzing the tribological behavior of retrieved hip femoral head by using a stylus profilometer (SP) and confocal microscopy (CM). These variations in the results can affect the results of the surface topography [13–15], so the measurement principle should be close to the physical functional behavior of the surface [2].

Additional works have analyzed the behavior of optical instruments. Thompson et al. [16] make a quantitative comparison of areal topography measurements by using four optical techniques on a selective laser melting manufactured part. These techniques are confocal microscopy (CM), coherence scanning interferometry (CSI), focus variation microscopy (FVM), and X-ray computed tomography (XCT). The authors analyze the profile discrepancy between instrument pairs, obtaining high values (near 50%), due to the poorer capture of smaller scale peaks and pits of the FVM instrument. In the same line of work, Feidenhans'l et al. [17] compares optical methods for surface roughness measurement, employing different scatterometers and confocal microscopy (CM). The results are compatible between instruments, but it is necessary to include a Gaussian smoothing function to compensate for the differences.

All these results are interesting, although only partially covering the type of parts that the present work addresses, that is to say, those that are manufactured and used in mechanical manufacturing processes, such as machining processes. Moreover, algorithms or measuring procedures are weakly or not described in all that previous work. For these reasons, this work analyses and compares the application of both techniques, stylus profilometer (SP) and confocal microscopy (CM) measurements, in the evaluation of a series of surfaces, which include machined surfaces and two roughness standards (type C1 and C4 spacing standard), by establishing a comparison between them, and not only considering the results obtained (roughness parameters and their associated uncertainty), but also the procedure and the requirements and performances that these techniques need and offer. The work also considers other aspects related with the measurement, such as the set-up operations of pre-measurement samples, the operating time, operational considerations, data storage requirements, and the cost of instruments and maintenance.

2. Evaluation Procedure of Roughness Parameters

2.1. Parameters Calculation

In order to perform the proposed comparison, a specific procedure to evaluate the roughness parameters of a profile from its coordinates (x, z) has been developed. This way, when the results are compared, only the difference due to the type of instrument, and especially the data acquisition procedure, will be evaluated, which is not affected by systematic effects due to the calculation software that is different for each instrument.

The procedure of obtaining results is defined by the following steps:

1. Obtaining the extracted profile measured by different instruments. The file contains the sampled x coordinates and the digitized z coordinates (x_r, z_r).
2. Form removal. Due to the impossibility of placing the measured profile fully parallel to the measurement base, or the presence of geometric errors on the surface of the measurand, it is necessary to eliminate this form by fitting the data to a nominal shape (line, polynomial, and circle). Correction can be done in two ways: applying a tilt correction or by subtraction of the mean. This one will be used when the angle rotated by the surface is very small. The coordinates (x_{rFr}, z_{rFr}) are obtained. When the fitting to a regression line is employed, its mathematical representation is:

$$z = ax + b \tag{1}$$

The coefficients a and b can be calculated using the least squares method. The line that best fits the set of coordinates (x_i, z_i) is:

$$z_i - ax_i - b + e_i \approx 0 \tag{2}$$

where e_i is the residual. It is possible to solve a linear system, obtaining the coefficients a and b, using:

$$\mathbf{P} = \left(\mathbf{A} \cdot \mathbf{A}^T\right)^{-1} \cdot \mathbf{A}^T \cdot \mathbf{B} \quad \text{where} \quad \mathbf{A} = \begin{bmatrix} x_1 & 1 \\ \vdots & \vdots \\ x_n & 1 \end{bmatrix} \quad \mathbf{B} = \begin{bmatrix} z_1 \\ \vdots \\ z_n \end{bmatrix} \quad \mathbf{P} = \begin{bmatrix} a \\ b \end{bmatrix} \tag{3}$$

Using the same procedure, it is possible to eliminate the form of surfaces that can be adjusted to quadratic polynomials.

3. λ_s Filter. The profile is filtered using a low pass filter with a λ_s cut-off. The primary profile is obtained with coordinates (x_{rFrLs}, z_{rFrLs}).
4. λ_c Filter. The profile is now filtered using a high pass filter with a λ_c cut-off according to the standard ISO 16610-21 [18]. This way, the waviness profile is eliminated, obtaining the roughness profile with coordinates $(x_{rFrLsLc}, z_{rFrLsLc})$.
5. Obtaining the roughness parameters, according to the standards ISO 4287 [19] and ISO 4288 [20]. In order to solve the calculation of the equations described in the aforementioned standards (roughness parameters R_a, R_q, R_{sk}, and R_{ku}), a method of integral analytical calculation, the trapezoidal method, has been used instead of the formula of discrete calculation (Equations (8)–(11)) to improve the accuracy of the results.

$$R_p = \max_{1 \le i \le m} z_{pi} \tag{4}$$

$$R_v = \max_{1 \le i \le m} z_{vi} \tag{5}$$

$$R_z = R_p + R_v \quad \Rightarrow \text{all calculated over a sampling length} \tag{6}$$

$$R_t = R_p + R_v \quad \Rightarrow \text{all calculated over the evaluation length} \tag{7}$$

$$R_a = \frac{1}{l}\int_0^l |z(x)|dx \quad \approx \quad R_a = \frac{1}{n}\sum_{i=1}^{n}|z_i| \tag{8}$$

$$R_q = \sqrt{\frac{1}{l}\int_0^l z^2(x)dx} \quad \approx \quad R_q = \frac{1}{n}\sum_{i=1}^{n} z_i^2 \tag{9}$$

$$R_{sk} = \frac{1}{R_q^3}\left[\frac{1}{l}\int_0^l z^3(x)dx\right] \quad \approx \quad R_{sk} = \frac{1}{R_q^3}\frac{1}{n}\sum_{i=1}^{n} z_i^3 \tag{10}$$

$$R_{ku} = \frac{1}{R_q^4}\left[\frac{1}{l}\int_0^l z^4(x)dx\right] \quad \approx \quad R_{ku} = \frac{1}{R_q^4}\frac{1}{n}\sum_{i=1}^{n} z_i^4 \tag{11}$$

Parameters R_t, R_z, R_p, and R_v measure the amplitude of the profile (peak and valley distances), R_a is used to calculate the average roughness, R_q measures the variance of the amplitude distribution function (ADF) of the profile, R_{sk} analyze the asymmetry of the ADF, and R_{ku} evaluates the spikiness of the profile. Depending on the type of instrument evaluated, different coordinates (x, z) will be generated, so it will not be necessary to execute all steps.

2.2. Model for Calculating Uncertainties

Following the calculation model of roughness parameters described in the previous section, for the calculation of the uncertainty of these parameters, the Monte Carlo method is used. This numerical resolution method has a high application in metrological fields [21–23].

The documents Supplement 1 to the "Guide to the expression of uncertainty in measurement (GUM)"—Propagation of distributions using a Monte Carlo method [24] and Supplement 2 to the "Guide to the expression of uncertainty in measurement (GUM)"—Extension to any number of output quantities [25] have been used for the development of the calculation algorithms.

The calculation algorithms are provided in the following steps:

1. Definition of the output quantities. Parameters R_t, R_z, R_p, R_v, R_a, R_q, R_{sk}, R_{ku}, and R_{Sm}.
2. Definition of input quantities. The sampled coordinates and the digitized coordinates, and scale division error of the instrument on the z-axis. In this work, only the sources that directly affect the coordinates of the captured profile, and therefore the variability of the measurements, have been

taken into account. All of these magnitudes analyze the variations generated by the instrument in the measurements: noise in the readings, imperfections in the reference of the instrument, sampling and digitizing process, and rounding-off of the coordinates and software calculations, as well as the horizontal and vertical resolution of the instrument and the idealization of the Gaussian filter [26,27].

3. Assignment of the probability density functions (PDF) to the input variables. For the input variables defined above, it is established that variability due to the process of obtaining the sampled coordinates responds to a normal distribution with a mean of the raw coordinate value and a standard deviation equal to 2% of the sampling step [28,29]. In order to analyze the variability of the z-coordinates, it was necessary to perform a repeatability study of the measurements [30]. That is to say, measuring the same profile a large number of times, by the same operator, with the same measurement procedure, same measuring system, and same operation conditions [31]. After doing different experiments, it has been observed that it is practically impossible to obtain the same position of the z-coordinate to be analyzed. Therefore, an analysis of the repeatability of the parameters P_p and P_v is planned, where the parameters P_p and P_v are the maximum profile peak height of the primary profile and maximum profile valley depth of the primary profile, respectively. For this type C1 spacing standard, grooves having a sine wave profile will be measured and the standard deviations S of the previous parameters will be determined. This study assumes that this variability corresponds to a *t*-distribution (employed when a series of indications are evaluated). The standard uncertainty associated with this variability can be estimated as:

$$u(x) = \sqrt{\frac{\nu_p}{\nu_p - 2}\frac{S}{\sqrt{n}}} \quad (12)$$

where S is the maximum standard deviation of the parameters P_p and P_v, obtained in the experiment, ν_p is the degrees of freedom of the obtained parameters, and n represents the numbers of measurements made in the roughness measurement (typically one). Therefore, the digitized coordinates respond to a *t*-distribution of the mean value of the raw coordinate z and a standard uncertainty equal to the value calculated by the previous equation. The experimental values obtained are shown in Section 4. The scale division errors of the instrument on the z-axis, responds to a rectangular distribution of limits $[-E, E]$, where E is the scale division on the z-axis.

4. Propagation. According to the recommendation of Supplement 1 to the GUM, in its Section 7.2.3, it is possible to use a small number of iterations (M) for complex models. Taking into account this recommendation, the standard deviation and the mean of the values obtained after performing the M iterations of the model, could be taken as $u(y)$ and y respectively, and can be assigned to a Gaussian PDF $g_Y(\eta) = N(y, u^2(y))$. In the simulations, 10,000 replications have been performed, requiring between 2 and 70 min of calculation on a computer with an i7-6700HQ-Intel(R) Core(TM) and 16 GB memory.

5. Results. The standard deviation of the resulting values obtained in the simulations, as well as their mean, is calculated. To determine the amplitude of the coverage interval, the minimum interval method is used [32,33]. Using the suggestions of Supplement 2 of the GUM, it is possible to determine the covariance matrix C of the calculated parameters (all parameters will be correlated to a certain degree, due to the presence of common input variables in all of them). From the covariance matrix it is possible to calculate the matrix of correlation coefficients r.

2.3. Algorithm Validation

In order to determine the validity and accuracy of the developed algorithms, both reference and synthetic data sets [34] have been used. These data are the coordinates (x, z) of a series of points linked to a reference profile. The reference data have been obtained from the Internet-based Surface

Metrology Algorithm Testing System of the National Institute of Standard and Technology (NIST) [29]. These simulation-generated profiles and F1-type software standards have been generated in accordance with the specifications of ISO 5436-2 [35].

One of the verifications performed in this work is shown in Table 1. The F1-type standard Mill.sdf file [29] shown in Figure 1 has been used and its roughness parameters have been evaluated using a cut-off (λ_c) of 0.8 µm and a Gaussian filter according to ISO 16610-21 [18] (steps 4 and 5).

Table 1. Reference values vs calculated values (NIST Mill measured profile).

Parameter	Reference Value	Calculated Value	Q_1 ($\times 10^{-6}$)	Percentage Difference (%)
R_t [µm]	1.09408	1.09408	1.5	0.0001
R_z [µm]	0.89833	0.89833	1.9	0.0002
R_p [µm]	0.46672	0.46672	4.0	0.0008
R_v [µm]	0.43161	0.43161	2.0	0.0005
R_a [µm]	0.16764	0.16765	10.3	0.0061
R_q [µm]	0.20479	0.20479	2.2	0.0011
R_{sk} [-]	0.1388	0.1388	3.5	0.0025
R_{ku} [-]	2.37947	2.37947	6.6	0.0027
R_{Sm} [µm]	255.82	255.83	6542.4	0.0025

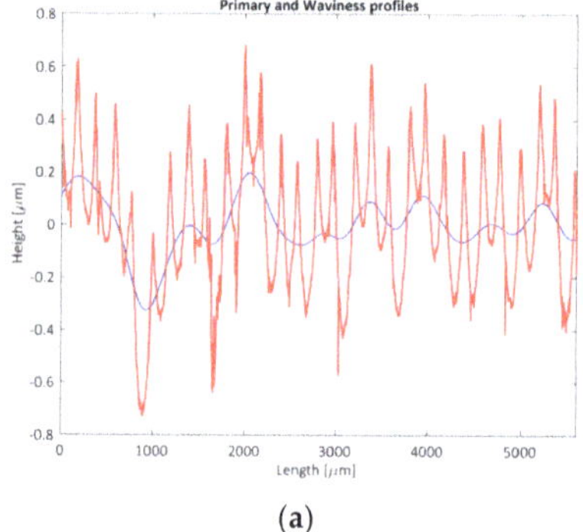

(**a**)

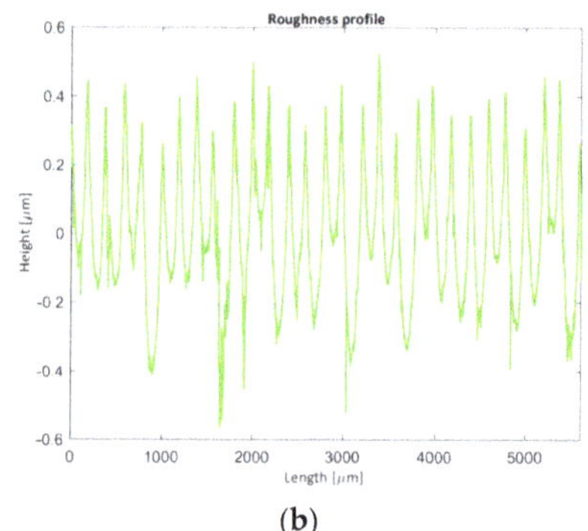

(**b**)

Figure 1. Mill F1-type (**a**) primary and waviness profile; (**b**) roughness profile.

In order to determine how good the developed algorithms are, the calculated results with the reference data are compared with NIST data. The difference, in absolute value, between the parameters obtained by the proposed algorithms and the reference parameters is used.

$$Q_1 = \left| R^{calculated} - R^{reference} \right| \tag{13}$$

In view of the results, it can be observed that the maximum percentage difference was 0.0061%, which is the reason why we claim the developed algorithms behave in a totally satisfactory manner. In addition, the algorithms of form elimination and low pass filtering for λ_s (steps 2 and 3) using the software provided by the Physikalisch-TechnischeBundesanstalt (PTB) "Software to Analyse Roughness of Profiles-Version 2.09" [36] have been validated, since they provide similar results to those shown. A Type D roughness standard has been measured, and the data of the extracted profile has been introduced in the previous software, obtaining the following results (Table 2).

Table 2. Reference values vs. calculated values (PTB's "Software to Analyse Roughness of Profiles").

Parameter	PTB Reference Value	Calculated Value	Q_1 ($\times 10^{-3}$)	Percentage Difference (%)
R_t [µm]	2.0468	2.0489	2.1	0.1026
R_z [µm]	2.0037	2.0051	1.4	0.0699
R_p [µm]	1.0109	1.0116	0.7	0.0692
R_v [µm]	0.9928	0.9935	0.7	0.0705
R_a [µm]	0.8943	0.8947	0.4	0.0447
R_q [µm]	0.9068	0.9071	0.3	0.0331
R_{sk} [-]	0.0107	0.0110	0.3	2.8037
R_{ku} [-]	1.0516	1.0511	0.5	−0.0475
R_{Sm} [µm]	80.96	80.89	70	−0.0865

3. Methodology of Experimental Study on Roughness

To perform the comparison of measurement characteristics between the image confocal microscope and some portable stylus measurement instrumentations, three specimens are measured. The first one is a milled part with the same tool and with different feed per tooth (measurand 1 to 7); The second one is a Type C1 spacing standard with grooves having a sine wave profile (measurand 8); The third one is a Type C4 spacing standard with grooves having an accurate profile (measurand 9). Figure 2 shows the three specimens, as well as their measurement areas.

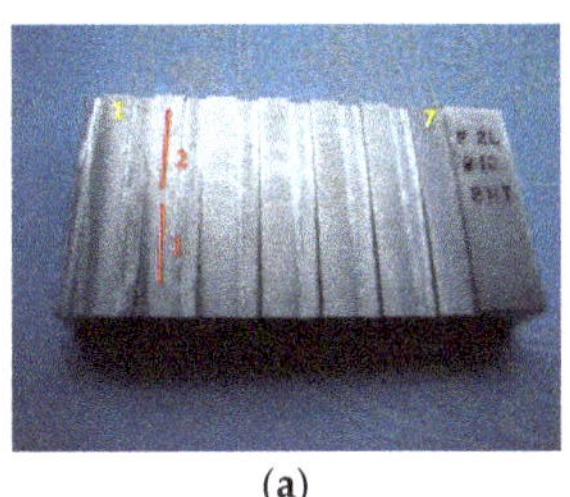
(a)

(b)

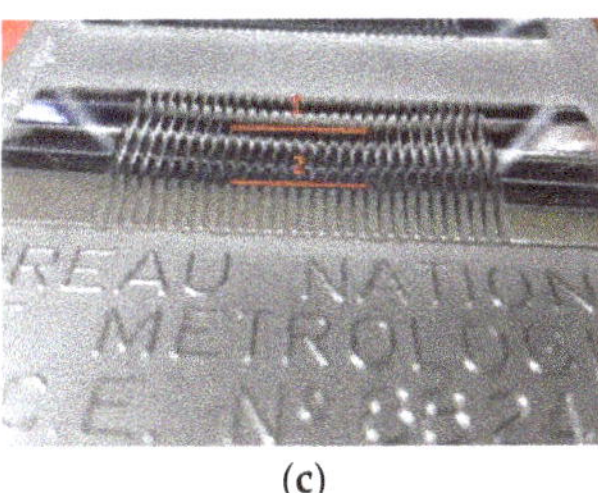
(c)

Figure 2. (**a**) Measurands 1 to 7; (**b**) Measurand 8; (**c**) Measurand 9.

The technical features of the three instruments used in the comparison are detailed below:

- Stylus profilometer 1 (*SP-1*): portable stylus profilometer with a mechanical probe, brand HOMMELWERKE, model TESTER T1000 WAVE with probe TKL 300L, stylus tip of 5 µm/90°. Measuring range: ±80 µm/±320 µm. Vertical resolution (Z): 0.01 µm/0.04 µm. Transverse length: 0.05–20 mm. X sampling: 0.583 µm (Figure 3a).
- Stylus profilometer 2 (*SP-2*): portable stylus profilometer with a mechanical probe, brand SM, model PROFILTEST SM7 with stylus tip of 5 µm. Measuring range: 320 µm. Vertical resolution (Z): 0.01 µm. X sampling: 2.5 µm (Figure 3b).
- Confocal microscope (*CM-3*) confocal microscope, brand Leica (Wetzlar, Germany), model DCM-3D. It uses episcopic illumination, with light source LED type of wavelength 460 nm. Image Acquisition Sensor: monochrome CCD for confocal applications. The equipment has five lenses, with amplification between 5× and 150×. A 50× lens was used in the measurements to provide agreement between amplification and data acquisition speed. This lens has the following characteristics: Measuring field: 254.64 × 190.90 µm. Pixel size: 0.332 µm and equal to the lateral resolution (XY). Vertical resolution (Z): <3 nm. Total measuring field of instrument: 114 × 75 × 40 mm (Figure 3c).

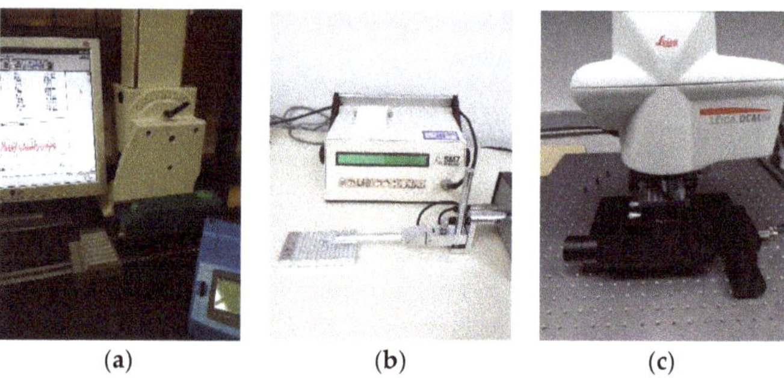

Figure 3. Experimental setup. (a) SP-1; (b) SP-2; (c) CM-3.

4. Analysis of Experimental Results

The main aspects, as defined in Section 1, are analyzed below and will serve as a comparison of the evaluated methods, tactile and optical.

First, the repeatability study of the measurements made with the three instruments used in the comparison is developed. Measurand 8 is measured 25 times. Table 3 establishes the values obtained in this study. The standard deviations are similar for the three instruments, although the confocal microscope (*CM-3*) has the higher vertical resolution.

Table 3. Results of the repeatability study.

Instrument	Standard Deviation R_p (μm)	Standard Deviation R_v (μm)
SP-1	0.0685	0.0584
SP-2	0.0539	0.0747
CM-3	0.0545	0.0707

4.1. Tactile Method

- Measuring ranges: In height (z-axis), the smallest measuring field allowed by each instrument has been selected. In length (x-axis), a profile of length of 5.6 mm ($7\times\ \lambda_c$) has been evaluated.
- Operating time: The effective measurement time, which included the movement of the probe on the surface, the calculation of parameters by the equipment, and the saving of the data files, has been less than 2 min. The preparation time of the sample, which includes the positioning and inclination adjustment by rotation of the *SP-1* probing system (Figure 3a), lasted 10 min at most.
- Environmental considerations: Regarding the *SP-1* and *SP-2* environmental requirements, the measurements were carried out at a controlled temperature of 20 ± 1 °C
- Measurand preparation: It included the cleaning of the measurand by a mixture of alcohol and ether, as well as the correct placement of the measurand on the basis of measurement (indicated in the previous point). Prior to measurement, the part had been thermally stabilized for at least 3 h.
- Numeric values of the parameters: Table 4 shows, as an example, the results of measurand 8 when it was measured with the *SP-1* equipment for a simulation performed for $M = 10^4$. Figure 4a,b show, by way of example, the histograms of parameters R_p and R_q. It was verified that these parameters can be reasonably approximated to a normal distribution. Analogous behavior was obtained for the other parameters. Table 5 shows the correlation coefficients matrix of the roughness parameters, verifying the presence of correlation between them. Some of them show high correlation coefficient values (>0.5), due to the fact that all the parameters employ the same input quantities, that is to say, raw coordinates (x_r, z_r). Table 6 (columns 2 and 4) and Table 7 (columns 2 and 4) show some of the values obtained in the measurements of the different

measurands after applying the algorithm used for the evaluation of roughness parameters. As it can be observed, there were discrepancies between the values of the equipment, higher for the amplitude parameters (peak and valley), presenting differences between 11% and 17% of the value analyzed and minimal when evaluating the amplitude parameter (average of ordinates), with differences in percentage between 0.2% and 4%. The difference in the spacing parameter was due to the different x sampling value of each piece of equipment. When comparing measurands, better amplitude parameter results (average of ordinates) were obtained when a standard was evaluated (measurand 8). When the piece was measured, differences up to 12% were obtained due to the impossibility of measuring the same profile with different equipment, and the presence of irregularities in the surface of the piece. The uncertainty of the parameters was similar when a standard or a piece was measured. The amplitude parameters (peak and valley) presented values of uncertainty of hundredths of a micrometer and were at least one or two orders of magnitude higher than the uncertainty of the amplitude parameters (average of ordinates). This was due to the process of obtaining these parameters.

- Data storage requirements: The data file, ASC/.txt extension, presented 9598/2241 rows of values, providing coordinates (x_r, z_r)/step and z_r/z coordinates, with a file size of 260/25 kB (*SP-1*/*SP-2*).
- Cost of instrumentation/maintenance: The acquisition cost of the equipment was 12,000/6000 €. The maintenance was practically non-existent (lubrication of the vertical slide guides of the probing head for the *SP-1* equipment).

Table 4. Results of the simulations of the eight measurand with the *SP-1* equipment.

Roughness Parameter	Parameter Estimation y	Standard Uncertainty $u(y)$	Shortest 95% Coverage Interval	
			Lower Limit	Upper Limit
R_t [µm]	3.433	0.032	3.375	3.496
R_z [µm]	3.367	0.016	3.336	3.399
R_p [µm]	1.679	0.012	1.658	1.703
R_v [µm]	1.688	0.012	1.667	1.711
R_a [µm]	0.9977	0.0009	0.9960	0.9993
R_q [µm]	1.1062	0.0009	1.1045	1.1080
R_{sk} [-]	0.0019	0.0016	−0.0012	0.0052
R_{ku} [-]	1.4860	0.0015	1.4829	1.4888
R_{Sm} [µm]	100.167	0.018	100.142	100.206

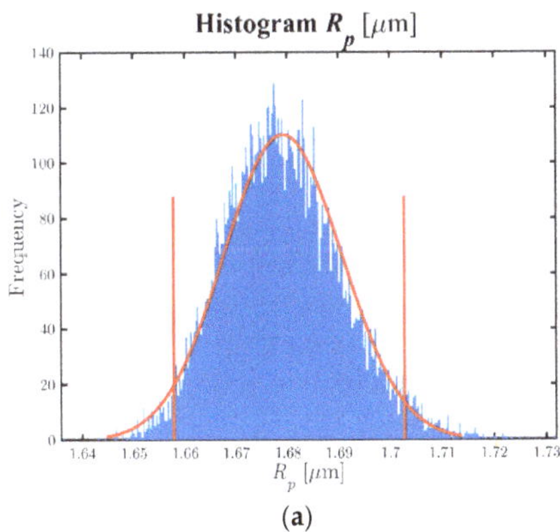

(a)

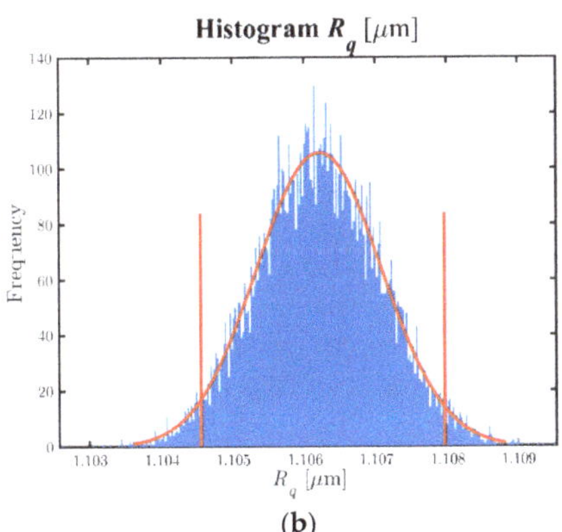

(b)

Figure 4. (**a**) R_p histogram; (**b**) R_q histogram.

Table 5. Correlation coefficient matrix of the roughness parameters.

Roughness Parameter	R_t	R_z	R_p	R_v	R_a	R_q	R_{sk}	R_{ku}	R_{Sm}
R_t	1.00	0.72	0.51	0.51	0.08	0.11	−0.01	0.14	0.00
R_z	0.72	1.00	0.71	0.71	0.12	0.17	−0.01	0.22	−0.01
R_p	0.51	0.71	1.00	0.00	0.09	0.12	0.12	0.14	−0.02
R_v	0.51	0.71	0.00	1.00	0.08	0.12	−0.14	0.17	0.00
R_a	0.08	0.12	0.09	0.08	1.00	0.91	0.00	−0.29	−0.04
R_q	0.11	0.17	0.12	0.12	0.91	1.00	0.01	0.01	−0.03
R_{sk}	−0.01	−0.01	0.12	−0.14	0.00	0.01	1.00	−0.01	−0.01
R_{ku}	0.14	0.22	0.14	0.17	−0.29	0.01	−0.01	1.00	0.01
R_{Sm}	0.00	−0.01	−0.02	0.00	−0.04	−0.03	−0.01	0.01	1.00

Table 6. Roughness values obtained with the different instruments and methods (measurand 8).

Instrument	*SP-1*		*SP-2*		*CM-3*	
	Parameter Estimation	Standard Uncertainty	Parameter Estimation	Standard Uncertainty	Parameter Estimation	Standard Uncertainty
R_t [µm]	3.433	0.032	4.015	0.074	3.831	0.034
R_z [µm]	3.367	0.016	3.781	0.036	3.686	0.017
R_p [µm]	1.679	0.012	1.866	0.025	1.947	0.012
R_v [µm]	1.688	0.012	1.915	0.026	1.739	0.011
R_a [µm]	0.9977	0.0009	1.0189	0.0019	1.0180	0.0007
R_q [µm]	1.1062	0.0009	1.1370	0.0020	1.1309	0.0007
R_{sk} [-]	0.0019	0.0016	−0.0455	0.0039	0.0406	0.0013
R_{ku} [-]	1.4860	0.0015	1.4828	0.0044	1.5104	0.0012
R_{Sm} [µm]	100.167	0.018	103.3108	0.027	101.808	0.017

Table 7. Roughness values obtained with the different instruments and methods (measurand 2).

Instrument	*SP-1*		*SP-2*		*CM-3 (With Outliers)*		*CM-3 (Outliers Removed)*	
	Parameter Estimation	Standard Uncertainty	Parameter Estimation	Standard Uncertainty	Parameter Estimation	Standard Uncertainty	Parameter Estimation	Standard Uncertainty
R_t [µm]	2.613	0.046	2.939	0.088	8.484	0.044	3.442	0.090
R_z [µm]	2.243	0.022	2.350	0.038	4.061	0.019	2.632	0.029
R_p [µm]	1.010	0.016	1.064	0.027	2.250	0.014	1.295	0.018
R_v [µm]	1.233	0.016	1.286	0.027	1.811	0.014	1.337	0.023
R_a [µm]	0.3300	0.0008	0.3697	0.0018	0.39844	0.00063	0.3732	0.0018
R_q [µm]	0.4186	0.0008	0.4661	0.0019	0.53093	0.00065	0.4660	0.0011
R_{sk} [-]	−0.2420	0.0084	−0.4338	0.0168	0.338	0.012	−0.375	0.010
R_{ku} [-]	3.0868	0.0191	3.0862	0.0415	12.79	0.13	3.107	0.027

4.2. Confocal Microscope

- Measuring ranges: The evaluation field of 100 µm was selected. In length (x-axis), a profile of length 5.6 mm (7× λ_c) was evaluated. Because the field of measurement of the objective was less than the total length of scanning, 25 measurement fields had to be evaluated and the overlapping images (stitch) had to be implemented. The option of "profile measuring" was used in the instrument, obtaining only coordinates (x_r, z_r), keeping the coordinate y_r constant.
- Operating time: The effective measurement time, which included the measurement of the 25 measurement fields on the surface, the calculation of parameters by the equipment and the saving of the data files, was about 10 to 15 min. Sample preparation time, which included the placement and inclination adjustment by a tilt table (Figure 3b), lasted between 15 and 20 min.
- Operational considerations: There was a significant probability of obtaining outliers or points not measured in the sample, so it was necessary to repeat the measurement a high number of times. Unmeasured points appeared when the Charge-Coupled Device (CCD) sensor of the confocal microscope did not receive enough light intensity to detect a peak position. This fact

could be caused by local slope effects, which meant that the reflected light was not picked up by the target, or that the light intensity selected for the measurement was not adequate (low). On the contrary, if the luminous intensity of the equipment was very high, the CCD sensor could became saturated, tampering the peak position, and obtaining incorrect z-coordinate values. These effects could produce sharp peaks and valleys which were not real, as Figure 5b shows. Also, due to the geometry of the roughness profile, the illumination beam could only solve slopes with a maximum angle of 90 degrees. Very fine adjustment of the light intensity used in the measurement, and even the creation of different levels of illumination according to the depth of the sample, were therefore necessary. [37]. Likewise, it was be necessary to correct the unmeasured points and outliers by using interpolation and filtering techniques, respectively.

- Environmental considerations: The equipment was installed in the *Laboratorio de Investigación de Materiales de Interés Tecnológico* (LIMIT) laboratory of *ETS de Ingeniería y Diseño Industrial* (ETSIDI). The measurements were carried out at a controlled temperature of 20 ± 1 °C.
- Measurand preparation: The same as in the tactile instrument.
- Numeric values of the parameters: Table 6 (columns 6 and 7) and Table 7 (columns 6 to 9) show the roughness parameters obtained when evaluating measurands 8 and 2 respectively. If the results are compared when measuring the same sample with the stylus profilometers (Section 4.1), there are discrepancies between both methods, due to the causes mentioned above (outliers and not-measured points). When the standard was measured, differences of values of the parameters of between 2% and 15% were obtained. These differences could reach 300% when the piece was measured. In both cases, the smallest differences between the roughness parameter were obtained when the amplitude parameter (average of ordinates) was compared. As other authors have verified, most of the roughness parameters measured with the confocal microscope show higher values than those obtained with the profilometers. The uncertainty of the parameters showed the same order the magnitude as those obtained with the tactile instruments. To solve the problem of discrepancies, a morphological profile filter was employed: scale space techniques according to ISO 16610-49:2015 [38] so that it was possible to smooth the profile using different circular disks [39] and the outliers would be eliminated. Figure 6a,b show the results obtained when applying this technique. Also, it was decided to use a robust Gaussian regression filter [39,40] in order to improve the results, defined in standard ISO 16610-31:2016 [41]. This way, the waviness profile would not be affected by possible outliers that were not eliminated in the previous step. Figure 6c shows the primary profile and the waviness profile, and Figure 6d shows the roughness profile. By applying these modifications, it was possible to observe how the results obtained improve considerably when implementing the previous techniques, Table 6 (column 9), obtaining a maximum difference, when the R_{sk} parameters evaluated at 50% of its value (without the filtering technique the difference was 240%). As an average value, the difference between parameters was 14% and 105% without the filtering technique.
- Data storage requirements: The data file had 16,885 points with coordinates (x_r, z_r) and a size of 310 kB.
- Cost of instrumentation/maintenance: The acquisition cost of the equipment was approximately 120,000 €. Maintenance was practically non-existent (replacement of LED light sources after about 10,000 h of operation).

Finally, the compatibility between the results of the roughness parameters obtained by the different measurement methods and instruments and the values provided by the calibration certificate of the type C1 spacing standard (measurand 2) were analyzed. The calibrated parameters of the standard give the next values: Parameter R_a = 1.003 µm with an expanded uncertainty (k = 2) of 3% of the value of the parameter; Parameter R_{Sm} = 101.3 µm with an expanded uncertainty (k = 2) of 3 µm. When comparing the previous values with the uncertainty of the obtained measurements, it could be affirmed that the results were coherent with the certified values (the measured values were contained in the interval characterized by the certified value and its expanded uncertainty). Figure 7a,b shows this consideration. It was verified that the values of the analyzed roughness parameters, obtained with the confocal microscope, always exceeded the certified values.

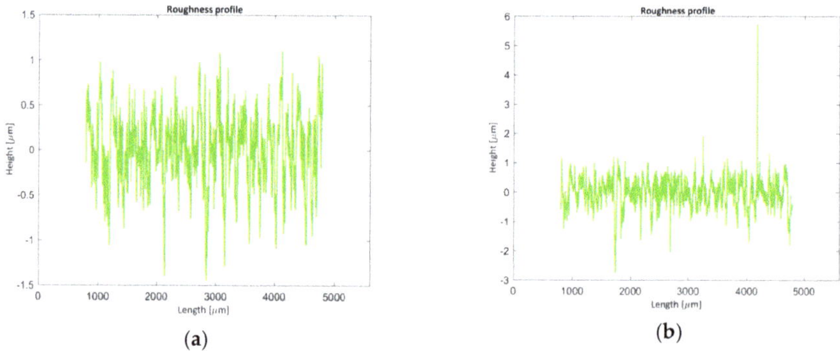

Figure 5. Roughness evaluated profile in measurand 2 (**a**) *SP-1*; (**b**) *CM-3*.

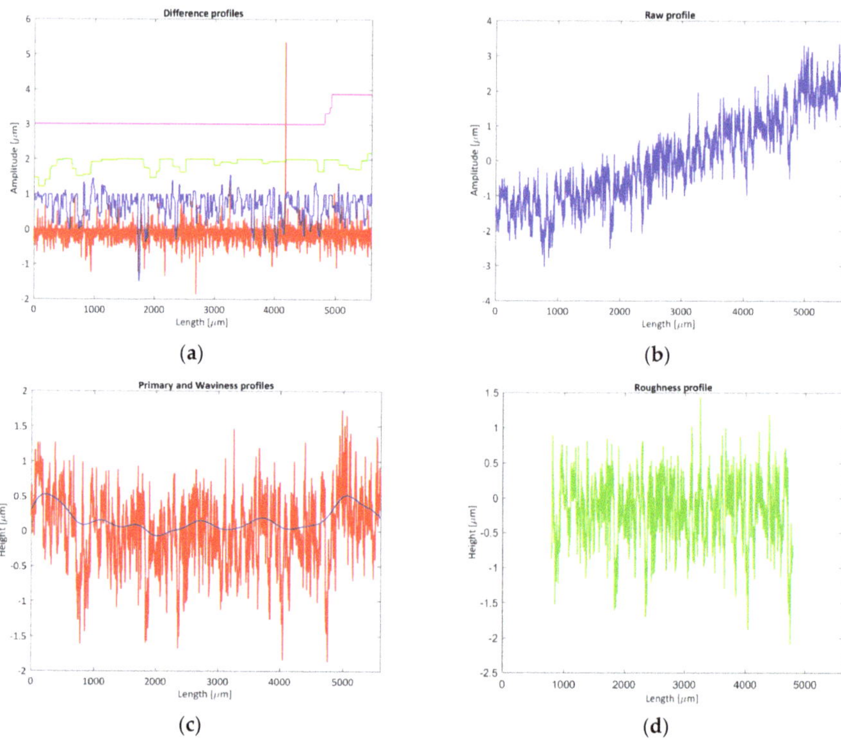

Figure 6. Roughness evaluated profile with a confocal microscope (outliers eliminated): (**a**) morphological filter application; (**b**) raw profile obtained after filtration; (**c**) waviness profile; (**d**) roughness profile.

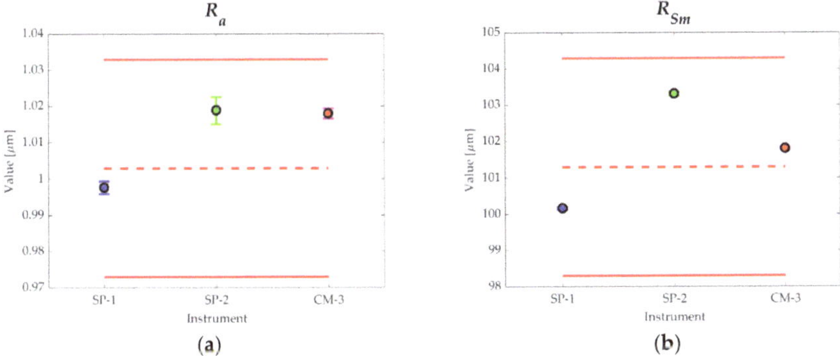

Figure 7. Comparison of results for: (**a**) R_a; (**b**) R_{Sm}.

It should be noted that the present work has used parts that could be measured by any of the two considered techniques, optical or tactile. Therefore, the size and location of the measured parts had not been taken into account. Both size and location were factors of great relevance, which in many situations determine the type of measurement to be performed. In this sense, contact measurement techniques were usually advantageous compared to optical techniques due to the wide variety of existing profilometers, their portability, and the possibility of working in less restrictive environmental conditions.

5. Conclusions

This work has developed a comparative study between two kinds of measurement techniques that can be used in surface metrology, namely, two portable stylus profilometers (tactile) and a confocal microscope (optical). Measurement results, needed requirements, and available performances of both techniques have been analyzed. Both the metrological characteristics (measurement precision and accuracy, complexity of algorithms for data processing, etc.) and the operational characteristics (measuring ranges, measurement speed, environmental and operational requirements, costs, etc.) were studied.

The algorithms and procedures for calculating the uncertainties associated with the roughness parameters have been developed by employing the Monte Carlo method. The correlations between parameters have been found.

Measurements of several measurands obtained with the portable stylus profilometer have been shown to be the most reliable method when compared with the certified values of the measured standard, in line with Leach conclusions [2]. The procedure required little preparation and measurement time, and the measurement algorithms were simple. As a disadvantage, the low vertical resolution, i.e., the z-value, of the two instruments employed when compared to optical methods, could be neglected. With respect to the low x sampling value of the portable stylus profilometer employed, if compared with the confocal microscope, it could also be neglected. This conclusion, for other types of stylus profilometers with better metrological characteristics r (higher lateral resolution) that were not analyzed in this paper, may not be true.

The use of a confocal microscope allows for improving the previous parameters (lower uncertainty) at the expense of greater measurement and preparation times, as well as a very high cost of equipment. Besides, it forced the development of more complex algorithms due to the presence of outliers and points not measured on the sample in order to obtain reliable results. From the experimental study, it had been observed that when the confocal microscope was employed, it was possible to reduce nearly 10 times the difference between parameters (compared with results obtained with a stylus profilometer) when the profile was filtered employing a morphological profile filter and a robust Gaussian regression filter.

Further studies would be suitable in order to include more terms that contribute to the uncertainty of the roughness parameters (standard contribution and sensor interaction (light or tip) among others).

Author Contributions: Conceptualization, A.S.L. and J.C.G.; Methodology, P.M.; Software, J.C.G.; Validation, C.W. and T.F.P.; Formal Analysis, A.S.L.; Investigation, P.M.; Resources, A.S.L.; Data Curation, T.F.P. and C.W.; Writing-Original Draft Preparation, J.C.G.; Writing-Review & Editing, J.C.G., P.M., and A.S.L.; Visualization, T.F.P.; Supervision, J.C.G.; Project Administration, A.S.L.; Funding Acquisition, A.S.L.

Funding: This work has been carried out within the framework of the Project DPI2016-78476-P "*Desarrollo colaborativo de patrones de software y estudios de trazabilidad e intercomparación en la caracterización metrológica de superficies*", belonging to the 2016 Call for R & D Projects corresponding to the State Program to Promote Scientific and Technical Research Excellence, State Subprogram of Generation of Knowledge.

Conflicts of Interest: The authors declare no conflict of interest. The funders had no role in the design of the study; in the collection, analyses, or interpretation of data; in the writing of the manuscript, and in the decision to publish the results.

References

1. Leach, R. Chapter 1: Introduction to Surface Topography. In *Characterisation of Areal Surface Texture*; Leach, R., Ed.; Springer: Berlin/Heidelberg, Germany, 2013; pp. 1–13, ISBN 978-3-642-36458-7.
2. Leach, R.; Weckenmann, A.; Coupland, J.; Hartmann, W. Interpreting the probe-surface interaction of surface measuring instruments, or what is a surface? *Surf. Topogr. Metrol. Prop.* **2014**, *2*, 035001. [CrossRef]
3. Mathiaa, T.G.; Pawlus, P.; Wieczorowski, M. Recent trends in surface metrology. *Wear* **2011**, *271*, 494–508. [CrossRef]
4. Leach, R. *Optical Measurement of Surface Topography*; Leach, R., Ed.; Springer: Berlin/Heidelberg, Germany, 2011; ISBN 978-3-642-12011-4.
5. Conroy, M.; Armstrong, J. A comparison of surface metrology techniques. *J. Phys. Conf. Ser.* **2005**, *13*, 458–465. [CrossRef]
6. Artigas, R. Chaper 11: Imaging Confocal Microscopy. In *Optical Measurement of Surface Topography*; Leach, R., Ed.; Springer: Berlin/Heidelberg, Germany, 2011; pp. 237–286, ISBN 978-3-642-12011-4.
7. Vorburger, T.V.; Rhee, H.-G.; Renegar, T.B.; Song, J.-F.; Zheng, A. Comparison of optical and stylus methods for measurement of surface texture. *Int. J. Adv. Manuf. Technol.* **2007**, *33*, 110–118. [CrossRef]
8. Poon, C.Y.; Bushan, B. Comparison of surface roughness measurements by stylus profiler, AFM and non-contact optical profiler. *Wear* **1995**, *190*, 76–88. [CrossRef]
9. Nouira, H.; Salgado, J.A.; El-Hayek, N.; Ducourtieux, S.; Delvallée, A.; Anwer, N. Setup of a high-precision profilometer and comparison of tactile and optical measurements of standards. *Meas. Sci. Technol.* **2014**, *25*, 044016. [CrossRef]
10. Piska, M.; Metelkova, J. On the comparison of contact and non-contact evaluations of a machined surface. *MM Sci. J.* **2014**, *2*, 476–480. [CrossRef]
11. Nieslony, P.; Krolczyk, G.M.; Zak, K.; Maruda, R.W.; Legutko, S. Comparative assessment of the mechanical and electromagnetic surfaces of explosively clad Ti-steel plates after drilling process. *Precis. Eng.* **2017**, *47*, 104–110. [CrossRef]
12. Merola, M.; Ruggiero, A.; De Mattia, J.S.; Affatato, S. On the tribological behavior of retrieved hip femoral heads affected by metallic debris. A comparative investigation by stylus and optical profilometer for a new roughness measurement protocol. *Measurement* **2016**, *90*, 365–371. [CrossRef]
13. Krolczyk, G.M.; Maruda, R.W.; Nieslony, P.; Wieczorowski, M. Surface morphology analysis of Duplex Stainless Steel (DSS) in Clean Production using the Power Spectral Density. *Mesurement* **2016**, *94*, 464–470. [CrossRef]
14. Krolczyk, G.M.; Maruda, R.W.; Krolczyk, J.B.; Nieslony, P.; Wojciechowski, S.; Legutko, S. Parametric and nonparametric description of the surface topography in the dry and MQCL cutting conditions. *Mesurement* **2018**, *121*, 225–239. [CrossRef]
15. Niemczewska-Wójcik, M. The influence of the surface geometric structure on the functionality of implants. *Wear* **2011**, *271*, 596–603. [CrossRef]

16. Thompson, A.; Senin, N.; Giusca, C.; Leach, R. Topography of selectively laser melted surfaces: A comparison of different measurement methods. *CIRP Ann.-Manuf. Technol.* **2017**, *66*, 543–546. [CrossRef]
17. Feidenhans'l, N.A.; Hansen, P.E.; Pilný, L.; Madsen, M.H.; Bissacco, G.; Petersen, J.C.; Taboryski, R. Comparison of optical methods for surface roughness characterization. *Meas. Sci. Technol.* **2015**, *26*, 085208. [CrossRef]
18. ISO 16610-21:2011. *Geometrical Product Specifications (GPS)—Filtration—Part 21: Linear Profile Filters: Gaussian Filters*; ISO: Geneva, Switzerland, 2011.
19. ISO 4287:1997. *Geometrical Product Specifications (GPS)—Surface Texture: Profile Method—Terms, Definitions and Surface Texture Parameters*; ISO: Geneva, Switzerland, 1997.
20. ISO 4288:1996. *Geometrical Product Specifications (GPS)—Surface Texture: Profile Method—Rules and Procedures for the Assessment of Surface Texture*; ISO: Geneva, Switzerland, 1996.
21. Cox, M.G.; Siebert, B.R.L. The use of a Monte Carlo method for evaluating uncertainty and expanded uncertainty. *Metrologia* **2006**, *43*, S178–S188. [CrossRef]
22. Sousa, J.A.; Ribeiro, A.S. The choice of method to the evaluation of measurement uncertainty in metrology. In Proceedings of the IMEKO XIX World Congress—Fundamental and Applied Metrology, Lisbon, Portugal, 6–11 September 2009; pp. 2388–2393.
23. Caja, J.; Gómez, E.; Maresca, P. Optical measuring equipments. Part I: Calibration model and uncertainty estimation. *Precis. Eng.* **2015**, *40*, 298–304. [CrossRef]
24. Evaluation of Measurement Data—Supplement 1 to the 'Guide to the Expression of Uncertainty in Measurement'—Propagation of Distributions Using a Monte Carlo Method. Available online: https://www.bipm.org/utils/common/documents/jcgm/JCGM_101_2008_E.pdf (accessed on 19 August 2018).
25. Evaluation of Measurement Data—Supplement 2 to the 'Guide to the Expression of Uncertainty in Measurement'—Extension to Any Number of Output Quantities. Available online: https://www.bipm.org/utils/common/documents/jcgm/JCGM_102_2011_E.pdf (accessed on 19 August 2018).
26. NIST Surface Roughness and Step Height Calibrations. Available online: https://www.nist.gov/sites/default/files/documents/pml/div683/grp02/nistsurfcalib.pdf (accessed on 28 June 2018).
27. Harris, P.M.; Leach, R.K.; Giusca, C. *Uncertainty Evaluation for the Calculation of a Surface Texture Parameter in the Profile Case*; NPL Report MS 8; Queen's Printer: Cambridge, UK, 2010; ISSN 1754-2960.
28. Haitjema, H.; van Dorp, B.; Morel, M.; Schellekens, P.H.J. Uncertainty estimation by the concept of virtual instruments. In Proceedings of the Recent Developments in Traceable Dimensional Measurements, Munich, Germany, 20–21 June 2001; Decker, J.E., Ed.; SPIE: Bellingham, WA, USA, 2001; pp. 147–157.
29. Internet Based Surface Metrology Algorithm Testing System. Available online: https://physics.nist.gov/VSC/jsp/index.jsp (accessed on 26 June 2018).
30. Hüser, D.; Hüser, J.; Rief, S.; Seewig, J.; Thomsen-Schmidt, P. Procedure to Approximately Estimate the Uncertainty of Material Ratio Parameters due to Inhomogeneity of Surface Roughness. *Meas. Sci. Technol.* **2016**, *27*, 085005. [CrossRef]
31. International Vocabulary of Metrology—Basic and General Concepts and Associated Terms (VIM). Available online: https://www.bipm.org/utils/common/documents/jcgm/JCGM_200_2012.pdf (accessed on 19 August 2018).
32. Fotowicz, P. An analytical method for calculating a coverage interval. *Metrologia* **2006**, *43*, 087001. [CrossRef]
33. Wübbeler, G.; Krystek, M.; Elster, C. Evaluation of measurement uncertainty and its numerical calculation by a Monte Carlo method. *Meas. Sci. Technol.* **2008**, *19*, 084009. [CrossRef]
34. Cox, M.G.; Harris, P.M. The design and use of reference data sets for testing scientific software. *Anal. Chim. Acta* **1998**, *380*, 339–352. [CrossRef]
35. ISO 5436-2:2012. *Geometrical Product Specifications (GPS)—Surface Texture: Profile Method; Measurement Standards—Part 2: Software Measurement Standards*; ISO: Geneva, Switzerland, 2012.
36. RPTB Version 2.09—Software to Analyse Roughness of Profiles. Available online: https://www.ptb.de/rptb/login (accessed on 26 June 2018).
37. Barajas, C. Caracterización geométrica de huellas de dureza Brinell mediante equipos ópticos. Modelo de microscopia confocal. Ph.D. Thesis, Technical University of Madrid, Madrid, Spain, December 2015.
38. ISO 16610-49:2015. *Geometrical Product Specifications (GPS)—Filtration—Part 49: Morphological Profile Filters: Scale Space Techniques*; ISO: Geneva, Switzerland, 2015.

39. Muralikrishnan, B.; Raja, J. *Computational Surface and Roundness Metrology*; Springer: London, UK, 2009; pp. 1–263, ISBN 978-1-84800-296-8.
40. Hüser, D. Selected Filtration Methods of the Standard ISO 16610, 5 Precision Engineering, PTB. 2016. Available online: https://www.ptb.de/cms/fileadmin/internet/fachabteilungen/abteilung_5/5.1_oberflaechenmesstechnik/DKD-Richtlinien/Selected_Filtration_Methods_of_ISO-16610.pdf (accessed on 26 June 2018).
41. ISO 16610-31:2016. *Geometrical Product Specifications (GPS)—Filtration—Part 31: Robust Profile Filters: Gaussian Regression Filters*; ISO: Geneva, Switzerland, 2016.

© 2018 by the authors. Licensee MDPI, Basel, Switzerland. This article is an open access article distributed under the terms and conditions of the Creative Commons Attribution (CC BY) license (http://creativecommons.org/licenses/by/4.0/).

Article

Product Lifecycle Management as Data Repository for Manufacturing Problem Solving

Alvaro Camarillo [1,2,*], José Ríos [2] and Klaus-Dieter Althoff [3,4]

1 Exide Technologies SAS, 5 allée des Pierres Mayettes, 92636 Gennevilliers, France
2 Mechanical Engineering Department, Polytechnic University of Madrid, Jose Gutierrez Abascal 2, 28006 Madrid, Spain; jose.rios@upm.es
3 German Research Center for Artificial Intelligence (DFKI), Trippstadter Straße 122, 67663 Kaiserslautern, Germany; klaus-dieter.althoff@dfki.de
4 Institut für Informatik, University of Hildesheim, Universitätsplatz 1, 31141 Hildesheim, Germany
* Correspondence: alvaro.camarillo@exide.com; Tel.: +33-141-21-23-67

Received: 2 July 2018; Accepted: 14 August 2018; Published: 18 August 2018

Abstract: Fault diagnosis presents a considerable difficulty to human operators in supervisory control of manufacturing systems. Implementing Internet of Things (IoT) technologies in existing manufacturing facilities implies an investment, since it requires upgrading them with sensors, connectivity capabilities, and IoT software platforms. Aligned with the technological vision of Industry 4.0 and based on currently existing information databases in the industry, this work proposes a lower-investment alternative solution for fault diagnosis and problem solving. This paper presents the details of the information and communication models of an application prototype oriented to production. It aims at assisting shop-floor actors during a Manufacturing Problem Solving (MPS) process. It captures and shares knowledge, taking existing Process Failure Mode and Effect Analysis (PFMEA) documents as an initial source of information related to potential manufacturing problems. It uses a Product Lifecycle Management (PLM) system as source of manufacturing context information related to the problems under investigation and integrates Case-Based Reasoning (CBR) technology to provide information about similar manufacturing problems.

Keywords: product lifecycle management (PLM); manufacturing problem solving (MPS); fault diagnosis; smart factory; process failure mode and effect analysis (PFMEA); case-based reasoning (CBR)

---

1. Introduction

A manufacturing failure is an event in which some part of a manufacturing system does not perform according to its operational specifications and therefore it occurs in a specific manufacturing context. Due to such a failure, production is disturbed to a certain extent. The consequence is that production targets may not be achieved. The gap between the resulting production state and the intended production state is thus a problem. A Manufacturing Problem Solving (MPS) procedure starts when such a situation arises. The first step in an MPS procedure is the fault diagnosis, which presents considerable difficulty to human operators in supervisory control of manufacturing systems [1,2]. Providing support to these operators, by means of knowledge-based software applications, has been identified as essential to achieve good results from the solving procedure [3–5].

The characterization of the manufacturing context, where a failure might occur, can be set by means of processes, process steps, machines, tooling, process parameters, and product manufacturing features. Therefore, the context of the failure can be described by means of Process-Product-Resource (PPR) data. When looking at data, historical failure data could be used to assist in the MPS process. The use of such historical data would require identifying similar failures, which, in turn, requires making use of manufacturing context data. Product Lifecycle Management (PLM) systems are considered the main source of PPR digital information, and therefore their use would facilitate any attempt to develop a knowledge-based solution to support an MPS process [5,6]. PLM systems are one of the main components of the digital manufacturing and Industry 4.0 strategy [7,8].

The technological vision of a Smart Factory [9,10] is also included in the strategy Industry 4.0 [8], and eventually aims to develop a factory model that is context-aware and assists people and machines in the execution of their tasks. Fault diagnosis can be considered as one of these tasks to be supported.

A typical approach to a Smart Factory [9] is the deployment of smart artifacts, along the value streams, that are able to collect a large quantity of data from their environment and to communicate with each other (Internet of Things) [8]. This approach implies large investments, which prevents some companies from taking their first steps into Industry 4.0 [11,12]. This work presents an approach that aims to contribute to the technological vision of a Smart Factory but observing the constraints of being based on existing facilities, existing company data repositories, and with low investment.

The proposed approach is a knowledge-based development that integrates two main types of software applications: PLM and Case-Based Reasoning (CBR) [5]. The developed prototype assists human operators during an MPS process. It guides the user through the MPS activity based on the problem-solving method 8D [13]. The user introduces a query describing a problem that occurs in a production line, and the developed prototype software proposes several potential causes to be checked in the line. Based on the user feedback related to these first proposed causes, the system will propose the next lower level of causes. The process is repeated until the root cause of the problem is identified. More detailed information can be found in Camarillo et al. [5].

As sources of information, the prototype integrates two repositories. First, the existing company Process Failure Mode and Effect Analysis (PFMEA) records [14] are stored in the case base of the Case-Based Reasoning (CBR) system [15]. Second, the company's Product Lifecycle Management (PLM) repository contains the PPR data. The Case-Based Reasoning (CBR) system is used to search for and retrieve information concerning similar problems. The similarity determination between the current problem and the problems contained in the case base depends on the PPR data stored in the PLM repository. The architecture of the system is based on SEASALT (Shared Experience using an Agent-based System Architecture LayouT) [16], which is a multi-case-based, domain-independent reasoning architecture for extracting, analyzing, sharing, and providing experiences. Camarillo et al. [17,18] show a description of a first prototype and initial results. Camarillo et al. [5] shows the MPS process model, the architecture of the developed prototype, the main ontology to support the MPS process, and the validation of the prototype. This paper provides an insight into the information and communication models of the developed prototype.

The next sections of this paper are structured as follows. Section 2 contains a review of research works with similar approaches. Sections 3 and 4 discuss the created information and communication models, which are implemented in the developed prototype. Section 5 describes the configuration of the developed system. Section 6 provides a brief description of an application case. The paper ends with the conclusions.

2. Related Works

The developed approach is grounded on three basic models: an MPS process model, an MPS knowledge representation model, and an MPS system architecture model [5].

The MPS process model defines the steps to be taken by the user to solve a problem with the support of the proposed system. This process model basically follows the steps defined in the 8D method [13] and specifies the kind of interaction expected at each step between the user and the system. The left side of Figure 1 shows the developed Graphical User Interface (GUI), the center part of Figure 1 shows the main steps of the MPS process, and the right side of Figure 1 shows the main systems of the developed prototype. The main steps of the MPS process are explained next and their links with the developed GUI are shown in Figure 1.

1. The user inputs a basic description of a manufacturing problem into the system (S1).
2. Based on the user input, the system searches and collects context information related to the problem from the PLM system repository. The result is shown to the user (S1).
3. Combining the input from the user and the data collected from the PLM repository, the system creates a global query to search for possible solutions (S2).
4. The system distributes the global query among its agents [16] that look for the most similar proposals out of their own case bases by applying CBR. The 10 most similar proposals coming from the agents are presented to the user. Initially, only the proposed containment actions and problem causes are displayed (S2).
5. The user must check the proposed failure modes or causes at the manufacturing location where the problem was identified and give feedback to the system. At this stage, the user may decide to refine his problem formulation and then go back to Step 1 (S2).
6. Once the possible root causes are identified, the system provides the related proposals for corrective and preventive actions (S3).
7. As part of the lessons learned step, the user gives feedback to be analyzed by a Knowledge Engineer. When appropriate, the Knowledge Engineer will update the CBR subsystems to extend the case bases (S4, S5, S6).

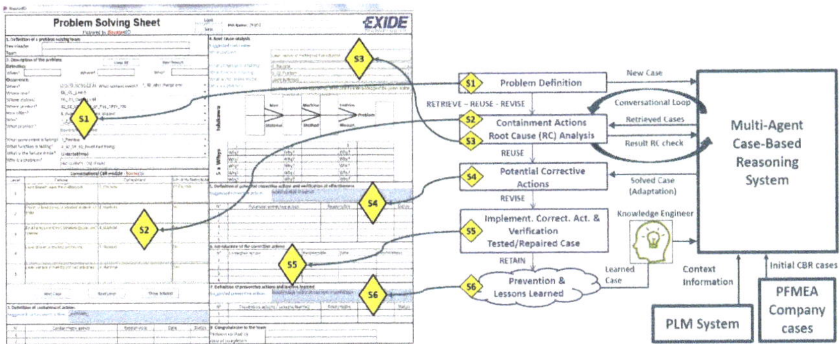

Figure 1. Process model of knowledge-based system. PLM, Product Lifecycle Management; PFMEA, Process Failure Mode and Effect Analysis; CBR, case-based reasoning.

The MPS knowledge representation model is based on an ontology that allows for the representation of any knowledge related to the MPS process [5]. It comprises the following main concepts: Problem, Component, Function, Failure, Context, and Solution (Figure 2). The relations among these six concepts, their associated taxonomies, and their parameters have been designed to fulfil several constraints: support a generic definition of a manufacturing process and its location, be compatible with the information structure of the PFMEA method, comprise concepts to describe different aspects of a manufacturing problem, and to allow case similarity determination.

The proposed ontology defines the concept "Problem" similarly as in FMEA (Failure Mode and Effect Analysis) [14], where a component performs a function, and the latter fails in a defined mode. Component, Function, and Failure form a unique trio. The concept "Component" is subdivided into six subtypes: Process, Man, Machine, Material, Method, and Environment. The concept "Context" allows for representing the setting of a problem, is subdivided into seven different types of contexts: Material, Process, Machine, Event, Method, Man, and Environment, and has an associated taxonomy represented through the relationship type "is part of" pointing to itself. Each subtype of Context has different types of attributes to specify each type of technical information in the context (e.g., pressures, temperatures, and dimensions). These attributes are used in the configuration of the PLM system to store PPR information explicitly associated with the problem. More detailed information about this ontology can be found in Camarillo et al. [5].

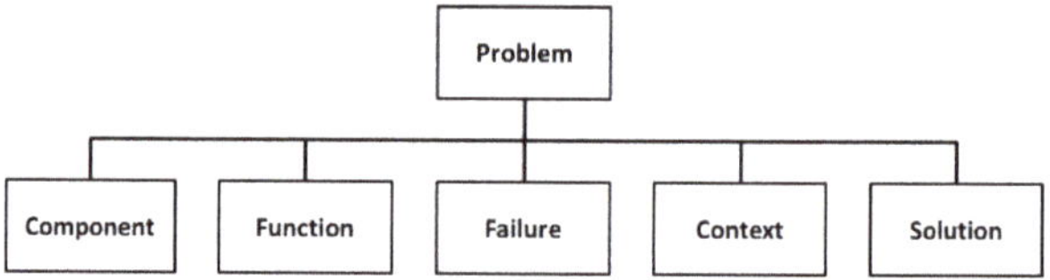

Figure 2. Manufacturing Problem Solving (MPS) top-level Ontology.

The proposed MPS system architecture model is based on SEASALT (Shared Experience using an Agent-based System Architecture LayouT) [16]. The developed architecture supports the deployment of the different agents across different manufacturing plants of a company. Within each plant, agents can be deployed across the areas with different manufacturing processes. In this way, each topic agent, hosted in a specific manufacturing process of a specific manufacturing plant, will be able to collect and to store knowledge related to its own area, becoming an expert of its process and plant. By means of a coordinator agent, a topic agent can communicate and interchange information with all the other topic agents hosted in different processes and/or plants through the company's intranet. Each topic agent has its own case base and uses CBR technology to find the most similar cases related to a user query. This information exchange supports the MPS process by providing the user with solutions for the most similar failures stored in any topic agent of the architecture [5].

In the literature review, several relevant works developed by other researchers were identified. Firstly, in relation to the modeling of PFMEA concepts. Dittmann et al. [19] presented an ontology to support FMEA concepts. The information model proposed in this work enhanced that ontology mainly by adding the concepts of Problem and Context [5]. In relation to the use of a PLM repository, Bertin et at. [6] propose, as part of a Lessons Learned System (LLS), the use of a PLM system as the central repository of data, but they put the focus on the Engineering Change Request (ECR) process of the company, whereas this work focuses on problem solving at production lines.

The work of Yang et al. [1] presents a fault diagnosis system for software intensive manufacturing systems and processes. They also profit from the stored information in the FMEA documents of the company and use CBR as an Artificial Intelligence (AI) tool. Nevertheless, they propose a second AI technology, deep-level Bayesian diagnosis network, to be used in cases of dynamic multi-fault diagnosis with uncertainty. The approach presented in this paper shares with them the use of FMEA and CBR but remains at a simpler AI level. However, the application scope of this work considers the sharing of knowledge among different manufacturing processes and plants (represented by topic agents). Also, contrary to the single-diagnosis suggestion proposed by Yang et al. [1], this proposed system uses an MPS method to guide the user step-by-step through the resolution of problems, which allows multiple cycles of problem redefinition, and that is fundamental when addressing very complex problems.

Finally, two relevant research works were identified in the field of fault diagnosis in aircraft maintenance: Chiu et al. [20] and Reus et al. [21]. Chiu et al. [20] propose the use of CBR together with genetic algorithms to enhance dynamic weighting and the design of non-similarity functions. With this approach, the proposed CBR system is able to achieve superior learning performance. As in the previous case, the approach presented in this paper remains at a simpler AI level, but proposes knowledge sharing among different MPS units. Reus et al. [21], as in this work, propose the use of SEASALT as a multi-agent architecture to share knowledge among multiple units. Nevertheless, the use of extended context-related information to enrich the similarity calculation is not addressed. Therefore, the link to a PLM system to enrich the similarity calculation and the search for solutions is outside its scope.

The next section introduces the developed information models and their link to the data sources, with special focus on the data related to PFMEA and the PPR concepts to be supported by the PLM system repository.

3. Information Models and Sources of Data

The proposed knowledge-based system uses data from four different sources (Figure 3):

- User query: through the developed Graphical User Interface (GUI), the user introduces a rough definition of the problem that occurs in the production facility. Taking the Kepner–Tregoe method [22] as a reference, the user should provide an answer to the questions 'What?' (a brief description of the problem), 'When?' (date and time), 'How often?' (frequency), 'Where?' (this question is divided into three different attributes related to the line and station where the problem happens and the product that is being produced), 'Who?' (operator name), and 'Why?' (a brief description of why it is a problem).
- Problem context: based on the user query, the system receives from the PLM system repository multiple details about the associated context of the problem.
- Cases from PFMEA: the system needs an initial set of cases (i.e., problems previously solved), stored in the case base of the CBR system, in order to propose solutions to the user. This work proposes the use of the company's available PFMEA documents to generate the initial case base.
- Knowledge engineer: an assigned employee in the company will analyze continuously the problems solved with the system and will create new cases to be stored in the case base (i.e., the reusing of created knowledge).

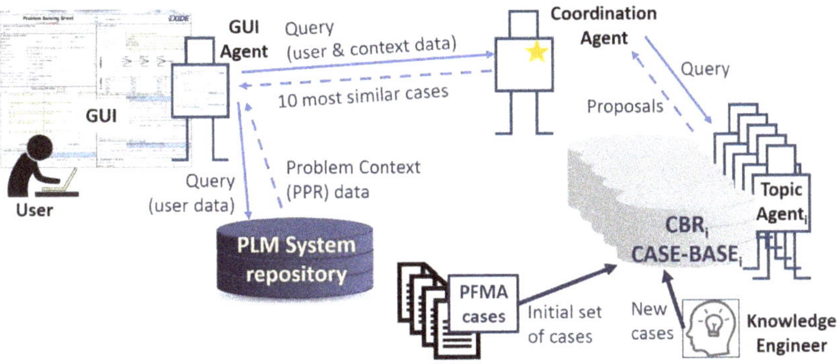

Figure 3. Top-level view of the proposed system data sources. GUI, graphical user interface; PPR, Process-Product-Resource; PFMA, Process Failure Mode Analysis.

Figure 3 shows the top-level view of the knowledge-based system. The user manages the system through the GUI. The GUI represents a Problem-Solving Sheet (PSS) divided into the corresponding areas of the 8D method. The user inputs the description of the problem through the GUI as described at the beginning of this section. Based on the input description, a GUI agent will send a query to the PLM system to receive the contextual information related to the data input by the user. Then, a comprehensive query, comprising both the user's input data plus the retrieved context data, is sent to the CBR system (Coordination Agent). The Coordination Agent will need to collect a first set of possible solved cases. The Coordination Agent communicates with the different topic agents to request proposals for similar problems. The case base of each topic agent needs to be populated with an initial set of cases (i.e., already solved problems). To do so, the company's PFMEAs are taken as the initial set of cases. The case base will be continuously extended with new solved problems. A solved problem is analyzed by an expert to decide whether it is included as a new case in the CBR case bases or not. A deeper insight to this process is given in the next sections.

The next subsections focus on the developed information models used to collect cases from PFMEA and to collect context information from the PLM system.

3.1. PFMEA Data

The initial source of cases for the knowledge-based system is the PFMEA documentation of the company. Figure 4 shows the information mapping between PFMEA and the knowledge model of the system [5]. As is shown in this figure, the PFMEA does not contain all the needed information to match totally the knowledge model of the system. Thus, the support of experts and the PLM system of the company are needed to translate the identified failures from the PFMEA into useful cases for the knowledge-based system. This situation derives from the fact that a PFMEA document represents a detailed analysis of all possible modes of failure associated with a specific production process [14]. However, the PFMEA document does not contain explicit technical information about the context of the process under analysis, and that information needs to be found in other documents associated with the PFMEA. In the proposed knowledge-based system, the created cases will be definitely detached from their original PFMEA documents, and the context information will have to be extracted from the PLM system of the company. This situation requires an expert to conduct a translation of failures into cases. A second issue to solve, in the translation activity, is that a PFMEA is a technique to prevent failures, but it is not designed to contain or to correct failures. Therefore, an expert has to fill these two fields, as well as define who is involved when the problem might happen (i.e., the parameter "Who"), a piece of information that is also not explicit in the PFMEA document.

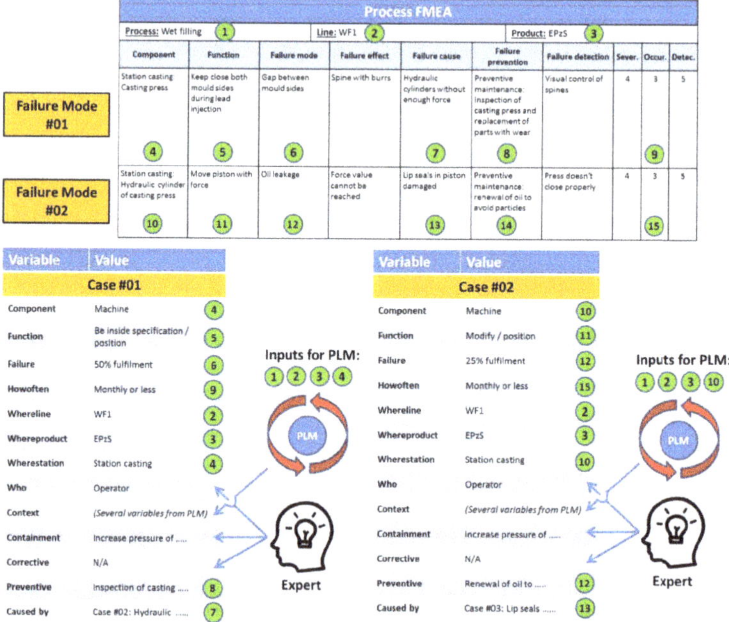

Figure 4. Information match between PFMEA and the knowledge model.

The fields in the PFMEA matrix (Figure 4) marked with 1, 2, 3 (i.e., the process, production facility, and product under analysis), and 4 (i.e., the component where the potential failure is identified) are used as input for the PLM to search for and retrieve the context data. The fields 4, 5, and 6 represent the core of the problem definition (i.e., Component, Function, and Failure). Field 9 (i.e., occurrence) defines the frequency of the problem, field 7 the preventive action, and field 8 indicates which failure causes the problem. The same process is repeated systematically with every new failure.

3.2. PLM Information Model and Manufacturing Context Data

The PLM system has the function of providing the context data related to the user query. For this, the PLM system has to be customized to fit into the defined ontological model, so the requested context information can be used by the knowledge-based system. This customization has two main areas of activity:

- PLM systems are a repository of data related to the product lifecycle, but not all the data are stored necessarily as data records in the system repository (e.g., technical reports written using a word processor application could contain product or process parameters, but for the PLM repository the file is a black box, and it only knows the metadata associated with the "Technical Report" type of file). Therefore, the parameters defined by experts as relevant for the similarity calculation (i.e., the parameters that distinguish one process from another one) have to be declared as parameters in the PLM repository.
- The context information is retrieved based on the query defined by the user. As was mentioned in Section 3, the user inputs data about the affected production line, the station, the product, the user, and the date when the problem happened. Taking these data, the PLM system needs to access all the context-related data. This means that the system needs to move automatically from element to element, collecting as much data as possible. For this reason, specific relationships to link all the related items in the PLM system have to be created.

In the case study presented in this work, the selected company had no prior PLM system available. The software Aras Innovator (version 11.0) was selected as a PLM system and its installation and customization were part of the case study. On top of the two customization activities defined above, an experimental novel PLM information model was developed with the aim of accelerating the retrieval process of context data.

The taxonomy of the class Context of the ontological model (Figure 5) has six main elements: Process, Machine, Material, Man, Method, and Environment. These six elements were declared as main elements in the PLM data structure ("Items" as per the standard terminology in Aras). Figure 5 shows these six elements with their relationships. These relationships will be the key to the process of information search as previously mentioned.

In the PFMEA methodology [14], on which the created ontology is partially based, the element Method represents the defined procedures or standards. Following this philosophy, the element Method contains the technical specification of each of the other component types in the PLM system.

A relevant issue in the management of technical information is that part of the information can be common for a whole family of components (e.g., a family of hex bolts with a specific diameter and thread where each one is distinguished from the others by length). In this sense, it is important to create a configuration in the PLM system that allows this type of information inheritance. This will ensure higher information consistency in the PLM system and a much easier update procedure. This was addressed by the creation of an additional element named "Parameter". A Parameter is defined as the smallest stored unit of information in the PLM repository. It contains information about a single attribute, and it is made of the type of the attribute (e.g., pressure or temperature), its limit type (i.e., max, min, nominal, or not applicable), its value (either numerical or a selection from a set of possible values), and its measurement unit. Each instantiation of a Parameter can be shared by multiple instantiations of Method. For example, an instance of the element Process called "Casting Plate 90" will have a link to an instance of an element Method called "Casting Plate 90 method", and this last one will contain a list of instances of the element Parameter, such as "Pressure/Max/10/bar" or "Temperature/Nom/300/°C". These parameters can be also used by the Method instantiation "Casting Plate 125 method" when the parameter Pressure does not change with the size of the grid.

Thus, this independency of Parameter-related Method allows the same parameter to belong to different methods. This inheriting process should not be confused with the type of inheritance between classes in typical Object-Oriented Programming (OOP). In the OOP case a child class inherits the attributes of its parent class, but in the proposed structure both an attribute and the specific values of such an attribute are inherited from an instance of the parent class to the instance of the child class. With this idea, each of the component subclasses can have links to multiple Method elements containing several types of technical data related to the family of the component (e.g., a method to define the material and supplier applicable to all hex bolts made of stainless steel, a method to define the thread and diameter applicable to all hex bolts with a standard thread and a diameter of 5/8", and a method to define the length applicable to a hex bolt with a length of 6").

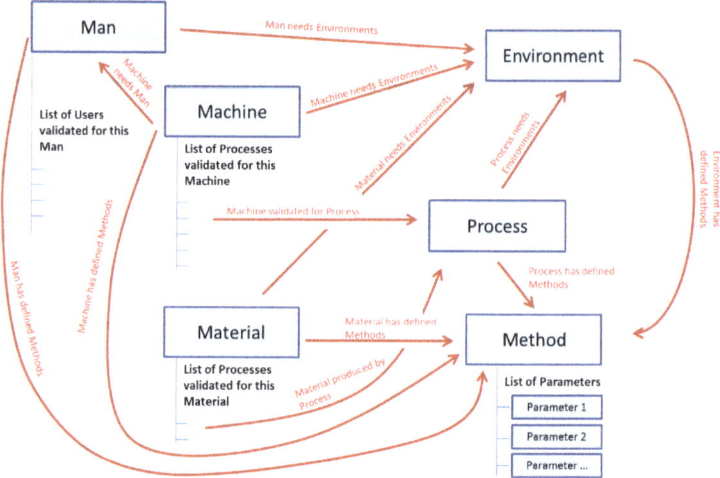

Figure 5. PLM structure of items.

The defined relationships and attributes for the PLM system are (Figure 6 shows an example):

- Machine attribute: each machine has a list of processes validated for it. This list can be shorter or longer depending on the complexity of the machine (e.g., pneumatic cylinder versus complete assembly line) and depending on the number of products that the machine can produce (each product can have specific processes).
- Machine to Man: a machine needs one or more human resources to control it. These human resources are part of "Man", and they can be of different subtypes (e.g., operator, process engineer, quality inspector . . .).
- Machine to Environment: a machine works in a defined environment, which comprises specific conditions or requirements, for example, a limited range of room temperature, or a stable power supply.
- Machine to Method: a machine has defined methods, which contain the technical specification of each of the other element types.
- Material attribute: a material has a list of processes validated for producing it. As in the case of Machine, this list can be longer or shorter.
- Material to Environment: a material works in an environment.
- Material to Method: a material has defined methods.
- Process to Environment: a process works in an environment.
- Process to Method: a process has defined methods.

- Process to Machine and Material: a process has two indirect relationships with material and machine. A process is validated for a pair composed of a machine and a material. Each existing material has validated processes. Each machine has validated processes. The common process will define the manufacturing process of that material under production in that machine.
- Man attribute: a man has a list of users validated for it. It is important to highlight the difference between Man and User. Man represents a type of human resource (e.g., an operator), and User represents a specific person (e.g., Mr. Müller, who belongs to the type of Man operator).
- Man to Environment: a man works in an environment.
- Man to Method: a man has defined methods that can execute.
- Environment to Method: an environment has defined methods.
- Method to Parameter: a method has a list of parameters.

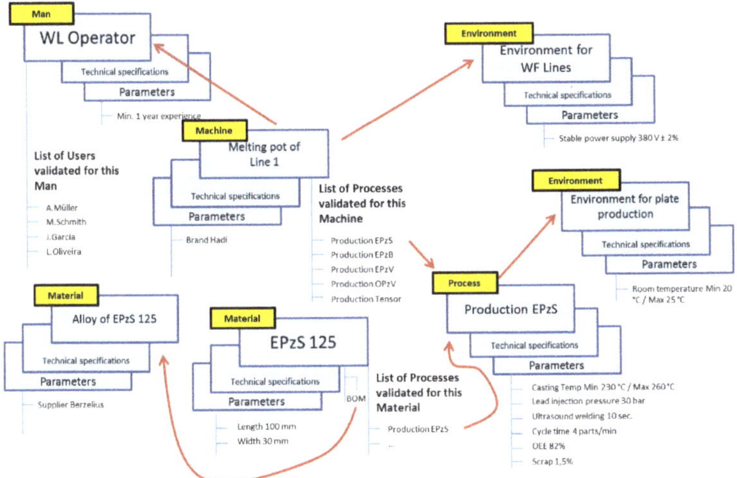

Figure 6. Example of the PLM structure of items (OEE: Overall Equipment Effectiveness).

Finally, each of the explained elements will have a set of standard attributes that contains the basic data of the element. These attributes are:

- The Name of the element.
- The Element reference number, which is the identification number in the PLM system.
- The Revision level.
- The State (i.e., released or not).
- The Effective date from which the item is released.
- The Classification (i.e., the position of the component in the Context taxonomy).
- The Item Nature. This is an attribute to indicate whether the component is real or abstract. This attribute allows the user of the knowledge-based system, during the introduction of the problem description, to select a general family of products or lines instead of a specific one (e.g., "Casting Lines" versus "Casting Line 3"). Since the system searches for context data in the PLM repository, the PLM system must support also these general elements to be able to return data. Specific elements are tagged with "Real", and general ones with "Abstract".

The next section presents the communication models developed for this work. These models define the communication among the main four actors of the proposed system: user, agents, PLM system, and CBR system.

4. Communication Model

4.1. Communication among Agents and Users

The communication model implemented in the developed system is a simplified version of SEASALT [16]. SEASALT is a multi-agent architecture made of three types of agents that communicate with each other to provide the user with solutions (Figure 7):

- GUI Agent: there are as many agents of this type as user access points in the network of the system. It manages the communication with the user, guiding him through the eight steps of the problem-solving method 8D. It manages also the communication with the PLM system (see Section 4.2) and the communication with the Coordination Agent.
- Topic Agent: there are as many agents of this type as production units in the network of the system. Each production plant should have at least one of these agents, and each production plant can have as many agents of this type as different production processes in the plant. Each Topic Agent has its own case base with solved cases of problems related to its production process, and it manages its own CBR engine to extract the most similar cases from its case base.
- Coordination Agent: there is a single agent of this type. It coordinates the communication among agents, and it is in charge of selecting the 10 most similar cases out of the multiple proposals coming from the topic agents.

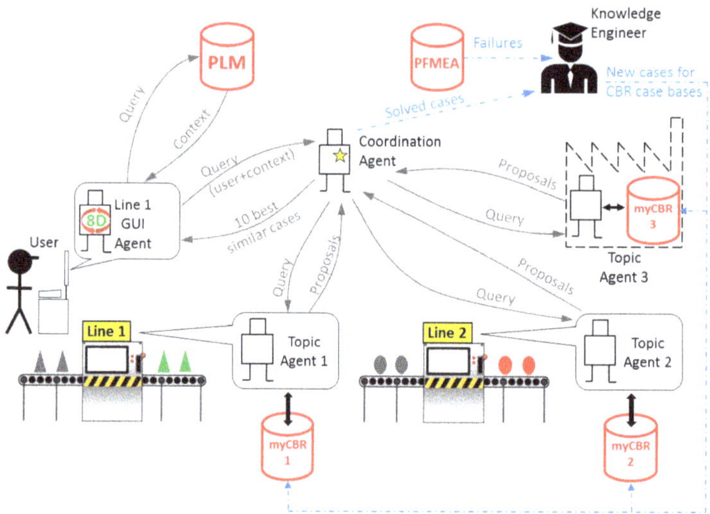

Figure 7. Communication model.

The system has to be maintained by one or more knowledge engineers that are responsible for two main tasks:

- Translating failures stored in PFMEA documents into cases following the model presented in Section 3.1. This function could be automated to a large extent, but it is not part of the scope of this work, and it will be considered in future work.
- Analyzing problems that were solved with the system and rating their relevancy as a case to be added to the case base of a topic agent.

4.2. Communication with the PLM

The GUI Agent is responsible for communicating with Aras Innovator, the PLM system implemented in the case study company. The communication language of Aras Innovator is AML (Aras Markup Language). AML is an XML (Extensible Markup Language) dialect and markup language that drives the Aras Innovator server. Clients submit AML documents to the Aras Innovator server via HTTP (Hypertext Transfer Protocol), and the server returns an AML document containing the requested information.

The central element that builds Aras Innovator is the ItemType. An ItemType is a business object that represents the template or definition for the items that are created from it. In OOP (Object-Oriented Programming) terms, the ItemType is similar to the class definition, and the items that are created from it are the class instances or objects. Almost everything in Aras Innovator is defined through an ItemType. For instance, ItemType defines properties, forms, or views available for this item, its lifecycle, the work-flows associated with the item, permissions, relationships, server and client methods and events to run on the item, etc. ItemType is designed to hold as little information as a name, or as much information as required for the most complex business objects.

The first step in any retrieval sequence is to get the ID (identification number) of the item that contains the requested information. Figure 8 shows an example of a message requesting Aras Innovator to search for an item of type Part (i.e., any mechanical design) and to send back all the data related to it and the returned message from Aras Innovator.

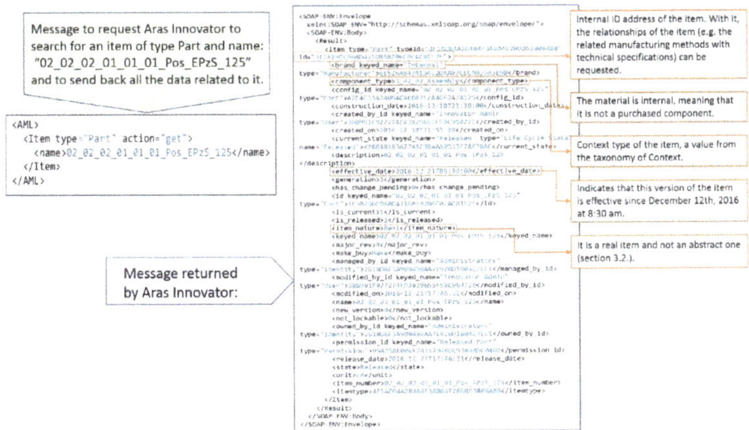

Figure 8. Example of communication with Aras Innovator.

A very important point in the retrieval process from the PLM system is that the extracted context information should be that which was released and effective when the problem happened. The date is the parameter "When" in the user query (see Section 3). For example, the user may investigate a quality claim about a product that was created several months ago. In this case, the current version of context information in the PLM repository might not be relevant, because between the current date and the date when the problem happened several items could have been changed (e.g., new suppliers, changes in the design, or improvements in the processes). Aras Innovator returns by default the item released at the time of the request. Nevertheless, in case the release date is later than the date of the problem, there are functions that return all the previously released versions of an item.

Once the ID of an item is known, a message can be sent to the PLM system to request all its relationships. In the next example, all the manufacturing methods related to an item are requested.

```
<AML>
  <Item type="Part" id="3F3829DC968D421DB58206C0CAC01F2F" action="get">
    <Relationships>
      <Item type="Manufacturing Method List_Part" action="get"></Item>
    </Relationships>
  </Item>
</AML>
```

Following the relationships among elements defined in the PLM information model (Section 3.2), and with the AML queries presented above, it is possible to extract all the context information of a problem from the limited amount of information introduced by the user in the query (i.e., 'What?', 'When?', 'How often?', 'Where?', 'Who?', and 'Why?').

4.3. Communication with the CBR System

Section 4.1 described how the Topic Agent hosts the CBR system. For this case study, the open source system myCBR was selected [18]. Since both myCBR and the multi-agent architecture (on which the MPS system is built) are developed in Java, the communication between both systems is done directly through internal Java communication. The classes of both myCBR and the agent are instantiated in the same source code, and they can interchange information through their public functions.

The Topic Agent goes through the following communication steps to retrieve from its case base a set of similar cases related the user query:

- Sending an order to myCBR to load the CBR project of the production area where the agent is located. This project contains the information model to describe problems (based on the top-level ontology presented in Section 2), the defined rules to calculate similarity, and the name of the case base where similar cases will be searched for. These concepts will be explained with more detail in Section 5.
- Sending an order to myCBR to obtain the main concept of the information model of the project that was opened in the step before. In this case, this is "Problem", the upper element of the MPS ontology (see Figure 1).
- Sending an order to myCBR to open the case base associated with the open project.
- Sending the retrieval parameters, which are the retrieval method and the problem attributes that are inside of the messaged received from the Coordination Agent.
- Launching the retrieval process.
- Receiving the 10 most similar proposals from of the open project.
- The proposals are sent to the Coordination Agent.

Once the way in which the CBR system receives a query and sends proposals is presented, the next section explains in detail how the CBR system manages the received data and how it calculates similarities in order to propose similar cases to the user.

5. CBR System Functioning Description

The CBR systems, which are hosted in each of the topic agents of the network, are responsible for seeking similar cases that could help the user to solve the current problem. For this function, they search in their own case base. The search is based on the information contained in the user query and the PLM context data received from the Coordination Agent. The user and PLM parameters are shown in the upper right corner of Figure 9. These parameters are presented also in Figure 10 with the format in which they are received by the topic agents. From the user inputs (numbers 1 to 12), some of the parameters (numbers 1, 2, 3, 4, and 6) are used in the request for context data in the PLM system as was explained in Section 3.2. The output of the PLM repository (number 13 to a previously undefined number "n" of attributes depending on each context case), together with some of the user query parameters (numbers 2, 3, 4, 5, 8, 9, 10, and 12), are used to create the query to be sent to the CBR system. This CBR query is shown in the upper left corner of Figure 9. Finally, there are some parameters in the user query (numbers 1, 7, and 11), shown in red in Figure 9, which are not used for the similarity calculation. The problem date (number 1) is used in the PLM system to find the released context information at the time that the problem happened. The parameters "What" and "Why" (numbers 7 and 11) are used by the knowledge engineer to understand better the solved problem, and to make a decision on whether the case will be included in the case base as a new case or not (see Section 4.1).

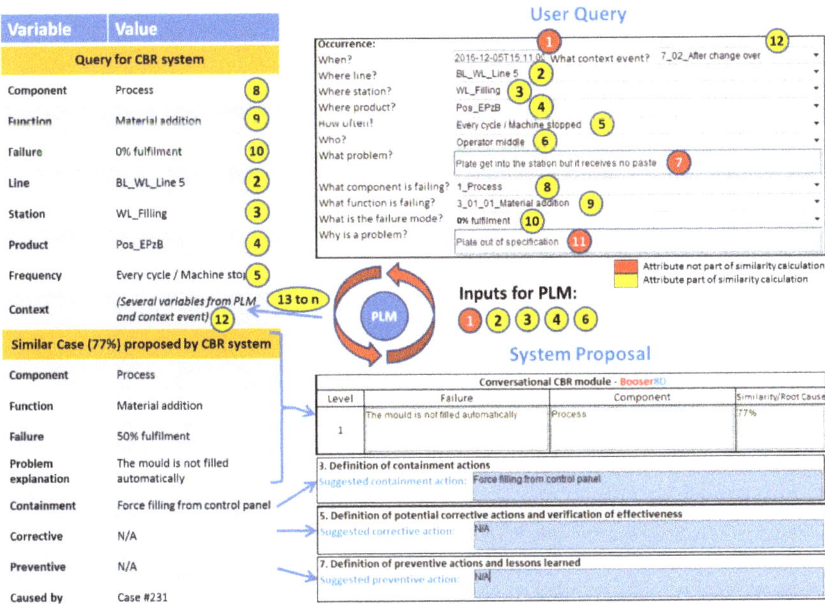

Figure 9. Information match between application user interface and the knowledge model.

The parameters delivered to the CBR system should match with the internal information model of myCBR [5], which is built on the general knowledge model of the system, and which is used to describe the reality under analysis. myCBR works with projects, which are the basic container of classes, attributes, a Similarity Measure Function (SMF), instances of the classes, and case bases. Each project can contain one or more classes, and each class contains attributes of different types. Each numerical attribute has to be configured with its range of valid values, and each taxonomy attribute type has to be configured with its list of valid values. Finally, each attribute receives an SMF.

SMFs are the functions associated with each attribute used by myCBR to calculate the final result, which will evaluate how similar is the received query in relation to each of the cases stored in the case base. Each attribute brings its own similarity contribution to the global similarity result, and the global similarity result is a pondered weight of all of them. Figure 11 shows the selected weights used in the case study of this work and the global formula applied in the project. Detailed information can be found in Camarillo et al. [5].

Query

```
<Query>
    User query
 1  <When>2016-12-05T15:11:00</When>
 2  <Whereline>BL_WL_Line 5</Whereline>
 3  <Wherestation>WL_Filling</Wherestation>
 4  <Whereproduct>Pos_EPzB</Whereproduct>
 5  <Howoften>8_Every cycle / Machine stopped</Howoften>
 6  <Who>Operator middle</Who>
 7  <Whatproblem>Plate get into the station but it receives no paste</Whatproblem>
 8  <Component>1_Process</Component>
 9  <Function>3_01_01_Material addition</Function>
10  <Failure>0% fulfilment</Failure>
11  <Whyproblem>Plate out of-specification</Whyproblem>
12  <Context_Event>7_02_After change over</Context_Event>
    PLM contribution to query
13  <2_1_Context Process>1_03_05_Filling</2_1_Context Process>
14  <2_Filling Pressure Max>...</2_Filling Pressure Max>
    ...
15  <2_2_Context Man>2_01_Operator</2_2_Context Man>
16  <2_Experience>9</2_Experience>
    ...
17  <2_3_Context Material>3_02_02_Assembly</2_3_Context Material>
18  <2_Material Brand>Internal</2_Material Brand>
    ...
19  <2_4_Context Machine>4_02_04_Modifying element</2_4_Context Machine>
20  <2_Machine Brand>Hadi</2_Machine Brand>
 n  ...
</Query>
```

Attribute not part of similarity calculation

Attribute part of similarity calculation

Figure 10. Example of query.

- 40% → Core problem description (i.e., component, function, and failure).
  8 9 10
- 40% → The parameters to define 'who' and 'how often', and the core context description (i.e., process, man, machine, material, environment, method, and event).
  5 6 12 13 15 17 19 21 22
- 10% → Line, station, and product.
  2 3 4
- 10% → The whole set of context technical parameters extracted from the PLM system.
  14 16 18 20

$$
\begin{aligned}
\text{Sim}(q,c) = {} & 0{,}4\cdot(\text{sim}_8(q_8,c_8)+\text{sim}_9(q_9,c_9)+\text{sim}_{10}(q_{10},c_{10}))/3+ \\
& 0{,}4\cdot(\text{sim}_5(q_5,c_5)+\text{sim}_6(q_6,c_6)+\text{sim}_{12}(q_{12},c_{12})+\text{sim}_{13}(q_{13},c_{13})+\text{sim}_{15}(q_{15},c_{15})+ \\
& \text{sim}_{17}(q_{17},c_{17})+\text{sim}_{19}(q_{19},c_{19})+\text{sim}_{21}(q_{21},c_{21})+\text{sim}_{22}(q_{22},c_{22}))/9+ \\
& 0{,}1\cdot(\text{sim}_2(q_2,c_2)+\text{sim}_3(q_3,c_3)+\text{sim}_4(q_4,c_4))/3+ \\
& 0{,}1\cdot(\text{sim}_{14}(q_{14},c_{14})+\text{sim}_{16}(q_{16},c_{16})+\text{sim}_{18}(q_{18},c_{18})+\text{sim}_{20}(q_{20},c_{20}))/4
\end{aligned}
$$

Figure 11. Weights for global similarity calculation.

As previously mentioned, myCBR works with numerical and taxonomy attributes. Numerical attributes (e.g., numbers 2, 3, 4, 5, 14, or 16 in Figure 10) contain any value between a defined minimum and maximum. Taxonomy attributes (e.g., numbers 8, 9, 12, 13, or 20 in Figure 10) contain values from a specific list of valid values. The Similarity Measure Function (SMF) of numerical attributes calculates the relative distance between the value at query and case of this attribute. For example, in the attribute number 16, "experience", if the query has the value 9 years, the case under analysis has the value 23 years, and the defined range for this attribute is from 0 to 45 years, the similarity contribution to the global value of this attribute would be:

$$sim(q,c) = 1 - |9 - 23| / |45 - 0| = 1 - 0.31 = 0.69.$$

The SMF of taxonomy attributes is calculated differently. The nodes of the taxonomy are associated with position factors, which are set based on experience and a trial-and-error process, until the results given by the system are satisfactory. The similarity value is calculated based on the value of the position factor of the common node related to both the query and the case. Figure 12 shows the example of the attribute "Function", which is noted with number 9 in the figures above. In this case, the upper node (i.e., "Function") was set with a position factor value of 0, the next one with value of 0.5, the following one with value of 0.75, and the lowest node with value of 1. To calculate the similarity, the SMF searches for the closest common node between query and case, which in this example is "Modify". This node has a position factor of 0.5, and therefore the similarity result of the example is 0.5 as is shown in the figure.

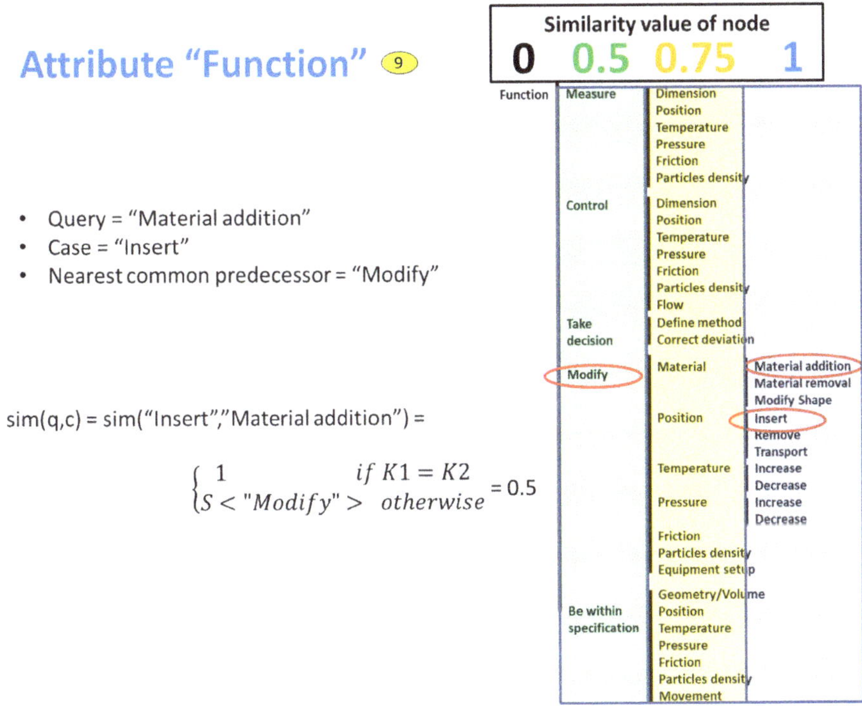

Figure 12. Example of individual similarity for taxonomy attributes.

A last relevant point related to similarity calculation is the range of supported manufacturing processes. In that sense, the developed system was designed to work for any kind of manufacturing process. Therefore, in the case of the context parameters, which are extracted from the PLM repository, a very wide range of possible parameter types should be expected. For some processes, a parameter of pressure might be relevant, but for other processes the relevant parameter could be a temperature. In the presented multi-agent architecture, each topic agent is specialized to the process where it is located. Therefore, when a topic agent receives a query, it will look for the parameters that it knows. Thus, all context parameters that are unknown by the topic agent will be ignored (i.e., no contribution in terms of similarity). It could also happen that not all of the parameters associated with its process are included in the query. In that case, the parameters that are not found will be assigned a null value, reducing the similarity result. In this way, cases from the same process will have always the highest similarity values followed by the cases from similar processes (i.e., processes with similar context parameters). The cases from very different processes will have the lowest similarity value.

Once the global similarity is calculated, the system displays the corresponding results. The bottom part of Figure 9 shows how the output in the system looks. In this case, the similarity value is 77%, and the proposal for the containment, corrective, and preventive actions is displayed to the user through the GUI.

The next section presents the details of the implementation of the developed system prototype in a case study company from the battery manufacturing sector.

6. Case Study in Battery Manufacturing

The developed system prototype was implemented as a case study in the company Exide Technologies, a global provider of stored electrical energy solutions (i.e., batteries and associated equipment and services) for transportation and industrial markets. Exide Technologies has several production plants running similar processes in Europe and the USA. For this reason, Exide could benefit from this research work.

For this case study, two plants of the company were selected: one located in Germany, and another one in Spain; both plants produce similar products with similar processes. They produce power batteries for the industrial market (i.e., forklifts or similar applications) with the process denominated Wet Filling to produce the positive plates. Gravity casting, with which the negative grids are produced, was also selected to test the performance of the system with a second manufacturing process.

The implementation of the case study can be divided in three phases: the definition of key PPR (Product-Process-Resource) attributes to ensure proper similarity calculation within the CBR systems, the collection of cases for the CBR case bases, and the test of the system on the shop floor.

As presented in Section 5, the similarity calculation is done based on a set of attributes that are selected out of the whole range of attributes that define the PPR reality of the case study. The selection criteria, and their individual weight in the similarity calculation, are related to the possible variation range of each attribute. For example, in a defined PPR scenario where the height of the product never changes across all existing part numbers, height would be a very bad candidate for similarity calculation, because it is not going to help to distinguish among cases. On the contrary, if the key differentiating characteristic of a range of products is the color, and there is a variety of colors, then color would be an excellent candidate for similarity calculation. To identify these key attributes for similarity, staff from production and engineering in the German plant of Exide were interviewed with a focus on the wet-filling process. The steps followed in these interviews were the identification of key PPR elements, the identification of the relationships among them, and finally the identification of their relevant attributes and their corresponding variability range. This information was used in the customization of both the PLM and CBR systems.

The next step in the case study was to collect enough cases to fill the CBR case bases of the topic agents. This activity was focused only on the wet filling production area of the German plant, leaving the Spanish plant for validation purposes. Four sources were used: the PFMEA of the wet-filling process, existing 8D reports coming from quality claims reported in the past, cases taken directly from the field in the German plant of the case study company, and some other cases from other manufacturing processes and companies, which helped also to test the capability and flexibility of the developed prototype. For the collection of cases directly on the shop floor, an unqualified individual without any kind of industrial background was engaged for two weeks, which helped to demonstrate that the proposed representation method for production problems is intuitive enough, and it is valid even for very unskilled operators. Table 1 shows a summary of all collected cases. In the table, the term "level" refers to any disruption that can be identified in production. The term "case" refers to a set of levels describing the whole chain of problems, from a visible one to the final root cause. The following example illustrates these concepts: a pump provides less pressure than defined, because the piston inside is worn out, because the lubrication oil has contamination particles, because the filter is broken. This example represents a single case with four levels.

Table 1. Case base of prototype application.

	Cases	Levels
Wet filling shop floor (German plant)	31	81
8D reports: Wet filling (German plant)	3	13
PFMEA: Wet filling (German plant)	16	60
Other processes/Other companies	16	72
TOTAL	**72**	**226**

For the last phase in the case study, several computers were installed in the selected production lines with access to the multi-agent architecture and to the PLM system. From these computers, the operators were able to access the problem-solving system to introduce their queries and receive recommendations from the system.

The group leaders of both plants were trained in the prototype system and they used it together with other operators to solve 10 problems occurring during their shifts. Since the cases in the prototype were expressed in English, the support of a translator was needed. The queries introduced by the users, the applicability of the results of the proposals coming out of the system, and the real solutions of the problems were all recorded for final analysis. The system was successfully tested, demonstrating the feasibility of the proposed approach [5].

7. Conclusions

This work has presented the details of the information and communication models of a developed system prototype to support a Manufacturing Problem Solving (MPS) process. The main contribution of the proposed system is that it integrates an MPS method with CBR on an agent-based distributed architecture and with a PLM system as a manufacturing context data repository. This novel approach had not been proposed in the reviewed literature.

The novelty from the modeling perspective resides in the information model created and implemented in a PLM system to facilitate the storage of and search for manufacturing context information that is used to calculate the similarity among production problems on the shop floor. The relationships among items in the system, and variables for the kind of explicit information related to each item, have been designed to facilitate the searching process associated with an MPS process, where the focus is placed on collecting as much contextual information as possible concerning the problem under investigation.

The approach is also an example of a low investment proposal that can be included in the conceptual frame of the technological vision Industry 4.0. Even if this case study is far from all advanced features envisaged today in the Smart Factory concept, it can be understood as a first small step to motivate some companies to start taking steps into Industry 4.0.

Author Contributions: Conceptualization, A.C., J.R., and K.-D.A.; Methodology, A.C.; Software, A.C.; Validation, A.C.; Supervision, J.R. and K.-D.A.; Writing (Original Draft Preparation), A.C. and J.R.; Writing (Review & Editing, J.R. and K.-D.A.

Funding: This research received no external funding.

Conflicts of Interest: The authors declare no conflict of interest.

References

1. Yang, S.; Bian, C.; Li, X.; Tan, L.; Tang, D. Optimized fault diagnosis based on FMEA-style CBR and BN for embedded software system. *Int. J. Adv. Manuf. Technol.* **2018**, *94*, 3441–3453. [CrossRef]
2. Ye, N.; Zhao, B.; Salvendy, G. Neural-networks-aided fault diagnosis in supervisory control of advanced manufacturing systems. *Int. J. Adv. Manuf. Technol.* **1993**, *8*, 200–209. [CrossRef]
3. Liu, D.R.; Ke, C.K. Knowledge support for problem-solving in a production process: A hybrid of knowledge discovery and case-based reasoning. *Expert Syst. Appl.* **2007**, *33*, 147–161. [CrossRef]
4. Lundgren, M.; Hedlind, M.; Kjellberg, T. Model driven manufacturing process design and managing quality. *Procedia CIRP* **2016**, *50*, 299–304. [CrossRef]
5. Camarillo, A.; Ríos, J.; Althoff, K.D. Knowledge-based multi-agent system for manufacturing problem solving process in production plants. *J. Manuf. Syst.* **2018**, *47*, 115–127. [CrossRef]
6. Bertin, A.; Noyes, D.; Clermont, P. Problem solving methods as Lessons Learned System instrumentation into a PLM tool. In Proceedings of the 14th IFAC Symposium on Information Control Problems in Manufacturing, Bucarest, Romania, 23–25 May 2012.
7. Stark, J. *Product Lifecycle Management*; Springer International Publishing: Cham, Switzerland, 2015.
8. Hermann, M.; Pentek, T.; Otto, B. *Design Principles for Industrie 4.0 Scenarios: A Literature Review*; Working Paper; Technische Universität Dortmund—Fakultät Maschinenbau—Audi Stiftungslehrstuhl Supply Net Order Management: Dortmund, Germany, 2015.
9. Wang, S.; Wan, J.; Li, D.; Zhang, C. Implementing smart factory of industrie 4.0: An outlook. *Int. J. Distrib. Sens. Netw.* **2016**, *12*, 3159805. [CrossRef]
10. Zuehlke, D. Smart Factory—Towards a factory-of-things. *Annu. Rev. Control* **2010**, *34*, 129–138. [CrossRef]
11. Witchalls, C.; Chambers, J. The Internet of Things Business Index: A Quiet Revolution Gathers Pace. The Economist Intelligence Unit. 2013. Available online: http://www.economistinsights.com/analysis/internet-things-business-index (accessed on 1 March 2018).
12. Lee, I.; Lee, K. The Internet of Things (IoT): Applications, investments, and challenges for enterprises. *Bus. Horiz.* **2015**, *58*, 431–440. [CrossRef]
13. Riesenberger, C.A.; Sousa, S.D. The 8D Methodology: An Effective Way to Reduce Recurrence of Customer Complaints. In Proceedings of the World Congress on Engineering, London, UK, 30 June–2 July 2010; Volume 3.
14. VDA. *Qualitätsmanagement in der Automobilindustrie—Qualitätsmanagement-Methoden Assessments*; Verband der Autoindustrie (VDA): Berlin, Germany, 2015.
15. Richter, M.M.; Weber, R.O. *Case-Based Reasoning: A Textbook*; Springer: Heidelberg, Germany, 2013.
16. Bach, K. Knowledge Acquisition for Case-Based Reasoning Systems. Ph.D. Thesis, University of Hildesheim, Hildesheim, Germany, 2012.
17. Camarillo, A.; Ríos, J.; Althoff, K.D. CBR and PLM applied to diagnosis and technical support during problem solving in the Continuous Improvement Process of manufacturing plants. *Procedia Manuf.* **2017**, *13*, 987–994. [CrossRef]
18. Camarillo, A.; Ríos, J.; Althoff, K.D. Agent Based Framework to support Manufacturing Problem Solving integrating Product Lifecycle Management and Case-Based Reasoning. In Proceedings of the IFIP 14th International Conference on Product Lifecycle Management, Seville, Spain, 10–12 July 2017; Springer: Cham, Switzerland; Volume 517, pp. 116–128.
19. Dittmann, L.; Rademacher, T.; Zelewski, S. Performing FMEA using ontologies. In Proceedings of the 18th International Workshop on Qualitative Reasoning, Evanston, IL, USA, 2–4 August 2004.
20. Chiu, C.; Chiu, N.H.; Hsu, C.I. Intelligent aircraft maintenance support system using genetic algorithms and case-based reasoning. *Int. J. Adv. Manuf. Technol.* **2004**, *24*, 440–446. [CrossRef]

21. Reuss, P.; Althoff, K.D.; Hundt, A.; Henkel, W.; Pfeiffer, M. Multi-Agent Case-Based Diagnosis in the Aircraft Domain. In Proceedings of the 23rd International Conference on Case-Based Reasoning Workshop, Frankfurt, Germany, 28–30 September 2015; pp. 43–52.
22. Kepner, C.H.; Tregoe, B.B. *The New Rational Manager—An Updated Edition for a New World*; Princeton Research Press: Princeton, NJ, USA, 2008.

© 2018 by the authors. Licensee MDPI, Basel, Switzerland. This article is an open access article distributed under the terms and conditions of the Creative Commons Attribution (CC BY) license (http://creativecommons.org/licenses/by/4.0/).

Article

Risk Management of Hazardous Materials in Manufacturing Processes: Links and Transitional Spaces between Occupational Accidents and Major Accidents

Francisco Brocal [1,*], Cristina González [2], Genserik Reniers [3,4], Valerio Cozzani [5] and Miguel A. Sebastián [2]

1 Department of Physics, Systems Engineering and Signal Theory, Escuela Politécnica Superior, Universidad de Alicante, Campus de Sant Vicent del Raspeig s/n, 03690 Sant Vicent del Raspeig, Alicante, Spain
2 Manufacturing and Construction Engineering Department, ETS de Ingenieros Industriales, Universidad Nacional de Educación a Distancia, Calle Juan del Rosal, 12, 28040 Madrid, Spain; cggaya@ind.uned.es (C.G.); msebastian@ind.uned.es (M.A.S.)
3 Faculty of Technology, Policy and Management, Safety and Security Science Group (S3G), TU Delft, 2628 BX Delft, The Netherlands; G.L.L.M.E.Reniers@tudelft.nl
4 Faculty of Applied Economics, Antwerp Research Group on Safety and Security (ARGoSS), University Antwerp, 2000 Antwerp, Belgium
5 Department of Civil, Chemical, Environmental, and Materials Engineering, Università di Bologna, Via Terracini, 28, 40131 Bologna, Italy; valerio.cozzani@unibo.it
* Correspondence: francisco.brocal@ua.es; Tel.: +34-96-590-9750

Received: 12 September 2018; Accepted: 5 October 2018; Published: 9 October 2018

Abstract: Manufacturing processes involving chemical agents are evolving at great speed. In this context, managing chemical risk is especially important towards preventing both occupational accidents and major accidents. Directive 89/391/EEC and Directive 2012/18/EU, respectively, are enforced in the European Union (EU) to this end. These directives may be further complemented by the recent ISO 45001:2018 standard regarding occupational health and safety management systems. These three management systems are closely related. However, scientific literature tackles the researching of these accidents independently. Thus, the main objective of this work is to identify and analyse the links and transitional spaces between the risk management of both types of accident. Among the results obtained, three transitional spaces can be pointed out which result from the intersection of the three systems mentioned. Similarly, the intersection of these spaces gives shape to a specific transitional space defined by the individual directives linked to Directive 89/391/EEC. These results are limited from a regulatory and technical perspective. Thus, the results are a starting point towards developing models that integrate the management systems studied.

Keywords: risk assessment; dangerous substance; Directive 89/391/EEC; Directive 2012/18/EU; ISO 45001:2018 standard; emerging risk; major accident; manufacturing; occupational accident; risk management

---

1. Introduction

In the past few years, manufacturing processes involving chemical substances have evolved to a great extent, from the new technologies applied to these processes to the new products and materials produced and their regulatory framework. The chemical industry is at the heart of the European Union (EU) manufacturing industry, representing approximately 7% of EU industrial production and a 1.1%

share of EU GDP. It supplies two-thirds of its production to other sectors within the manufacturing industry [1].

In regard to the regulatory framework in the EU, two relatively-recent regulations stand out at first glance. Regulation (EC) No 1907/2006 on Registration, Evaluation, Authorisation and Restriction of Chemicals (REACH Regulation), aims to protect human health and the environment by ensuring greater safety in the production and use of chemical substances. The REACH Regulation, which entered into force in 2007, applies to all chemical substances and thus has an impact on many businesses [2]. The classification and labelling of hazardous chemicals is governed by Regulation (EC) No 1272/2008 on classification, labelling and packaging of substances and mixtures (CLP Regulation). The CLP Regulation entered into force in 2009 and it enables identification of dangerous substances by means of classification and labelling, and informing users about their hazards through standard symbols and phrases [3].

In 2018, approximately 145,000 substances were classified according to the CLP Regulation [4]. Also in 2018, more than 21,000 substances were registered in the European Economic Area under the REACH Regulation, of which more than 12,000 are used in manufacturing processes. Figure 1 shows the evolution in the number of chemical substances registered in the European Economic Area under REACH [5] starting in January 2009, when records first exist, until August 2018. To this end, taking into account the uses and exposure to these substances, a distinction is made between their global life cycle and the life cycle applied to manufacturing.

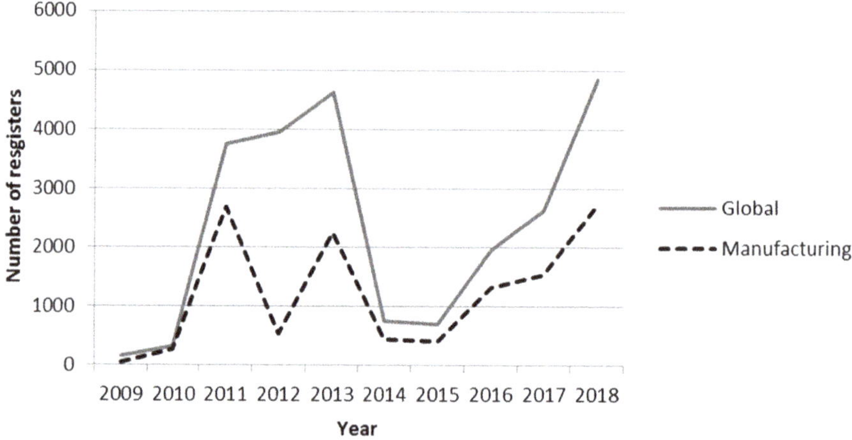

Figure 1. Evolution in the number of chemical substances registered in the European Economic Area under the Registration, Evaluation, Authorisation and Restriction of Chemicals (REACH) regulation. January 2009–August 2018 [5].

This context of change generates both opportunities and challenges in many fields of knowledge. Among these fields, risk management is particularly important, both from a systemic point of view and from more specific perspectives. Among these specific perspectives, chemical risk management should be noted in order to prevent both occupational accidents and major accidents in manufacturing environments where hazardous materials are used. For this, two separate and solid legislative frameworks exist in the EU.

In the case of the risk management of occupational accidents, Directive 89/391/EEC on the introduction of measures to encourage improvements in occupational safety and health (OSH) should be pointed out [6]. This directive has been developed through a broad set of specific directives, with Directive 98/24/EC on OSH in the field of chemical agents [7] standing out in the context of hazardous materials.

In the case of the risk management of major accidents and hazards involving dangerous substances, Directive 2012/18/EU is applicable [8]. This directive states that its provisions should be applied without prejudice to the provisions of Union law relating to OSH and the working environment, and, in particular, without prejudice to Council Directive 89/391/EEC [6].

Hence, these risk management frameworks are closely related [9] and, as a result, are open to being studied from an integrative point of view. However, despite the importance of this relationship, scientific literature addresses the study of risk management of occupational accidents and major accidents involving dangerous materials in an independent and practically excluding manner.

In this regard, using the ScienceDirect database [10], the search for the keyword 'Directive 2012/18/EU' anywhere in an article, returns 13,584 papers. Similarly, the keyword 'Directive 89/391/EEC' returns 473 papers. However, combining both keywords with the AND operator returns three articles, which are: Rasmussen et al. [11], Besserman and Mentzer [12], and Li and Guldenmund [13].

This circumstance defines a closed border between these management systems, making it difficult to develop integrative techniques and methodologies that favour the reduction of risk due to hazardous substances. In addition to the described relationship between directives, the recent opportunity defined by the ISO 45001:2018 standard on OSH management systems [14] should be pointed out.

Thus, the main objective of this study is to identify and analyze the links and transitional spaces between the risk management of occupational accidents and major accidents involving hazardous materials. To achieve this goal, the methodology followed by this study is based on researching and carrying out a comparative analysis of the legal and standardised context described above. To this end, this work is organised as follows: (1) study of the legal context within the framework of the EU; (2) comparative analysis between management systems derived from Directive 89/391/EEC [6] and Directive 2012/18/EU [8]; (3) comparative analysis between management systems derived from the ISO 45001:2018 standard [14] and Directive 2012/18/EU [8]; (4) analysis of transitional spaces between risk management of hazardous materials in manufacturing processes; (5) discussion of results; and (6) conclusions.

2. Legal Context in the European Union (EU)

Directive 2012/18/EU (hereinafter, Directive Seveso III) is applicable in regard to the management of major accidents and hazards involving dangerous substances. This directive states that its provisions should be applied without prejudice to the provisions of Union law relating to OSH and the working environment, and, in particular, without prejudice to Council Directive 89/391/EEC.

The object of Directive 89/391/EEC (hereinafter, Framework Directive) is to introduce measures to encourage improvements in OSH. To this end, it contains general principles concerning the prevention of occupational risks, the protection of safety and health, the elimination of risk and accident factors, the informing, consultation, balanced participation in accordance with national laws and/or practices and training of workers and their representatives, as well as general guidelines for the implementation of said principles.

The Framework Directive serves as basis for more specific directives covering all the risks connected with safety and health in the workplace. Thus, 20 specific directives have been enacted to date since 1989, as listed in Table 1. Thus, Directive Seveso III is applicable to serious accidents and the Framework Directive to occupational risk prevention in a wide sense: that is, considering prevention of occupational accidents, illnesses and other dangers to the safety and health of workers.

Considering that the objectives of this study are linked to major accidents and occupational accidents, both concepts will be defined below. Then, the main relationships that exist between the directives involved in the legal context set out here will be analysed.

Table 1. Individual directives within the meaning of Article 16 (1) of Directive 89/391/EEC on occupational safety and health (OSH).

Nr	Individual Directives	Topic	Year (First Publication)
1	Directive 89/654/EEC [15]	Workplace	1989
2	Directive 2009/104/EC [16]	Work equipment	1989
3	Directive 89/656/EEC [17]	Personal protective equipment	1989
4	Directive 90/269/EEC [18]	Manual handling of loads	1990
5	Directive 90/270/EEC [19]	Display screen equipment	1990
6	Directive 2004/37/EC [20]	Carcinogens or mutagens at work	1990
7	Directive 2000/54/EC [21]	Biological agents at work	1990
8	Directive 92/57/EEC [22]	Temporary or mobile construction sites	1992
9	Directive 92/58/EEC [23]	Safety and/or health signs	1992
10	Directive 92/85/EEC [24]	Pregnant workers	1992
11	Directive 92/91/EEC [25]	Mineral-extracting industries; drilling	1992
12	Directive 92/104/EEC [26]	Mineral-extracting industries	1992
13	Directive 93/103/EC [27]	Work on board fishing vessels	1993
14	Directive 98/24/EC [7]	Risks related to chemical agents at work	1998
15	Directive 99/92/EC [28]	Risks from explosive atmospheres	1999
16	Directive 2002/44/EC [29]	Vibration	2002
17	Directive 2003/10/EC [30]	Noise	2003
18	Directive 2004/40/EC [31]	Electromagnetic fields	2004
19	Directive 2006/25/EC [32]	Artificial optical radiation	2006
20	Directive 2013/35/EU [33]	Electromagnetic fields	2013

2.1. Major Accident and Occupational Accident

Directive Seveso III defines 'major accident' as an occurrence such as a major emission, fire, or explosion resulting from uncontrolled developments in the course of the operation of any establishment covered by this directive, and leading to serious danger to human health or the environment, immediate or delayed, inside or outside the establishment, and involving one or more dangerous substances.

The Framework Directive and its specific directives do not define occupational accidents. To this end, others sources are required. For example, Eurostat defines an occupational accident (or accident at work) as a discrete occurrence during the course of work which leads to physical or mental harm [34]. The ISO 45001:2018 standard defines 'incident' as an occurrence arising out of, or in the course of, work that could or does result in injury and ill health. An incident where injury and ill health occurs is sometimes referred to as an 'accident' [14].

Comparing the definitions for major accidents and occupational accidents, it could be said that a major accident could also be considered an occupational accident whenever there is harm to workers (injury and ill health). Among the specific directives developed following the Framework Directive, Directive 98/24/EC on chemical agents [7] is the one which, in principle, is more closely linked to Directive Seveso III. Taking this directive into account, it could be said that an occupational accident involving chemical substances is an occurrence arising out of, or in the course of, work that could or does result in injury and ill health for workers. All of this is regardless of the level of severity or seriousness of damage.

In any case, there are other directives besides Directive 98/24/EC on chemical agents [7] that are closely linked to Directive Seveso III. In order to identify these directives, the definition of major accident included in Directive Seveso III will be taken into account, considering to this end the occurrence of a major emission, fire, or explosion.

These events are covered by the industrial safety technologies (IST) studied by Sebastián and Brocal [35], which can be defined as follows: a set of instruments and industrial processes that enable the practical use in analysis, evaluation and control of specific risks being able to be classified into: work equipment; places and workplaces; handling, storage and transport; electricity; fires; and chemicals.

Sebastián and Brocal [35] studied the relationship that exists between IST and specific, applicable directives, the (adapted) results of which are included in Table 2. Considering this result, a preliminary approximation is then offered in regard to the technical–legal relationship that exists between the individual directives and Directive Seveso III.

Table 2. Industrial safety technologies (IST) linked to individual directives within the meaning of Article 16 (1) of Directive 89/391/EEC (adapted from [35]).

CLASSIFICATION OF IST	DIRECTIVE 89/654/EEC WORKPLACE [15]	DIRECTIVE 2009/104/EC WORK EQUIPMENT [16]	DIRECTIVE 773/1997 PERSONAL PRO. EQUIPMENT [17]	DIRECTIVE 90/269/EEC MANUAL HANDLING LOADS [18]	DIRECTIVES 2004/37/EC AND 98/24/EC CHEMICAL [7]	DIRECTIVE 92/58/EEC SIGNAL. [9]	DIRECTIVE 99/92/EC EXPLOSIVE ATMOS. [28]
WORK EQUIPMENT	○	•	○	–	•	○	•
PLACES AND WORKPLACES	•	○	○	○	•	○	•
HANDLING, STORAGE AND TRANSPORT	•	•	○	•	•	○	•
ELECTRICITY	•	•	○	–	•	○	•
FIRES	•	•	○	–	•	○	•
CHEMICALS	•	•	○	○	•	○	•

• Direct link. ○ Cross link.

2.2. *Activity Involving Chemical Agents*

Directive 98/24/EC defines 'Chemical agent' [7] as any chemical element or compound, on its own or admixed, as it occurs in the natural state or as produced, used or released, including release as waste, by any work activity, whether or not produced intentionally and whether or not placed on the market. In addition, 'hazardous chemical agent' means: (a) any chemical agent which meets the criteria for classification as hazardous within any physical and/or health hazard classes laid down in the CLP Regulation, whether or not that chemical agent is classified under that regulation; (b) any chemical agent which, whilst not meeting the criteria for classification as hazardous in accordance with point (a) may, because of its physicochemical, chemical or toxicological properties and the way it is used or is present in the workplace, present a risk to the safety and health of workers, including any chemical agent that is assigned an occupational exposure limit value under Article 3 of this directive. Directive Seveso III defines 'dangerous substance' as a substance or mixture covered by Part 1 or listed in Part 2 of its Annex I, including in the form of a raw material, product, by-product, residue or intermediate.

The main difference observed on comparing the definitions of 'dangerous substance' (Seveso III) and 'hazardous chemical agent' (Directive 98/24/EC [7]), lies in the former referencing specific substances and amounts (Annex I) while the latter has a much broader and general definition, considering any substance that could give rise to an occupational risk, regardless of whether they meet the classification criteria laid down in the CLP Regulation [3].

Thus, Directive 98/24/EC [7] will be applicable to any work with dangerous substances according to Seveso III, given that these dangerous substances will also be hazardous chemical agents.

Additionally, Directive 98/24/EC defines 'Activity involving chemical agents' as any work in which chemical agents are used, or are intended to be used, in any process, including production, handling, storage, transport or disposal and treatment, or which result from such work [7]. Directive Seveso III defines 'presence of dangerous substances' as the actual or anticipated presence of dangerous substances in the establishment, or of dangerous substances which it is reasonable to foresee may be generated during loss of control of the processes, including storage activities, in any installation within the establishment, in quantities equal to or exceeding the qualifying quantities set out in Part 1 or Part 2 of its Annex I.

The main difference observed in regard to criteria regarding the presence of dangerous substances again lies in the variables collected in Annex I of Directive Seveso III. Thus, Directive 98/24/EC [7] will be applicable to any occupational activity involving dangerous substances included in Seveso III.

Directive 98/24/EC on chemical agents [7] is complemented by Directive 2004/37/EC on carcinogens or mutagens at work [20], as collected in Table 2. The CLP Regulation [3] may be considered the connection point between both directives. Similarly, this regulation is also linked closely to Directive Seveso III.

With the aim of establishing these links, Table 3 collects 7 dangerous substances. These 7 substances have been selected as follows: (a) of the 48 substances collected in Annex I, Part 2 of Seveso III, those with a CAS number have been selected, amounting to 35; (b) for each of these 35 substances with a CAS number, the INFOCARQUIM database [36] has been used to determine those which are carcinogens or mutagens (1A/1B) according to the CLP Regulation; (c) from the 35 substances above, 7 substances have been identified as carcinogens or mutagens; (d) for each of these 7 substances, their H statements have been identified with examples of manufacturing processes, also by using the INFOCARQUIM database; (e) for the 7 substances above, their threshold limit values (VLA, *Valor Límite Ambiental*) have been identified according to the document on the limits of chemical agents for professional exposure in Spain [37].

VLA are reference values for chemical agent concentration in the air and represent the conditions for which it is believed that, based on current knowledge, most workers may be exposed to on a daily basis throughout their work life without suffering adverse effects on their health [37].

Thus, the 7 substances mentioned will be subject to the implementation of Directive Seveso III according to the figure collected in columns 2 and 3. Additionally, these substances fall within the scope of Directive 98/24/EC [7] and Directive 2004/37/EC [6] whenever they are found in workplaces. When these substances may be inhaled by workers, applicable threshold limit values must be considered. For Spain, these values are collected in Table 3.

Although not contained in Table 3, the applicability of other directives and regulations should be analysed for every manufacturing process that is studied; for instance, the directives collected in Table 1. To this end, one of the essential sources of information is the H statement collected in Table 3.

Table 3. Carcinogenic or mutagenic substances from Annex I, Part 2 of Directive Seveso III. Environmental value limits and related manufacturing processes.

DIRECTIVE SEVESO III: PART 2 [8]				**CLP Regulation [36]**		**Examples of Manufacturing Processes [36]**	**VLA 2018 [37]**	
Column 1		Column 2	Column 3					
Dangerous substances	CAS number	Qualifying quantity (tonnes) for the application of		Carcinogens or mutagens	Phrases H (Hazard)		VLA-ED ® (Reference value for Daily Exposure)	VLA-EC ® (Reference value for Short-term Exposure)
		Lower-tier requirements	Upper-tier requirements					
Arsenic pentoxide, arsenic (V) acid and/or salts	1303-28-2	1	2	Carc. 1A	H301,H331, H350, H410	Special glass manufacturing (light bulbs and tubes, optic glass, glass for liquid-crystal displays (LCD), etc.)	0.01 mg/m^3	-
Arsenic trioxide, arsenious (III) acid and/or salts	1327-53-3	-	0,1	Carc. 1A	H300,H314, H350, H410	Manufacturing of low-melting glass and manufacturing of electronic components	0.01 mg/m^3	-
Ethyleneimine	151-56-4	10	20	Carc. 1B, Muta. 1B	H225, H300, H310, H314, H330, H340, H350, H411	Manufacturing of ethyleneimine and aziridine polymers	0.2 ppm; 0.36 mg/m^3	-
Formaldehyde (concentration $\geq$ 90%)	50-00-0	5	50	Carc. 1B	H301, H311,H314, H317, H331, H341, H350	Metal cutting, machining and grinding processes	0.3 ppm; 0.37 mg/m^3	0.6 ppm; 0.74 mg/m^3
Ethylene oxide	75-21-8	5	50	Carc. 1B, Muta. 1B	H220,H315,H319, H331, H340, H350	Glycol (ethylene glycol), polyglycol and polyol synthesis used in manufacturing fibres, coolants and foams	1 ppm; 1.8 mg/m^3	-
Propylene oxide	75-56-9	5	50	Carc. 1B, Muta. 1B	H224, H302,H311, H319, H331, H335, H340, H350	Manufacturing of flexible or rigid polyurethane foams	2 ppm; 4.8 mg/m^3	-
4, 4′-Methylene bis (2-chloraniline) and/or salts, in powder form	101-14-4	-	0,01	Carc. 1B	H302, H350, H410	Manufacturing of jet engine turbine blades, radar systems and home appliances	0.01 ppm; 0.1 mg/m^3	-

2.3. Workplace

Directive 89/654/EEC [15] lays down minimum requirements for safety and health at the workplace. For the purposes of this directive, 'workplace' means the place intended to house workstations on the premises of the undertaking and/or establishment and any other place within the area of the undertaking and/or establishment to which the worker has access in the course of his employment.

On the other hand, Directive Seveso III shall apply to establishments. Establishment means the whole location under the control of an operator where dangerous substances are present in one or more installations, including common or related infrastructures or activities; establishments are either lower-tier establishments or upper-tier establishments. Thus, establishments under Seveso III are also workplaces under the definition of Directive 89/654/EEC [15].

2.4. Installation

Directive Seveso III defines 'installation' as a technical unit within an establishment and whether at or below ground level, in which dangerous substances are produced, used, handled or stored; it includes all the equipment, structures, pipework, machinery, tools, private railway sidings, docks, unloading quays serving the installation, jetties, warehouses or similar structures, floating or otherwise, necessary for the operation of that installation.

Also, Directive 2009/104/EC defines 'work equipment' as follows: any machine, apparatus, tool or installation used at work [16]. Installations considered to be work equipment are for example: surface treatment installations, painting installations, installations composed of a combination of machines that work interdependently, etc. [38]. As for general service or protection installations, such as electrical installations, gas or fire protection, annexed to the workplace, that are considered as an integral part thereof, then Directive 89/654/EEC on workplaces is applicable [38].

As a result, the concept of installation, as defined by Directive Seveso III, may be considered to be part of the scope of Directive 2009/104/EC on work equipment [16], as well as Directive 89/654/EEC on the workplace [15]. More specifically, Directive 98/24/EC [7] indicates that work equipment and protective systems provided by the employer for the protection of workers shall comply with the relevant EU provisions on design, manufacture and supply with respect to health and safety. Likewise, the employer shall take measures to provide sufficient control of plant, equipment and machinery or provision of explosion suppression equipment or explosion pressure relief arrangements.

Furthermore, Directive 98/24/EC [7] specifically points out the need to adopt measures in view of explosions linked to work equipment and installations. Thus, a direct link is established between Directive 99/92/EC [28] where 'explosive atmosphere' means a mixture with air, under atmospheric conditions, of flammable substances in the form of gases, vapours, mists or dust in which, after ignition has occurred, combustion spreads to the entire unburned mixture. However, for example this directive shall not apply to the manufacture, handling, use, storage and transport of explosives or chemically unstable substances.

3. Comparative Analysis between Management Systems Derived from the Framework Directive and Directive Seveso III

Directive Seveso III indicates that member states shall require the operator to draw up a document in writing setting out the major-accident prevention policy (MAPP) and to ensure that it is properly implemented. The MAPP shall be implemented by appropriate means, structures and by a safety management system, in accordance with Annex III of this directive, and it will be proportionate to the major-accident hazards, and the complexity of the organization or the activities of the establishment. Table 4 shows the structure of the safety management system according to such Annex III. This Annex III is linked, in turn, to Annex II regarding minimum data and information to be considered in the safety report referred to in Article 10 of Directive Seveso III.

Table 4. Correspondence between structures of risk management systems: Directive Seveso III, Framework Directive and ISO 45001:2018 Standard.

	Directive Seveso III: Safety management system–Issues [8]	Framework Directive–Articles [6]	ISO 45001:2018 Standard–Sections [14]
(i) Organization and personnel	The roles and responsibilities of personnel involved in the management of major hazards at all levels in the organization, together with the measures taken to raise awareness of the need for continuous improvement.	• Article 5. General provision • Article 6. General obligations on employers • Article 7. Protective and preventive services • Article 11. Consultation and participation of workers • Article 13. Workers' obligations	• Section 5.3 Organizational roles, responsibilities and authorities • Section 10.3 Continual improvement
	The identification of training needs of such personnel and the provision of the training so identified.	• Article 6 (6.3.d). General obligations on employers • Article 12. Training of workers	• Section 5.4 (e.4) Consultation and participation of workers
	The involvement of employees and of subcontracted personnel working in the establishment which are important from the point of view of safety.	• Article 6 (6.4). General obligations on employers	• Section 8.1.4.3 Outsourcing
(ii) Identification and evaluation of major hazards	Adoption and implementation of procedures for systematically identifying major hazards arising from normal and abnormal operation including subcontracted activities where applicable and the assessment of their likelihood and severity.	• Article 6 (6.3.a). General obligations on employers • Article 9 (9.1.a). Various obligations on employers	• Section 6.1.2 Hazard identification and assessment of risks and opportunities
(iii) Operational control	Adoption and implementation of procedures and instructions for safe operation, including maintenance, of plant, processes and equipment, and for alarm management and temporary stoppages.	• Article 6 (6.2). General obligations on employers • Article 9 (9.1.b). Various obligations on employers • Article 10. Worker information	• Section 8.1.1 Operation planning and control • Section 8.1.2 Eliminating hazards and reducing occupational safety and health (OSH) risks
	Taking into account available information on best practices for monitoring and control, with a view to reducing the risk of system failure.	• Article 9 (9.1.c). Various obligations on employers • Article 10. Worker information	
	Management and control of the risks associated with ageing equipment installed in the establishment and corrosion: inventory of the establishment's equipment, strategy and methodology for monitoring and control of the condition of the equipment; appropriate follow-up actions and any necessary countermeasures.	• –	• –
(iv) Management of change	Adoption and implementation of procedures for planning modifications to, or the design of new installations, processes or storage facilities.	• Article 6 (6.1 and 6.3.d). General obligations on employers	• Section 8.1.3 Managing of change
(v) Planning for emergencies	Adoption and implementation of procedures to identify foreseeable emergencies by systematic analysis, to prepare, test and review emergency plans to respond to such emergencies and to provide specific training for the staff concerned. Such training shall be given to all personnel working in the establishment, including relevant subcontracted personnel.	• Article 8. First aid, fire-fighting and evacuation of workers, serious and imminent danger	• Section 8.2 Emergency preparedness and response
(vi) Monitoring performance	Adoption and implementation of procedures for the ongoing assessment of compliance with the objectives set by the operator's major-accident prevention policy (MAPP) and safety management system, and the mechanisms for investigation and taking corrective action in case of non-compliance. The procedures shall cover the operator's system for reporting major accidents or 'near misses', particularly those involving failure of protective measures, and their investigation and follow-up on the basis of lessons learnt. The procedures could also include performance indicators such as safety performance indicators (SPIs) and/or other relevant indicators.	• Article 9 (9.1.c and 9.1.c). Various obligations on employers	• Section 9.1 Monitoring, measurement, analysis and performance evaluation • Section 10.2 Incident, nonconformity and corrective action
(vii) Audit and review	Adoption and implementation of procedures for periodic systematic assessment of the MAPP and the effectiveness and suitability of the safety management system.The documented review of performance of the policy and safety management system and its updating by senior management, including consideration and incorporation of necessary changes indicated by the audit and review.	• –	• Section 9.2 Internal audit

The Framework Directive does not explicitly develop a safety management system. However, it contains general principles concerning the prevention of occupational risks, the protection of safety and health, the elimination of risk and accident factors, the informing, consultation, balanced participation in accordance with national laws and/ or practices and training of workers and their representatives, as well as general guidelines for the implementation of the said principles.

In themselves, the aforementioned general principles form the basis for a management system; in this case, a system to manage the safety and health of workers. To develop this, each member state of the European Union must transpose the Framework Directive into their national legal systems.

By way of example, Law 31/1995 on occupational risk prevention transposes the Framework Directive into Spanish law [39]. This Law explicitly states that occupational risk prevention must be integrated into an undertaking's general management system, across all of the activities and across the hierarchy thereof, by means of implementing and putting into practice an occupational risk prevention plan. This occupational risk prevention plan should include the organisational structure, responsibility, roles, practices, procedures, processes and resources necessary to carry out risk prevention activity in the undertaking. The management and implementation instruments that are essential to the risk prevention plan are: the assessment of occupational risk and the planning of preventive action.

Also, Royal Decree 39/1997, which validates the Regulation on Prevention Services, develops those aspects that make it possible to integrate occupational risk prevention management into the undertaking's activities and across the hierarchical levels thereof, based on a plan that includes work techniques, organisation and conditions [40]. Thus, as shown in Table 4, correspondence may be established between the structure of the Framework Directive and the management system included in Annex III of Directive Seveso III.

Correspondence may be classified as strong, weak or non-existent. Correspondence is strong when an issue in Annex III has its direct equivalence (major risk management vs. occupational risk management) with one or more articles of the Framework Directive. Correspondence is weak when an issue only has partial equivalence. And, obviously, correspondence is non-existent whet there is no equivalence.

Considering this, there is strong correspondence in regard to the following sections of Annex III: (i) Organization and personnel, (ii) Identification and evaluation of major hazards, (iv) Management of change, and (v) Planning for emergencies.

Correspondence is weak in sections (iii) Operational control and (vi) Monitoring performance. In regard to section (iii), the Framework Directive does not explicitly state: the management and control of risks associated with ageing equipment installed in the establishment and corrosion. In regard to section (vi), the Framework Directive does not explicitly state: the adoption and implementation of procedures for the ongoing assessment of compliance with the objectives set by the operator's MAPP and safety management system; near misses. There is no (explicit) correspondence in regard to section (vii) regarding Audit and review.

Furthermore, the Framework Directive contains specific guidelines in regard to chemical agents. These guidelines may be understood in terms of the relevance of the risk of such agents on occupational risk overall. Specifically, article 6.2 of the directive states as one of the general principals of preventive activity: replacing the dangerous by the non-dangerous or the less dangerous. Also, article 6.3.a: evaluate the risks to the safety and health of workers, inter alia in the choice of work equipment, the chemical substances or preparations used, and the fitting-out of workplaces. In order to carry out such assessment regarding chemical substances, Directive 98/24/EC [7] and Directive 2004/37/EC [6], on chemical agents and on carcinogens and mutagens at work, respectively, are essential.

Considering that these directives are part of the development of the Framework Directive, they share with it the general principles of risk management. As a result, the minimum provisions for the protection of workers contained in each directive must be managed under the said general management principles. Thus, a correspondence may be observed between the said principles and provisions, from the general to the specific.

Such correspondence makes it possible, in turn, to relate the sections of Annex III of Directive Seveso III with the specific provisions of each directive. This correspondence with the aforementioned directives may be seen in Table 5. The level of detail is lower than that found in Table 4, as in this case, the aim is to offer a global overview of the common thread described, from the general (Table 4) to the specific (Table 5).

Table 5. Correspondence between structures of risk management systems of Directive Seveso III, Directive 98/24/CE on chemical agents and Directive 2004/37/EC on carcinogens or mutagens.

Directive Seveso III [8]	Directive 98/24/CE on chemical agents at work [7]	Directive 2004/37/EC on carcinogens or mutagens at work [20]
(i) Organization and personnel	• Article 8. Information and training for workers	• Article 11. Information and training of workers
(ii) Identification and evaluation of major hazards	• Article 4. Determination and assessment of risk of hazardous chemical agents	• Article 3. Scope–determination and assessment of risks
(iii) Operational control	• Article 5. General principles for prevention of risks associated with hazardous chemical agents and application of this Directive in relation to assessment of risks • Article 6. Specific protection and prevention measures	• Article 4. Reduction and replacement • Article 5. Prevention and reduction of exposure • Article 8. Foreseeable exposure • Article 9. Access to risk areas • Article 10. Hygiene and individual protection
(iv) Management of change	• Article 4. (4.2 and 4.5). Determination and assessment of risk of hazardous chemical agents • Article 8. Information and training for workers	• Article 3. Scope–determination and assessment of risks • Article 11. Information and training of workers
(v) Planning for emergencies (vi) Monitoring performance	• Article 7. Arrangements to deal with accidents, incidents and emergencies	• Article 5. Prevention and reduction of exposure • Article 7. Unforeseen exposure
(vii) Audit and review	• Article 4. Determination and assessment of risk of hazardous chemical agents	• Article 3. Scope–determination and assessment of risks

Thus, comparing the results of Tables 4 and 5 offers the following main learnings: (a) section (i) Organization and personnel of Directive Seveso III corresponds closely with the Framework Directive, while this correspondence is weaker with the individual directives analysed, limited mainly to training requirements; (b) section (ii) to (vi) of Directive Seveso III correspond closely with the individual directives. This correspondence is the result of going from the general in the Framework Directive to the specific on issues regarding chemical agents and carcinogens and mutagens pursuant to Directive 98/24/EC [7] and Directive 2004/37/EC [20], respectively, (c) in regard to section (vii) on Audit and review, which has no correspondence with the Framework Directive, reviewing the risk assessment whenever necessary is explicitly considered, such as when working conditions change, new scientific knowledge is achieved on the effects of chemical agents or limit values, etc.

4. Comparative Analysis between Management Systems Derived from the ISO 45001:2018 Standard and Directive Seveso III

Implementing an OSH management system conforming to the ISO 45001:2018 standard enables an organization to manage its OSH risks and improve its OSH performance [14]. According to this voluntary standard, an OSH management system can assist an organization to fulfil its legal and other requirements. The implementation and maintenance of an OSH management system, its effectiveness and its ability to achieve its intended outcomes are dependent on a number of key factors which can include the integration of the OSH management system into the organization's business processes and compliance with its legal and other requirements.

As indicated by the ISO 45001:2018 standard, its adoption in a given organization, however, will not in itself guarantee prevention of work-related injury and ill health to workers, provision of safe and healthy workplaces and improvement of OSH performance [14]. So, in the context of this work, that is from the perspective of a risk management system, the adoption or implementation of the ISO 45001:2018 standard [14] may be understood as a tool that may assist an organisation in complying with the Framework Directive and Directive Seveso III.

As with the previous section, which carried out a comparative analysis between management systems derived from the Framework Directive and Directive Seveso III, this section contains a similar analysis, using the same criteria, comparing between management systems derived from the ISO 45001:2018 standard [14] and Directive Seveso III.

Hence, as Table 5 also shows, correspondence may be established between the structure of the ISO 45001:2018 standard [14] and the management system included in Annex III of Directive Seveso III. Consequently, the result obtained makes it possible to establish a similar correspondence between the said standard and the Framework Directive, since Directive Seveso III works as a common denominator.

The results from such analysis establish a strong correspondence with all sections from Annex III except with section (iii) Operational control, where aspects related to management and control of the risks associated with ageing equipment installed in the establishment and corrosion are not explicitly stated by the ISO 45001:2018 standard [14]. This is also true in regard to the Framework Directive, Directive 98/24/CE on chemical agents [7] and Directive 2004/37/EC on carcinogens or mutagens [20].

As a result of this correspondence, the ISO 45001:2018 standard [14] is linked closely, not only to Directive Seveso III but also to the Framework Directive and its individual directives, especially Directive 98/24/EC [7] and Directive 2004/37/EC [20], on chemical agents and carcinogens or mutagens at work, respectively.

On the other hand, the OSH management system approach defined by the ISO 45001:2018 standard [14] is founded on the concept of Plan-Do-Check-Act (PDCA). This standard indicates that it can be applied to a management system and to each of its individual elements, as follows:

- Plan: determine and assess OSH risks, OSH opportunities and other risks and other opportunities, establish the OSH objectives and processes necessary to deliver results in accordance with the organization's OSH policy;
- Do: implement the processes as planned;
- Check: monitor and measure activities and processes with regard to the OSH policy and OSH objectives, and report the results;
- Act: take actions to continually improve OSH performance towards achieving the intended outcomes;

The ISO 45001:2018 standard incorporates the PDCA concept into a new framework [14]. This framework can be integrated with the framework defined by Seveso III, considering the results shown in Table 4. The result of such integration is shown in Figure 2.

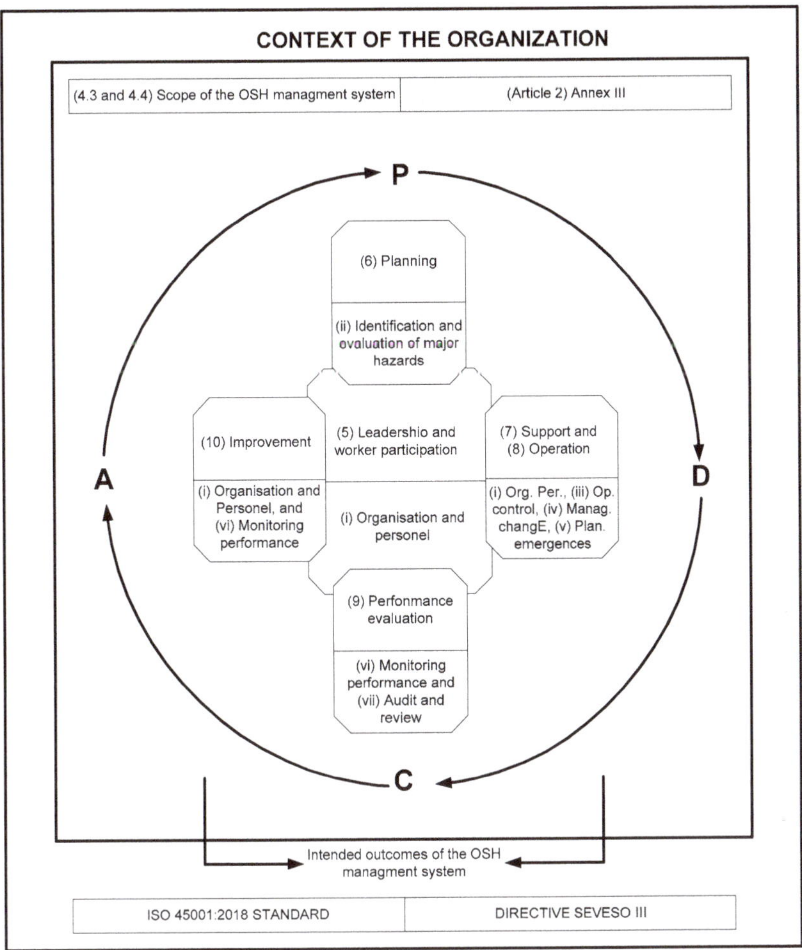

Figure 2. Integration of the Plan-Do-Check-Act (PDCA) framework defined by the ISO 45001:2018 standard [14] with the management system derived from Directive Seveso III [8].

5. Analysis of Transitional Spaces Between the Risk Management of Hazardous Materials in Manufacturing Processes

An analysis is carried out below of the main transitional spaces that arise from the results obtained in the previous sections. Each of these transitional spaces may be understood as the intersection between the correspondences identified between the different sets analysed: that is, the legal context and management systems derived from Directive Seveso III, the Framework Directive and the ISO 45001:2018 standard [14].

5.1. Legal Context

As a result of the analysis carried out in the legal context section, Figure 3 shows the main correspondence observed between the directives studied; that is, between Directive Seveso III and the individual directives within the meaning of Article 16 (1) of the Framework Directive.

To show these links summarily, the same thread has been followed that configures the structure of the legal context section. To this end, the basic outline of a general manufacturing process has been considered, involving chemical agents (Directive 98/24/EC [7] and Directive 2004/37/EC [20]). Thus, this manufacturing process may take place in one or more workplaces (Directive 89/654/EEC [15]) where, among other chemical risks, there may be the risk of explosion (Directive 99/92/EC [28]). Such workplaces may be considered as establishments with installations in the scope of implementation of Directive Seveso III, other types of establishments (not falling within Directive Seveso III) or other types of workplaces. In all cases, these workplaces will have installations and work equipment (Directive 2009/104/EC [16]) that will configure the corresponding manufacturing process.

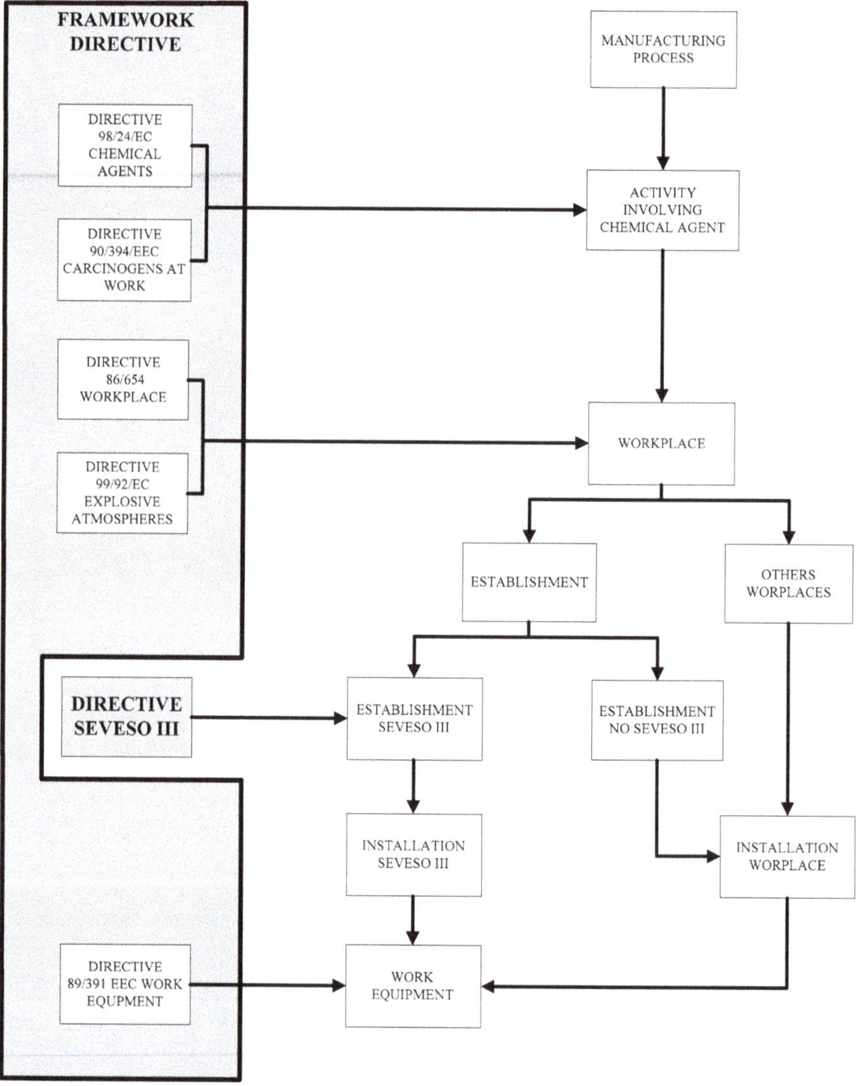

Figure 3. Main links formed between the individual directives (Framework Directive [6]) and Directive Seveso III [8] within the general structure of a manufacturing process.

Analysis of directives related to the safety and health of persons other than the individual directives within the meaning of Article 16 (1) of the Framework Directive, are not within the remit of this study. However, it should be pointed out that there may be connections to other regulations, such as the REACH Regulation [2] and CLP Regulation [3], or Directive 2006/42/EC on machinery [41]. These examples may be considered as fundamental regulations in the field of safety and health.

The main transitional spaces between these links, as shown in Figure 3, are describe below, considering in this regard the intersections between Directive Seveso III and the individual directives within the meaning of Article 16 (1) of the Framework Directive:

- *Activity involving chemical agents:* Directive 98/24/EC on chemical agents [7] will be applicable to any work activity linked to a manufacturing process that involves the use of any of the dangerous substances collected in Directive Seveso III. Additionally, other individual directives may be applicable depending on the characteristics of the manufacturing process in each specific case, as well as the chemical agents used; for instance, Directive 2004/37/EC on carcinogens or mutagens at work [20], as shown for the substances collected in Table 3.
- *Workplaces:* Directive 98/654/EEC on Workplaces [15] is applicable to any work activity linked to a manufacturing process integrated in an establishment, following Directive Seveso III. Additionally, other individual directives may also be applicable depending on the characteristics of the manufacturing process in each specific case, as well as the chemical agents used; for instance, Directive 99/92/EC on explosive atmospheres [28]. This could be the case, for instance, in the case of manufacturing processes that use any of the flammable substances collected in Table 3, such as: ethyleneimine (H225: very flammable liquid and vapours), Ethylene oxide (H220: extremely flammable gas) and propylene oxide (H224: extremely flammable liquid and vapours).
- *Installation:* Directive 2009/104/EC on work equipment [16], as well as Directive 89/654/EEC on workplaces [15], will be applicable to any work activity linked to any manufacturing process that involves one or more installations pursuant to Directive Seveso III. In general, a manufacturing process will be configured by elements that fall within the definition of work equipment and workplace, which may completely or partially configure an installation pursuant to Directive Seveso III.

5.2. Management Systems

The four transitional spaces (TS) represented conceptually in Figure 4 are derived from the results obtained in the two comparative analyses carried out between risk management systems (the links shown in Tables 4 and 5): that is, between Directive Seveso III and the Framework Directive, and between Directive Seveso III and the ISO 45001:2018 standard [14].

Transitional space TS 1 is configured by the correspondence that exists between the structure of the Framework Directive and the management systems derived from Directive Seveso III. Similarly, TS 2 is configured by the correspondence that exists between the structure of the ISO 45001:2018 standard and the management system derived from Directive Seveso III.

Transitional space TS 3 may be considered a 'natural space' between the Framework Directive and the ISO 45001:2018 standard [14] since the goal of both systems is to manage safety and health at work. However, studying this transitional space falls outside the aim of this study as such a study would form part of an analysis process regarding the implementation of the ISO 45001:2018 standard [14] by an organisation.

In any case, given that the result of implementing the ISO 45001:2018 standard [14] may be understood as a tool that can help an organisation to meet such legal requirements as complying with the Framework Directive and Directive Seveso III, the three systems may share an intersectional space, giving rise to transitional space TS 4.

Additionally, given that the provisions of Directive 98/24/EC on chemical agents [7] and Directive 2004/37/EC on carcinogens or mutagens at work [20] are in line with the general management guidelines set by the Framework Directive, the correspondence between these directives and Annex III of Directive Seveso III has been studied and the results are collected in Table 5.

The said results may be considered as a specific correspondence transferred from transitional space TS 1 to transitional space TS 4 when the intersection of the three systems takes place. Thus, TS 4 may be considered to include, at least, the provisions of Directive 98/24/EC [7] and Directive 2004/37/EC [20]. In other words, transitional space TS 4 may be considered to be a specific transitional space resulting from the intersection of three general transitional spaces.

Even if this specific transitional space has been studied in regard to the aforementioned directives, it should be pointed out that other directives, from those considered in Table 1, may also be of interest.

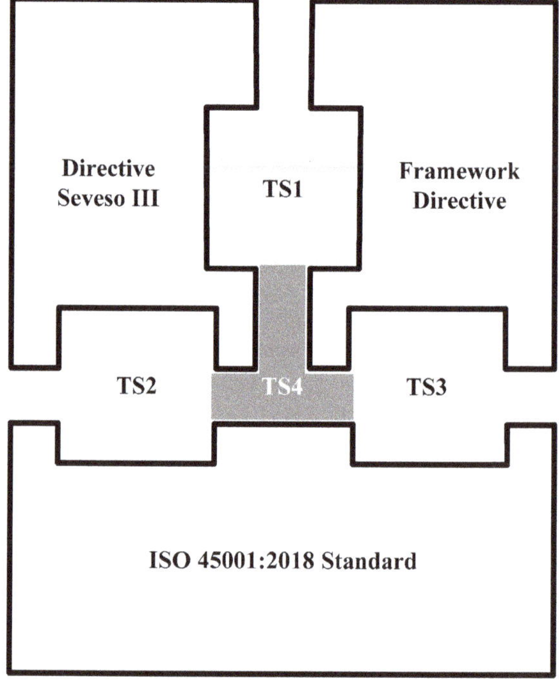

TS1, TS2, TS3: the transitional spaces configured by the correspondences identified in the Table 4

TS 4: the transitional space resulting from the intersection of TS1, TS2 and TS3.

Figure 4. Transitional spaces between the following management systems: Directive Seveso III [8], Framework Directive [7] and ISO 45001:2018 standard [14].

6. Discussion

There are signs of a revitalizing interest in foundational issues in risk assessment and management, which is welcome and necessary for meeting the challenges currently faced by the field of risk. These are related to societal problems and complex technological and emerging risks [42] which can exist in manufacturing processes alongside traditional risks [43].

The second European Survey of Enterprises on New and Emerging Risks reveals that dangerous substances (or biological substances) are most prevalent in the European Union in certain sectors such as manufacturing (51.7%) [44]. As a result, new challenges for the management of dangerous substances in the workplace are emerging, for example, in the area of green jobs (bio-energy production, new types of energy storage) and in relation to the use of innovative materials (e.g. nanomaterials) and technologies with currently unknown health risks (such as 3D printing) and substances recognised as endocrine disrupters [45].

In this context, Brocal et al. [46] considered that the relationship between the prevention of occupational accidents and major accidents is especially important. Based on this integrative perspective, Zio [47] points out to the realization that, to manage risk in a systematic and effective way, it is necessary to consider all phases of potential accident scenarios together.

There is a general research focus on dynamic risk assessment and management rather that static or traditional risk assessment [42], which can consider the dynamic evolution of conditions, both internal and external to the system, affecting risk assessment [48]. The effectiveness in the application of a dynamic risk management framework in collecting and considering evidence of emerging risks relies on the continuous development of dynamic techniques for hazard identification and risk assessment, joined with a proper safety culture [49].

However, this study does not distinguish between the risk management of both types of risks; that is, between traditional risk and emerging risk. This is due to the general characteristics of the legal and standardised context in which this study has been carried out, towards meeting the main objective set out.

In any case, this objective may be extended in future research in the direction of emerging risk management. In this regard, the CWA 16649:2013 standard on managing emerging technology-related risks may be considered to be a reference management system [50]. Such a direction may be justified, among others, by considering the data in Figure 1, which shows that the number of chemical substances used in manufacturing processes has increased in the past few years. Furthermore, by considering the hypothesis that, in general, such substances are a source of risk, implementing the TICHNER (Technique to Identify and CHaracterize NERs) technique developed by Brocal et al. [51], it could be stated that the situation described in manufacturing processes could constitute an emerging risk.

6.1. Major Accident and Occupational Accident

In regard to the definitions for 'accident' considered in this study, 'major accident' is defined by Directive Seveso III, while 'occupational accident' is not defined by the directives studied. Therefore, other sources are required. Besserman and Mentzer [12] consider that each country defines their statistics differently including different definitions for lost time incidents, non-fatal injuries, and what constitutes the manufacturing/chemical industry, among others. Furthermore, such authors indicate that there is minimal reporting of true process safety metrics resulting from the loss of containment of a hazardous substance.

In any case, a major accident can also be considered to be an occupational accident if harm to workers exists. Inversely, an occupational accident can also be a major accident when it meets the requirements set out in Directive Seveso III. The causality relationship between both types of accident has not been studied here, although its interest in the sphere of safety is evident. In the 1970s, the effect of human actions and organizational factors on accident occurrence was recognized, but it took until the mid- 1980s before management became aware that they were key to achieving a good level of both occupational and process safety [52]. The incorporation of Bayesian networks into risk assessment may be another interesting focus for both research and industrial purposes, because it allows a systemic approach considering human error and management influences [49].

6.2. Links and Transitional Spaces

The main correspondence between risk management in both types of accidents is the presence in the manufacturing process of one or more of the hazardous substances included in Annex I of Directive Seveso III, for which directives derived from the Framework Directive are also applicable. The said directives may be understood as the deployment of this correspondence into further links that are more specific, giving rise to two types of transitional spaces: functional transitional spaces and regulatory transitional spaces.

Functional transitional spaces (that is, activities involving chemical agents, workplaces and installations) are interconnected as briefly outlined in Figure 3. Such interconnection must be understood as a basic outline that can be extended in different directions in the field of risk management as, for instance: the safety change management studied by Gerbec [53]; management of exposure to nanomaterials studied by Hunt et al. [54]; machine safety, which can be generally defined through Directive 2006/42/EC and the ISO 12100: 2010 and ISO/TR 14121-1: 2012 standards on safety of machinery [55,56]

As regarding regulatory transitional spaces, they are defined by the intersections between the risks management systems studied through the Framework Directive, Directive Seveso III and the ISO 45001:2018 standard.

The links shown in Table 4 between management systems derived from the Framework Directive and Directive Seveso III, have been classified as: four are strong (Annex III sections (i), (ii), (iv) and (v)), two are weak (Annex III sections: ((iii) and (vi)), and one is non-existent (Annex III section (vii)).

This result is coherent under the perspective of the Framework Directive, which compiles a set of general guidelines that are further developed by specific directives, as shown in the results in Table 5. Therefore, the set established by the Framework Directive and individual directives completes and strengthens the seven links above, configuring the transitional space TS 1.

As regards the correspondence between management systems derived from the ISO 45001:2018 standard and Directive Seveso III, shown also in Table 4, 6 of these links have been classified as strong and one as relatively strong (sections Annex III: (iii)), configuring transitional space TS 2.

These results are also coherent with the aim of an OSH management system as defined by the ISO 45001:2018 standard. Moreover, these results make it possible to draw up a coherent correspondence between the structures of the ISO 45001:2018 standard [14] and the Framework Directive, configuring transitional space TS 3.

As regards TS 4, given that it may be considered to be a specific transitional space resulting from the intersection of three general transitional spaces, it is key to defining and channelling the transitions between systems by means of the corresponding individual directives in each case, including at least Directive 98/24/EC on chemical agents [7]. This consideration is still valid when the intersection occurs between two systems, and the linking and transitional role played by the individual directives is equally important.

The risk assessment techniques are a key structuring element between the functional and regulatory transition spaces, as shown in the results of Table 4 and, especially, of Table 5. However, according to Brocal et al. [9] it is necessary to deepen through future research on the analysis of differentiating and applicative criteria between the techniques used in the field of safety occupational and safety linked to major accidents.

These links and transitional spaces may facilitate system integration. In this regard, Li and Guldenmund [13] point out that according to the literature, an integrated management system is more advanced than independent safety systems, as safety is just one of the comprehensive organization management objectives.

In the process of integrating systems, besides the existence of transitional spaces, non-traditional spaces also exist as a result of the specific aspects of each system that have no direct correlation with the other systems. These non-transitional spaces have not been studied here, yet may be equally important in any integration process, since they define the frontiers necessary to avoid unwanted interference with the transitional spaces.

7. Conclusions

The main objective of this study has been met through identifying and analysing the links and transitional spaces between the risk management of occupational accidents and major accidents that involve hazardous substances in manufacturing processes. To this end, the risk management systems derived from Directive Seveso III, the Framework Directive and the ISO 45001:2018 standard [14] have been analysed, obtaining three main results.

The first result of this analysis, the main link identified between the risk management of both types of accidents, is the presence in a manufacturing process of any of the hazardous substances included in Annex I of Directive Seveso III, for which the directives derived from the Framework Directive are also applicable, and the principles and guidelines of the ISO 45001:2018 standard are applicable on a voluntary basis [14].

As a second result, the intersection of Directive Seveso III, the Framework Directive and the ISO 45001:2018 standard, configures three general transitional spaces (TS 1, TS 2 and TS 3).

In turn, and as a third result, the intersection of these three general transitional spaces configures a specific transitional space (TS 4), which is key to defining and channelling the transition between systems by means of the individual directives that may correspond in each case, which will, at least, include the directive on chemical substances. This will enable integration processes between the systems considered.

The above results are limited from a regulatory and technical perspective. In regard to the regulatory perspective, the context is limited to the EU, as well as Directive Seveso III, the Framework Directive and any individual directives that further develop it. Other directives and regulations that are especially relevant in the sphere of safety, as for instance the REACH Regulation [2] and CLP Regulation [3], are open to further analysis to enable their integration into the management systems. From a technical perspective, no distinction has been made between traditional risk management and emerging risk management. Similarly, no distinction has been made between static and dynamic approaches.

The results and limitations stated above may point towards future paths of research, from which models to integrate the risk management of occupational accidents and major accidents may be developed based on real experiences and data.

By way of final conclusion, the identification and analysis of the links and transitional spaces carried out by this study aspire to being a starting point that will inspire other researchers to continue and further develop this study with the end goal of efficiently integrating risk management systems related to accidents derived from dangerous substances in manufacturing processes.

Author Contributions: Conceptualization, F.B., C.G. and M.A.S.; Investigation, F.B., C.G. and M.A.S; Methodology, F.B., C.G.; Supervision, F.B, C.G., M.A.S., G.R. and V.C.; Writing—original draft, F.B., C.G. and M.A.S.; Writing—review and editing, F.B, C.G., M.A.S., G.R. and V.C.

Funding: This research was funded by the Ministry of Economy and Competitiveness of Spain, title: 'Analysis and Assessment of technological requirements for the design of a New Emerging Risks standardized management SYStem (A2NERSYS)' with reference DPI2016-79824-R.

Conflicts of Interest: The authors declare no conflict of interest.

References

1. European Commission. *Growth–Internal Market, Industry, Entrepreneurship and SMEs*; European Commission: Brussels, Belgium, 2018; Available online: https://ec.europa.eu/growth/sectors/chemicals_en (accessed on 24 August 2018).
2. European Union. *Regulation (EC) No 1907/2006 of the European Parliament and of the Council of 18 December 2006 Concerning the Registration, Evaluation, Authorisation and Restriction of Chemicals (REACH), Establishing a European Chemicals Agency, Amending Directive 1999/45/EC and Repealing Council Regulation (EEC) No 793/93 and Commission Regulation (EC) No 1488/94 as well as Council Directive 76/769/EEC and Commission Directives 91/155/EEC, 93/67/EEC, 93/105/EC and 2000/21/EC*; Official Journal of the European Union: Brussels, Belgium, 2006.
3. European Union. *Regulation (EC) No 1272/2008 of the European Parliament and of the Council of 16 December 2008 on Classification, Labelling and Packaging of Substances and Mixtures, Amending and Repealing Directives 67/548/EEC and 1999/45/EC, and Amending Regulation (EC) No 1907/2006*; Official Journal of the European Union: Brussels, Belgium, 2008.

4. European Chemical Agency (ECHA). Classification and Labelling Inventory Database. Available online: https://echa.europa.eu/information-on-chemicals/cl-inventory-database (accessed on 24 August 2018).
5. European Chemical Agency (ECHA). Registered Substances Database. Available online: https://echa.europa.eu/information-on-chemicals/registered-substances (accessed on 24 August 2018).
6. European Communities. *Directive 89/391/EEC of 12 June 1989 on the Introduction of Measures to Encourage Improvements in the Safety and Health of Workers at Work—'Framework Directive' Directive 89/654/EEC—Workplace Requirements of 30 November 1989 Concerning the Minimum Safety and Health Requirements for the Workplace (First Individual Directive within the Meaning of Article 16 (1) of Directive 89/391/EEC)*; Official Journal of the European Communities: Brussels, Belgium, 1989.
7. European Union. *Directive 98/24/EC–Risks Related to Chemical Agents at Work of 7 April 1998 on the Protection of the Health and Safety of Workers from the Risks Related to Chemical Agents at Work (Fourteenth Individual Directive within the Meaning of Article 16(1) of Directive 89/391/EEC)*; Official Journal of the European Union: Brussels, Belgium, 1998.
8. European Union. *Directive 2012/18/EU of the European Parliament and of the Council of 4 July 2012 on the Control of Major-Accident Hazards Involving Dangerous Substances, Amending and Subsequently Repealing Council Directive 96/82/EC Text with EEA Relevance*; Official Journal of the European Union: Brussels, Belgium, 2012.
9. Brocal, F.; Sebastián, M.A.; González, C.; Reniers, G.; Paltrinieri, N. Standardized risk assessment techniques: A review in the framework of occupational safety. In *Safety and Reliability—Safe Societies in a Changing World, Proceedings of the 28 TH International European Safety and Reliability Conference (ESREL 2018), Trondheim, Norway, 17–21 June 2018*; Haugen, S., Barros, A., Gulijk, C.V., Kongsvik, T., Vinnem, J., Eds.; CRC Press: London, UK, 2018.
10. ScienceDirect Databases. Available online: https://www.sciencedirect.com/search/advanced (accessed on 6 August 2018).
11. Rasmussen, K.; Sokull-Klüttgen, B.; Yu, I.J.; Kanno, J.; Hirose, A.; Gwinn, M.R. Chapter 7–Regulation and Legislation. In *Adverse Effects of Engineered Nanomaterials*, 2nd ed.; Fadeel, B., Pietroiusti, A., Shvedova, A.A., Eds.; Academic Press: Cambridge, MA, USA, 2017; pp. 159–188. [CrossRef]
12. Besserman, J.; Mentzer, R.A. Review of global process safety regulations: United States, European Union, United Kingdom, China, India. *J. Loss Prevent Proc. Ind.* **2017**, *50*, 165–183. [CrossRef]
13. Li, Y.; Guldenmund, F.W. Safety management systems: A broad overview of the literature. *Saf. Sci.* **2018**, *103*, 94–123. [CrossRef]
14. *ISO 45001–Occupational Health and Safety Management Systems. Requirements with Guidance for Use*; International Organization for Standardization (ISO): Geneva, Switzerland, 2018.
15. European Union. *Directive 89/654/EEC of 30 November 1989 Concerning the Minimum Safety and Health Requirements for the Workplace (First Individual Directive within the Meaning of Article 16 (1) of Directive 89/391/EEC)*; Official Journal of the European Union: Brussels, Belgium, 1989.
16. European Union. *Directive 2009/104/EC–Use of Work Equipment of 16 September 2009 Concerning the Minimum Safety and Health Requirements for the Use of Work Equipment by Workers at Work (Second Individual Directive within the Meaning of Article 16(1) of Directive 89/391/EEC)*; Official Journal of the European Union: Brussels, Belgium, 2009.
17. European Union. *Directive 89/656/EEC–Use of Personal Protective Equipment of 30 November 1989 on the Minimum Health and Safety Requirements for the Use by Workers of Personal Protective Equipment at the Workplace (Third Individual Directive within the Meaning of Article 16 (1) of Directive 89/391/EEC)*; Official Journal of the European Union: Brussels, Belgium, 1989.
18. European Union. *Directive 90/269/EEC–Manual Handling of Loads of 29 May 1990 on the Minimum Health and Safety Requirements for the Manual Handling of Loads where There Is a Risk Particularly of Back Injury to Workers (Fourth Individual Directive within the meaning of Article 16 (1) of Directive 89/391/EEC)*; Official Journal of the European Union: Brussels, Belgium, 1990.
19. European Union. *Directive 90/270/EEC–Display Screen Equipment of 29 May 1990 on the Minimum Safety and Health Requirements for Work with Display Screen Equipment (Fifth Individual Directive within the Meaning of Article 16 (1) of Directive 89/391/EEC)*; Official Journal of the European Union: Brussels, Belgium, 1990.
20. European Union. *Directive 2004/37/EC–Carcinogens or Mutagens at Work of 29 April 2004 on the Protection of Workers from the Risks Related to Exposure to Carcinogens or Mutagens at Work (Sixth individual Directive within the meaning of Article 16(1) Directive 89/391/EEC)*; Official Journal of the European Union: Brussels, Belgium, 2004.

21. European Union. *Directive 2000/54/EC–Biological Agents at Work of the European Parliament and of the Council of 18 September 2000 on the Protection of Workers from Risks Related to Exposure to Biological Agents at Work (Seventh Individual Directive within the Meaning of Article 16(1) of Directive 89/391/EEC)*; Official Journal of the European Union: Brussels, Belgium, 2000.
22. European Union. *Directive 92/57/EEC–Temporary or Mobile Construction Sites of 24 June 1992 on the Implementation of Minimum Safety and Health Requirements at Temporary or Mobile Construction Sites (Eighth Individual Directive within the Meaning of Article 16 (1) of Directive 89/391/EEC)*; Official Journal of the European Union: Brussels, Belgium, 1992.
23. European Union. *Directive 92/58/EEC–Safety and/or Health Signs of 24 June 1992 on the Minimum Requirements for the Provision of Safety and/or Health Signs at Work (Ninth Individual Directive within the Meaning of Article 16 (1) of Directive 89/391/EEC)*; Official Journal of the European Union: Brussels, Belgium, 1992.
24. European Union. *Directive 92/85/EEC–Pregnant Workers of 19 October 1992 on the Introduction of Measures to Encourage Improvements in the Safety and Health at Work of Pregnant Workers and Workers who Have Recently Given Birth or Are Breastfeeding (Tenth Individual Directive within the Meaning of Article 16 (1) of Directive 89/391/EEC)*; Official Journal of the European Union: Brussels, Belgium, 1992.
25. European Union. *Directive 92/91/EEC–Mineral-Extracting Industries–Drilling of 3 November 1992 Concerning the Minimum Requirements for Improving the Safety and Health Protection of Workers in the Mineral-Extracting Industries through Drilling (Eleventh Individual Directive within the Meaning of Article 16 (1) of Directive 89/391/EEC)*; Official Journal of the European Union: Brussels, Belgium, 1992.
26. European Union. *Directive 92/104/EEC–Mineral-Extracting Industries of 3 December 1992 on the Minimum Requirements for Improving the Safety and Health Protection of Workers in Surface and Underground Mineral-Extracting Industries (Twelfth Individual Directive within the Meaning of Article 16 (1) of Directive 89/391/EEC)*; Official Journal of the European Union: Brussels, Belgium, 1992.
27. European Union. *Directive 93/103/EC–Work on Board Fishing Vessels of 23 November 1993 Concerning the Minimum Safety and Health Requirements for Work on Board Fishing Vessels (Thirteenth Individual Directive within the Meaning of Article 16 (1) of Directive 89/391/EEC)*; Official Journal of the European Union: Brussels, Belgium, 1993.
28. European Union. *Directive 99/92/EC–Risks from Explosive Atmospheres of 16 December 1999 on the Minimum Requirements for Improving the Safety and HEALTH protection of Workers Potentially at Risk from Explosive atmospheres (15th individual Directive within the meaning of Article 16(1) of Directive 89/391/EEC)*; Official Journal of the European Union: Brussels, Belgium, 1999.
29. European Union. *Directive 2002/44/EC–Vibration of 25 June 2002 on the Minimum Health and Safety Requirements Regarding the Exposure of Workers to the Risks Arising from Physical Agents (Vibration) (Sixteenth Individual Directive within the Meaning of Article 16(1) of Directive 89/391/EEC)*; Official Journal of the European Union: Brussels, Belgium, 2002.
30. European Union. *Directive 2003/10/EC–Noise of 6 February 2003 on the Minimum Health and Safety Requirements Regarding the Exposure of Workers to the Risks Arising from Physical Agents (Noise) (Seventeenth Individual Directive within the Meaning of Article 16(1) of Directive 89/391/EEC)*; Official Journal of the European Union: Brussels, Belgium, 2003.
31. European Union. *Directive 2004/40/EC of the European Parliament and of the Council of 29 April 2004 on the Minimum Health and Safety Requirements Regarding the Exposure of Workers to the Risks Arising from Physical Agents (Electromagnetic Fields) (18th Individual Directive within the Meaning of Article 16(1) of Directive 89/391/EEC)*; Official Journal of the European Union: Brussels, Belgium, 2004.
32. European Union. *Directive 2006/25/EC–Artificial Optical Radiation of 5 April 2006 on the Minimum Health and Safety Requirements Regarding the Exposure of the Workers to Risks Arising from Physical Agents (Artificial Optical Radiation, 19th Individual Directive within the Meaning of Article 16(1) of Directive 89/391/EEC)*; Official Journal of the European Union: Brussels, Belgium, 2006.
33. European Union. *Directive 2013/35/EU–Electromagnetic Fields of 26 June 2013 on the Minimum Health and Safety Requirements Regarding the Exposure of Workers to the Risks Arising from Physical Agents (Electromagnetic Fields) (20th Individual Directive within the Meaning of Article 16(1) of Directive 89/391/EEC) and Repealing Directive 2004/40/EC*; Official Journal of the European Union: Brussels, Belgium, 2013.
34. Eurostat. Statistics Explained. Accidents at Work Statistics. Available online: https://ec.europa.eu/eurostat/statistics-explained/index.php/Accidents_at_work_statistics (accessed on 28 June 2018).

35. Sebastián, M.A.; Brocal, F.M.A. Analysis of Open Resources from INSHT for Application to University Teaching of Industrial Safety Technology. *Procedia Eng.* **2015**, *132*, 228–235. [CrossRef]
36. INSHT (National Institute for Safety and Health at Work). INFOCARQUIM (Information about Carcinogens Chemicals) Database. Available online: http://infocarquim.inssbt.es/Forms/About.aspx (accessed on 6 August 2018).
37. *Límites de exposición profesional para agentes químicos en España (Occupational Exposure Limits for Chemical Agents in Spain)*; INSSBT (National Institute for Safety, Health and Welfare at Work): Madrid, Spain, 2018.
38. *Guía Técnica para la evaluación y prevención de los riesgos relativos a la utilización de equipos de trabajo (Technical Guide for the assessment and prevention of risks related to the use of work equipment) (Royal Decree 1215/1997—BOE nº 188)*, 2nd ed.; INSHT (National Institute for Safety and Health at Work): Madrid, Spain, 2011.
39. Spain. Law 31/1995 of 8 November on Prevention of Occupational Risks. (BOE, No. 269, 10.11.1995). Available online: https://www.boe.es/buscar/pdf/1995/BOE-A-1995-24292-consolidado.pdf (accessed on 9 October 2018).
40. Spain. Royal Decree 39/1997, January 17, adopting Regulations on Prevention Services. (BOE, No. 27, 31.01.1997). Available online: https://www.boe.es/buscar/pdf/1997/BOE-A-1997-1853-consolidado.pdf (accessed on 9 October 2018).
41. European Union. *Directive 2006/42/EC of the European Parliament and of the Council of 17 May 2006 on machinery, and Amending Directive 95/16/EC (recast)*; Official Journal of the European Union: Brussels, Belgium, 2006.
42. Aven, T. Risk Assessment and Risk management: Review of Recent Advances on Their foundation. *Eur. J. Oper. Res.* **2016**, *253*, 1–13. [CrossRef]
43. Brocal, F.; Sebastián, M.A.; González, C. Theoretical framework for the new and emerging occupational risk modeling and its monitoring through technology lifecycle of industrial processes. *Saf. Sci.* **2017**, *99*, 178–186. [CrossRef]
44. EU-OSHA (European Agency for Safety and Health at Work) (Jul. 24, 2018). 2014. European Survey of Enterprises on New and Emerging Risks (ESENER-2) [Interactive Survey]. EU-OSHA, Bilbao. Available online: https://osha.europa.eu/en/surveys-and-statistics-osh/esener/2014 (accessed on 6 August 2018).
45. Helathy Workplaces. *Manage Dangerous substances. Campaign Guide*; EU-OSHA (European Agency for Safety and Health at Work): Bilbao, Spain, 2018.
46. Brocal, F.; Sebastián, M.A.; González, C. Chapter 2: Advanced manufacturing processes and technologies. In *Management of Emerging Public Health Issues and Risks. Multidisciplinary Approaches to the Changing Environment*; Benoit, R., Weiss, K., Thireau, V., Eds.; Academic Press (Elsevier): Cambridge, MA, USA, 2018; ISBN 9780128132906, in press.
47. Zio, E. The future of risk assessment. *Reliab. Eng. Syst. Safe.* **2018**, *177*, 176–190. [CrossRef]
48. Paltrinieri, N.; Scarponi, G. Addressing dynamic risk in the petroleum industry by means of innovative analysis solutions. *Chem. Eng. Trans.* **2014**, *36*, 451–456.
49. Villa, V.; Paltrinieri, N.; Khan, F.; Cozzani, V. Towards dynamic risk analysis: A review of the risk assessment approach and its limitations in the chemical process industry. *Saf. Sci.* **2016**, *89*, 77–93. [CrossRef]
50. *Managing Emerging Technology-related Risks. CWA 16649:2013*; European Committee for Standardization (CEN): Brussels, Belgium, 2013.
51. Brocal, F.; González, C.; Sebastián, M.A. Technique to identify and characterize new and emerging risks: A new tool for application in manufacturing processes. *Saf. Sci.* **2018**, *109*, 144–156. [CrossRef]
52. Jain, P.; Pasman, H.J.; Waldram, S.P.; Rogers, W.J.; Mannan, M.S. Did we learn about risk control since Seveso? Yes, we surely did, but is it enough? An historical brief and problem analysis. *J. Loss Prev. Process Ind.* **2017**, *49*, 5–17. [CrossRef]
53. Gerbec, M. Safety change management—A new method for integrated management of organizational and technical changes. *Saf. Sci.* **2017**, *100*, 225–234. [CrossRef]
54. Hunt, G.; Lynch, I.; Cassee, F.; Handy, R.D.; Fernandes, T.F.; Berges, M.; Kuhlbusch, T.A.J.; Dusinska, M.; Riediker, M. Towards a Consensus View on Understanding Nanomaterials Hazards and Managing Exposure: Knowledge Gaps and Recommendations. *Materials* **2013**, *6*, 1090–1117. [CrossRef] [PubMed]
55. *ISO 12100 Preview Safety of Machinery–General Principles for Design–Risk Assessment and Risk Reduction*; International Organization for Standardization (ISO): Geneva, Switzerland, 2010.

56. *ISO/TR 14121-2. Preview Safety of Machinery–Risk Assessment–Part 2: Practical Guidance and Examples of Methods*; International Organization for Standardization (ISO): Geneva, Switzerland, 2012.

© 2018 by the authors. Licensee MDPI, Basel, Switzerland. This article is an open access article distributed under the terms and conditions of the Creative Commons Attribution (CC BY) license (http://creativecommons.org/licenses/by/4.0/).

Article

Improving Pallet Mover Safety in the Manufacturing Industry: A Bow-Tie Analysis of Accident Scenarios

Karolien van Nunen [1,2,*], Paul Swuste [2,3], Genserik Reniers [2,3], Nicola Paltrinieri [4], Olga Aneziris [5] and Koen Ponnet [6]

1. Research Chair Vandeputte, Law Enforcement, Faculty of Law, University of Antwerp, 2000 Antwerp, Belgium
2. Antwerp Research Group on Safety and Security (ARGoSS), Engineering Management, Faculty of Applied Economics, University of Antwerp, 2000 Antwerp, Belgium; P.H.J.J.Swuste@tudelft.nl (P.S.); G.L.L.M.E.Reniers@tudelft.nl (G.R.)
3. Safety and Security Science, Faculty of Technology, Policy and Management, Delft University of Technology, 2628 CD Delft, The Netherlands
4. Department of Mechanical and Industrial Engineering, Norwegian University of Science and Technology—NTNU, 7491 Trondheim, Norway; nicola.paltrinieri@ntnu.no
5. National Centre for Scientific Research "Demokritos", 15310 Agia Paraskevi, Greece; olga@ipta.demokritos.gr
6. Media, ICT & Interpersonal Relations in Organisations & Society (MIOS), Faculty of Political and Social Sciences, University of Antwerp, 2000 Antwerp, Belgium; koen.ponnet@uantwerpen.be

* Correspondence: karolien.vannunen@uantwerpen.be; Tel.: +31-498-522-100

Received: 22 August 2018; Accepted: 9 October 2018; Published: 12 October 2018

Abstract: A Belgian manufacturing company uses pallet movers for internal transport. Despite the company's efforts to improve occupational safety, accidents with pallet movers remain noteworthy. In order to control occupational accidents, it is crucial to have a clear view of the potential accident scenarios that are present in a company. The bow-tie method is a way to capture and visualize these accident processes in an integrative way. Included in the bow-tie are safety barriers (both technical as organizational and human) and management delivery systems that can intervene in these accident processes. Once bow-ties are composed, they are an excellent point of departure to assign indicators to the safety barriers and management delivery systems in order to control (i.e., prevent or mitigate) accident scenarios. Two types of indicators can be distinguished. Firstly, there are general indicators that are assigned to management delivery systems interrupting multiple accident scenarios, which can yield a higher safety gain (as they intervene in multiple accident scenarios). Secondly, there are scenario-specific indicators targeting one specific accident scenario, which can be valuable as they target a specific problem in the company. For the development of the bow-ties, a multi-method design with the inclusion of different data sources was used, leading to a comprehensive overview. This makes the bow-tie analysis of internal transport with pallet movers transferable to other settings where pallet movers are used for internal transport.

Keywords: manufacturing industry; bow-tie analysis; pallet mover accidents; accident analysis; safety barriers

---

1. Introduction of the Study

Internal transport represents a well-known occupational hazard in many modern industrial environments. Pallet movers (synonyms for 'pallet mover' are pallet jack, walkie-rider, pallet truck, transpallet) are used for internal transport in several industries, and these machines are easy to operate, compared to, for example, forklift trucks. However, pallet movers are inherently dangerous machines.

They often operate close to pedestrian workers and, charged with a load, their total mass can be well above two tons. Often, their load is not secured to the machine, leading to instability when gravity gets a grip of the load.

The company under investigation in this study is a manufacturing plant located in Belgium, which is part of an American multinational, producing consumer products all over the world. Pallet movers are used frequently for internal transport. Like many major American companies, this company also pays a lot of attention to the safety of their employees. The company uses the so-called 6W-2H and why-why techniques to analyze its recordable accidents without or with lost work time. The 6W-2H problem analysis produces a description of the context of accidents (what, where, which, when, who, to whom, how, how much). The why-why analysis is performed to get to the so-called root causes of accidents and is based on mapping of what happened during the accident process and why this happened. Countermeasures are formulated based on the results of these accident analyzes. The entire accident analysis is the responsibility of the supervisors, but are hereby supported by the Health and Safety (HSEQ) staff.

According to figures provided by the European safety manager of the abovementioned multinational, the 6W-2H and why-why techniques did not help in reducing occupational accidents substantially. As an alternative, a bow-tie analysis is proposed, providing a detailed and comprehensive insight into potential accident scenarios, including possible safety barriers (both technical and non-technical) and management delivery systems which can prevent or mitigate the accident processes. In Section 4, the choice for the bow-tie method is supported.

To test the bow-tie analysis, a pilot project was formulated, focussing on accidents during internal transport with pallet movers. Accidents with pallet movers represent a significant share of the total number of accidents. In the years 2015 and 2016, about ten percent of all recordable accidents with lost work time that occurred at the European plants of the multinational involved a pallet mover.

The research question of this study is 'Which accident scenarios are possible during internal transport with pallet movers and which safety barriers (both technical as organizational and human) and management delivery systems can influence (i.e., prevent or mitigate) these accident scenarios?'.

2. Background of the Plant under Investigation

At the Belgian plant, approximately 300 people are employed. Two out of three employees are permanent ('own company employees') and one out of three is a contractor.

The production process is manifested on several floors. The upper floors are responsible for supplying raw materials, the production of intermediate products, and supplying these intermediate products to lower floors, where product finishing and packing take place. The ground floor has production lines for finishing the products and packing lines. These lines are implanted in an existing space which was initially not designed for that purpose.

The entire plant is characterized by a lot of load manipulation, which means that products are stocked in many temporary buffers. For example, on the ground floor, the packing materials are transported from the warehouse to a first large buffer (by a forklift truck), followed by transportation from the first large buffer to a second smaller buffer (by a pallet mover), and from the second smaller buffer to a buffer in front of the production or packing line (by a pallet mover).

The transportation routes of forklift trucks are separated from transportation routes of pallet movers, leading to (almost) no possible contact between the two types of internal transport.

Figure 1 shows two types of pallet movers being used at the plant: Standard electrical pallet movers and electrical stackers which can lift loads to approximately 1.8 m. Pallet movers can transport a load of 1.2 to 2 tons, have a driving speed up to 6 km/h and a standard emergency stop. Maintenance and repair of pallet movers are performed by the company itself on a regular basis, while inspections are performed by an external company every three months.

Approximately 75% of the operators handling pallet movers are provided by one steady contractor company. The contractor staff is characterized by a frequent rotation.

Figure 2 shows the most frequent loads transported at the plant. Finished products are not transported with pallet movers, only raw materials, intermediate products, packing material, and off-quality products.

During the period from 2015 to 2016, eight recordable accidents occurred at the Belgian location. In three of these accidents, a pallet mover was involved. Two of these pallet mover accidents happened with a contractor employee, and one with a company employee.

Figure 1. The types of pallet movers—(**a**): standard pallet mover and (**b**): stacker.

Figure 2. The most frequent loads transported with pallet movers.

3. Research Methodology

To compose the bow-ties of accidents with pallet movers, a multi-method design was used.

A literature search was conducted using electronic databases of the library of the Delft University of Technology, the British Health and Safety Executive (HSE), the American Occupational safety and Health Administration (OSHA), and the American National Institute for Occupational Safety and Health (NIOSH). Search terms were the following: 'pallet mover', 'pallet jack', 'walkie-rider', 'pallet truck, 'transpallet', and 'accident'. Safety related articles on pallet movers were rather scarce in the literature. Therefore, articles on forklift truck accidents (using the search term 'forklift') were also included in the literature study, for as far as the accident processes had similarities with the ones of pallet movers.

Belgian and Dutch national data on pallet mover accidents were requested. For Belgium, data on pallet mover accidents are obtained from Fedris, the Belgian federal agency for occupational risks. For The Netherlands, Storybuilder is used, which is a software tool developed for the Dutch Ministry of Social Affairs and Employment [1]. Both agencies have provided data on accidents reported to and investigated by the labor inspectorate. In both databases, the type of equipment involved in the accidents can be selected. This equipment classification is based on the classification by ESAW (European Statistics on Accidents at Work). Pallet movers fall under the code 11.04 ('mobile handling devices, handling trucks (powered or not)—barrows, pallet trucks, etc.'). It should be noted that the data obtained does not only cover accidents with pallet movers, but also other equipment falling under the same code. Hence, national numbers of pallet mover accidents cannot be given. However, the databases also contain information on types of accident scenarios, root causes, and failing safety barriers. The latter qualitative information was used to complement the development of the bow-ties.

Documents and data concerning pallet mover safety available at the Belgian plant (and by extension at all European plants if available) were analyzed: The material of the pallet mover training (presentation, syllabus), an observation checklist used to evaluate the use of pallet movers (HSEQ staff and supervisors use this checklist to evaluate the behavior of operators during pallet mover use), the minutes of monthly safety meetings, a pallet mover maintenance checklist, the most recent pallet mover inspection overview, a checklist for interims regarding pallet mover use, and the safety notifications regarding pallet movers in the incident registration system. During 2015–2016, 127 safety notifications regarding pallet movers were available at the Belgian plant comprising information on accidents (n = 9), near-misses (n = 9), unsafe conditions (n = 93), and positive feedback to the operators (n = 16). All available accident analyzes (the 6W-2H and why-why techniques) performed after a recordable accident with a pallet mover were also taken into account. In Appendix A, an example of such a recordable accident is given, containing a short description of the accident and the actions taken in response to the accident.

At the Belgian plant, observations were being held at the workplaces of the pallet mover operators. A personal introduction of the researchers and the purpose of the study were given before the observations, and an introduction was given to the contractors during their daily team meetings. Observations were complemented with interviews with operators (n = 25), team leaders (n = 5), the HSEQ staff (n = 3), and management (n = 2). Interviews with operators and team leaders were performed on-the-job and took approximately fifteen minutes per person. The following aspects were addressed during the interviews:

- A job description of their tasks involving a pallet mover
- Problems and obstacles encountered during the use of pallet movers
- Facilitating aspects regarding the use of pallet movers
- Accidents, near-misses, or unsafe conditions with pallet movers
- Accidents that are most likely to occur with pallet movers
- Suggestions for improvement regarding the use of pallet movers
- Additional questions based on the observations

Interviews with the HSEQ staff and management were performed in a meeting room and took approximately one hour per person, discussing topics as the safety management system and findings during the fieldwork. The staff of the external company that provides the training of the pallet mover operators was also interviewed.

4. The Bow-Tie Model

The safety metaphor used in this study is the so-called bow-tie (Figure 3). The bow-tie model originates from the engineering domain and combines a hazard and safety barrier concept, dating as far back as DeBlois (1926), Gibson (1961) and Haddon (1963), together with a scenario concept, known from the Swiss cheese model of Reason (1997) [2–8]. A bow-tie model is comprised of a fault tree (the left-hand side of the model), which represents the risk factors of a failure, and an event tree (the right-hand side of the model), which represents the consequences of a failure [9].

The bow-tie metaphor illustrates an accident process, starting with a hazard on the left-hand side. A hazard (or energy) is a source or a condition with the potential for causing harm. Various accident scenarios, pictured as left-right arrows, can migrate to the center point of the metaphor, the central event. This central event represents a state where the hazard (energy) has become uncontrollable and, thus, becomes an undesirable event with a potential for harm or damage. The central event proceeds the consequences at the right-hand side of the metaphor, such as causing harm to people or damage to assets or environment.

The strength of the metaphor is its relationship between accident scenarios, technical safety barriers, non-technical safety barriers, and management delivery systems. A scenario is a sequence of events and conditions necessary for an accident to occur. Looking at the scenarios, two types of events can be distinguished. There are pre-event accident scenarios presented at the left-hand side of the central event (leading to the central event), and there are post-event accident scenarios depicted at the right-hand side of the central event (leading to the consequences). The technical safety barriers, represented as the black boxes in the scenarios, are technical entities that can interrupt the accident scenario. An example of a technical safety barrier is an emergency stop on a pallet mover. The non-technical (or organizational and human) safety barriers are represented as the white boxes in the scenarios, being non-technical entities that can interrupt the accident scenario. An example of a non-technical safety barrier is the removal of leaking cubitainers (which interrupts the pre-event accident scenario of losing control over the pallet mover due to leaked products on the floor). The upwards arrows in Figure 3 represent the influence of management delivery systems. Management delivery systems influence the quality (in terms of reliability and availability) of the technical and non-technical safety barriers. For example, maintenance of the emergency stop on the pallet mover does not interrupt the accident scenario in a direct way, but is a management delivery system influencing the reliability of the technical safety barrier 'emergency stop on pallet mover'. Another example of a management delivery system is the training of pallet mover operators on removing leaking cubitainers, which influences the reliability of the non-technical safety barrier 'removal of leaking cubitainers'.

There are two types of safety barriers that can be distinguished. There are safety barriers to prevent the occurrence of the central event, which are presented in the pre-event accident scenarios, and there are safety barriers to control or to mitigate the consequences, which are presented in the post-event accident scenarios.

The bow-tie model has a hidden time factor. Less than adequate safety barriers or management delivery systems can be manifested over a long period of time. If a hazard becomes uncontrollable and reaches the central event, scenarios reaching their consequences will usually unroll very quickly. Pre-event accident scenarios may take days, week, months, or even longer, while post-event accident scenarios develop in hours, minutes, or even shorter.

After bow-ties are developed, the next step is to assign indicators to the safety barriers and the management delivery systems [10]. Indicators are able to visualize possibilities for improvement, to indicate safety improvement or safety decline over time, and create benchmarking (e.g., between different plants). Once the indicators are developed, targets and limits should be assigned to every indicator (what is acceptable or unacceptable as a result or for instance tolerable with leeway for improvement). Additionally, responsibilities have to be set up in order to achieve goals and to define actions when targets are not met. According to the needs of the company, indicators can be prioritized (which indicators are for instance more important or more feasible to implement).

Traditionally, the scope of accident and incident investigations, whether performed internally or externally, is usually limited to investigating the immediate causes and decision making processes related to the accident sequence. Important factors contributing to the accident are hereby often overlooked. However, since the method used to analyze incident data and accident information influences the proposed prevention measures, it should be of no surprise that those investigations don not directly guide one towards the most effective improvements and solutions. On the other extreme, methods and models to investigate the socio-technical system, such as those proposed by Rasmussen ('drift to danger' model) [11], Leveson (the STAMP method) [12] or Hollnagel (the FRAM method) [13], are often too general for the application intention, and this way surpass their goal. The bow-tie method as employed in this research finds a way between both situations, and thoroughly investigates a (possible) accident without being too high-level.

The majority of accident modeling techniques has been designed to address process safety. However, most of these models tend to be also applicable to the field of occupational safety [14]. The same applies for the bow-tie model, which has entered the field of occupational safety through the European Workgroup for development of the Occupational Risk Model (WORM), which started with the aim of decreasing the occupational accident rate in the Netherlands by 10–15% [14,15]. The value of this model lies mainly in its suitability for qualitative analysis. In addition, bow-ties have been proposed for quantifying occupational risk in the framework of the WORM research project. An initial [16] and a more general form of the model have been presented in Reference [17]. Bowties for quantifying occupational risk have been presented in the following cases: Falls from heights [18], falling objects [19], contact with moving parts of machines [20], activities near moving vehicles [21], fires [22] and hazardous substances [23].

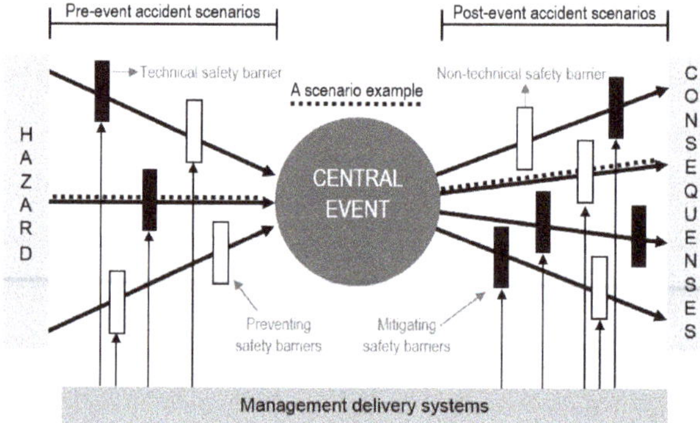

Figure 3. The bow-tie model.

5. Results

5.1. The Outcomes of the Current Accident Analysis in the Plant under Investigation

At the Belgian plant, pallet mover accidents are being analyzed with the 6W-2H and why-why techniques. This accident analysis allows us to focus on technical, organizational and human aspects. However, in the accident analysis of the plant, a trend of 'blaming the victim' can be identified. For example, looking at the pallet mover accident as described in the Appendix A, the following aspects can be indicated as important contributors for the accident: The area was very crowded creating a narrow maneuvering space, the load that had to be picked up was not standing in the designated zone and the production pressure was high. The accident analysis shows that the involved operator was personally blamed, as disciplinary measures against the operator were taken by the contractor company.

The countermeasures that are taken based on the results of the accident analysis are also mainly directed towards the operators, such as giving personal warnings and the retraining of pallet mover operators. The observation checklist to evaluate the behavior of operators during the use of pallet movers is often used after an incident occurred, meaning that it is implicitly assumed that the behavior of the operator is one of the root causes of incidents. Organization-oriented countermeasures mainly focus on the adaptation of the working environment to create more maneuvering space and a better overview.

5.2. Bow-Ties of Accident Processes with Pallet Movers

Table 1 presents an overview of the left side of the bow-tie: Possible hazards, pre-event accident scenarios, preventing technical and non-technical safety barriers, preventing management delivery systems, and central events of accident processes with pallet movers. Table 2 comprises central events, mitigating technical and non-technical safety barriers, mitigating management delivery systems, post-event accident scenarios, and consequences.

Tables 1 and 2 were composed based on the findings of the literature study [24–40], information available through Belgian and Dutch national accident databases, the analysis of documents and data available at the Belgian plant regarding pallet movers, and the results of the fieldwork (observations and interviews).

Table 1. The elements on the left-hand side of the bow-ties of accident processes with pallet movers based on: (1) Literature; (2) Belgian and Dutch National data; (3) Document analysis; (4) Safety notifications in the incident registration system of the Belgian plant; (5) Accident analysis; (6) Fieldwork: Observations and interviews.

Hazard	Pre-Event Accident Scenario	Preventing Technical and Non-Technical Safety Barriers and Management Delivery Systems	Central Event
• Load (1,2,3,4,5,6)	• Limited sight due to height of load (1,2,3,5,6) • Load not secured/not well secured (e.g., no cage, not strapped) (e.g., due to production pressure) (1,2,3,4,5,6) • Load wrongly loaded (e.g., not centered, too highly loaded, forks under width pallet instead of length, containers not straight on each other, too heavy) (2,3,4,5,6)	• Camera on pallet mover (1) • Mirror in workplace (1,2,6) • Traffic management: floor markings (1,2,3,5,6) • Use horn on pallet mover (1,2,3,4,5,6) • clothing employees with reflective clothing (2) • Design pallet mover (less heavy) (1) • Bump detection (1,3,6) • Loading correctly (e.g., no load on double pallets) (2,3,4,6) • Secure the load (e.g., stretching, strapping, cage) (1,2,4,5,6) • Provision of material to secure the load (6) • Using the material to secure the load (2,4,6) • Clear and accessible guidelines/procedures for loading (3,6) • Training pallet mover operator (1,3,4,5,6) • Sensitization/communication on loading (3,4,5,6) • Planning production and staffing (sufficient time and people to perform the job) (1,5,6)	• Unstable load (1,2,3,4,5,6) • Loss of pallet mover control (1,2,3,4,5,6)
• Speed pallet mover (1,2,3,4,6) • Acceleration pallet mover (1,6)	• Driving or accelerating too fast (1,2,3,4,6) • Due to production pressure • Wanting to take a break faster • Due to gross negligence (play behavior, act tough)	• Speed limitation (1) • Acceleration limitation (1,6) • Emergency stop pallet mover (1,6) • Provision of a sufficient number of pallet movers (3) • Clear and accessible guidelines/procedures on speed and acceleration (3,5,6) • Training pallet mover operator (1,3,4,5,6) • Sensitization/communication on speed and acceleration (3,4,5,6) • Planning production and staffing (sufficient time and people to perform the job) (1,5,6)	• Unstable load (1,2,3,4,5,6) • Loss of pallet mover control (1,2,3,4,5,6)
• Design workspace (1,2,3,4,5,6)	• Limited sight due to insufficient light (1,2,6) • Distraction due to hard/insufficient noise (1,5) • Too narrow of a maneuvring/parking space (1,2,3,4,5,6) • Limited sight at crossings, corners, entrances/exits (1,3,4,5,6) • Ridges, sharp endings at infrastructure (e.g., at pallet lift, diked areas) (4,5,6) • Too crowded working area/buffers (e.g., too much load, too much traffic) (2,4,5,6) • Pedestrian or other internal transport in the zone of the pallet mover transport (2,4,5,6) • Unsafe location to recharge thebattery (4,6) • Unsuitable floor material (too rough or too smooth) (2,6)	• sup>· Sufficient light (1,6) • sup>· Ear protection (6) • Camera on pallet mover (1) • Mirror in workplace (1,2,6) • Sandblasting floor (6) • Traffic management: Floor markings (1,2,3,5,6), stop sign or other road signs (1,2,4,6), separate pathways and lanes (1,2,6) • Use horn on pallet mover (1,2,3,4,5,6) • Bump detection (1,3,6) • (Reorganisation) layout (e.g., sufficient pre-defined zones for full and empty pallets or cubitainers, as least load manipulation as possible) (1,2,3,4,5,6) • Increasing the frequency of emptying buffers (3,4,5,6) • Shielding ridges, sharp endings at infrastructure (6) • Clear and accessible guidelines/procedures (e.g., for maneuvring in a narrow space) (3,5,6) • Training pallet mover operator (1,3,4,5,6) • sup>· Provision of predefined locations to recharge the battery (6)	• Unstable load (1,2,3,4,5,6) • Loss of pallet mover control (1,2,3,4,5,6)

Table 1. *Cont.*

Hazard	Pre-Event Accident Scenario	Preventing Technical and Non-Technical Safety Barriers and Management Delivery Systems	Central Event
• Condition workspace (1,2,3,4,5,6)	• Floor in bad condition (e.g., holes, loose floor plates) (1,2,3,4,6) • Insufficient housekeeping (e.g., loads outside predefined spaces, pallets not stacked correctly) (2,3,4,5,6) • Insufficient cleanliness (e.g., wet floor, the products on the floor) (2,3,4,5,6) • Incoming rain, humidity of the workspace due to condensation (6)	• Supervising housekeeping and cleanliness (3,4,5,6) • Cleaning wet/dirty floor (4,5,6) • Shield of damages and repair as soon as possible (6) • Sandblasting floor (6) • Lift the forks 10–20 cm while driving to avoid irregularities in the surface (3) • Clear and accessible guidelines/procedures (e.g., on housekeeping) (3,5,6) • Training pallet mover operator (1,3,4,5,6) • Sensitization/communication (e.g., on importance of notifications of damaged floor, on importance housekeeping) (3,4,5,6)	• Unstable load (1,2,3,4,5,6) • Loss of pallet mover control (1,2,3,4,5,6)
• Condition material (equipment and load) (1,2,3,4,5,6)	• Poor condition pallet mover due to absent maintenance (e.g., malfunctioning emergency stop) (1,2,3,4,6) • Not inspected pallet movers (3,4,6) • Damaged load (e.g., damaged pallets, leaking cubitainers) (1,3,4,6) • Misfit between pallet mover and load (1,4) • Substantial mass of pallet mover (1) • Pallet mover or load not fit for the job (e.g., pallet mover too wide for the workspace) (3,4,5)	• Sandblasting floor (6) • Purchasing pallet movers that are fit for the job (3,4,5) • Purchasing less heavy pallet movers (1) • Purchasing loads that are fit for the job (3,4) • Timely and adequate maintenance of pallet mover (3,4,6) • Timely and adequate inspection of pallet movers and loads (3,4,5,6) • Timely replacement of outdated pallet movers (3,5,6) • Removing damaged loads (1,3,6) • Provide inspection mark that is not obstructive during work (e.g., sticker instead of a strap) (4,6) • Communication with the supplier of damaged loads (3) • Performing start-up check of pallet mover (e.g., of an emergency stop) (3,4,5,6) • Providing clear/readable start-up checklist that is easy to fill in (3,4,6) • Clear and accessible guidelines/procedures (e.g., on start-up check pallet mover) (3,5,6) • Training pallet mover operator (1,3,4,5,6) • Sensitization/communication (e.g., on not using damaged pallets, on signalising defects at pallet movers) (3,4,5,6)	• Unstable load (1,2,3,4,5,6) • Loss of pallet mover control (1,2,3,4,5,6) • Breakdown pallet mover (1,2,3,4,5,6)

Table 1. *Cont.*

Hazard	Pre-Event Accident Scenario	Preventing Technical and Non-Technical Safety Barriers and Management Delivery Systems	Central Event
• Operating the pallet mover (1,2,3,4,5,6)	• Not following the safe practices for pallet mover use (2,3,4,5,6) • Due to ignorance or inexperience (3,5,6) • Due to production pressure (3,4,5,6) • Due to distraction operator (e.g., using a phone, reading during pallet mover use) (2,4,5) • Due to gross negligence (intoxication, play behavior, act tough) (2,3) • Not following the traffic rules (e.g., ignoring a stop sign) (4,5) • Losing control due to a physical/medical problem (e.g., fatigue, bad eyes, concentration impaired by medication (2,3) • Operator in a no-go area for pallet movers (1,4,5,6) • Unauthorised use of pallet mover (no or an expired licence) (3,4,5,6) • Difficult maneuvring situations (e.g., pallet needs to be turned first) (6) • Non-routine circumstances (e.g., maintenance machines) whereby the usual routes cannot be taken (4,6)	• Pre-selection of pallet mover operators (3,5,6) • Training pallet mover operator (1,3,4,5,6) • On-the-job assistance/training of new employees (5,6) • Clear and accessible guidelines/procedures for pallet mover use (also for non-routine circumstances) (3,5,6) • Sensitization/communication (e.g., on no-go areas) (3,4,5,6) • Leadership supervision, coaching, and feedback on unsafe practices (3,4,5,6) • Planning production and staffing (sufficient time and people to perform the job) (1,5,6) • Sufficient predefined parking spaces for all pallet movers (3,4,5,6,6) • Provide personal keys (3,4,6) • Remove key from pallet mover when not in use (2,3,4,6) • Take measures so it cannot be bypassed to start/drive pallet mover without keys (2,4,6) • Always carry driving license for pallet mover (3,4,6) • Visual and physical demarcation of no-go areas for pallet movers (6)	• Unstable load (1,2,3,4,5,6) • Loss of pallet mover control (1,2,3,4,5,6)

Table 2. The elements on the right-hand side of the bow-ties of the accident processes with pallet movers based on the [1] Literature; [2] Belgian and Dutch National data; [3] Document analysis; [4] Safety notifications in the incident registration system of the Belgian plant; [5] Accident analysis; [6] Fieldwork: observations and interviews.

Central Event	Mitigating Technical and Non-Technical Safety Barriers and Management Delivery Systems	Post-event Accident Scenario	Consequence
• Unstable load (1,2,3,4,5,6)	• sup>· Application of first aid [2,4,5] • On time [2] • With the correct diagnose and response action [2] • sup>· Personal Protective Equipment (PPEs) [2,5] • Provision of PPEs [2] • Using the PPEs [2] • Maintenance of PPEs [2] • Safety features on pallet mover (e.g., tilt protection, bump detection) [2,3,6] • Protection of objects/infrastructure (e.g., with foam, with steel bar) [3,5,6]	• Loss of load [1,2,3,4,5,6]	• Injury [1,3,4,5,6] • Damage [1,3,4,5,6] • Stop production process [4,5,6]
• Loss of control pallet mover [1,2,3,4,5,6] • Breakdown pallet mover [1,2,3,4,5,6]	• Traffic management: Separate pathways and lanes [1,2,6] • Emergency stop pallet mover [1,6] • Increased visibility and/or audibility of pallet mover [2] • Application of first aid [2,4,5] • On time [2] • With the correct diagnose and response action [2] • sup>· Personal Protective Equipment (PPEs) [2,5] • Provision of PPEs [2] • Using the PPEs [2] • Maintenance of PPEs [2] • Safety features on the pallet mover (e.g., tilt protection, bump detection) [2,3,6] • Protection of objects/infrastructure (e.g., with foam, with steel bar) [3,5,6]	• Pallet mover hits pedestrian/other internal transport [1,2,3,4,5,6] • Pallet mover hits operator pallet mover [1,2,3,4,5,6] • Pallet mover hits object/infrastructure [1,2,3,4,5,6]	• Injury [1,3,4,5,6] • Damage [1,3,4,5,6] • Economic loss (because of a stop of the production process) [4,5,6]

6. Assigning Indicators to the Bow-Ties

Once the bow-ties are composed, they are an excellent point of departure to assign indicators to the technical and non-technical safety barriers and to the management delivery systems in order to control (i.e., prevent or mitigate) the accident scenarios. Indicators can support a company's safety management and provide information on a preferred safety goal.

It is important that the indicators focus on aspects that are applicable to the operating environment. For instance, it could be an improvement to replace all present pallet movers with new, less heavy equipment. However, in the context of the company, it is possible that this is not feasible due to budgetary constraints. Another example is the reorganization of the layout of the floors, which is an important aspect in order to create more maneuvering space. However, in the plant under investigation, installations are integrated into an existing space which was initially not designed for that purpose, leading to space constraints. Additionally, in the past years, several improvements have already been made regarding the layout of the working space and the work floor, and there is, of course, a limitation to the possibilities in creating more maneuvering space. A question arising from this space constraint is whether pallet movers are the best equipment to use at particular spaces in a production facility with limited maneuvering space and if it would not be better to search for an alternative way to transport the material. This is related to the inherent safety of a company [41], which will be discussed at the end of this section.

Besides the applicability of indicators in the company environment, the focus of the indicators can also be chosen based on the presence of management delivery systems. For instance, as can be seen in Tables 1 and 2, certain management delivery systems are present in multiple scenarios. It concerns the following management delivery systems: 'training of pallet mover operators', 'sensitization and communication', 'guidelines and procedures', and 'planning of production and staffing'. Indicators assigned to these management delivery systems can be considered as general indicators, as they are linked to multiple accident scenarios. As an example, the management delivery systems 'training of pallet mover operators' is further elaborated below.

Next to the general indicators, scenario-specific indicators can also be developed. Decisions on what scenarios should be focused on can be based on the plant-specific risks regarding pallet mover use. Two examples of company-specific pre-scenarios will be further elaborated below, being 'narrow maneuvering space' and 'leaking cubitainers'.

6.1. General Indicators: Training of Pallet Mover Operators

In Table 3, the possible indicators for the frequently occurring management delivery system 'training of pallet mover operators' have been elaborated. An indicator of the content of the (re)training is proposed, and an indicator of the quality control of the (re)training. Additionally, there are indicators regarding the coverage ratio of the training and the retraining.

Table 3. Possible indicators for the management delivery system 'training of pallet mover operators'.

Sequentiality in follow-up	Content (re)training	Evaluation of the content of the (re)training every two years: Is the (re)training completely tailored to the needs of the company? (yes/no)Aspects to take into consideration: - Use of examples of specific risks and possible accident scenarios at the company? E.g., narrow maneuvring spaces, wet floors, too crowded buffers ... - Sufficient rules and guidelines for the target audience?
	Quality control (re)training	Yearly evaluation of the percentage of participants of the (re)training evaluating the training as positively (i.e., a score of 7 out of 10 or higher)
	Coverage ratio training	Monthly evaluation of the percentage of starting pallet mover operators that are trained for pallet mover use
	Coverage ratio retraining	Yearly evaluation of percentage pallet mover operators receiving a retraining every five years

It should be noted that a certain sequentiality is present in the follow-up of the indicators. For instance, a high coverage ratio of the (re)training is negligible if the content of the (re)training is not tailored to the needs of the company and if the quality of the (re)training is evaluated as substandard.

6.2. Scenario-Specific Indicators: Narrow Maneuvering Space

A specific risk at the plant under investigation is the narrow maneuvering space. Therefore, this scenario was chosen to be further elaborated into scenario-specific indicators. Figure 4 shows the bow-tie of the pre-event accident scenario of a narrow maneuvering space. Only a selection of possible safety barriers and management delivery systems has been included in the bow-tie. Two non-technical safety barriers have been included: 'no pallet mover use at too narrow spaces' and 'correct pallet mover use at narrow spaces'. One technical safety barrier has been included: 'sandblasting floor'. Sandblasting of the transportation routes with pallet movers (Figure 5) leads to a better grip and a shorter breaking distance. The following management delivery systems have been included: 'traffic management' and 'training pallet mover operators'.

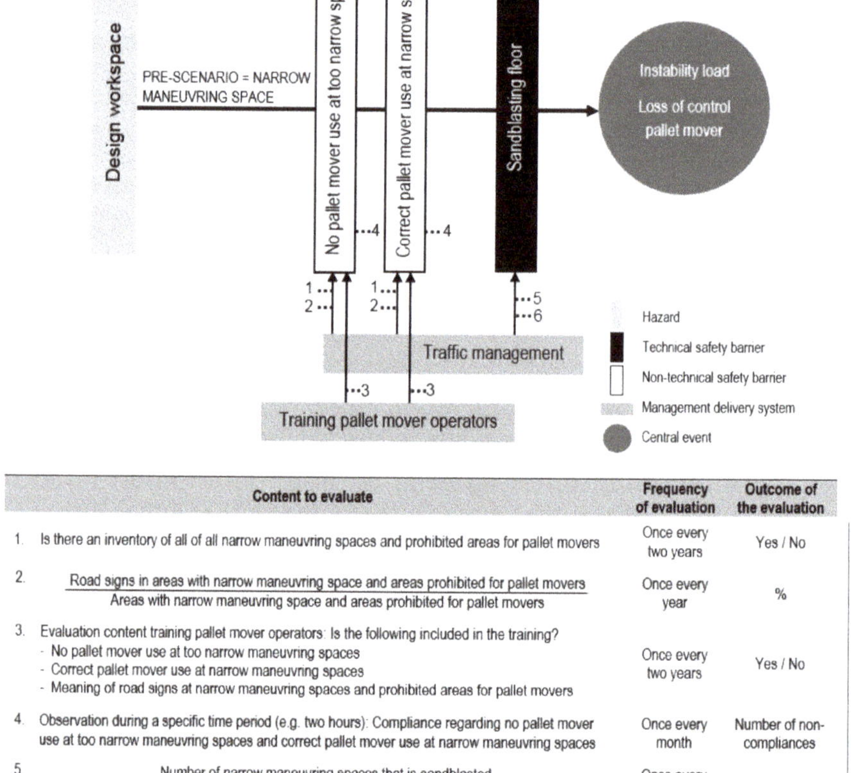

	Content to evaluate	Frequency of evaluation	Outcome of the evaluation
1.	Is there an inventory of all of all narrow maneuvring spaces and prohibited areas for pallet movers	Once every two years	Yes / No
2.	Road signs in areas with narrow maneuvring space and areas prohibited for pallet movers / Areas with narrow maneuvring space and areas prohibited for pallet movers	Once every year	%
3.	Evaluation content training pallet mover operators: Is the following included in the training? - No pallet mover use at too narrow maneuvring spaces - Correct pallet mover use at narrow maneuvring spaces - Meaning of road signs at narrow maneuvring spaces and prohibited areas for pallet movers	Once every two years	Yes / No
4.	Observation during a specific time period (e.g. two hours): Compliance regarding no pallet mover use at too narrow maneuvring spaces and correct pallet mover use at narrow maneuvring spaces	Once every month	Number of non-compliances
5.	Number of narrow maneuvring spaces that is sandblasted / Number of narrow maneuvring spaces	Once every two years	%
6.	Number of timely maintained sandblasted floors / Number of sandblasted floors	Once every year	%

Figure 4. The possible indicators for the pre-scenario of 'narrow maneuvring spaces'.

Figure 5. An example of a sandblasted and not sandblasted pallet mover transportation route.

As with the general indicators, a sequentiality is also present in the follow-up of the scenario-specific indicators. For example, when a company does not include correct pallet mover use at narrow maneuvering spaces in the training, it does not make sense to evaluate non-compliances on this topic.

6.3. Scenario-Specific Indicators: Leaking Cubitainer

Another specific risk at the plant under investigation is the worn drain valves of cubitainers, leading to spills of liquid content (Figure 6). Products on floors lead to a longer breaking distance and a higher chance of losing control over the pallet mover. Figure 7 presents the bow-tie of the pre-event accident scenario of damaged loads, and more specific leaking cubitainers, complemented with possible indicators for this specific scenario. Again, only a selection of possible safety barriers and management delivery systems has been included in the bow-tie, namely, the non-technical safety barriers 'purchasing cubitainers fit for the job', 'removal of leaking cubitainers', and 'removal of leaked product', the technical safety barrier 'sandblasting floor', and the management delivery systems 'training pallet mover operators' and 'traffic management'.

Figure 6. A cubitainer with leaking drain valves leading to a spill of the liquid content.

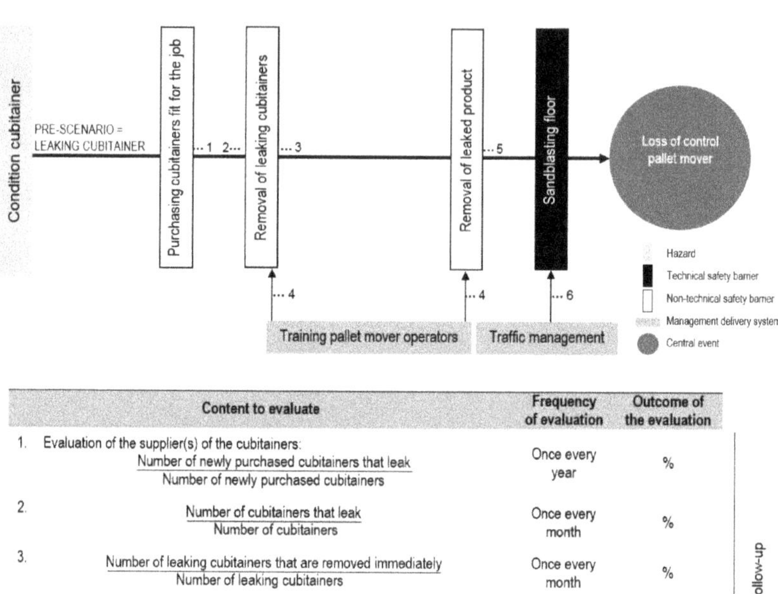

Content to evaluate	Frequency of evaluation	Outcome of the evaluation
1. Evaluation of the supplier(s) of the cubitainers: $\frac{\text{Number of newly purchased cubitainers that leak}}{\text{Number of newly purchased cubitainers}}$	Once every year	%
2. $\frac{\text{Number of cubitainers that leak}}{\text{Number of cubitainers}}$	Once every month	%
3. $\frac{\text{Number of leaking cubitainers that are removed immediately}}{\text{Number of leaking cubitainers}}$	Once every month	%
4. Evaluation content training pallet mover operators: Is the following included in the training? - Not using leaking cubitainers - Signalizing leaking cubitainers - Immediately cleaning leaked product on the floor	Once every two years	Yes / No
5. Observation during a specific time period (e.g. two hours): Compliance with immediately cleaning leaked product on the floor	Once every month	Number of non-compliances
6. $\frac{\text{Number of transportation routes with cubitainers that is sandblasted}}{\text{Number of transportation routes with cubitainers}}$	Once every two years	%

Figure 7. The possible indicators for the pre-scenario of 'leaking cubitainer'.

6.4. Evaluation of the Indicators

Once the indicators have been determined, they have to be evaluated. Based on the specificity of the indicator and the needs of the company, this evaluation can take place on a yearly basis, or if needed with a lower or a higher frequency.

Additionally, after the development of the indicators, targets and limits have to be assigned to every indicator. This means that the company has to decide what is acceptable as a result and what is not. In the example of the coverage ratio of the training, a target could be that 100% of all pallet mover operators should be trained for pallet mover use. In the example of the number of leaking cubitainers, a target could be that <5% of the cubitainers is leaking.

An important aspect is that responsibilities have to be indicated: Who does what and when in order to reach the goals of the indicators. In the example of the (re)training, responsibility has to be indicated for the subscriptions for the (re)training. The same applies to responsibilities to take actions when a target is not achieved.

Based on the elaborated examples, it could seem that a lot of collection and registration is needed for all indicators. However, it should be noted that the given examples and their accompanying indicators are very specific. Once the entire set of indicators has been developed, it will become clear that many of them can be collected and registered under the same heading.

In order to facilitate an adequate monitoring of the indicators, a system should be set up to report and to collect the required data. Such systems are often already (partly) present in a company.

To be complete, something should be said on the necessity of the indicators, which is a reflection that should be made before developing the indicators. After all, it should first be analyzed if the processes that are present in the company are as inherently safe as possible [41]. In the given example regarding the narrow maneuvering spaces, whether a pallet mover is the safest equipment to use at these places (and by extension at all places) should be considered. In the example of the leaking cubitainers, whether cubitainers are necessary and if the process cannot be designed in such a way that the transportation of the liquid is minimalized (and by extension that a large amount of manipulation of all loads at the plant is minimalized) should be considered. Kletz's [42] article entitled "what you don't have, can't leak" is a good resume of this matter.

7. Discussion and Conclusions

The outcomes deriving from the current accident analysis that is used at the Belgian plant under investigation (6W-2H and why-why technique) does not seem sufficient to take adequate measures in order to prevent accidents with pallet movers. When preventive measures are taken based on accident analyses, several shortcomings can be identified. Firstly, preventive measures based on accident analyzes only focus on accidents that already occurred, leaving all other potential accident scenarios out of scope. A method that includes not only company specific data, but also generic data sources such as the literature and national accident data, generates information on the entire accident process, including aspects that have not (yet) occurred at a specific plant. Additionally, it can be concluded that the recommendations resulting from the current accident analysis that is used at the plant under investigation, are mostly individual-oriented. This is however not an intrinsic problem of the 6W-2H and why-why technique, as this technique focusses on both technical, organizational as human aspects. Hence, the focus on the human aspects is not a consequence of the technique itself, but of the way the technique is applied.

To address the shortcomings of the current accident analysis, an alternative method—i.e., the bow-tie method—is chosen in order to address pallet mover safety. The bow-tie was chosen for several reasons. Firstly, the bow-ties were composed using a multi-method design. This multi-method design leads to a better comprehensiveness of the entire accident process of pallet mover use and gives a detailed overview of what could possibly go wrong with a pallet mover. In the bow-ties, the possible causes and consequences of potential accidents are identified. Additionally, the bow-tie includes the influence of safety measures (safety barriers and management delivery systems) on the evolution of accident scenarios [43].

Due to the use of a multi-method design to compose the bow-ties, not only company specific data were included, but also generic data sources such as the literature and national accident data. This leads to information on the entire accident process, including aspects that have not (yet) occurred at a specific plant.

Because of the comprehensive character of the bow-tie method, the results are easily transferable to other production facilities where pallet movers are used for internal transport, assuming that the hazards are the same. This means that, if this study was conducted in another production facility with similar hazards and similar a working environment, composition of the bow-ties would have led to a similar outcome as in Tables 1 and 2 (this does not mean that the process of linking indicators to the bow-ties is the same, as this is very company specific). In other words, the bow-tie method leads to a general model that is transferable and applicable in every setting where, in this case, pallet movers are being used. However, the indicators may be different.

Another advantage of the bow-tie method is that it allows us to make a clear distinction between preventing and mitigating safety barriers and management delivery systems.

Seven hazards regarding pallet mover use could be identified based on the composition of the bow-ties: Load, the speed of the pallet mover, acceleration of the pallet mover, the design of the workplace, conditions of the workplace, conditions of materials (load and pallet mover), and operating the pallet mover. Through several identified pre-event accident scenarios, these hazards can lead to different central events: Instability of the load, loss of control over the pallet mover, and a breakdown of the pallet mover. At their turn, these central events can lead through several post-accident scenarios to different consequences: Injury, damage, or economic loss. Several technical and non-technical safety barriers and management delivery systems to prevent or mitigate the central event could be linked to the accident scenarios.

The identified safety barriers and management delivery systems mainly focus on organizational aspects, and, to a lesser extent, on the individual behavioral aspects of operators. The pitfall of 'blaming the victim', which is often present in other methods of accident analysis where there is primarily focused on the individual behavior of the operators, is therefore reduced when using the bow-tie method.

Once bow-ties are composed and safety barriers and management delivery systems have been identified, indicators should be developed and monitored consequently. These indicators should be composed based on their applicability in the company, meaning what is possible given a specific company environment. When developing indicators, an important distinction can be made. Firstly, there are general indicators. In the bow-ties, certain management delivery systems can be linked to many of the accident scenarios. When indicators are developed for frequently occurring management delivery systems, these indicators can be considered as general because they are not linked to only one scenario. Secondly, there are scenario-specific indicators. This means that indicators are linked to specific scenarios that require attention in a plant. With both the general and the scenario-specific indicators, a certain sequentiality should be acknowledged in the follow-up of the indicators. For all indicators, it is therefore important to set priorities.

To conclude, indicators are an important result of a bow-tie analysis. When a company reaches a consensus on a set of indicators to be monitored, a unique insight is obtained on the status and development of potential accident scenarios. Management can intervene adequately to ensure a safe production.

Author Contributions: Conceptualization, K.v.N., P.S. and G.R.; Methodology, K.v.N., P.S. and G.R.; Validation, K.v.N., P.S. and G.R.; Formal Analysis, K.v.N and P.S.; Investigation, K.v.N and P.S.; Data Curation, K.v.N., P.S. and G.R.; Writing-Original Draft Preparation, K.v.N.; Writing-Review & Editing, P.S., G.R., N.P., O.A. and K.P.; Visualization, K.v.N.; Supervision, G.R.; Project Administration, K.v.N.; Funding Acquisition, G.R."

Acknowledgments: The authors would like to thank the company under investigation in this article for their cooperation and for the possibility to conduct this research.

Conflicts of Interest: The authors declare no conflict of interest.

Appendix A

Table A1. An example of a recordable accident with a pallet mover at the Belgian plant (input based on the 6W-2H and why-why analysis).

Description Accident Based on the Accident Analysis	Actions in Response to the Accident
• An operator (contractor) got his foot clamped under the pallet mover, causing him to trip and pull the pallet mover over his foot, resulting in a contusion of his foot  • A load was urgently needed in order to avoid a line stop, so the operator broke with normal procedures and did not first remove the loads stocked in front of the load that had to be picked up • Therefore, he had to pass through a narrow space and did not follow the safe practice of walking beside and not in front of the pallet mover • The operator also did not follow the safe practice of keeping the pallet mover at arm's length • There were too many loads stocked by the forklifts in the area, creating the narrow space (the forklift truck driver responsible for the stocking of the area was not consulted during the accident analysis). • The load that had to be picked up was not standing in the designated zone	• Disciplinary measures against the operator were taken by the contractor company • Communication on the accident and safe practices • Retrain pallet mover operators • Reorganization layout of the zone: a part of the loads is now stocked elsewhere, in order to create more maneuvring space

References

1. Bellamy, L.; Ale, B.; Whiston, J.; Mud, M.; Baksteen, H.; Hale, A.; Papazouglou, L.; Bloemhoff, A.; Damen, M.; Oh, J. The software tool storybuilder and the analysis of horrible stories of occupational accidents. *Saf. Sci.* **2008**, *46*, 186–197. [CrossRef]
2. DeBlois, L. *Industrial Safety Organization for Executives and Engineer*; McGraw-Hill Book Company: New York, NY, USA, 1926.
3. Gibson, J. The contribution of experimental psychology to the formulation of the problem or safety—A letter for basic research. Behavioural Approaches to Accident Research. In *Accident Research, Methods and Approaches*; Haddon, W., Suchman, E., Klein, D., Eds.; Harper & Row: New York, NY, USA, 1961.
4. Haddon, W. A note concerning accident theory and research with special reference to motor vehicle accidents. *Ann. N. Y. Acad. Sci.* **1963**, *107*, 635–646. [CrossRef] [PubMed]
5. Reason, J. *Managing the Risks of Organisational Accidents*; Alderschot: Ashgate, UK, 1997.
6. Swuste, P.; Gulijk, C.; van Zwaard, W. Safety metaphors and theories, a review of the occupational safety literature of the US UK and The Netherlands, till the first part of the 20th century. *Saf. Sci.* **2010**, *48*, 1000–1018. [CrossRef]
7. Swuste, P.; Gulijk, C.; van Zwaard, W.; Oostendorp, Y. Occupational safety theories, models and metaphors in three decades since World War II, in the United States, Britain, and the Netherlands: A literature review. *Saf. Sci.* **2014**, *62*, 16–27. [CrossRef]
8. Swuste, P.; Gulijk, C.; van Zwaard, W.; Lemkowitz, S.; Oostendorp, Y.; Groeneweg, J. Developments in the safety science domain in the fields of general and safety management between 1970–1979, the year of the near disaster on Three Mile Island, a literature review. *Saf. Sci.* **2016**, *86*, 10–26. [CrossRef]
9. Lu, L.; Liang, W.; Zhang, L.; Zhang, H.; Lu, Z.; Shan, J. A comprehensive risk evaluation method for natural gas pipelines by combining a risk matrix with a bow-tie model. *J. Nat. Gas Sci. Eng.* **2015**, *25*, 124–133. [CrossRef]
10. Swuste, P.; Theunissen, J.; Schmitz, P.; Reniers, G.; Blokland, P. Process safety indicators, a review of literature. *J. Loss Prev. Process Ind.* **2016**, *40*, 162–173. [CrossRef]
11. Rasmussen, J. Risk management in a dynamic society: A modelling problem. *Saf. Sci.* **1997**, *27*, 183–213. [CrossRef]
12. Leveson, N. A new accident model for engineering safer systems. *Saf. Sci.* **2004**, *42*, 237–270. [CrossRef]
13. Hollnagel, E.; Goteman, O. The functional resonance accident model. *Proc. Cognit. Syst. Eng. Process Plant* **2004**, *2004*, 155–161.
14. Jacinto, C.; Silva, C. A semi-quantitative assessment of occupational risks using bow-tie representation. *Saf. Sci.* **2010**, *48*, 973–979. [CrossRef]
15. Hale, A.; Ale, B.; Bellamy, L.; Whiston, J.; Mud, M.; Papazoglou, I.; Bloemhof, B.; Post, J.; Oh, J. Best practice in risk assessment: Work accidents. In Proceedings of the European Consumer Safety Association Conference, Edinburgh, UK, 20–21 April 2005.
16. Papazoglou, I.A.; Ale, B.J.M. A logical model for the quantification of occupational risks. *Reliab. Eng. Syst. Saf.* **2007**, *92*, 785–803. [CrossRef]
17. Papazoglou, I.A.; Aneziris, O.N.; Bellamy, L.J.; Ale, B.J.M.; Oh, J. Quantitative occupational risk model: Single hazard. *Reliab. Eng. Syst. Saf.* **2017**, *160*, 162–173. [CrossRef]
18. Aneziris, O.N.; Papazoglou, I.A.; Baksteen, H.; Mud, M.; Ale, B.J.; Bellamy, L.J.; Hale, A.R.; Bloemhoff, A.; Post, J.; Oh, J. Quantified Risk Assessment for Fall from Heights. *Saf. Sci.* **2008**, *46*, 198–220. [CrossRef]
19. Aneziris, O.N.; Papazoglou, I.A.; Mud, M.; Damen, M.; Bellamy, L.J.; Manuel, H.J.; Oh, J. Occupational risk quantification owing to falling objects. *Saf. Sci.* **2014**, *69*, 55–70. [CrossRef]
20. Aneziris, O.N.; Papazoglou, I.A.; Konstandinidou, M.; Baksteen, H.; Mud, M.; Damen, M.; Bellamy, L.J.; Oh, J. Quantification of occupational risk owing to contact with moving parts of machines. *Saf. Sci.* **2013**, *51*, 382–396. [CrossRef]
21. Aneziris, O.N.; Papazoglou, I.A.; Bellamy, L.J.; Mud, M.; Damen, M.; Manuel, H.J.; Oh, J. Occupational Risk Management for activities performed near vehicles. In Proceedings of the 25th European Safety and Reliability Conference (ESREL 2015), Zurich, Switzerland, 7–10 September 2015.

22. Aneziris, O.N.; Papazoglou, I.A.; Konstandinidou, M.; Baksteen, H.; Mud, M.; Damen, M.; Kuiper, J.; Bellamy, L.J.; Oh, J. Occupational Risk Management for fires. In Proceedings of the European Safety and Reliability Conference (ESREL 2010), Rhodes, Greece, 5–9 September 2010.
23. Papazoglou, I.A.; Konstandinidou, M.; Aneziris, O.N.; Baksteen, H.; Mud, M.; Damen, M.; Kuiper, J.; Bellamy, L.J.; Oh, J. Occupational Risk Management for release of hazardous substances from closed containments. In Proceedings of the European Safety and Reliability Conference (ESREL 2010), Rhodes, Greece, 5–9 September 2010.
24. Stout, N. Characteristics of work related injuries involving forklift trucks. *J. Occup. Accid.* **1987**, *18*, 179–190.
25. Miller, B. Forklift safety by design. *Prof. Saf.* **1988**, *33*, 18–21.
26. Lifschultz, B.; Donoghue, E. Deaths due to forklift truck accidents. *Forensic Sci. Int.* **1994**, *65*, 121–134. [CrossRef]
27. Larsson, T.; Rechnitzer, G. Forklift trucks analysis of severe and fatal; occupational injuries, critical incidents and priority for prevention. *Saf. Sci.* **1994**, *17*, 275–289. [CrossRef]
28. Darcy, C.; Lovell, M.; Metcalfe, J. Injuries from forklift trucks. *Injury* **1995**, *26*, 285. [CrossRef]
29. Mack, K.; Haslegrave, C.; Gray, M. Usability of manual handling aids for transporting materials. *Appl. Ergon.* **1995**, *26*, 353–364. [CrossRef]
30. Born, C.; Ross, S.; Aron, B.; DeLong, W.; Iannacone, W. Patterns of Injury and Disability Caused by Forklift Trucks. *J. Trauma* **1996**, *40*, 636–639. [CrossRef] [PubMed]
31. UK-HSE. *Safety of Industrial Lift Trucks*; HMSO: Norwich, UK, 2003.
32. UK-HSE. *Warehousing and Storage Keep It Safe*; Health and Safety Executive INDG 412; Health and Safety Executive: Bootle, UK, 2011.
33. UK-HSE. *Pallet Safety*; Guidance note PM15; Health and Safety Executive: Bootle, UK, 2014.
34. Harris, G.; DeRosia, J. Occupant Protection and Standup Forklift (Lift Truck) Dynamics. In Proceedings of the 2003 ASME International Mechanical Engineering Congress (IMEC'03), Washington, DC, USA, 15–21 November 2003.
35. Larsson, T.; Lambeth, J.; Wilde, M.; Tully, G.; Askew, H.; Skinner, W.; Carter, B.; Martin, T.; Kenningham, L. Industrial forklift trucks, dynamic stability and the design of safe logistics. *Saf. Sci. Monit.* **2003**, *7*, 1–14.
36. Horberry, T.; Larsson, T.; Johnston, I.; Lambert, J. Forklift safety traffic engineering and intelligent transport systems: A case study. *Appl. Ergon.* **2004**, *35*, 575–581. [CrossRef] [PubMed]
37. St-Vincent, M.; Denis, D.; Imbeau, D.; Laberge, M. Work factors affecting manual materials handling in a warehouse superstore. *Int. J. Ind. Ergon.* **2005**, *35*, 33–46. [CrossRef]
38. Berry, T. An analysis of pallet truck incidents. In Proceedings of the ASME 2008 International Mechanical Engineering Congress and Exposition, Boston, MA, USA, 31 October–6 November 2008.
39. Sanders, D. Controlling the direction of walkie type forklifts and pallet jacks on sloping ground. *Assem. Autom.* **2008**, *28*, 317–324. [CrossRef]
40. Kleinert, S.; Overmeyer, L. Integration of 3D camera systems on forklift trucks. *Logist. J.* **2013**, 1–7. [CrossRef]
41. Kletz, T. *Plant Design for Safety—A User-Friendly Approach*; Hemisphere: New York, NY, USA, 1991.
42. Kletz, T. What You Don't Have, Can't Leak. *Chem. Ind.* **1978**, *6*, 287–292.
43. De Dianous, V.; Fievez, C. Aramis project: A more explicit demonstration of risk control through the use of bow-tie diagrams and the evaluation of safety barrier performance. *J. Hazard. Mater.* **2006**, *130*, 220–233. [CrossRef] [PubMed]

© 2018 by the authors. Licensee MDPI, Basel, Switzerland. This article is an open access article distributed under the terms and conditions of the Creative Commons Attribution (CC BY) license (http://creativecommons.org/licenses/by/4.0/).

MDPI
St. Alban-Anlage 66
4052 Basel
Switzerland
Tel. +41 61 683 77 34
Fax +41 61 302 89 18
www.mdpi.com

Materials Editorial Office
E-mail: materials@mdpi.com
www.mdpi.com/journal/materials

www.ingramcontent.com/pod-product-compliance
Lightning Source LLC
LaVergne TN
LVHW071608170726
843515LV00008B/2350

* 9 7 8 3 0 3 9 3 6 7 6 6 5 *